PHYSICS OF THE SOLAR CORONA AND TRANSITION REGION

Part II

Proceedings of the Monterey Workshop, held in Monterey, California, August 1999

Edited by

ODDBJORN ENGVOLD

Institute of Theoretical Astrophysics, Oslo

and

JOHN W. HARVEY

National Solar Observatory, Tucson

In collaboration with

CAROLUS J. SCHRIJVER and NEAL E. HURLBURT

Lockheed-Martin Advanced Technology Center, Palo Alto, CA

Reprinted from *Solar Physics*
Volume 193, Nos. 1–2 (2000)
Volume 195, No. 1 (2000)
Volume 195, No. 2 (2000)
Volume 197, No. 2 (2000)

KLUWER ACADEMIC PUBLISHERS
DORDRECHT / BOSTON / LONDON

A C.I.P. Catalogue record for this book is available from the Library of Congress.

ISBN 0-7923-70430

Published by Kluwer Academic Publishers,
P.O. Box 17, 3300 AA Dordrecht, The Netherlands

Sold and distributed in North, Central and South America
by Kluwer Academic Publishers,
101 Philip Drive, Norwell, MA 02061, U.S.A.

In all other countries, sold and distributed
by Kluwer Academic Publishers,
P.O. Box 322, 3300 AH Dordrecht, The Netherlands

Cover illustration: AR9077: Solar Magnetic Arcade. This striking close-up of AR9077 was made by the orbiting TRACE satellite shortly after the flare erupted. It shows million-degree hot solar plasma cooling down while suspended in an arcade of magnetic loops. *Astronomy Picture of the Day, http://antwrp.gsfc.nasa.gov/apod/ap000720.html*

Printed on acid-free paper

Printed in the Netherlands

TABLE OF CONTENTS

PREFACE

Solar Physics publishes up to two Topical Issues per year that focus on areas of especially vigorous and active research. The present Topical Issue contains papers of recent results on the solar corona, as well as on the transition region and low solar wind. The majority of these papers, which were all refereed in accordance with the standards of Solar Physics, were presented in August 1999 at a workshop held in Monterey, California.

The Sun's magnetic field is responsible for the spectacularly dynamic and intricate phenomenon that we call the corona. The past decade has seen an enormous increase in our understanding of this part of the solar outer atmosphere, both as a result of observations and because of rapid advances in numerical studies. The *Yohkoh* satellite has observed the Sun now for over eight years, producing spectacular sequences of images that convey the complexity of the corona. The imaging and spectroscopic instruments on SOHO have added information on the cooler part of the corona. And since April of 1998 TRACE has given us very high resolution images of the 1–2 MK corona, at cadences that allow detailed observations of field oscillations, loop evolution, mass ejecta, etc.

The papers of this Topical Issue revolve around one key theme: the entire outer atmosphere of the Sun is intrinsically dynamic, evolving so rapidly that even the concept of a single local temperature for a single fluid often breaks down. Moreover, the corona is an intrinsically nonlinear and non-local medium. These aspects are discussed in this Topical Issue, including both papers that review recent developments (both based on observations and on theoretical/numerical modeling), and original research papers based on observations from many different observatories.

We are very grateful to the many referees who were given little time to respond, and to the staff of Kluwer for the production of the topical issues and their reprints. The papers accepted for this Topical Issue add up to such a volume that they have to be distributed over two Topical Issues of Solar Physics (December 1999 and April 2000), which are reprinted in two bound volumes, of which this is the second.

THREE-DIMENSIONAL SEPARATOR RECONNECTION – HOW DOES IT OCCUR?

K. GALSGAARD[1], E. R. PRIEST[1] and Å. NORDLUND[2]
[1]*School of Mathematics and Statistics, University of St. Andrews, KY16 9SS, Scotland*
(e-mails: klaus@mcs.st-and.ac.uk; eric@mcs.st-and.ac.uk)
[2]*NBIfAFG, Astronomical Observatorium, Julie Maries vej 30, 2100 Kbh Ø Denmark*
(e-mail: aake@astro.ku.dk)

(Received 5 October 1999; accepted 18 February 2000)

Abstract. In two dimensions magnetic energy release takes place at locations where the magnetic field strength becomes zero and has an x-point topology. The x-point topology can collapse into two y-points connected by a current sheet when the advection of magnetic flux into the x-point is larger than the dissipation of magnetic flux at the x-point. In three dimensions magnetic fields may also contain singularities in the form of three-dimensional null points. Three-dimensional nulls are created in pairs and are therefore, at least in the initial stages, always connected by at least one field line – the separator. The separator line is defined by the intersection of the fan planes of the two nulls. In the plane perpendicular to a single separator the field line topology locally has a two dimensional x-point structure. Using a numerical approach we find that the collapse of the separator can be initiated at the two nulls by a velocity shear across the fan plane. It is found that for a current concentration to connect the two nulls along the separator, the current sheet can only obtain two different orientations relative to the field line structure of the nulls. The sheet has to have an orientation midway between the fan plane and the spine axis of each null. As part of this process the spine axes are found to lose their identity by transforming into an integrated part of the separator surfaces that divide space into four magnetically independent regions around the current sheet.

1. Introduction

Two-dimensional magnetic reconnection has been investigated for many years and today we have a good understanding of the theory describing it (see Priest and Forbes (2000) for a recent review). When a 2D x-point is stressed by oppositely directed in-flowing flux an electric current is generated at the null. The magnitude of the electric current and the value of the magnetic resistivity, determine how fast the magnetic field lines can reconnect at the null point. If the reconnection is insufficient to process all of the incoming magnetic flux then the x-point collapses into a current sheet ending at two y-points (Syrovatsky, 1981).

Over the last decade interest has turned towards three-dimensional magnetic reconnection. The processes that govern three-dimensional magnetic reconnection are not yet fully understood. The increased complexity however allows reconnection to take place not only at singular points – three-dimensional nulls – but also in regions where the magnetic field is non-vanishing (Hesse and Schindler, 1988).

The physics of the solar corona and transition region. Part II.
Reprinted from Solar Physics **193,** 2000.

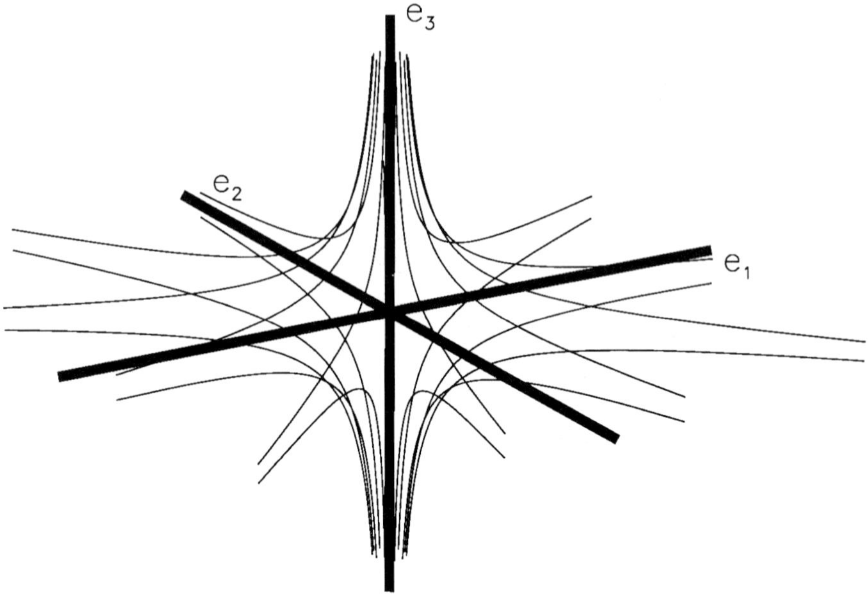

Figure 1. Single null point topology indicated by magnetic field lines, the eigenvectors e_1 and e_2 define the fan plane and e_3 the spine axis.

This paper investigates separator reconnection that occurs in a double null point configuration. The topology of a single null can be described by a first-order Taylor expansion of the magnetic field around the null. In this expansion the Jacobian matrix of the magnetic field contains information on the linear field line topology. However, the zero divergence of the magnetic field, the trace of the Jacobian matrix, severely restricts the possible topologies of the nulls (Parnell *et al.*, 1996). Assuming that the eigenvalues of the Jacobian matrix are all real and non-zero, then there will be two of one sign (assumed positive) and one of the opposite sign (assumed negative). Then the eigenvectors of the positive eigenvalues define a plane in which the magnetic field lines diverge from the null point – the fan plane (Priest and Titov, 1996). The eigenvector associated with the negative eigenvalue determines an axis along which a single field line reaches the null – the spine axis (see Figure 1).

When magnetic null points are dynamically formed in a region of space that initially does not contain magnetic singularities, then their formation occurs as a bifurcation process in which at least two null points are formed. In this process there will always be at least one field line, the separator, connecting the two nulls. This field line is defined by the intersection of the two null point fan planes (Figure 2).

The magnetic field line structure along the separator resembles that of a two-dimensional x-point, although with the difference that in the vicinity of the separator the component of the magnetic field along the separator direction dominates the

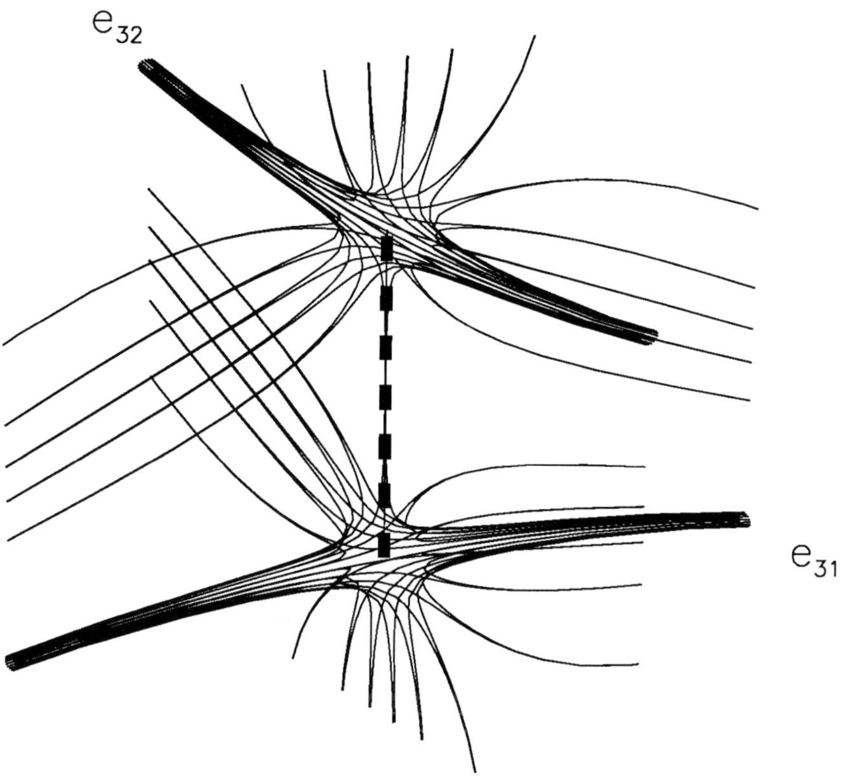

Figure 2. Double null point. The eigenvectors e_{31} and e_{32} define the spine axes and the thick dashed line shows the location of the separator.

magnetic energy. This provides a magnetic scalar pressure that oppose the collapse of the x-point topology along the separator.

We designed a numerical experiment to investigate the behaviour of a complicated magnetic field containing several magnetic nulls in the domain. Our aim is to understand how and where the collapse of length scales occurs in a topologically complicated magnetic field. This is important to understand, since current concentrations are locations in space where magnetic reconnection may occur. This change in magnetic field-line topology can release free magnetic energy that can heat the plasma and is therefore important in coronal heating.

The structure of the paper is as follows. In Section 2 the numerical experiment is described. Section 3 discusses the reasons for and locations of the current collapse. In Section 4 different scaling relations are obtained, and finally Section 5 summarises the results.

2. Experiment

The numerical experiment is carried out in a 3D Cartesian domain which is taken to be 2D periodic and bounded in the third direction. The boundary conditions are taken to be impenetrable and perfectly conducting. This conserves the total mass in the domain and an imposed boundary velocity can be used to stress the magnetic field inside the domain. The magnetic field is stressed by imposing a sinusoidal shear velocity with a wavelength equal to the domain size on the two boundaries (for more details see Galsgaard and Nordlund, 1997). The initial magnetic field is a constant-α force-free field defined by

$$\mathbf{B} = (\cos(ky) - \sin(kz), \cos(kz) - \sin(kx), \cos(kx) - \sin(ky)) \ , \qquad (1)$$

where $k = 2\pi$ and the domain is taken to be a unit box. The magnetic field contains 8 null points in the unit box and each null point connects to eight other null points. These connections divide into two main classes, where six are fan-fan separator lines and two represent spine-spine lines. The fan-fan connections may further be divided into two classes depending on the distance between the nulls and the behaviour of the local field around the separator line connecting them (the near and remote nulls in the following). The magnetic field topology in the domain is therefore very complex as the fan planes divide the domain into many magnetically independent flux regions. The spine-spine lines are structurally unstable in the sense that a general non-ideal perturbation destroys them. Initially, a constant density and plasma energy are assumed together with a zero velocity field. The plasma energy is chosen such that the average plasma β in the domain is $\frac{1}{6}$. Figure 3 indicates the initial magnetic field topology. The isosurfaces give the locations of the magnetic nulls, while the field lines indicate the locations of the spine axes and the fan planes.

3. The Collapse of the Separators

Before describing the collapse of the magnetic field structure, a short introduction to the global dynamical evolution of the magnetic field is presented (a more detailed discussion is given in Galsgaard and Nordlund (1997)). This is followed by a detailed analysis of the data with the aim of understanding the process that leads to the collapse of the separator lines connecting nearest neighbour null points.

When the magnetic field is perturbed by boundary motions, the disturbance propagates into the domain with at least the Alfvén velocity. A simple way to monitor the field reaction is to follow the development of the isosurfaces of weak magnetic field strength (the weak field tunnels in the following). The level of these may be chosen such that they show the locations of the nulls and their connections to the nearest neighbouring nulls (Figure 4).

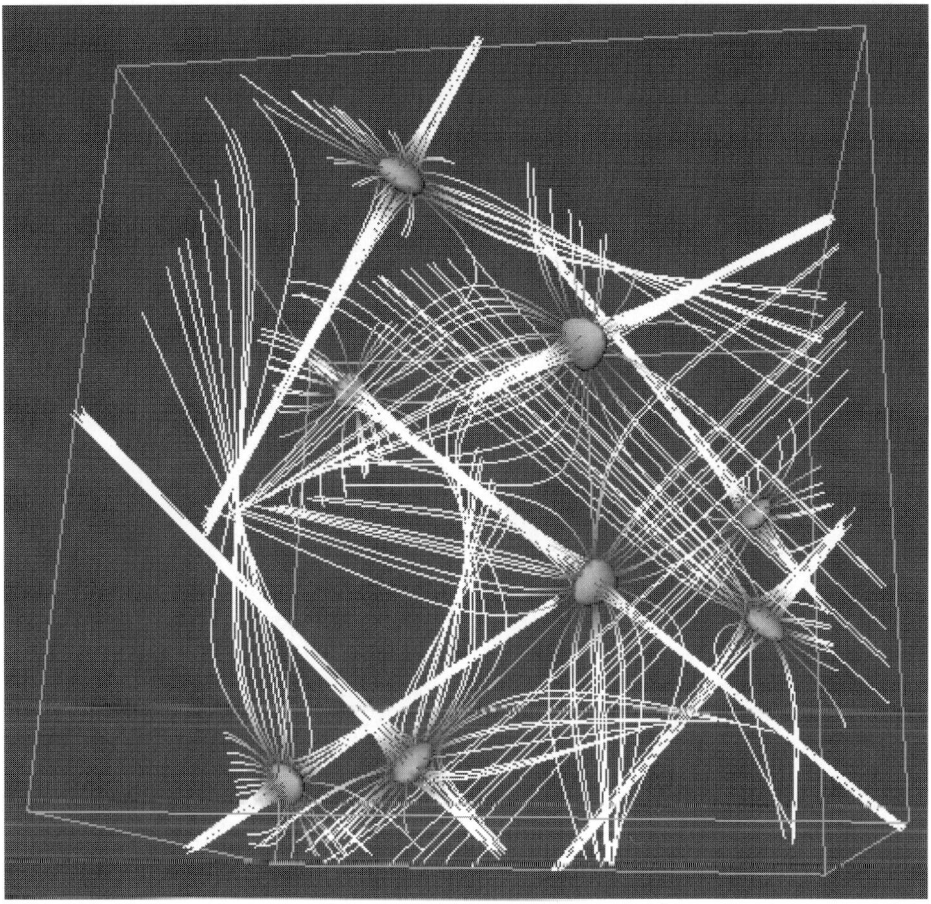

Figure 3. The initial field-line topology is indicated by showing the locations of the magnetic nulls by isosurfaces and the spine and fan planes by tracing magnetic field lines.

As the perturbation propagates into the domain the structure of the weak field tunnels changes. The panels in Figure 4 clearly show the collapse of one of these, while several of the other connections have only minor perturbations. The time sequence showing the collapse of this part of the isosurface surface indicates that magnetic field lines with different spatial orientation are being compressed across the collapse direction. Initially, the field is frozen into the plasma and the collapse generates a rapid change in the tangential component of the field lines across the isosurface. As a result of this there is a significant increase in the current strength. By repeating the experiments with different phases and directions of the boundary driving it was found that it is only the 'weak field tunnels' that collapse and form current concentrations in the domain. Two obvious questions arises from this: Why is it that only the weak field tunnels collapse and generate current accumulations

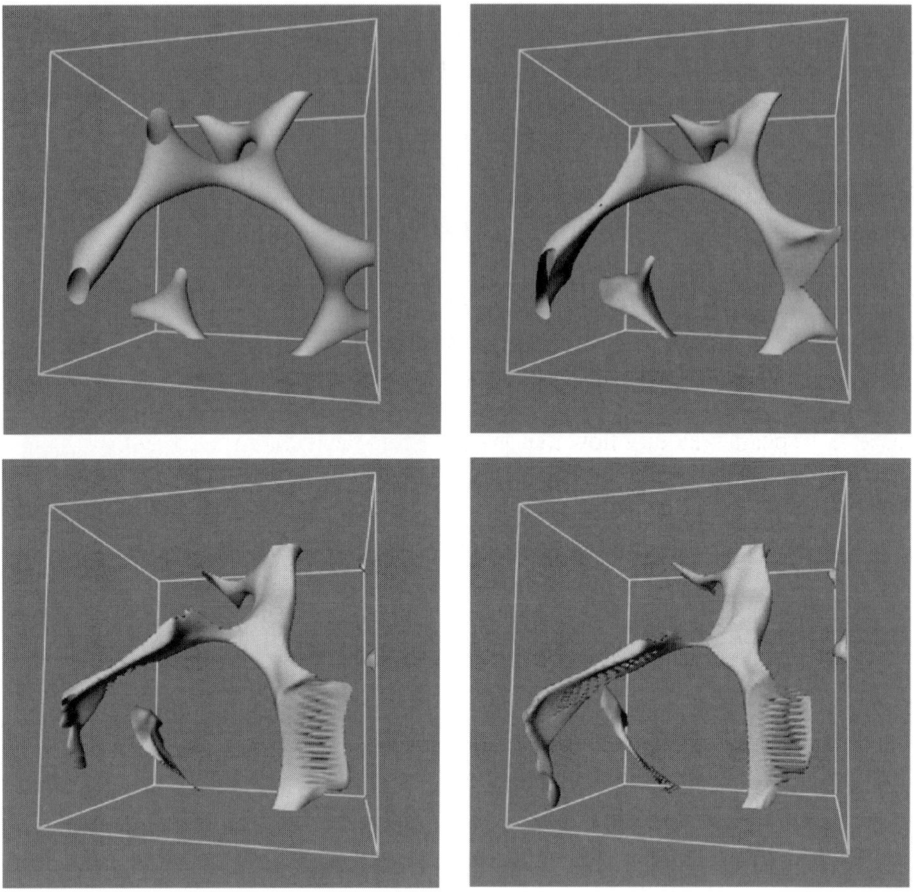

Figure 4. Isosurfaces of weak magnetic field strength showing the location of the nulls and their connections to the three nearest null points. The driving is imposed on the left- and right-hand side boundaries of the box. The panels represent the time development of the isosurface of constant magnetic field. The time for the panels, given in average Alfvén crossing times of the full domain, are 0, 1.0, 2.0, and 3.0. A movie of the time-dependent evolution is available on the www (*www-solar.mcs.st-and.ac.uk/~klaus*).

and not the remote fan-fan connections or the spine axes? What drives the collapse and determines the topology of the sheet formation?

When investigating the field-line topology along the two types of separator lines (the near and remote neighbours) it becomes obvious that the field line topology in planes perpendicular to the separators is slightly different for the two cases. The topology along the separator line connecting the nearby neighbours has a projected field-line topology that looks like a 2D x-point with a 90-deg angle between the two separator lines. In 2D this topology is unstable to any perturbation and the x-point can collapse into a sheet. In 3D this is not necessarily the case, because there is a field component along the separator which is dominating in a finite region around the separator line. This component then acts as a magnetic scalar pressure that may slow down or even stop the collapse from occurring when the perturbation is below a given amplitude. Along the separator connecting the remote neighbours the field lines are predominantly parallel to the separator and they only have a weak component showing an x-type topology in the perpendicular plane. The magnetic energy contained in this plane is therefore much smaller than that of the component along the separator and an x-type collapse along this class of separator lines is not likely to occur. The field-line topology along the spine axis is rotationally symmetric and either converging to or diverging from the spine axis depending on the location between the two nulls. From analytical and numerical investigations (Priest and Titov, 1996; Galsgaard, Rickard, and Reddy, 1997; Rickard and Titov, 1996; Craig and Fabling, 1996) it is known that to excite a current accumulation along the spine axis a very special velocity flow that is centred on the spine axis is required. This is unlikely to occur when the magnetic field is perturbed by 'random' perturbations. The most likely location for current accumulation in a double null point system is along the separator line, just as seen in these experiments, because the fan planes provide a much larger surface for trapping perturbations of the magnetic field than a single field line. It is, therefore, not as sensitive to the exact shape of the perturbation as spine reconnection.

To understand the collapse of the weak field tunnels we have to investigate the structural changes along the separator and its relation to the stress imposed by the boundary motions. For this purpose we concentrated on one particular null point pair in the domain. This pair represents the separator line that collapsed first in Figure 4 and was chosen because it is close to one of the driving boundaries. This makes the mapping between the field-line topology and the imposed boundary flow easier to follow. Investigating the development of the other separator lines in the domain shows that they become unstable to the same type of perturbation, but at different times.

Figure 5 shows the field-line topology of the two nulls and the structure of the imposed boundary motions. From the figure it is seen that the null point is orientated in a skew angle with respect to the driving boundary. The fan planes and spine axes are therefore penetrating the driving boundary at an angle. As the boundary motion stresses the field the perturbation propagates along the field lines. By looking at Figure 5 it is clear that the magnetic fields on either side of the fan plane are exposed to different types of stress. On one side (to the left of the fan plane of the nearest null) the boundary motion pushes the field towards the fan plane and up towards the null. This forces the null structure on this side of the fan plane to move away from the boundary and further into the domain. On the other side of the fan plane (on the right-hand side of the fan plane) the boundary motion drags the magnetic field away from the null point and down towards the driving boundary. Together this produces a shear motion across the fan plane and the spine axis on either side of the fan plane is moved in opposite directions. This shear of the null leads to the formation of a current concentration passing through

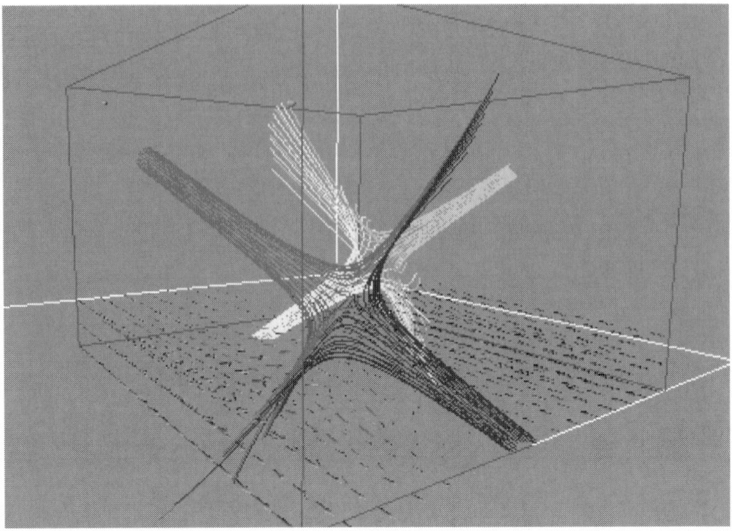

Figure 5. The initial field-line topology of the investigated null point pair is shown together with the structure of the imposed boundary velocity. The field lines indicate the magnetic field line structure and the vectors give the sinusoidal structure of the imposed boundary motions, with the velocity out of the paper in the left part of the box and into the paper in the right part of the box. The white lines indicate the locations of the full domain.

the null point, destroying its identity. Snapshots of the temporal development are shown at different times in Figure 6 and as a movie on the www (*www-solar.mcs.st-nd.ac.uk/˜ klaus*).

If a null point is isolated in space, then this type of collapse may occur at any orientation around the spine axis. In a double null this is not the case. The fact that the null point is a part of a pair of nulls imposes strong restrictions on how the collapse may take place, if the current has to accumulate along all of the separator line. The topology of the nulls means that the fan planes are orientated at an angle of nearly 90 deg to each other – the weak rotation along the separator is due to the initial force free current along the separator line, with the nulls having an angle of 90 deg between the fan plane and the spine axis. Because the initiation of the current sheet formation is driven by a shear flow across the fan plane, a single current sheet passing through both nulls is restricted to only two different orientations relative the structure of the nulls. In all cases it will have an orientation that penetrates the null at an angle of ± 45 deg relative the spine axis. In the case of Figure 5 the current sheet connecting the two nulls can either be horizontal or vertical. The choice of orientation is determined by the perturbation of the null system. This way the formation of the current sheet along the separator structurally looks like the collapse of a 2D x-point into two y-points connected by a current sheet. It is found that the fan plane and spine axis are changed into two separator surfaces extending from the edge of the current sheet. The identity of the nulls

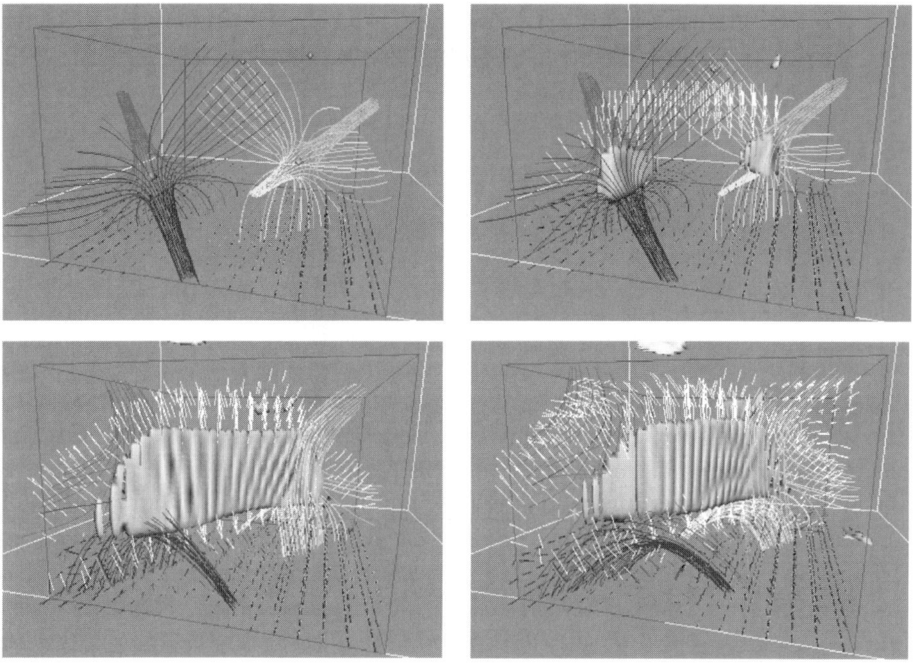

Figure 6. The panels show the collapse of the separator line. Indicating the degeneration of the null structure as a function of time. The panels represent the development at the times 0, 1.0, 2.0, and 3.0 in time units of the initial average Alfvén crossing time of the full domain. The field lines represent the magnetic field-line topology around the nulls. The starting points for the field-line traces are advected with the velocity flow as a function of time. This is done to indicate the change in field line connectivity as a function of time. The vectors at the base indicate the driving velocity, while the white vectors represent the high velocity outflow jets driven by the reconnection. The isosurface shows the location of the current accumulation. A movie of the time-dependent evolution is available on the www (*www-solar.mcs.st-and.ac.uk/~ klaus*).

and especially the identity of the spine axis is destroyed as a consequence of the separator collapse.

In the ideal limit the current sheet is formed because of the shear of the null points. This shear spreads along the separator and forms the current sheet where the peak current strength depends on how much the boundaries are moved. When the magnetic resistivity is non-zero then the magnetic Reynolds number decreases and eventually becomes of order unity in the current sheet and magnetic reconnection becomes important for the continued dynamical evolution. In the resistive phase the dynamics around the separator change dramatically. The initial shear flow is replaced by a stagnation flow that is initiated by the outflow jet sheets that develop along the sheet. The plasma in the current sheet is accelerated towards the edges by the tension force of the reconnected field lines. As the plasma is transported out of the sheet, then the plasma pressure and the magnetic pressure decrease at the centre of the current sheet and this initiates an inflow into the sheet. The new flux

TABLE I
Main parameters for the experiments

Number	t_d	v_d	$\langle\beta\rangle$	N^3	η
3	2	2	$\frac{1}{6}$	88^3	nu
4	8	0.5	$\frac{1}{6}$	88^3	nu
5	4	1	$\frac{1}{3}$	88^3	nu
6	4	1	$\frac{1}{12}$	88^3	nu
8	4	1	$\frac{1}{6}$	88^3	$2 \times$ nu
9	4	1	$\frac{1}{6}$	88^3	$0.5 \times$ nu
10	4	1	$\frac{1}{6}$	88^3	nu
11	4	1	$\frac{1}{6}$	88^3	0.001
12	4	1	$\frac{1}{6}$	88^3	0.01
15	4	1	$\frac{1}{6}$	88^3	0.003
16	4	1	$\frac{1}{6}$	88^3	0.005
17	4	1	$\frac{1}{6}$	88^3	0.0003
18	4	1	$\frac{1}{6}$	176^3	0.003
19	4	1	$\frac{1}{6}$	176^3	0.001
20	4	1	$\frac{1}{6}$	176^3	0.0003

reconnects and is expelled by the jet and the stagnation flow is started. The process continues until there is no more magnetic energy to be released and the current sheet then fades away.

4. Scaling Relations

Experiments have been carried out with different values of the boundary driving velocity, the average plasma β and the magnetic resistivity, with the intention of investigating the dependence of the dynamical behaviour on these parameters. The initial parameters of the experiments are given in Table I, where the columns represent the experiment number, the driving time, the driving velocity, the average plasma β, the numerical resolution and the type/value of the magnetic resistivity. For the magnetic resistivity 'nu' means that a fourth-order method combined with a discontinuous capture mechanism to provide the highest possible spatial resolution for the given numerical resolution has been used. In the rows with only a numerical value, this value is used as a constant magnetic resistivity.

The results of the experiments are summarised in Table II, where the time for the peak current, the peak current, the peak width of the current sheet in units of

TABLE II
Peak value results for the experiments.

Number	t_I	I_p	width	t_v	v
3	1.07	140.9	0.25	1.07	1.19
4	3.91	97.1	0.13	2.55	0.56
5	1.55	122.1	0.15	1.85	0.81
6	1.85	120.9	0.19	1.76	0.87
8	2.25	103.1	0.17	1.67	0.79
9	2.22	137.6	0.18	1.75	0.96
10	1.76	122.1	0.17	1.76	0.87
11	1.62	98.5	0.188	1.73	0.84
12	1.80	7.49	0.132	1.48	0.65
15	1.35	42.5	0.174	1.35	0.68
16	1.23	24.1	0.128	1.36	0.58
17	1.59	114.2	0.200	1.78	1.22
18	1.35	43.4	0.18	1.35	0.71
19	1.46	109.9	0.187	1.61	0.90
20	1.87	228.1	0.197	1.97	1.11

the unit box, the time for peak outflow velocity, and the peak outflow velocity are given.

Figure 7 shows the time dependent behaviour of the peak current, the sheet width and the outflow velocity as functions of magnetic resistivity and the maximum shear stroke ($v_d\, t$) for experiments 11–17. This shows that the dynamical evolution is time dependent and that a steady state is never reached. The reason is that the magnetic field structure around the current sheet is constantly changing after the reconnection is initiated. The relative importance of the field line component along the separator decreases with time as new flux is advected into the current sheet with different field orientations and strength.

A question often raised regards the speed of magnetic reconnection. Is the process fast enough to release enough magnetic energy in a high Reynolds-number plasma such as the solar corona? Can this really account for the energy required in a large solar flare? An attempt to provide scaling relations has been made. By taking the peak values from Table II plots of the dependence of the peak current, the maximum sheet width, and the peak outflow velocity on the constant resistivity are shown in Figure 8.

The left panel shows that the peak current has a near power law like dependence on the value of the magnetic resistivity, ν. The stars represents the low-resolution experiment while the diamonds are from the high-resolution experiments. The

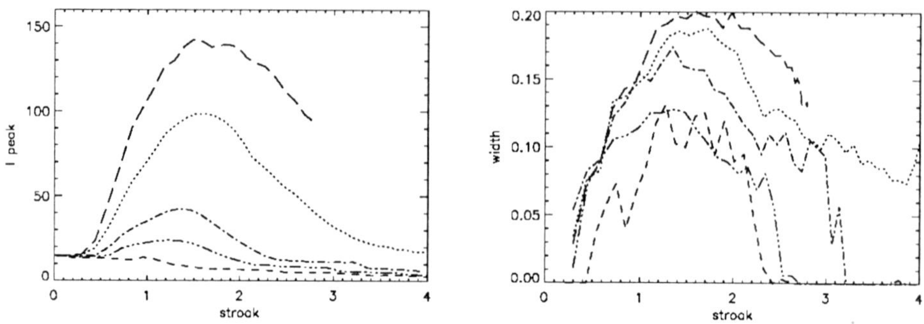

Figure 7. The panels represent the time-dependent development of the peak current, and the sheet width as functions of the boundary shear stroke for the experiments 11–17. The lines represents different values of the magnetic resistivity, η. The dotted line represents experiment 11 ($\eta = 0.001$), the short dashed line is experiment 12 ($\eta = 0.01$), the dot-dashed line is experiment 15 ($\eta = 0.003$), the dot-dot-dot-dashed line is experiment 16 ($\eta = 0.005$), and the long dashed line is experiment 17 ($\eta = 0.0003$).

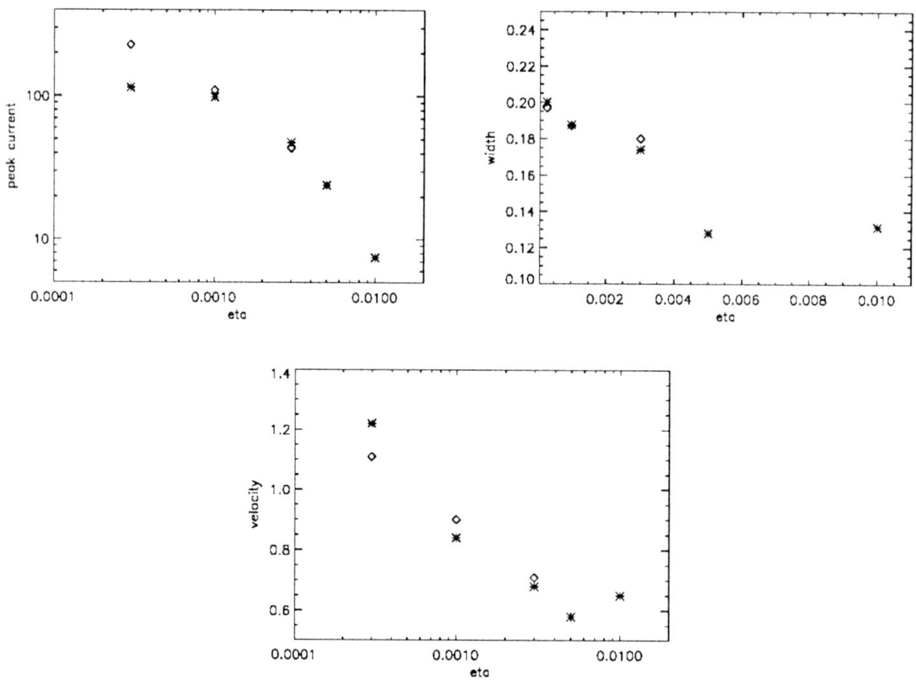

Figure 8. The panels show the scaling of the peak current, the width of the current sheet, and the outflow velocity as functions of η for experiments 11–20. The stars are used for the low-resolution experiments and the diamonds are used for the high-resolution experiments.

turnoff at the lowest value of η for both resolutions shows that the peak current is sufficiently resolved for η vales below the turnoff point for the given numerical resolution. Taking this into account, then the spread in η values with well resolved current concentrations is too small to provide a scaling relation for the peak current with any confidence. The peak values are reached well before the driving on the boundaries is changed. The peak values are therefore representing real peak values in the fully dynamical evolution of the magnetic field topology.

The middle panel of Figure 8 shows the peak jet velocity out of the current sheet. The velocities are well determined over the same range of η as the peak current. The trend indicates an exponential decrease of the jet velocity with η. Again the velocity for the lowest η value seems to be affected by lack of numerical resolution.

The width of the current sheet is found to decrease linearly with η and seems to reach a maximum extent of 22% of the domain for zero magnetic resistivity, the right panel. This corresponds to a width/length ratio of about 0.62 as the length of the current sheet is at all times close to the initial distance between the nulls. The values become even larger when the driving velocity is increased further, see below. In all of the experiments it is found that the thickness of the current sheet is always collapsing to the minimum numerical resolution – except for the highest values of η.

In experiments 1–10, different plasma β values and driving speeds were tried. The results are shown in Figure 9. The left panels show the dependence on the plasma β and the right panels show the effect of the imposed driving velocities. The top panels show that the peak current is nearly independent of the value of the average plasma β. The β value mentioned in Table I is an average value for the full domain, therefore in the vicinity of the separator the local plasma β is much larger than unity because the magnetic field strength is by definition low along this field line. Therefore, even changes of larger than unity of the β values may not change the dynamical evolution significantly. On the other hand, it is found that the peak current increases with the driving velocity. This may be explained by the higher energy content that is available per time unit for the same stroke distance when the driving speed is increased. From the middle row it is seen that the outflow velocity is somewhat dependent on the plasma β, while more dependent on the driving velocity. Especially it is seen that the time for the peak value is delayed in terms of the stroke distance as the driving speed is increased. In the lower panels the width of the current sheet is seen to behave in a similar way to the outflow velocity. The strong dependence of the sheet width on driving velocity is a result of the finite reconnection speed of the sheet. More magnetic flux is advected into the current sheet as the driving speed is increased and the sheet has to increase its width to increase its reconnection speed.

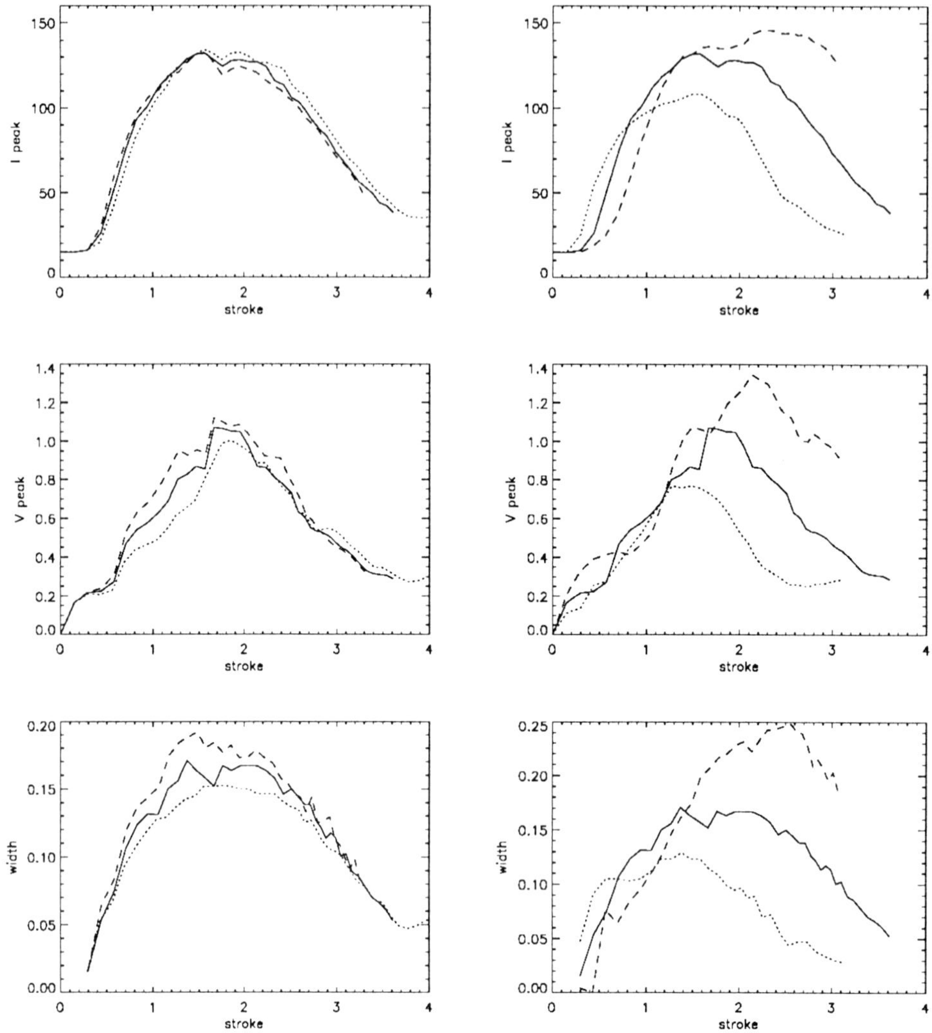

Figure 9. The left panels show the dependence on plasma β and the right panels the dependence on driving velocity. The top row represent the peak current, the middle row the outflow velocity and the bottom row the width of the current sheet. All data are from experiments 1–10. The full line represents the standard case, case number 10. In the left panels the long-dashed line case 6 and the dotted line case 5. In the right panels the long-dashed line case 3 and the dotted line case 4.

5. Conclusion

We have performed numerical experiments to investigate the current sheet formation along a magnetic separator defined by the intersection of the fan planes of two magnetic null points. These experiments show that the current accumulation is generated by a shear flow across the fan plane of the two nulls. To generate a current sheet all the way along the separator there are only two possible orientations for the collapse. These are at angles of about ± 45 deg of the spine axis. The spine axes no longer seem to be important for this particular shear. The continuous existence of the fan planes and spine axes through the null points is disrupted by the development of the separator current sheet. The fan plane is cut into two halves by the developing current sheet, with each half of the fan plane being attached to either edge of the sheet. In the same process the spine axis on either side of the fan plane connects to different ends of the current sheet. As part of the disruption process the spine axis looses its identity as a singular line reaching the null and is integrated into the separator surfaces defining the x-type topology along the separator line. The formation of the separator current sheet is, in contrast to the similar process in 2D, formed by a shear flow across the separator, and not until magnetic reconnection starts will a stagnation flow centred in the sheet develop.

The nature of reconnection in 3D and in particular whether or not it is fast is a complicated issue (Priest and Forbes, 2000), which we shall discuss in a future analysis of the data of this experiment.

In the discussion of null point reconnection until now it has not been decided which of the three possible types (spine, fan, and separator) is the most important for the dynamical evolution of a magnetic field. This experiment strongly suggests that separator reconnection is the most important type of null reconnection. This conclusion was also reached by Galsgaard, Rickard, and Reddy (1997), who investigated a more restricted set of double null situations. From theoretical considerations this may not be that surprising. The fan surfaces have a much larger interaction surface than the spine axis, and will therefore be much easier to perturb in a way that information is propagated towards the separator where reconnection is then able to take place. It would therefore be of great interest to find a well observed energy release event on the Sun, where we can say with confidence that null points are involved and then try to understand the process that leads to the energy release in this situation.

Acknowledgements

KG and EP would like to thank PPARC for financial support. The work of ÅN was supported in part by the Danish Research Foundation, through its establishment of the Theoretical Astrophysics Center. We acknowledge the use of the PPARC funded Compaq MHD Cluster in St. Andrews.

References

Craig, I. J. D. and Fabling, R. B.: 1996, *Astrophys. J.* **462**, 969.
Galsgaard, K. and Nordlund, Å: 1997, *J. Geophys. Res.* **102**, 231.
Galsgaard, K., Rickard, G., and Reddy, R.: 1997, *Solar Phys.* **176**, 299.
Hesse, H. and Schindler, K.: 1988, *J. Geophys. Res.* **93**, 5559.

Parnell, C. E., Smith, J. M., Neukirch, T., and Priest, E. R.: 1996, *Phys. Plasma* **3**, 759.
Priest, E. and Forbes, T.: 2000, *Magnetic Reconnection: MHD Theory and Applications.* Cambridge University Press, Cambridge, in press.
Priest, E. R. and Titov, V. S.: 1996, *Phil. Trans. R. Soc. London* **A354**, 2951.
Rickard, G. J. and Titov, V. S.: 1996, *Astrophys. J.* **472**, 840.
Syrovatsky, S. I.: 1981, *Ann. Rev. Astron. Astrophys.* **19**, 163.

THE STRUCTURE OF FORCE-FREE MAGNETIC FIELDS

E. A. EVANGELIDIS, L. L. VAUGHAN and G. J. J. BOTHA*
Centre for Physics and Mathematics, 2 Solonos St., Volos, Greece
**Department of Pure and Applied Physics, Queen's University, Belfast BT7 1NN, U.K.*

(Received 17 September 1999; accepted 10 February 2000)

Abstract. Incontrovertible evidence is presented that the force-free magnetic fields exhibit strong stochastic behavior. Arnold's solution is given with the associated first integral of energy. A subset of the solution is shown to be non-ergodic whereas the full solution is shown to be ergodic. The first integral of energy is applied to the study of these fields to prove that the equilibrium points of such magnetic configurations are saddle points. Finally, the potential function of the first integral of energy is shown to be a member of the Helmholtz family of solutions. Numerical results corroborate the theoretical conclusions and demonstrate the robustness of the energy integral, which remains constant for arbitrarily long computing times.

1. Introduction

The concept of magnetic configurations producing no effect on ionized matter was introduced in astrophysics by Lüst and Schlüter (1954) in order to interpret an eruption on the Sun. The explicit solution of the governing equation was given by Chandrasekhar (1956a,b), Chandrasekhar and Kendall (1957) in terms of the Hansen (1935) solutions of the vectorial Helmholtz equation. The theoretical development led to Woltjer (1958) formulating the following result: *for any magnetic configuration with a magnetic field* **B** *derivable from a vector potential* **A** *constant on some boundary enclosing a volume V, the integral $\int_V \mathbf{A}\cdot\nabla\wedge\mathbf{A}dV = \int_V \mathbf{A}\cdot\mathbf{B}\,\mathrm{d}V$ is an ideal MHD invariant.* This expression was then used by Chandrasekhar and Woltjer (1958) to show that the variation of the integral $\int_V[B^2/8\pi - (\alpha/8\pi)\mathbf{A}\cdot\nabla\wedge\mathbf{A}]\,\mathrm{d}V$, with α a constant Lagrange multiplier and **A** constant on the boundary, leads to the force-free field expression

$$\nabla\wedge\mathbf{B} = \alpha\mathbf{B}\,. \tag{1}$$

It is noticed that the only requirement on α is that it should be a constant. With the advent of projects aiming at the magnetic confinement of thermonuclear plasma to produce energy from fusion, the force-free field configuration has been used extensively in spheromacs and reversed field pinches. An exhaustive review of this state of affairs can be found in Taylor (1986).

Today we know that the solar atmosphere is structured on a wide range of scales and that it is extremely dynamic (Golub and Pasachoff, 1997). The first evidence

The physics of the solar corona and transition region. Part II.
Reprinted from Solar Physics **193,** 2000.

for this was obtained from ground-based observations during the 1950s and 1960s, but it was only with the advent of space-based observations that the true nature of the solar atmosphere was discovered. Observations from *Skylab* during the 1970s (Poletto *et al.*, 1975) and more recently from the *Yohkoh*, SOHO, and TRACE missions show a solar plasma dominated by an ubiquitous magnetic field where force-free fields play a major role (Priest, 1984; Zirker *et al.*, 1997; Démoulin, 1999). Since there is an extensive literature on this subject we concentrate on the discussion of the structure of the force-free field equation.

The present analysis begins by presenting Arnold's (1965) solution of the force-free field equation in Cartesian coordinates and the first integral of energy is derived, which has gone undetected so far. A subset of solutions, with one of the components of the magnetic field zero, is proved theoretically and confirmed numerically to be non-ergodic (Section 3). Then, in Sections 4 and 5, the equilibrium points of the full solution are shown to be saddle points of hyperboloidal surfaces in phase space, thus proving that the force-free field configuration is unstable. Finally, the analytic solution is generalized and the associated potential function of the energy integral is shown to be an eigenfunction of the Helmholtz equation of eigenvalue $\sqrt{2}$.

2. The Dynamical Structure of the Force-Free Magnetic Fields

The solution of the force-free field equation in Cartesian coordinates offers an insight into the nature of such magnetic configurations. It is a simple matter to prove that the following system of equations,

$$\begin{aligned}
\frac{\mathrm{d}x}{\mathrm{d}s} &= B_x = A \sin z + C \cos y \,, \\
\frac{\mathrm{d}y}{\mathrm{d}s} &= B_y = B \sin x + A \cos z \,, \\
\frac{\mathrm{d}z}{\mathrm{d}s} &= B_z = C \sin y + B \cos x \,,
\end{aligned} \tag{2}$$

satisfies the force-free field equation in the form $\mathbf{B} \wedge (\nabla \wedge \mathbf{B}) = 0$. In system (2) A, B, and C are constants and x, y, and z are some generalized angle coordinates. s is an affine parameter, for instance it could represent distance along a streamline, or time in the case of a velocity field. This solution was first given by Arnold (1965) in the context of topology. Figure 1(a) follows one streamline in three-dimensional Cartesian space, using a Runge–Kutta numerical integration of system (2). The pitch angle can be calculated from the expression $\mathrm{d}z/\mathrm{d}y$. The solution is extremely sensitive to the parameters and initial position, as shown by Figure 1(b). Figure 1(b) also shows that the streamlines twist around each other as they move through space.

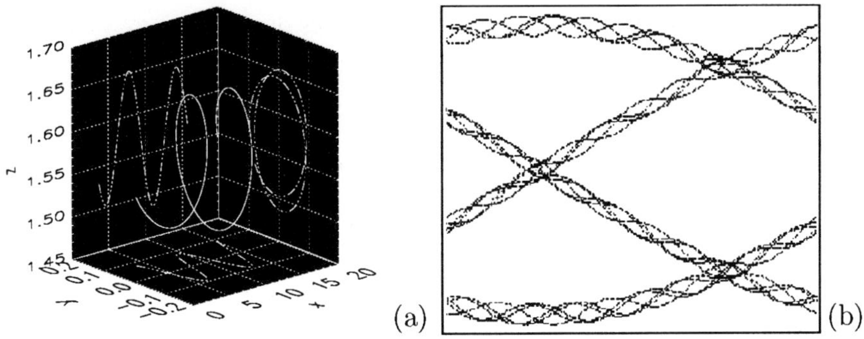

Figure 1. Streamlines of Arnold's solution with (a) $A = 1.1$, $B = 0.3$, $C = 1.3$ and initial position $x = 3.78$, $y = 0.08$, and $z = 1.53$; (b) with variation of typical parameters and initial conditions.

These numerical results were obtained with an energy conservation to machine accuracy. The general case of solution (2) with all constants nonzero, will be discussed first, since this will facilitate the understanding of the analysis of the special cases.

Taking the derivatives of expressions (2) the system follows

$$\frac{d^2x}{ds^2} = \alpha \cos x \cos z - \beta \sin x \sin y \ ,$$

$$\frac{d^2y}{ds^2} = \beta \cos y \cos x - \gamma \sin y \sin z \ , \tag{3}$$

$$\frac{d^2z}{ds^2} = \gamma \cos z \cos y - \alpha \sin z \sin x \ ,$$

with $\alpha = AB$, $\beta = BC$, and $\gamma = CA$. These are the equations of motion of a three-dimensional oscillator each component of which is under the influence of a combination of periodic forces described by the expressions on the right-hand sides. It is remarkable that the simplest possible equation of magneto-hydrodynamics has been transformed into a complicated problem of dynamical mechanics. These results can be written in the succinct form

$$\frac{d^2x_i}{ds^2} = -\frac{\partial U}{\partial x_i} \tag{4}$$

with

$$-U = \alpha \sin x \cos z + \beta \sin y \cos x + \gamma \sin z \cos y \ . \tag{5}$$

The existence of a potential function allows the derivation of the Hamiltonian of the system, which in this occasion reads

$$H = \tfrac{1}{2}(\dot{x}^2 + \dot{y}^2 + \dot{z}^2) - \alpha \sin x \cos z - \beta \sin y \cos x - \gamma \sin z \cos y \ , \tag{6}$$

$$H = \tfrac{1}{2}(A^2 + B^2 + C^2) . \tag{7}$$

It is noticed that the results presented so far correspond to the special case of the force-free field equation $\nabla \wedge \mathbf{B} = k\mathbf{B}$ when $k = 1$. The generalization of Arnold's solution to cases when $k \neq 1$ will be given in Section 6.

3. Particular Cases of Arnold's Solution

If Arnold's solution, system (2), is considered with any two of its constants equal to zero, then the motion it describes is non-ergodic. For instance, when $B = C = 0$ the motion is the straight line $y = \alpha_0 x + \alpha_1$ with α_0 and α_1 constant.

If one of the constants in Arnold's solution is zero (say $A = 0$), then the system of equations can be reduced to its normal form of two independent oscillators

$$\begin{aligned} \frac{\mathrm{d}^2\xi}{\mathrm{d}t^2} &= \beta \cos\xi , \\ \frac{\mathrm{d}^2\eta}{\mathrm{d}t^2} &= \beta \cos\eta , \end{aligned} \tag{8}$$

with

$$2x = \xi - \eta, \qquad 2y = \xi + \eta . \tag{9}$$

The corresponding energy levels are

$$\tfrac{1}{2}\dot{\xi}^2 - \beta \sin\xi = e_1, \qquad \tfrac{1}{2}\dot{\eta}^2 - \beta \sin\eta = e_2 , \tag{10}$$

whereas the integral of energy is

$$E = \frac{1}{2}\left(\frac{\dot{\xi}^2 + \dot{\eta}^2}{2} + \dot{z}^2\right) - \beta \sin\frac{\xi + \eta}{2} \cos\frac{\xi - \eta}{2} . \tag{11}$$

The solution of Equations (10) can be given in terms of elliptic functions and the problem is reduced to the motion of a point under the influence of two forces perpendicular to and independent of each other. Hence, the motion in the (ξ, η) plane is in general a complicated Lissajous figure. The z-component can be written now as

$$\dot{z} = \sqrt{2E - (e_1 + e_2)} \equiv D, \quad \text{constant.} \tag{12}$$

Equations (10) are valid for $e_1 - BC \geq 0$, $e_2 - BC \geq 0$, so that the motion in the (ξ, η) plane takes place inside the rectangle defined by the turning points of these equations. The decomposition (10) shows that the reduced problem with one of the constants A, B, or C zero, is non-ergodic, since a point lying initially on manifold e_1 will never visit manifold e_2 and *vice versa*.

Another way of studying the reduced Arnold's solution is the following. The system of equations

$$\dot{x} = C \cos y, \qquad \dot{y} = B \sin x \tag{13}$$

has the particular integral

$$B \cos x + C \sin y = D\,, \tag{14}$$

a result directly derivable from the z-component of (3), which confirms that the velocity $\dot{z}$ is constant and therefore $z = Dt + z(0)$. The simplified form of Arnold's solution for $B = C$, $D = 0$ can be integrated immediately to give

$$\tan\frac{x}{2} = e^{Ct}\,, \qquad \tan\frac{y}{2} = \frac{e^{Ct} - 1}{e^{Ct} + 1}\,, \tag{15}$$

which shows that the solution settles to the point attractor $x = \pi$, $y = \pi/2$ as $t \to \infty$. Arnold's solution for $A = 0$, $B, C \neq 0$ (see Figures 2 and 3) is reduced to quadrature upon deriving

$$\frac{\dot{x}^2}{C^2} + \frac{D^2}{C^2}\left(1 - \frac{B}{D}\cos x\right)^2 = 1\,, \tag{16}$$

$$\frac{\dot{y}^2}{B^2} + \frac{D^2}{B^2}\left(1 - \frac{C}{D}\sin y\right)^2 = 1. \tag{17}$$

These expressions are obtained by squaring each component of system (2) and substituting Equation (14) into them. Equations (16) and (17) will be used to draw two- and three-dimensional phase portraits $(\dot{x}, x)$, $(\dot{y}, y)$, or $(x, y, \dot{x})$. It is rather obvious that they represent segments of circles in phase space of radii C and B respectively. For instance, the circle for (16) is defined by the transformation $X = D(1 - B\cos x/D)$. The integrated form of (16) reveals a palindromic motion. In order to integrate (16) it is rewritten in the form

$$C\,\mathrm{d}t = \frac{\mathrm{d}x}{\sqrt{1 - \dfrac{D^2}{C^2}\left(1 - \dfrac{B}{D}\cos x\right)^2}}\,, \tag{18}$$

the denominator of which leads to the inequality

$$\frac{D - C}{B} < \cos x < \frac{D + C}{B} \tag{19}$$

which provides the upper and lower limits of the integration. By expanding the right-hand side of (18) and applying the integration limits (19), the integrated form of (16) gives

$$\frac{T_x}{2} = \frac{1}{C}\Bigg\{\left[\cos^{-1}\left(\frac{D-C}{B}\right) - \cos^{-1}\left(\frac{D+C}{B}\right)\right]\left[1 + \frac{D^2}{2C^2} + \frac{B^2}{4C^2}\right] +$$

$$+\frac{BD}{4C^2}\left[\left(3-\frac{C}{D}\right)\sqrt{1-\frac{(D+C)^2}{B^2}} - \right.$$

$$\left. - \left(3+\frac{C}{D}\right)\sqrt{1-\frac{(D-C)^2}{B^2}}\right]\Bigg\} \tag{20}$$

accurate to first order, where T_x is the period of the palindromic motion. Similarly the motion in $(\dot{y}, y)$ space has two turning points and the motion takes place within the limits

$$\frac{D-B}{C} < \sin y < \frac{D+B}{C}\,. \tag{21}$$

The period of the $(\dot{y}, y)$ motion is obtained from (20) by interchanging the constants B and C. Similar results hold true for the particular cases of Arnold's solution when one of B or C is zero and the relevant equations are summarized immediately below:

$$B = 0: \tag{22}$$

$$\dot{x} = C\cos y + A\sin z \equiv F,$$

$$\frac{\dot{y}^2}{A^2} + \frac{F^2}{A^2}\left(1 - \frac{C}{F}\cos y\right)^2 = 1, \quad \frac{\dot{z}^2}{C^2} + \frac{F^2}{C^2}\left(1 - \frac{A}{F}\sin z\right)^2 = 1.$$

$$C = 0: \tag{23}$$

$$\dot{y} = A\cos z + B\sin x \equiv G,$$

$$\frac{\dot{z}^2}{B^2} + \frac{G^2}{B^2}\left(1 - \frac{A}{G}\cos z\right)^2 = 1, \quad \frac{\dot{x}^2}{A^2} + \frac{G^2}{A^2}\left(1 - \frac{B}{G}\sin x\right)^2 = 1.$$

The corresponding integrals of energy, similar to (11), can be found by equating to zero the corresponding coefficient in (6). Therefore, the study of this subset of special force-free configurations leads to the conclusion that in the cross section z = constant there develop areas in which motion along streamlines becomes trapped in 'boxes', within which it performs a crescent like motion. On this motion is further superimposed the downstream motion. The similarity of this behavior with that of the 'banana' orbits of magneto-hydrodynamics is striking. Since the period of the palindromic motion depends only on the constants B, C, and D (for

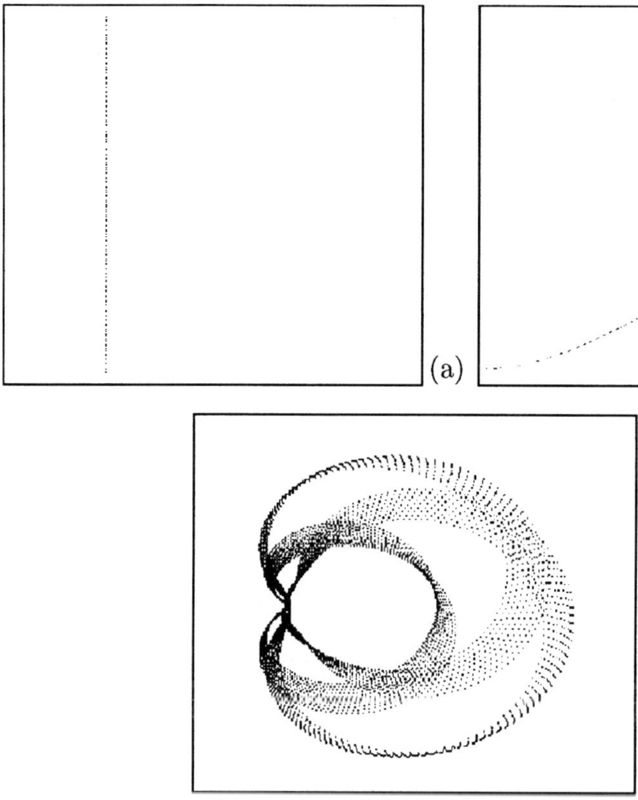

Figure 2. Non-ergodic Poincaré surface sections: (a) on the $x = 0$ plane and (b) on the $z = 0$ plane, with (c) the associated attractor, obtained by projecting the solution onto a periodic cylinder and plotting the (r, θ) plane with the z axis contracted. The parameters are $A = 0$, $B = 2$, $C = 5$ and the initial position is $x = 0$, $y = 0.201$, $z = 3$.

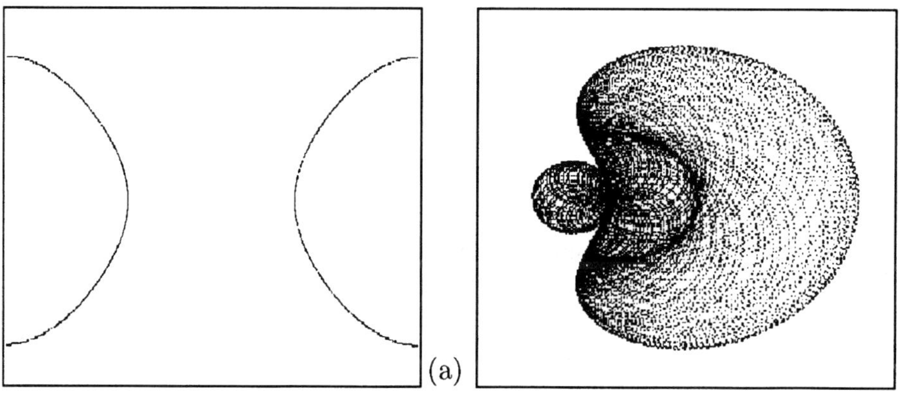

Figure 3. Non-ergodic (a) Poincaré surface section on the $z = 0$ plane with (b) the associated attractor. The parameters are $A = 0$, $B = 1.7321$, $C = 1.333$ and the initial position is $x = 4.867$, $y = 0.862$, $z = 0$.

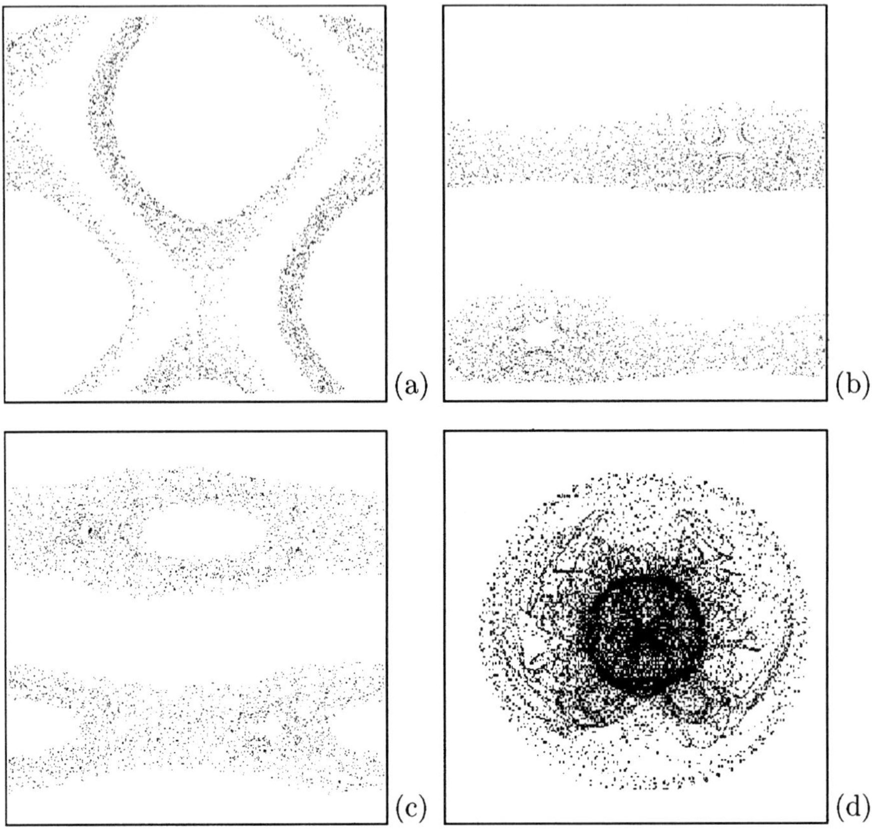

Figure 4. Ergodic Poincaré surface sections with their attractor. The parameters are $A = 1.111$, $B = 0.3115$, $C = 1.333$ and the initial position is $x = 1.57$, $y = 1.57$, $z = 0$. (a) The $x = 0$ plane. (b) The $y = 0$ plane. (c) The $z = 0$ plane. (d) The attractor.

$A = 0$), there is an ∞^3 of points partaking in the motion for any given set. At the same time the downstream motion along the z-direction tends to produce thinner and thinner filaments of any initial magnetic field, due to the incompressibility of such fields. This behavior corresponds to the classical description of an attractor and therefore, there are a total of ∞^5 such objects (Landau and Lifshitz, 1993).

4. Stability Analysis of Force-Free Magnetic Configurations

Figure 4 shows numerically the ergodic behavior of force-free magnetic fields when $A, B, C \neq 0$ for Arnold's solution (2). These numerical results were obtained with an energy conservation to machine accuracy. The ergodic behavior was also shown by Hénon (1966) using a purely numerical approach. In Section 2 it was shown that Arnold's solution leads naturally to the formulation of the potential function (5) and to the first integral of energy

$$E = \tfrac{1}{2}(\dot{x}^2 + \dot{y}^2 + \dot{z}^2) - \alpha \sin x \cos z - \beta \sin y \cos x - \gamma \sin z \cos y. \tag{24}$$

The existence of this first integral allows the discussion of ergodicity to proceed in terms of the decomposability of the energy manifold (Birkhoff, 1966; Chintsin, 1964). The existence of a potential function reduces the discussion of stability of the dynamical problem to the discussion of its extrema (Lagrange criterion). Indeed, if $\overline{\mathbf{x}}$ is a maximum of the potential function then for a perturbation leading to movement away from the maximum, $\dot{\mathbf{x}}^2/2 = U(\overline{\mathbf{x}}) - U(\mathbf{x}) > 0$. For a real potential this relation can always be satisfied. For direction of motion towards a maximum the relation $\dot{\mathbf{x}}^2/2 = U(\mathbf{x}) - U(\overline{\mathbf{x}}) < 0$ shows that such motion is physically impossible for real potentials.

Let $(\overline{x}, \overline{y}, \overline{z})$ be such an equilibrium point of system (2). Then from the defining expressions

$$A \sin \overline{z} = -C \cos \overline{y}, \qquad B \sin \overline{x} = -A \cos \overline{z}, \qquad C \sin \overline{y} = -B \cos \overline{x}\,, \tag{25}$$

the solution

$$\cos 2\overline{x} = \frac{C^2 - A^2}{B^2}, \quad \sin^2 \overline{x} = \frac{A^2 + B^2 - C^2}{2B^2}, \quad \cos^2 \overline{x} = \frac{B^2 + C^2 - A^2}{2B^2},$$

$$\cos 2\overline{y} = \frac{A^2 - B^2}{C^2}, \quad \sin^2 \overline{y} = \frac{B^2 + C^2 - A^2}{2C^2}, \quad \cos^2 \overline{y} = \frac{C^2 + A^2 - B^2}{2C^2},$$

$$\cos 2\overline{z} = \frac{B^2 - C^2}{A^2}, \quad \sin^2 \overline{z} = \frac{C^2 + A^2 - B^2}{2A^2}, \quad \cos^2 \overline{z} = \frac{A^2 + B^2 - C^2}{2A^2} \tag{26}$$

follows. One verifies easily that the derivatives of the potential function (5), that is the equations of motion (3), become zero at the stationary points (26). According to a theorem of real analysis:

A critical point c of a function f is a relative strict maximum, if the second derivative satisfies the relation $D_{ij} f(c) w^i w^j < 0$ for all directions w (Bartle, 1976).

Indeed, let us expand the potential in the vicinity of the critical point $\overline{\mathbf{x}}$ as defined by (25). It is

$$U(\mathbf{x}) = U(\overline{\mathbf{x}}) + (\mathbf{x} - \overline{\mathbf{x}})_i D_i U(\overline{\mathbf{x}}) + \tfrac{1}{2}(\mathbf{x} - \overline{\mathbf{x}})_i (\mathbf{x} - \overline{\mathbf{x}})_j D_{ij} U(\overline{\mathbf{x}}). \tag{27}$$

We have already proved that the first derivatives $D_i U(\overline{\mathbf{x}})$, that is the equations of motion, are zero for the stationary points of the force-free field equations. Recalling that at equilibrium points the velocity is zero, it follows from (24) that $U(\overline{\mathbf{x}}) = E$ and since $U(\mathbf{x}) - E = -u^2/2$, we are left with the equation

$$(\mathbf{x} - \overline{\mathbf{x}})_i (\mathbf{x} - \overline{\mathbf{x}})_j D_{ij} U(\overline{\mathbf{x}}) + u^2 = 0\,. \tag{28}$$

The second term can be calculated in the vicinity of the critical point from (2). It is

$$u^2 = A^2 + B^2 + C^2 + 2\alpha(x - \overline{x}) + 2\beta(y - \overline{y}) + 2\gamma(z - \overline{z}) \tag{29}$$

when terms up to first order are kept. Hence we find

$$\begin{aligned}
&B^2(x - \overline{x})^2 + C^2(y - \overline{y})^2 + A^2(z - \overline{z})^2 - \\
&-2(x - \overline{x})(y - \overline{y})\sqrt{(E_0 - C^2)(E_0 - B^2)} - \\
&-2(y - \overline{y})(z - \overline{z})\sqrt{(E_0 - A^2)(E_0 - C^2)} - \\
&-2(z - \overline{z})(x - \overline{x})\sqrt{(E_0 - B^2)(E_0 - A^2)} - \\
&-2\alpha(x - \overline{x}) - 2\beta(y - \overline{y}) - 2\gamma(z - \overline{z}) - 2E_0 = 0\,,
\end{aligned} \tag{30}$$

where $2E_0 = A^2 + B^2 + C^2$. Expression (30) can be further transformed into its canonical form upon using a well-established algebraic theorem, namely that: *Every quadratic form* $A(x, x) = \sum_{r=1}^{n} \sum_{s=1}^{n} \alpha_{rs} x_r x_s$ *in n variables and of rank r, with coefficients in a given field F, can be transformed by a non-singular transformation, with coefficients in F, into the form* $\alpha_1 X_1^2 + \cdots + \alpha_r X_r^2$ *(canonical form), where* $\alpha_1, \ldots, \alpha_r$ *are numbers in F and no one of them is equal to zero* (Ferrar, 1957).

For $n = 3$ the six distinct quadratic surfaces of geometry are retrieved. The actual expressions effecting the canonical transformation can be found in the Appendix, but they are of no importance in the qualitative discussion of the stability here. Naturally, not all of these six surfaces are acceptable solutions. Since they represent potentials of the force-free magnetic fields, their second derivatives must satisfy the constraints mentioned above. We summarize the situation in Table I.

The relief maps of the potential function in Figures 5 and 6 demonstrate clearly that the equilibrium points are saddle points, confirming the results of the theoretical description.

5. Stability of the Linear Mapping

Similar conclusions can be reached upon examination of the eigenvalues of the linear mapping of Arnold's solution (2). Expanding the right hand sides of those expressions in the neighborhood of the equilibrium point $(\overline{x}, \overline{y}, \overline{z})$, using double angle formulas and small angle approximations for the trigonometric functions, it is found that

$$\begin{aligned}
\dot{X} &= ZA\cos\overline{z} - YC\sin\overline{y}\,, \\
\dot{Y} &= XB\cos\overline{x} - ZA\sin\overline{z}\,, \\
\dot{Z} &= YC\cos\overline{y} - XB\sin\overline{x}\,,
\end{aligned} \tag{31}$$

TABLE I
Quadratic surfaces of geometry

1	$U = \alpha^2x^2 + \beta^2y^2 + \gamma^2z^2 - 1$. Ellipsoid.	$D^2f(c)(w)^2 = 2(\alpha^2x^2 + \beta^2y^2 + \gamma^2z^2) > 0$. Always positive. Physically unacceptable.
2	$U = \alpha^2x^2 + \beta^2y^2 - \gamma^2z^2 - 1$. One-sheet hyperboloid.	$D^2f(c)(w)^2 = 2(\alpha^2x^2 + \beta^2y^2 - \gamma^2z^2)$. A saddle point in general. In the present context $D^2f(c)(w)^2 < 0$ for $\alpha^2x^2 + \beta^2y^2 < \gamma^2z^2$ is the only physically acceptable branch. Branch $\alpha^2x^2 + \beta^2y^2 > \gamma^2z^2$ is inadmissible.
3	$U = \alpha^2x^2 - \beta^2y^2 - \gamma^2z^2 - 1$. Two-sheet hyperboloid.	Critical points are saddles. Only branch $\alpha^2x^2 < \beta^2y^2 + \gamma^2x^2$ is acceptable.
4	$U = \alpha^2x^2 + \beta^2y^2 - z$. Elliptic hyperboloid.	No critical point. Therefore, it is not a solution of the present dynamical system.
5	$U = \alpha^2x^2 - \beta^2y^2 - z$. Hyperbolic paraboloid.	No critical point. Inadmissible.
6	$U = \alpha^2x^2 + \beta^2y^2 - \gamma^2z^2$. Quadratic cone.	Trivial case, realizable for $A = B = C$. $(U(0) = 0)$

where $X = x - \overline{x}$, $Y = y - \overline{y}$ and $Z = z - \overline{z}$. Then the normal solution of the form

$$X = (X, Y, Z) = (X_i(0), Y_i(0), Z_i(0))e^{\lambda_i t} \tag{32}$$

can be found from system (31)'s characteristic equation (Dombre *et al.*, 1986)

$$\lambda^3 - \lambda\Lambda_1 - \Lambda_0 = 0 \tag{33}$$

with

$$\Lambda_1 = -A\cos\overline{z}B\sin\overline{x} - C\sin\overline{y}B\cos\overline{x} - A\sin\overline{z}C\cos\overline{y}\,, \tag{34}$$

$$\Lambda_0 = ABC(\cos\overline{x}\cos\overline{y}\cos\overline{z} - \sin\overline{x}\sin\overline{y}\sin\overline{z})\,. \tag{35}$$

Employing further expressions (25) we find

$$\Lambda_1 = \tfrac{1}{2}(A^2 + B^2 + C^2) = E_0\,, \tag{36}$$

$$\Lambda_0 = 2ABC\cos\overline{x}\cos\overline{y}\cos\overline{z}. \tag{37}$$

The last expression can be further transformed by means of solution (26) and $2E_0 = A^2 + B^2 + C^2$ into the form

$$\Lambda_0 = \pm 2\sqrt{(E_0 - A^2)(E_0 - B^2)(E_0 - C^2)}\,, \tag{38}$$

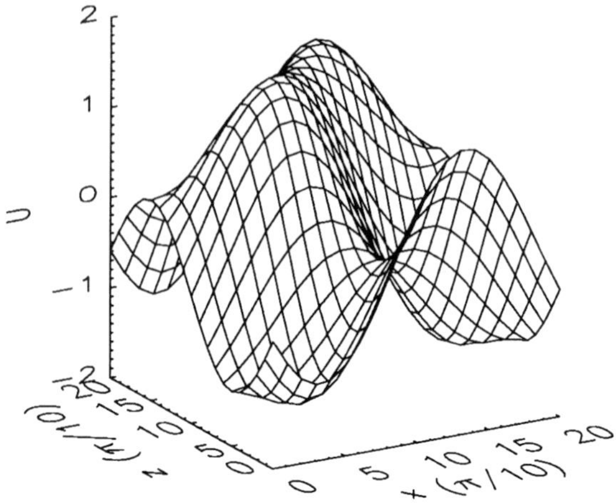

Figure 5. The potential function U on the (x, z) plane with $y = \pi/5$ and $\alpha = \beta = \gamma = 1$.

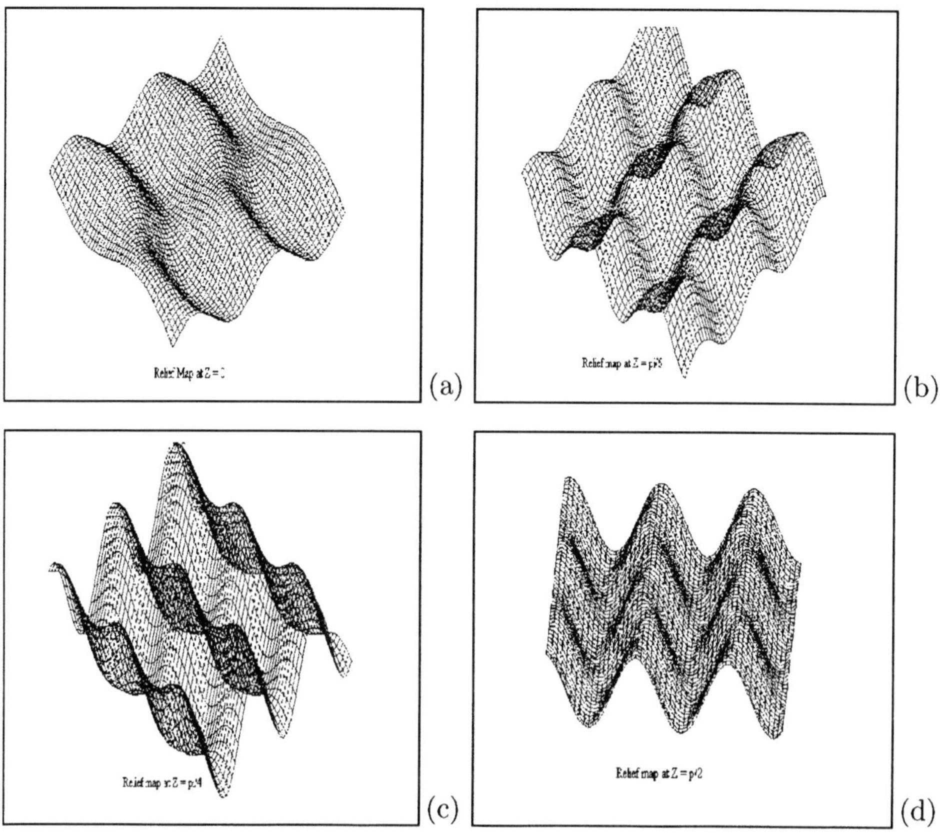

Figure 6. Relief maps of the potential U for typical values of α, β and γ on the (x, y) plane for (a) $z = 0$, (b) $z = \pi/6$, (c) $z = \pi/4$, (d) $z = \pi/2$.

so that the eigenvalues are given by the roots of

$$\lambda^3 - E_0\lambda \mp 2\sqrt{(E_0 - A^2)(E_0 - B^2)(E_0 - C^2)} = 0 \,. \tag{39}$$

The behavior of the general solution depends on whether the discriminant

$$\frac{b^2}{4} - \frac{E_0^3}{27} \quad \text{with} \quad b \equiv 2\sqrt{(E_0 - A^2)(E_0 - B^2)(E_0 - C^2)} \tag{40}$$

is larger, equal, or smaller than zero. The particular case where E_0 assumes the value of any amplitude A^2, B^2 or C^2 leads to the simplified equation

$$\lambda(\lambda^2 - E_0) = 0 \tag{41}$$

with eigenvalues

$$\lambda = 0, \quad \lambda = \pm\sqrt{E_0}. \tag{42}$$

In this case solution (32) is

$$X = X_1(0) + X_2(0)e^{+\sqrt{E_0}t} + X_3(0)e^{-\sqrt{E_0}t} \qquad (E_0 = A^2, B^2, C^2) \,. \tag{43}$$

The existence of the positive eigenvalue shows clearly that these particular equilibrium points are unstable for all real values of the parameters A, B, C in this particular case, except when the initial value of $X_2(0)$ is zero. The special case of the subset of solutions with $A = 0$, say, leads to exactly the same expression with $E_0 = (B^2 + C^2)/2$.

6. A Remark on the Potential Function of the Force-Free Magnetic Fields

As can be seen from the rotation of the force-free magnetic field equation (1),

$$\nabla^2\mathbf{B} + k^2\mathbf{B} = 0 \tag{44}$$

for any divergence free field, i.e., $\nabla \cdot \mathbf{B} = 0$. It follows that all solutions of Equation (1) should be members of the family of solutions to the Helmholtz equation (44), although the reverse is not true. Therefore, the question arises as to whether Arnold's solution is a member of the Helmholtz family of solutions. To that effect Arnold's solution is generalized to cases with $k \neq 1$. The generalized solution is easily found to be

$$\dot{x} = A\sin kz + C\cos ky \,, \tag{45}$$

$$\dot{y} = B\sin kx + A\cos kz \,, \tag{46}$$

$$\dot{z} = C\sin ky + B\cos kx \,, \tag{47}$$

$$-U = \alpha\sin kx\cos kz + \beta\sin ky\cos kx + \gamma\sin kz\cos ky \,, \tag{48}$$

with the first integral of energy

$$E = \tfrac{1}{2}(\dot{x}^2 + \dot{y}^2 + \dot{z}^2) - \alpha \sin kx \cos kz - \beta \sin ky \cos kx - \gamma \sin kz \cos ky \ . \tag{49}$$

The second partial derivatives of the generalized potential (48) give

$$\frac{\partial^2 U}{\partial x^2} + \frac{\partial^2 U}{\partial y^2} + \frac{\partial^2 U}{\partial z^2} = -2k_m^2 U \ . \tag{50}$$

The subscript has been added to indicate the eigenvalue of the magneto-static problem. The eigenvalue of Arnold's solution ($k = 1$) is $\sqrt{2}$. For the generalized case the relation $k_H = \sqrt{2}k_m$ holds true. Therefore, Arnold's solution is a member of the Helmholtz family of solutions.

7. Concluding Remarks

Following Arnold's considerations about fields described by magnetic force-free or vorticity-free type of equations, that they are generally chaotic, we have been able to derive rigorously explicit expressions for the source of this chaos. That is, we have proved that the equations of motion of these fields are described by an autonomous (time-independent) potential, whose equilibrium points are saddle points. The analysis has been particularly successful in defining explicitly ergodic and non-ergodic regions in parameter space and numerical integrations have borne out to any degree of accuracy the analytical proofs.

Force-free magnetic fields have attracted interest in space plasma research, where it is thought that this type of magnetic configuration might hold the key to the interpretation of observations. The idea has evolved that force-free fields represent steady state minima. This study shows that these fields are extremely unstable, which may play a significant part in explaining plasma behavior in space.

Appendix

We proceed here with the detailed derivation of the canonical expressions for the potential describing the force-free magnetic fields in the vicinity of their equilibrium point(s). For

$$\begin{aligned} b_{12} &= -\sqrt{(E_0 - C^2)(E_0 - B^2)}\ , & b_{11} &= B^2\ , \\ b_{23} &= -\sqrt{(E_0 - A^2)(E_0 - C^2)}\ , & b_{22} &= C^2 \\ b_{13} &= -\sqrt{(E_0 - B^2)(E_0 - A^2)}\ , & b_{33} &= A^2\ , \end{aligned} \tag{51}$$

Equation (30) can be written as

$$\begin{aligned} B(x,x) &= \sum_{i,j=1}^{3} b_{ij}x_i x_j = \sum_{r=1}^{3} b_{rr}x_r^2 + 2\sum_{r<s=1}^{3} b_{rs}x_r x_s \\ &= 2E_0 + 2\alpha x + 2\beta y + 2\gamma z \end{aligned} \tag{52}$$

with $x_1 = x,\ x_2 = y$ and $x_3 = z$. By making the transformation

$$X_1 = x_1 + \frac{b_{12}}{b_{11}}x_2 + \frac{b_{13}}{b_{11}}x_3, \qquad X_s = x_s, \quad s = 2,\ 3\ , \tag{53}$$

$$\beta_{22} = b_{22} - \frac{b_{12}^2}{b_{11}}, \qquad \beta_{33} = b_{33} - \frac{b_{13}^2}{b_{11}}, \qquad \beta_{23} = b_{23} - \frac{b_{12}b_{13}}{b_{11}}\ , \tag{54}$$

followed by a second transformation

$$X = X_1, \qquad Y = X_2 + \frac{\beta_{23}}{\beta_{22}}X_3, \qquad Z = X_3\ , \tag{55}$$

$$\alpha_{11} = b_{11}, \qquad \alpha_{22} = \beta_{22}, \qquad \alpha_{33} = \beta_{33} - \frac{\beta_{23}^2}{\beta_{22}}\ , \tag{56}$$

the bilinear form of Equation (52) becomes

$$B(x,x) = \alpha_{11}X^2 + \alpha_{22}Y^2 + \alpha_{33}Z^2\ . \tag{57}$$

Using transformations (53)–(56), the initial variables are now expressed in terms of the canonical variables of the quadratic form. They are

$$x = X - c_{12}Y + c_{13}Z, \qquad y = Y - c_{23}Z, \qquad z = Z \tag{58}$$

with

$$c_{12} = \frac{b_{12}}{b_{11}}, \qquad c_{13} = \frac{1}{b_{11}}\left(b_{12}\frac{\beta_{23}}{\beta_{22}} - b_{13}\right), \qquad c_{23} = \frac{\beta_{23}}{\beta_{22}}. \tag{59}$$

Substituting expressions (57) and (58) in (52), completing the squares and collecting terms produce the required expression

$$\delta_1 x_1^2 + \delta_2 x_2^2 + \delta_3 x_3^2 - 1 = 0 \tag{60}$$

used in the text. The symbols δ_i are abbreviations for the quantities

$$\delta_1 = \frac{\alpha_{11}}{\Delta}, \qquad \delta_2 = \frac{\alpha_{22}}{\Delta}, \qquad \delta_3 = \frac{\alpha_{33}}{\Delta}\ , \tag{61}$$

where

$$\Delta = \frac{\alpha^2}{\alpha_{11}} + \frac{\gamma_{12}^2}{\alpha_{22}} + \frac{\gamma_{13}^2}{\alpha_{33}} + 2E_0\ , \tag{62}$$

$$\gamma_{12} = \beta - \alpha c_{12}, \qquad \gamma_{13} = \alpha c_{13} - \beta c_{23} + \gamma\ , \tag{63}$$

and also

$$x_1 = X - \frac{\alpha}{\alpha_{11}}, \qquad x_2 = Y - \frac{\gamma_{12}}{\alpha_{22}}, \qquad x_3 = Z - \frac{\gamma_{13}}{\alpha_{33}}\,. \tag{64}$$

With these expressions we have completed our task of deriving the canonical magnetic surfaces of the force-free fields in terms of the initial constant magnetic fields of the problem.

Acknowledgements

Thanks are due to A. E. Evangelidis for producing numerous relief maps of the potential function U, some of which are reproduced in Figure 6.

References

Arnold, V. I.: 1965, *Comptes Rend. Acad. Sci., Paris* **261**, 17.
Bartle, R. G: 1976, *The Elements of Real Analysis*, John Wiley & Sons, New York, p. 379.
Birkhoff, G.: 1966, *Dynamical Systems*, American Mathematical Society, Vol. 106.
Chandrasekhar, S.: 1956a, *Proc. Natl. Acad. Sci.* **42**, 1.
Chandrasekhar, S.: 1956b, *Astrophys. J.* **126**, 232.
Chandrasekhar, C. and Kendall, P. C.: 1957, *Astrophys. J.* **126**, 457.
Chandrasekhar, S. and Woltjer, L.: 1958, *Proc. Natl. Acad. Sci.* **44**, 285.
Chintsin, A. J.: 1964, *Mathematische Grundlagen der Statistischen Mechanik*, Bibliographisches Institut, Mannheim.
Démoulin, P.: 1999, *J. Atmospheric Solar-Terrest. Phys.* **61**, 101.
Dombre, T., Frisch, U., Greene, J. M., Hénon, M., Mehr A., and Soward, A. M.: 1986, *J. Fluid Mech.* **167**, 353. After submitting our paper we discovered this paper with results similar to Equation (33).
Ferrar, W. L.: 1957, *Algebra*, Oxford University Press, Oxford, p. 148.
Golub, L. and Pasachoff, J. M.: 1997, *The Solar Corona*, Cambridge University Press, Cambridge.
Hansen, W. W.: 1935, *Phys. Rev.* **47**, 139.
Hénon, M.: 1966, *Comptes Rend. Acad. Sci., Paris* **262**, 312.
Jahnke, E. and Emde, F.: 1945, *Tables of Functions*, Dover Publications, New York, p. 168.
Landau, L. D. and Lifshitz, E. M.: 1993, *Fluid Mechanics*, Pergamon Press, Oxford, p. 13.
Lüst, R. and Schlüter, A.: 1954, *Z. Astrophys.* **34**, 263.
Poletto, G., Vaiana, G. S., Zombeck, M. V., Krieger, A. S., and Timothy, A. F.: 1975, *Solar Phys.* **44**, 83.
Priest, E. R.: 1984, *Solar Magnetohydrodynamics*, D. Reidel Publ. Co., Dordrecht, Holland.
Taylor, J. B.: 1986, *Rev. Mod. Phys.* **58**, 741.
Woltjer, L.: 1958, *Proc. Natl. Acad. Sci.* **44**, 489, 833.
Zirker, J. B., Martin, S. F., Harvey, K., and Gaizauskas, V.: 1997, *Solar Phys.* **175**, 27.

MODELS OF DYNAMIC CORONAL LOOPS

GIOVANNI PERES
Dip. di Scienze Fisiche ed Astronomiche, Sez. di Astronomia, Piazza del Parlamento 1, 90134 Palermo, Italy

(Received 23 September 1999; accepted 15 January 2000)

Abstract. This paper reviews the basic ideas underlying one-dimensional fluid dynamic models of coronal loops and presents some of their most recent applications. These models are an important theoretical support to explore the new scenario provided by the data of *Yohkoh*, SOHO, and TRACE, and are useful to interpret observations, when supplemented by appropriate spectral synthesis codes. Possible developments are also discussed.

1. Introduction

Since the beginning of X-ray coronal observations it has been recognized that the X-ray and XUV emitting corona is entirely made of plasma magnetically confined in loops (Vaiana, Krieger, and Timothy, 1973).

The data from the new satellites – mainly *Yohkoh*, SOHO, and TRACE – are showing us the extent to which the corona is variable, on many time and spatial scales. It is worth noting that, while many of the loops observed are variable and dynamic, the vast majority appear relatively steady, on the basis of their characteristic times, as discussed below in more detail.

The interesting new scenario given us by the recent data has revived interest in models of dynamic coronal loops. The scope of the applications of such models is quite broad and includes large dynamic events (e.g., flares) and small ones (microflares, loop brightenings, etc.), the slow evolution of loop plasma and studies of stability; they can be applied to a large fraction of the solar coronal phenomena.

The models of dynamic loops have been, and are, used both to explore theoretically the complex dynamics and thermal evolution of the confined solar corona, and as tools for an accurate interpretation of the data, a non-obvious task given the intricate interplay of the various physical effects.

In the following I will concentrate on the time dependent fluid dynamic models of the coronal loops. Section 2 of this paper will address the fundamental concepts, Section 3 will contain a short classification of the models, Section 4 will describe some examples of recent results, indicative of the broad scope of activity in this field, including a detailed modeling of a loop brightening observed with TRACE, and Section 5 will draw the conclusions.

The physics of the solar corona and transition region. Part II.
Reprinted from Solar Physics **193,** 2000.

2. Why 1-D Loop Fluid Models?

The mean free path of particles in the coronal plasma ($\approx 10^7$ cm) is typically much smaller than the loop length from one footpoint to the other, allowing a fluid treatment. Also, the plasma is frozen to the magnetic field and since the corona and transition region are typically at a low $\beta = p/(B^2/8\pi)$ (i.e. the magnetic pressure is much higher than the plasma thermal pressure), the plasma dynamics is dominated by the magnetic field and the plasma macroscopic motion can occur practically only along the magnetic field lines. Also thermal conduction, due to the electron diffusion, occurs only along the field lines – being much less effective across them (Spitzer, 1962). Therefore each plasma loop is independent from the neighboring ones (Rosner, Tucker, and Vaiana, 1978) and the dynamics and thermal evolution of the coronal plasma therein is one-dimensional, along the field lines.

For most of the loops, or the systems of loops, made of magnetically confined plasma the Alfvén travel time $t_A = L/V_A$ (where L is the characteristic field line length and $V_A = B/\sqrt{4\pi\rho}$ the Alfvén speed, with B the magnetic field and ρ the plasma mass density) is much shorter than the plasma evolution time (e.g., the radiative or conductive cooling times, or just the observed evolution time) implying that in most loops the magnetic field structure does not change during the plasma evolution. For instance $t_A \approx 10$ s for a loop with $L = 10^9$ cm, $B = 100$ G, and $\rho = 2 \times 10^{-14}$ g cm^{-3}, i.e., typical coronal values.

These considerations lead us to consider each magnetic flux tube as a rigid and steady container of the plasma, unless in the presence of magnetic field changes occurring within few Alfvén times; in practice almost each loop behaves as a rigid pipe filled with coronal plasma.

These model assumptions can even apply to events characterized by an evolving magnetic field, *after* the structures have stabilized; possible examples are reconnecting loops, after the reconnection is over, or forming loop arcades after morphological changes are over.

On the other hand since each magnetic loop, or even each magnetic 'fiber' forming a loop, is independent from the others, the single loop model can be properly applied to those loops in which the various fibers appear to behave coherently; for more complex cases one may consider multiple loop models (e.g., Hori *et al.*, 1997). In this respect the structure of the solar atmosphere, across the field line is of fundamental importance, being dictated possibly by the mechanisms of energy release, as recognized by Litwin and Rosner (1993).

The model, in order to treat satisfactorily the fluid behavior of the plasma, has to include the mass, momentum and energy conservation equations for the fluid plasma, plus the gas equation of state:

$$\frac{\partial n}{\partial t} + \nabla \cdot (n\mathbf{v}) = 0\,, \tag{1}$$

$$\frac{\partial (n\mathbf{v})}{\partial t} + \nabla \cdot (n\mathbf{v}\,\mathbf{v}) = -\frac{1}{m}\nabla p + n\mathbf{g}\,, \tag{2}$$

$$\frac{\partial u}{\partial t} + \nabla \cdot \left[(u + p)\mathbf{v} \right] = mn\mathbf{g} \cdot \mathbf{v} - n^2 P(T) + \mathcal{H}(\mathbf{r}) + \nabla F_c , \tag{3}$$

where n is the number density of hydrogen, $\mathbf{v}$ the macroscopic velocity of the plasma, p the plasma pressure, m the average particle mass, g the gravitational acceleration, u the plasma energy density, $P(T)$ the plasma radiative losses per unit emission measure, $\mathcal{H}(\mathbf{r})$ the heating, here prescribed per unit volume, and ∇F_c is the divergence of the heat flux; t and $\mathbf{r}$ are time and the spatial coordinate.

Beyond fluid dynamics, including pressure gradient and gravity components along the field lines, other fundamental physical effects are the thermal conduction, the optically thin radiative losses prescribed according to a spectral model (for example, Raymond and Smith, 1977; Cook *et al.*, 1989; Mewe, Lemen, and van den Oord, 1986; Mewe, Kaastra, and Liedahl, 1995, and references therein) and the effect of the heating.

The heating – which maintains the corona at a temperature of a million degrees and whose variation may be responsible for many of the observed changes – is a fundamental component of any coronal model and its formulation may depend on the specific problem. In the light of our ignorance of the heating mechanism, it may be prescribed as a function of the spatial coordinate and time: for instance, it can be assumed uniform in the corona or concentrated in some parts of the loop, to mimic different mechanisms, as in the Antiochos *et al.* (1999) and by Reale *et al.* (2000b) discussed below. The heating can be assumed to evolve in various ways, for instance fluctuating in time to simulate microflares (Peres, Reale, and Serio, 1993) or rapidly raising in time and decaying after some time. There are also attempts to model in detail specific mechanisms; for instance, in order to model flaring loops, time-dependent electron beam heating has been adopted according to the formulation, as in Nagai and Emslie (1984).

Some effects not always taken into account in the models are the effect of hydrogen ionization, which may be important in the energetics and dynamics of the lower transition region, and the plasma compressional viscosity. Although viscosity is not crucial in determining the large scale dynamics of the confined plasma, it is important for the plasma dynamics in the presence of shocks, such as those which develop in flaring loops, and for high-resolution spectra synthesized from flaring loops, such as the Ca XIX complex around 3.18 Å (Peres and Reale, 1993).

In some cases these models have been used to explore the relevance of plasma effects, such as departure from the Spitzer's formulation of thermal conduction (Karpen and De Vore, 1987; Peres, Rosner, and Serio, 1987), in determining the confined plasma dynamics. Incidentally, the above effect appears to be important mostly during the early flare phases which however are characterized by a low emission measure, making any related signature relatively hard to detect.

The numerical solution of the system equations is fundamental to treat realistic cases because the loop plasma evolution depends on several physical effects, all important and at work simultaneously, and many of them are non linear. Additional complexity comes from the formulation of the coronal radiative losses, sometimes approximated as piecewise power laws as, for instance, in Rosner, Tucker, and Vaiana (1978) or available as a lookup table. Furthermore, one typically needs the freedom to be able to use alternative heating formulations. All the above makes an analytical solution of the equations not viable.

The various authors who have undertaken loop plasma fluid modeling have used different numerical schemes. Certainly a very accurate numerical scheme is in order if one is looking for rather subtle effects, e.g., small-scale details of plasma dynamics or the fine structure of propagating waves, since the results may easily be affected by a coarse numerical scheme; however, it may happen that a model uses a highly refined numerical scheme with a coarse atmospheric model or a coarse treatment of the physical effects. An accurate model requires an accurate treatment of the physical effects and of the initial atmosphere as much as an accurate numerical scheme.

The initial conditions of the confined atmosphere may depend on the specific problem. Many studies address the evolution starting with stationary (or even static) initial atmosphere, either generated by a dedicated code or obtained through the relaxation of an approximately hydrostatic initial configuration (see, for instance, Antiochos *et al.*, 1999, in the following).

Also, the presence of the chromosphere and its evolution are important for the loop dynamic model, even if one is just interested in the coronal plasma evolution. The chromosphere, in a rather broad sense, acts as a mass source and as a heat flux sink for the corona and transition region. Furthermore it stabilizes the entire confined atmosphere as was ascertained through the work by Antiochos (1979), Habbal and Rosner (1979), and Peres *et al.* (1982). In this respect, one cannot put the loop lower boundary in the lower corona or in the transition region.

The chromosphere is optically thick to its own radiation; however, on the basis of accurate models of the chromosphere and of its radiative losses, it has been possible to ascertain that, if one is interested only in the energy balance and dynamics of the chromosphere, and not in its detailed spectrum, then its radiative losses can be conveniently approximated as effectively optically thin, considerably simplifying the calculations. A piecewise power law formulation of such losses was derived by Peres *et al.* (1982), on the basis of the Vernazza, Avrett, and Loeser (1981) models of the chromosphere; McClymont and Canfield (1983) proposed an analogous formulation of the chromospheric losses, also on the basis of Vernazza, Avrett, and Loeser (1981). More recently Kankelborg and Longcope (2000) have proposed another formulation based on the more updated chromosphere models of Fontenla, Avrett, and Loeser (1990).

Another aspect of the hydrodynamic modeling of loops, and for general for astrophysical plasma, is the adequate comparison with observations. The ever increasing quality and quantity of the data, the high level of refinement in the models, and the need to produce more and more detailed results on the coronal plasma dynamics requires the synthesis of the emission in specific lines or bands of interest from the computed evolution of the temperature, density and velocity of the plasma (e.g., Peres *et al.*, 1987; Reale *et al.*, 1988; Priest, 1998; Hori *et al.*, 1998). The approach by Maggio *et al.* (1994) is one of the most systematic in this respect; it has produced an environment, written in IDL, open to include new spectral codes and features of new observing instruments. It allows the synthesis of the results of models in the desired format, including that of the relevant observations.

3. A Simplified Classification of the Models

Over the years several groups have developed several one-dimensional fluid dynamics loop models and tackled different problems of coronal loop physics. Treating in detail the several codes and the whole scope of their studies is beyond the limits of this paper. However, rather than ignoring them, I will try a rapid overview and a coarse classification, with a list of references.

There are 0-dimensional models which limit themselves to calculating the averaged temperature and density in the coronal loop. The entire corona is described by these two numbers as functions of time. The exchange of mass and heat with the chromosphere is treated through the boundary conditions. The disadvantage of these codes is that it is impossible to predict most of the observed features but the advantage is that their simplicity allows to compute many cases quite simply and rapidly. These models have been used to study microflaring loops (Kopp and Poletto, 1993; Cargill, 1994; Cargill and Klimchuk, 1997)

Then there are models which treat in detail the chromosphere and the photosphere. Studying in detail a flare in the low atmosphere and predicting the relevant spectra requires to calculate both the time-dependent hydrodynamics and the radiative transfer in the evolving chromosphere. The group which included Canfield, McClymont and Fisher, along with other collaborators, then at the University of California, developed a code and published several papers, mainly McClymont and Canfield (1983a, b), Canfield, Fisher, and McClymont (1983), An *et al.* (1983), Fisher, Canfield, McClymont (1985a–c), on the modeling of chromosphere in flaring coronal loops, using a probabilistic escape treatment of radiative transfer. More recently Abbett and Hawley (1999) have developed a new model, discussed below in more detail.

Most of the codes, however, have been dedicated to study the corona and the transition region, often treating the chromosphere in an approximate way. Any attempt to review the results produced would require more space than can be taken by this review. I will try to credit the scientist involved, to mention some work, from which the interested scholar can find other references. The first to publish a model of this kind was Nagai (1980), who showed the feasibility of the chromospheric evaporation. A considerable effort in this field has been invested by the group at the Naval Research Lab., including Cheng, Doschek, and Karpen (1984), Mariska, Emslie, and Li (1989), Antiochos *et al.* (1999), Klimchuk (Ofman, Klimchuk, and Davila, 1998; Klimchuk and Mariska, 1988). The Palermo group (Peres *et al.*,

1982, 1987; Reale *et al.*, 1988; Betta *et al.*, 1997; Peres, 1997) collaborating with the groups of R. Rosner, now in Chicago, and L. Golub at the Center for Astrophysics, in Cambridge has addressed several aspects of loop plasma dynamics and solar and stellar coronal diagnostics. More recently Gan and Fang (1990) and Gan, Fang, and Zhang (1991), have developed a hydrodynamic loop code, and Hori *et al.* (1997) have devised a multi-loop code. Walsh, Bell, and Hood (1997) have paid attention to effects of the pulsed heating in coronal loops; Kankelborg and collaborators (this issue) are developing a model including recent developments on thermal conduction and radiative losses in the transition region.

Incidentally, let us note that just as important as the study of dynamical evolution is that of the stationary conditions of loops. A good knowledge of stationary loop conditions is important per se and because it provides the initial conditions, crucial for a correct determination of dynamic evolution. Static loop models were first studied by Rosner, Tucker, and Vaiana (1978) and Vesecky, Antiochos, and Underwood (1979). Then Serio *et al.* (1981) studied loops in hydrostatic conditions, higher than their pressure scale height and subject to non-uniform heating and found the relevant scaling laws. A recent treatment along this line is in Steele and Priest (1994). A more complete model is that of siphon flows in loops, to some extent related to the work of Antiochos *et al.* (1999) discussed below. For a recent and very detailed model of siphon flow loop we refer to Orlando, Peres, and Serio (1995a; b) and references therein.

4. Examples of Recent Results

In the following, three examples of loop hydrodynamic modeling will be shown to illustrate the range of physical problems addressed with these models. The first example studies the evolution of the chromosphere and of its spectrum during a flare; the second deals with the formation of a quiescent prominence knot, namely of cold plasma in the corona; the third studies a loop brightening, i.e., a phenomenon of the hot corona, and the detailed fitting of the related TRACE observations.

4.1. Modeling the Chromosphere and Photosphere

Abbett and Hawley (1999) have just published a paper on their new loop model which deals with the time-dependent hydrodynamics and radiative transfer in the lower atmosphere, plus due attention to the corona and transition region.

Abbett and Hawley study three cases of electron beam heated flaring chromosphere, the fundamental difference being the input energy flux: 10^9, 10^{10}, and 10^{11} erg s^{-1} cm^{-2}, respectively. The beam of non-thermal electrons has a power law energy distribution with a lower energy cutoff at 20 keV and an index of -5.

The case of intermediate energy flux carries all the important physical features the authors want to show and is briefly discussed here. Figure 1 shows the temperature, density of electrons and of mass along with the electron beam heating and the

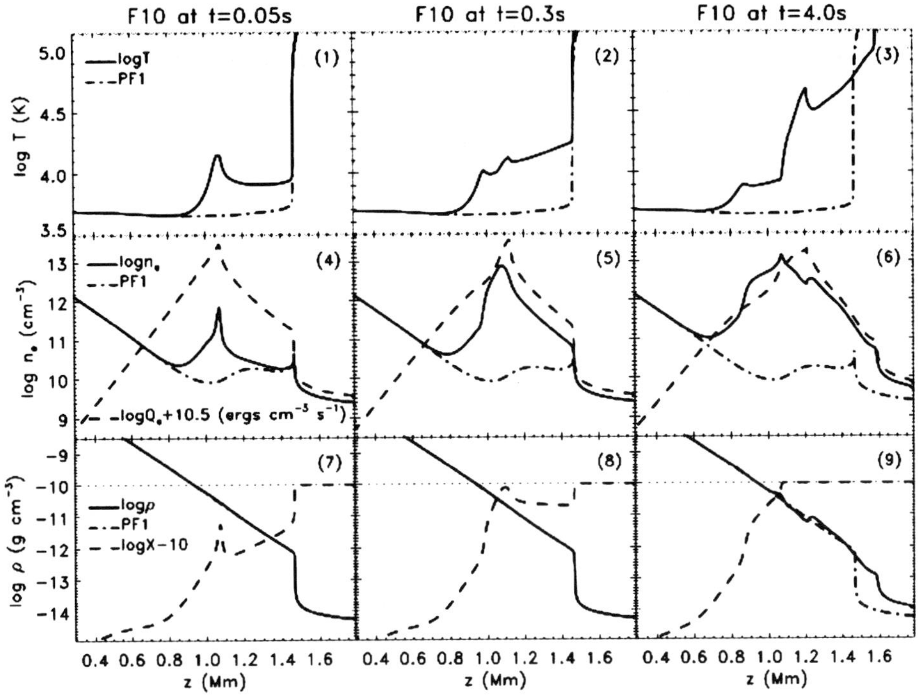

Figure 1. The low solar atmosphere during the early, gentle phase of electron beam heated flare evolution. The figures show the temperature (*top row*), electron density and electron beam heating per unit volume (*middle row*), the mass density and the hydrogen ionization fraction (*bottom row*) at selected times vs height in the chromosphere. In each panel the dot-dashed profile (PF1) is the initial atmospheric profile and the dotted line marks the level to which hydrogen is fully ionized (from Abbett and Hawley, 1999).

hydrogen ionization fraction vs height in the chromosphere at three moments of the early flare evolution. The initial flare phase is characterized by a gentle evolution; at later phases, approximately 25 s since the heating has started, the atmosphere cannot radiate the input power any more and evolves explosively. Figure 2 shows the explosive phase of the flare.

The authors study in detail the lines and continuum formation in the atmosphere. Figure 3 shows the Hα and the Ca II K spectral lines at selected times of the flare evolution, synthesized from the chromosphere model. The line shapes remain symmetric during the early, gentle, phase but become strongly non-symmetric in the later explosive phase characterized by a very dynamic and complex atmospheric evolution.

The case with a flux ten times smaller than the one discussed above never gets explosive, whereas the case with a flux ten times larger explodes in a fraction of a second. It is also worth noting that the X-ray heating is never important in each of the cases Abbett and Hawley have studied.

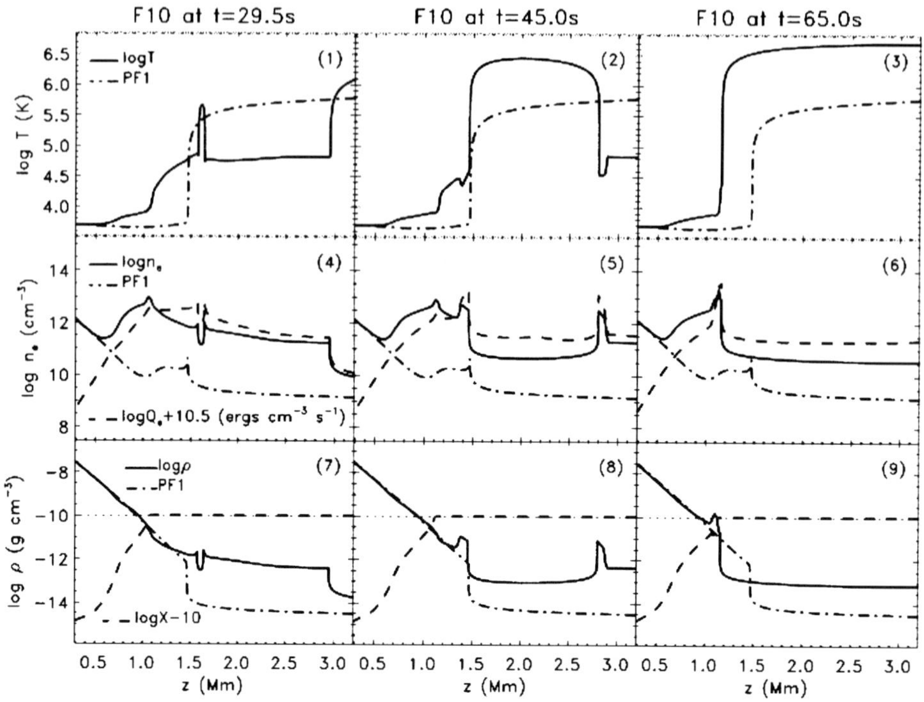

Figure 2. The atmosphere during the later, explosive phase of evolution. The notation is the same of Figure 1. (From Abbett and Hawley, 1999.)

4.2. Formation of a Quiescent Prominence

Antiochos *et al.* (1999) have simulated the formation of a prominence condensation in a coronal loop. More specifically they have studied the evolution of a prominence knot, which is just part of a prominence, inside a loop. The idea is that the whole prominence is formed by a set of loops each having a knot in the highest part and by modeling the formation of one such knot one can understand, in principle, the formation of whole prominences.

The topmost part of the loop consists of a rather long, almost horizontal branch (in fact slightly dipped at the very center of the loop); the rest of the loop on each side has the shape of a quarter of a circle rooted in the lowest atmosphere. The initial conditions consist of a hydrostatic configuration (relaxed from an initial trial configuration).

The key aspect of the simulation is a heating mostly concentrated at the footpoints: more specifically the volumetric heating, prescribed as a function of position, is $E = E_0 + E_1(s,t)$ where $E_0 = 1.5 \times 10^{-5}$ erg cm^{-3} s^{-1} and $E_1 = 10^{-3}$ erg cm^{-3} s^{-1} below s_1, the highest point of the chromosphere, and $E_1 = 10^{-3} e^{-(s-s_1)/\lambda}$ for $s > s_1$, and $\lambda = 10^8$ cm. E_1 is ramped up linearly in time over the first 1000 s and then left constant in time.

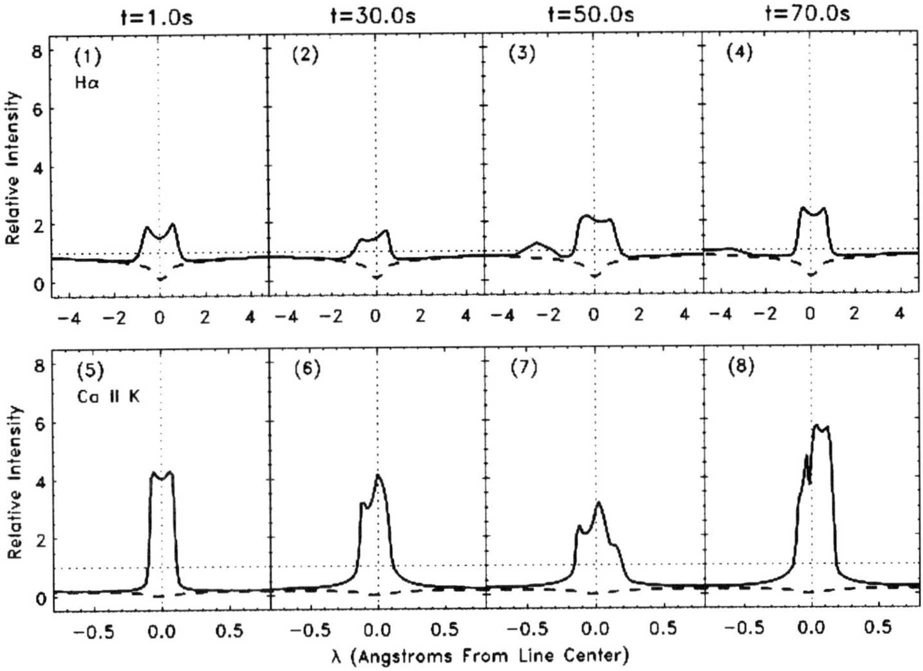

Figure 3. Evolution of the H_α (*top row*) and the Ca II K (*bottom row*) spectral lines during the flare. The intensity is relative to the continuum and the dashed line marks the preflare line profile. (From Abbett and Hawley, 1999.)

The loop is assumed to be symmetric and therefore only half of the loop is studied.

Under the effect of the heating, the loop first heats entirely, then the top part starts to cool and condense, as shown in Figure 4.

Figure 5 shows the very onset of the condensation and the following evolution, with the formation of shock fronts and steep transition regions between the condensation and the rest of the corona. The physical explanation is that the heating concentrated at the loop footpoint drives the plasma evaporation from chromosphere to the corona, where the heating is insufficient and hence the loss of equilibrium leads to the condensation.

Antiochos *et al.* have been able to solve this problem thanks to a new code with second order Godunov scheme, and a particular adaptive mesh scheme which allowed them to follow the evolution of the sharp fronts due the condensation.

It is worth noting, also, that Serio *et al.* (1981) already suggested an analogous mechanism as responsible for the prominence formation, during the long-term evolution of magnetic loops from the compact magnetic configuration of active regions to the more extended one of old regions hosting filaments. According to Serio *et al.* (1981) such an evolution would be energetically more convenient, with respect to

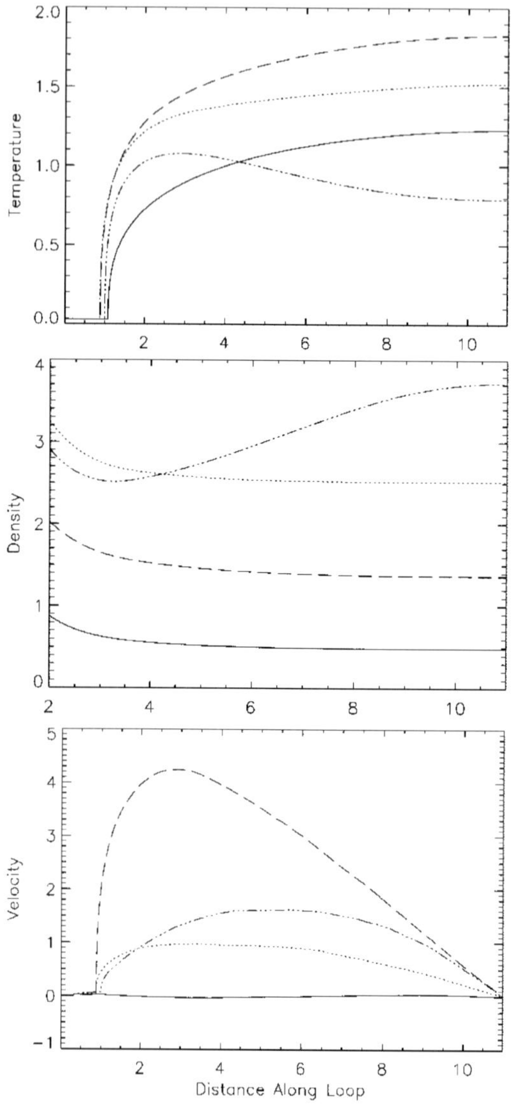

Figure 4. Temperature (*top panel*) in units of 10^6 K, number density (*middle*) in units of 10^9 cm^{-3}, and velocity (*bottom*) in units of 10^6 cm s^{-1}, vs distance from the loop base (in units of 10^9 cm), at selected times during the evolution before the formation of the condensation. The solid curves correspond to $t = 0$ s, the dashed curves to $t = 10\,000$ s, the dotted curves to $t = 30\,000$ s, and the dash-dotted curves to $t = 50\,000$ s. Note that the middle panel has a different spatial scale from the other two, showing just the part of the loop above the chromosphere. (Adapted from Antiochos *et al.*, 1999.)

TABLE I
Parameters of the simulations with transient constant heating

Name	Location	H_0 (10^{-3} erg cm^{-3} s^{-1})	s_0 (10^9 cm)	σ (10^9 cm)	Duration (s)
A	Footpoint	10	0.3	0.05	3000
B	Asymmetric	0.4	3	0.5	3000
C[a]	Asymmetric	0.4	3	0.5	3000

[a]As case B, but constant for the first 100 s and then decaying exponentially with an *e*-folding time of 300 s.

the case with maximum temperature at the loop apex, if the heating is concentrated at the loop base.

4.3. Modeling as a Diagnostic of the Heating

Along the same line of thought, Reale *et al.* (2000a, b) show that the detailed hydrodynamic/thermodynamic modeling of a loop brightening, observed with TRACE at a high angular and temporal resolution, leads to crucial insight on the location and timing of the heating driving the loop brightening. TRACE observed AR 8253, then very close to the solar disk center, on 26 June, 1998, taking one frame every 30 s over a few hours, in the 171 Å band; the pixel size is half an arc sec.

The loop gets bright initially at the footpoints. Then from each footpoint, bright, sharp fronts expand gradually along the loop; first from the northern footpoint, $\approx$ 10 min since the initial brightening and further 7 min later from the southern footpoint. The brightness fronts propagate at an apparent speed of 100 km s^{-1}. The surface brightness along the loop is always definitely asymmetric, with the northern footpoint and leg brighter than the southern ones. Figure 6 shows the region where the loop brightened with the loop clearly visible.

Reale *et al.* (2000b) inferred from the images that the loop, inclined on the solar surface by 60 deg, is approximately semi-circular, the footpoints separation being $\approx 10^{10}$ cm, with an almost equal distance between the center of the footpoints and the apex.

The first part of the sequence shows that the loop is not visible in the 171 Å band, which detects preferentially plasma around 10^6 K. The authors, therefore, assumed that the plasma is initially at 5×10^4 K and the pressure is 0.01 dyn cm^{-2}, i.e., the loop plasma is at rather low temperature, pressure and density.

Reale *et al.* (2000a) have used the Palermo-Harvard 1-D fluid dynamic code (Peres *et al.*, 1982; Betta *et al.*, 1997) to model the brightening; a semi-circular loop geometry and a constant cross section have been assumed for the loop; both geometrical assumptions being appropriate to the observed loop. In practice the geometry never changes during the evolution; while many loop subfilaments or threads brighten simultaneously, at the beginning, they all follow the same evolu-

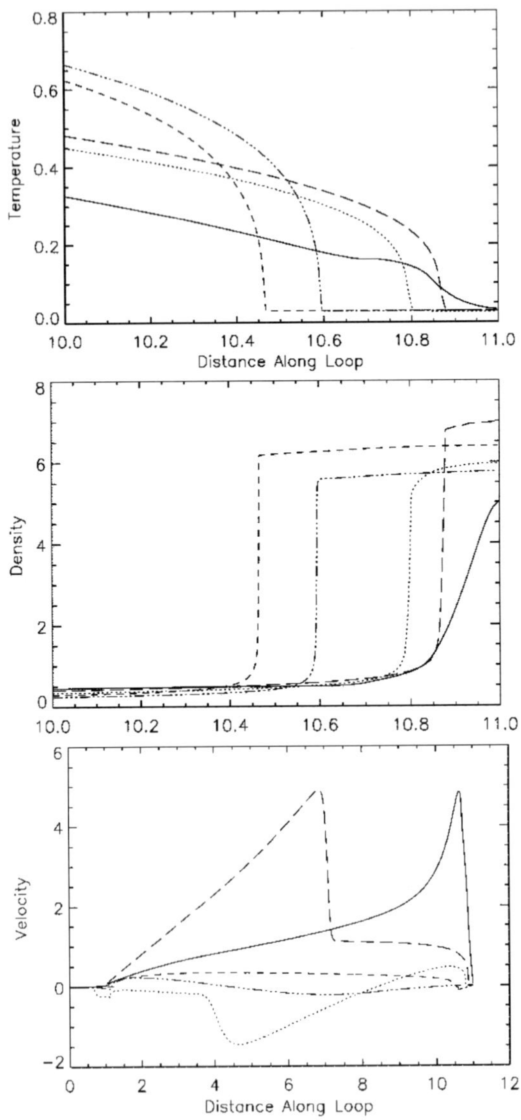

Figure 5. Analogous to Figure 4 but after the condensation phase has begun; the two topmost panels refer just to the part near the loop midpoint, the bottom panel to the entire half-loop. The solid curves correspond to $t = t_0 = 54\,000$ s, when the condensation first appears; the long dashed curves to $t = t_0 + 500$ s, the dotted curves to $t = t_0 + 1000$ s, the dash-dotted curves to $t = t_0 + 10\,000$ s, and the short dashed curves to $t = t_0 + 30\,000$ s. (Adapted from Antiochos *et al.*, 1999.)

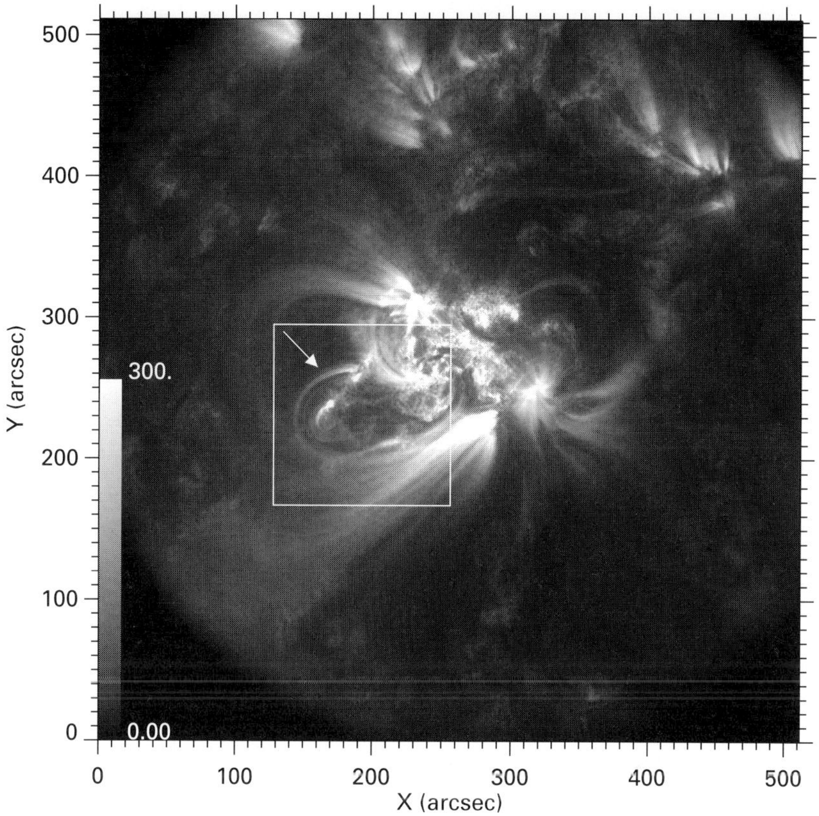

Figure 6. Whole field of view of the TRACE observation in the 171 Å band, made at 13:54:17 UT. The exposure time is 23 s and the brightness scale units are digital numbers. The square encloses the region where the loop appeared. The loop itself, already clearly visible in this image, is marked with an arrow. (From Reale *et al.*, 2000a.)

tion during the whole rising phase ($\approx$ 3000 s); later the various subfilaments appear to behave differently. Reale *et al.* (2000b) have modeled just the rising phase, to which the model can be applied correctly.

The heating term has been modeled as a separable function of position along the loop and time: the spatial distribution is Gaussian and different locations and widths along the loop have been tried, as shown in Table I. The impulsive heating is assumed to turn on and remain constant during the simulation; as described in Table I, the last simulation of the set considers heating with an exponentially decaying tail.

The observation is fitted with the model as follows: first, the loop geometry is derived from the observations, then, the initial conditions of the coronal loop plasma are inferred. Having defined the initial conditions, the intensity, spatial distribution and evolution of the heating are chosen and the loop plasma evolution

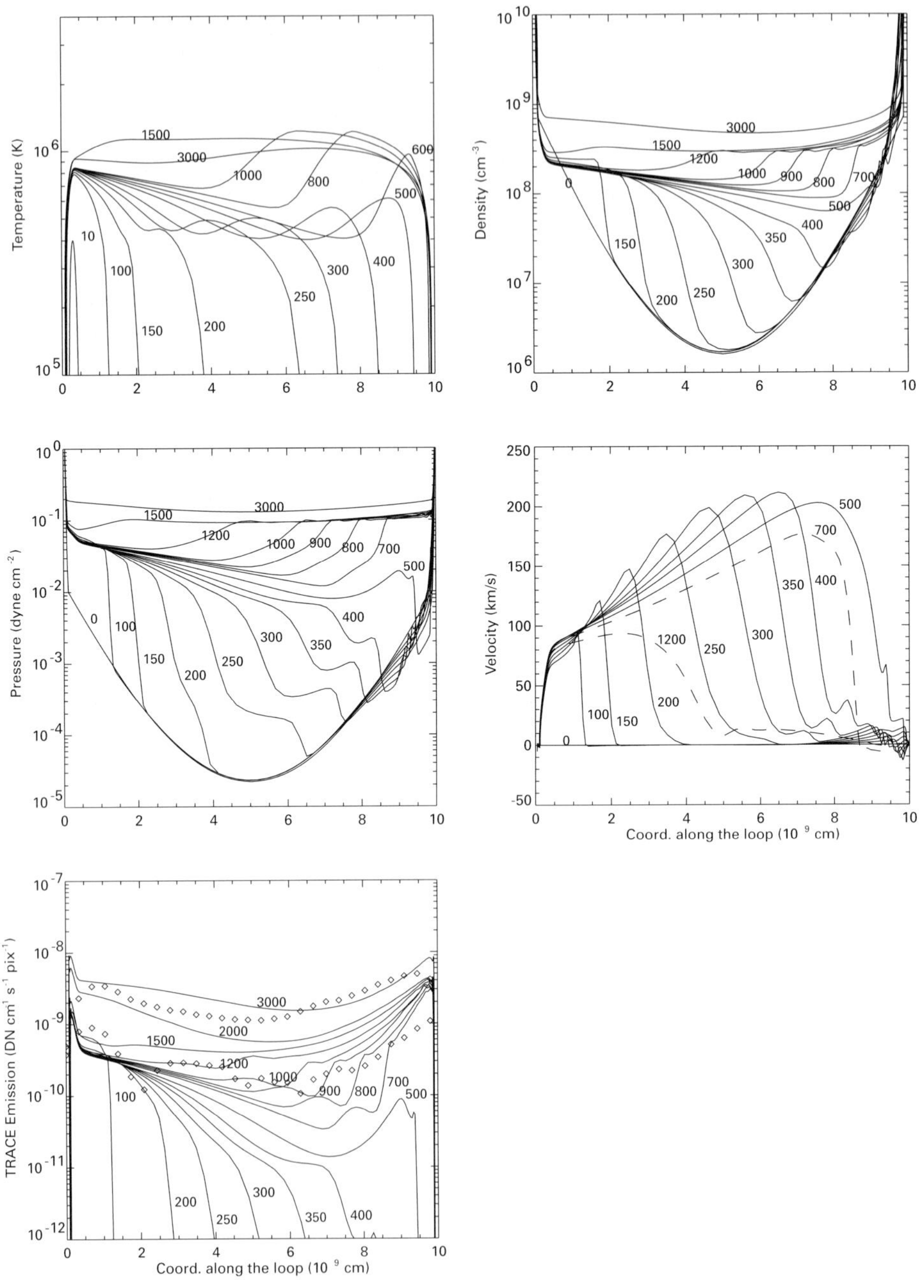

Figure 7. Plasma temperature, density, pressure, velocity and emission per unit volume in the TRACE 171 Å band, vs the spatial coordinate and time, for the model loop heated at the left footpoint (Model A in Table I). Time (in s since the start of the heating) is shown by each curve. To avoid confusion, velocity curves after 500 s are dashed. The lowest left panel shows the observed brightness distributions (diamonds) at $t = 1875$ s (*lower*) and at $t = 3374$ s (*upper*). (From Reale *et al.*, 2000b.)

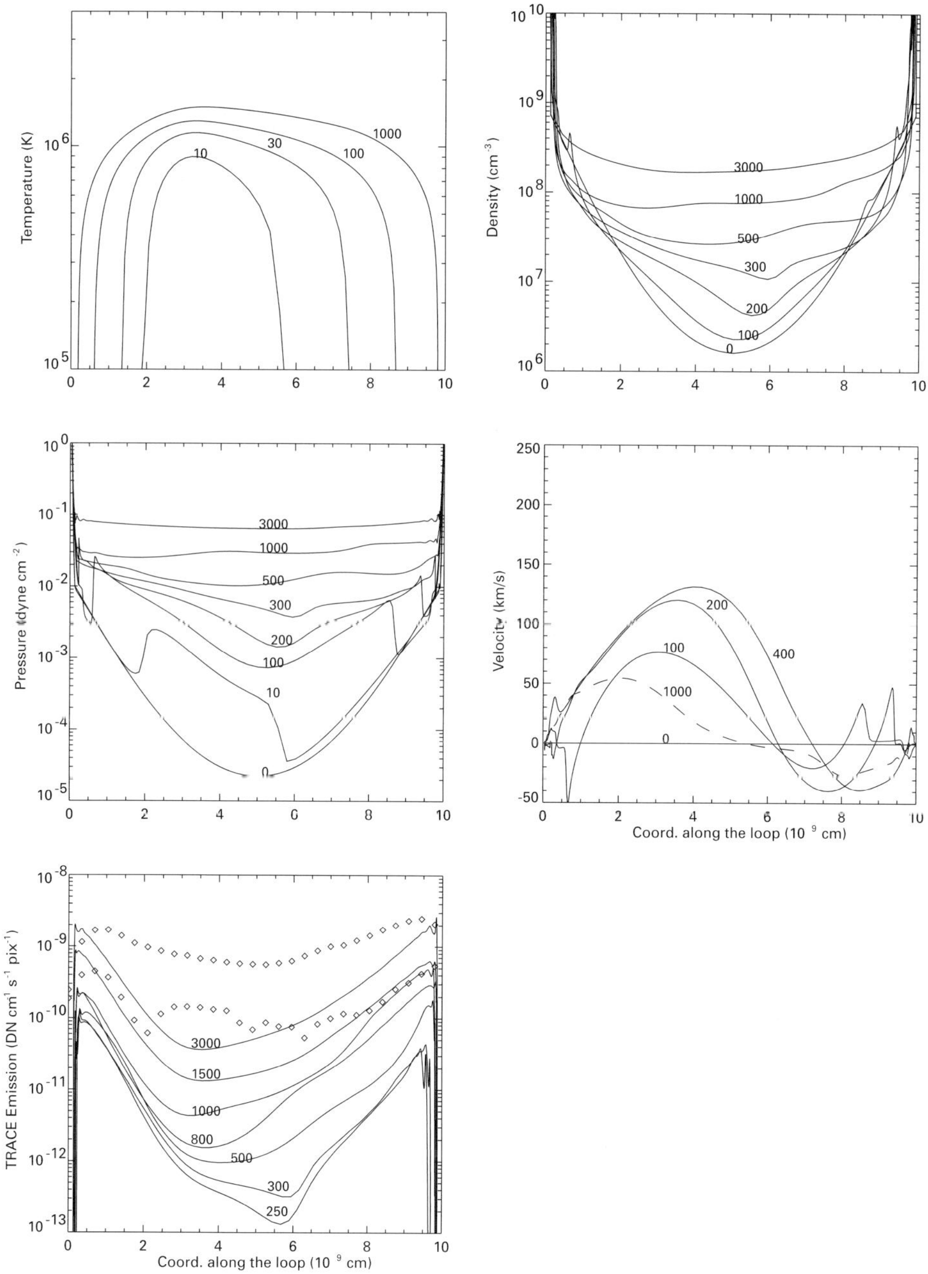

Figure 8. As Figure 7 for the model loop heated asymmetrically on the right leg (Model B in Table I). (From Reale *et al.*, 2000b.)

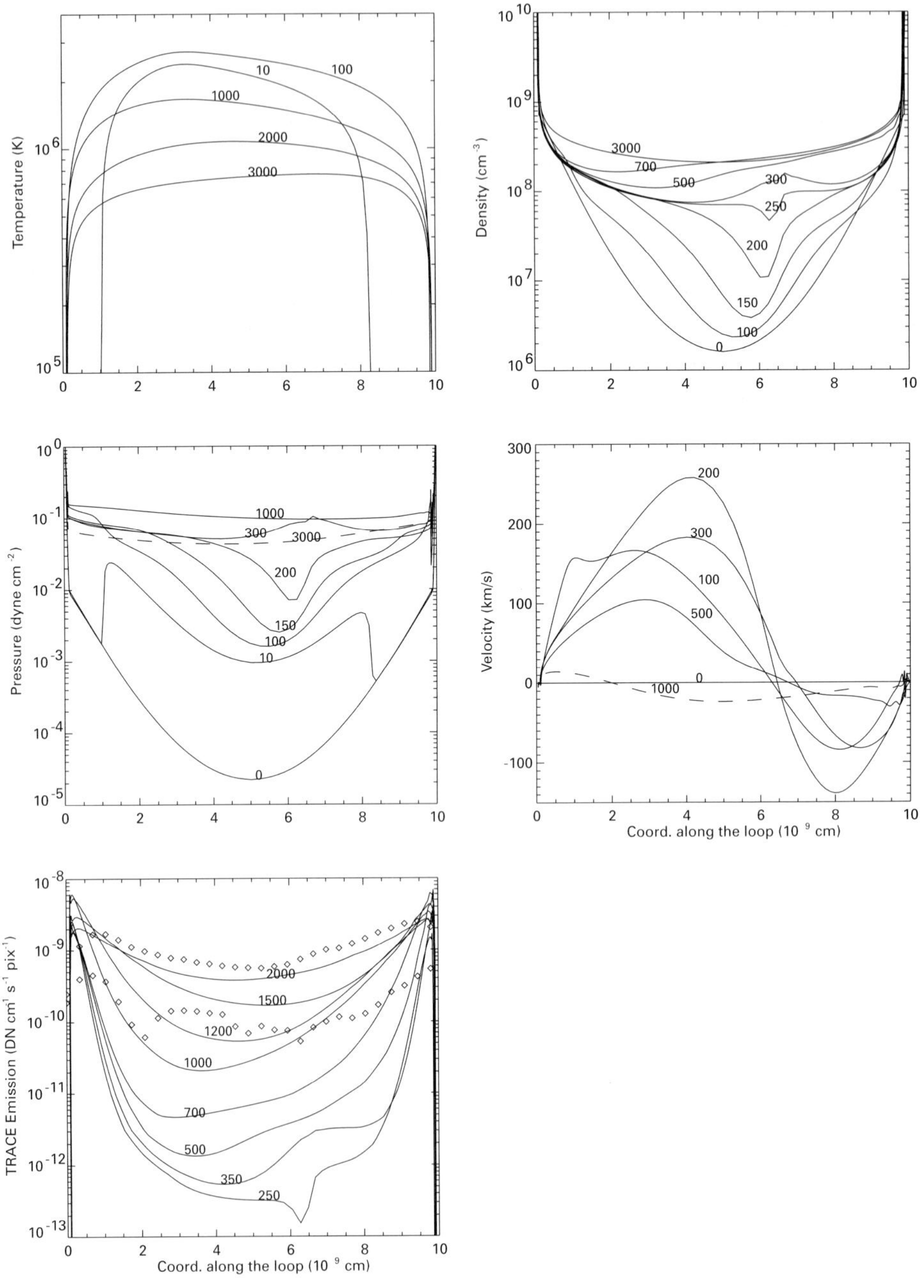

Figure 9. As Figure 7 for model loop C: the heating, on the right leg, is high and constant for 100 s and then exponentially decaying with an *e*-folding time 300 s. For a better readability the pressure at $t = 3000$ s is dashed. (From Reale *et al.*, 2000b.)

is then computed. The emission in the 171 Å band is then derived from the plasma density and temperature distribution along the loop during the whole evolution, taking into account also the folding with the instrument spectral response to allow a direct comparison with the focal plane data. The evolution of the synthesized emission is finally compared with observations, considering several characteristics including the brightness contrast between the two footpoints and the time lag between their brightening. From the comparison one can decide if and how to adjust the heating parameters to better fit the observations. The task is to best reproduce the observations.

In this context, Reale *et al.* (2000b) have paid particular attention to the contrast between footpoints and loop apex, and to the relative timing of the brightenings at the footpoints and at the apex. When interpreting the results it should be considered that the 171 Å spectral band (as all the TRACE coronal ones) is relatively narrow and the plasma is more easily detected in a narrow temperature range around 10^6 K.

The first model calculation (case A in Table I) consisted of heating one footpoint, since the first brightening occurs at a footpoint; the relevant results are in Figure 7. The simulations show that the opposite footpoint brightens first because that is the place where the plasma is compressed and the emission measure increases. This case shows how a too -straightforward- interpretation of observations to derive information on the heating can lead to wrong results. It is also very important to note that this simulation fails to reproduce observations, because the model predicts that the two footpoints and the two halves of the loop brighten alternatively, while the two footpoints brighten with only a slight time difference in the observations.

If the heating were located exactly at the loop apex the footpoints would, obviously, brighten simultaneously (at variance with observations). The heating cannot, therefore, be either at the footpoints or at the apex.

Therefore, on the basis of the relative brightness contrast between the two footpoints and the timing of their brightening Reale *et al.* (2000b) infer that the heating should be somewhere in between, most likely at, approximately, one third of the loop's length, reckoned from one footpoint, and opposite to the footpoint which brightens first. The evolution of the surface brightness in the 171 Å band, synthesized from this model (case B in Table I), agrees with observations much better than the simulations discussed before (Figure 8).

As for the timing of the heating, in all the simulations so far analyzed, heating is turned on suddenly and remains thereafter constant. Also, at late phases of the observed evolution, the observed brightness profile along the loop is flatter than the one obtained in case B. Reale *et al.* (2000b) then tried a heating which is suddenly turned on, stays on constantly for 100 s, and then decays exponentially with an e-folding time of 300 s. They obtain an even better agreement with observations (cf., Figure 9) and definitely the best fitting, among those tried by Reale *et al.* (2000b), of the spatial and temporal distribution of the emission along the loop.

5. Conclusions

The data collected by the recent coronal instruments are spurring the use of dynamic models of coronal loops and the development of new ones. These models are of fundamental importance to understand the macroscopic dynamics and thermal evolution of confined plasma under the combined effect of several physical mechanisms (including thermal conduction, heating and radiative losses) which act in a complex, non-linear way. As such they serve both as theoretical tools and as a means of accurately interpreting observations.

Within the limited space of this presentation, a sketchy reference to previous work has been presented. Also three very recent studies have been reviewed, in the perspective to show both the theoretical and the interpretative use of the models, as well as the range of application covering three different phenomena and physical conditions occurring in loops: the modeling of the optically thick chromospheric flare (Abbett and Hawley, 1999), the model of prominence formation (cold plasma forming in the corona) by Antiochos *et al.* (1999), and the modeling of a loop brightening observed by TRACE, yielding insight on the timing and location of the heating (Reale *et al.*, 2000a, b), at least for one specific event.

Each team involved in loop modeling is now addressing some problem in the relevant field and still a lot has to be said on each problem tackled so far. So in the near future one may expect a sizable amount of activity along lines already drawn in the recent past. One may also expect new developments of loop models, e.g., through the exploitation of more advanced numerical techniques (to approximate better the differential equations, to improve the numerical mesh, etc.) or refining the physical effects in the model.

The scope of loop modeling is rather broad and one may foresee many new developments. Just to mention a few of them, TRACE observations clearly show that loops are formed by many filaments (with a cross-field structure down to the resolution limit) which, in some cases, evolve differently; this may require the development of multi-loop models like the one by Hori *et al.* (1997), also, since observations can help to constrain them, it could be time to include in more detail specific heating mechanisms. Furthermore, chemical abundance stratification, or differentiation, in the solar atmosphere is a hot topic and Lenz, Low, and Rosner (1998) have provided a recent view on the problem, amenable to treatment in the loop model context; it is important to determine how the stratification, if present, influences the spectra and, possibly, the confined plasma dynamics through the radiative losses. It is equally important, on the other hand, to understand the extent to which dynamic loop phenomena may alter any gravitational chemical settling and bring back abundances to their standard average value.

There is a large number of problems in loop physics which have not been solved, observations provide new challenges, and the ever growing computing power grants that we can keep developing more detailed and accurate loop models to tackle the problems. And human ingenuity can lead easily to welcome surprises.

Acknowledgements

I acknowledge stimulating discussions with, and help by, F. Reale. This work was partially supported by Agenzia Spaziale Italiana and by M.U.R.S.T.

References

Abbett, W. P. and Hawley, S.: 1999, *Astrophys. J.* **521**, 906.
An, C. H., Canfield, R. C., Fisher, G. H., McClymont, A. N.: 1983, *Astrophys. J.* **267**, 421.
Antiochos, S. K.: 1979, *Astrophys. J.* **232**, L125.
Antiochos, S. K., MacNeice, P. J., Spicer, D. S., and Klimchuk, J. A.: 1999, *Astrophys. J.* **512**, 985.
Betta, R., Peres, G., Reale, F., and Serio, S.: 1997, *Astron. Astrophys. Suppl.* **122**, 585.
Canfield, R. C., Fisher, G. H., and McClymont, A. N.: 1983, *Astrophys. J.* **265**, 507.
Cargill, P. J.: 1994, *Astrophys. J.* **422**, 381.
Cargill, P. J. and Klimchuk, J. A.: 1997, *Astrophys. J.* **478**, 799.
Cheng, C.-C., Doschek, G. A., and Karpen, J. T.: 1985 *Astrophys. J.* **286**, 787.
Cook, J. W., Cheng, C.-C., Jacobs, V. L., and Antiochos, S. K.: 1989, *Astrophys. J.* **338**, 1176.
Fisher, G. H., Canfield, R. C., and McClymont, A. N.: 1985a, *Astrophys. J.* **289**, 414.
Fisher, G. H., Canfield, R. C., and McClymont, A. N.: 1985b, *Astrophys. J.* **289**, 425.
Fisher, G. H., Canfield, R. C., and McClymont, A. N.: 1985c, *Astrophys. J.* **289**, 434.
Fontenla, J. M., Avrett, E. H., and Loeser, R.: 1990, *Astrophys. J.* **355**, 700.
Gan, W. Q. and Fang, C.: 1990, *Astrophys. J.* **358**, 328.
Gan, W. Q., Fang, C., and Zhang, H. Q.: 1991, *Astron. Astrophys.* **241**, 618.
Habbal, S. R. and Rosner, R.: 1979, *Astrophys. J.* **234**, 1113.
Hori, K., Yokoyama, T., Kosugi, T., and Shibata, K.: 1997, *Astrophys. J.* **489**, 426.
Hori, K., Yokoyama, T., Kosugi, T., and Shibata, K.: 1998, *Astrophys. J.* **500**, 492.
Kankelborg, C. and Longcope, D.: 2000, *Solar Phys.* **190**, 59.
Karpen, J. T. and DeVore, C. R.: 1987, *Astrophys. J.* **320**, 904.
Klimchuk, J. A. and Mariska, J. T.: 1988, *Astrophys. J.* **328**, 334.
Kopp, R. A. and Poletto, G.: 1993, *Astrophys. J.* **418**, 496.
Lenz, D. D., Lou, Y.-C., and Rosner, R.: 1998, *Astrophys. J.* **504**, 1020.
Litwin, C. and Rosner, R.: 1993, *Astrophys. J.* **412**, 375.
Maggio A., Reale F., Peres G., and Ciaravella, A.: 1994, *Comp. Phys. Commun.* **81**, 105.
Mariska, J. T., Emslie, A. G., and Li, Peng: 1989, *Astrophys. J.* **341**, 106.
McClymont, A. N. and Canfield, R. C.: 1983a, *Astrophys. J.* **265**, 483.
McClymont, A. N. and Canfield, R. C.: 1983b, *Astrophys. J.* **265**, 497.
Mewe, R., Gronenschild, E. H. B. M., and van den Oord, G. H. J.: 1985, *Astron. Astrophys. Suppl. Ser.* **62**, 197.
Mewe, R., Kaastra, J. S., and Liedahl, D. A.: 1995, *Legacy* **6**, 16.
Mewe, R., Lemen, J. R., and van den Oord, G. H. J.: 1986, *Astron. Astrophys. Suppl. Ser.* **65**, 511.
Nagai, F.: 1980, *Solar Phys.* **68**, 351.
Nagai, F. and Emslie, A. G.: 1985, *Astrophys. J.* **279**, 896.
Ofman, L., Klimchuk, J. A., and Davila, J. M.: 1998, *Astrophys. J.* **493**, 474.
Orlando, S., Peres, G., and Serio, S.: 1995a, *Astron. Astrophys.* **294**, 861.
Orlando, S., Peres, G., and Serio, S.: 1995b, *Astron. Astrophys.* **300**, 549.
Peres, G.: 1997, *ESA SP-404*, p. 55.
Peres, G. and Reale, F.: 1993, *Astron. Astrophys.* **267**, 566.
Peres, G., Reale, F., and Serio, S.: 1993, in *J. Linsky and S. Serio (eds.) Physics of Solar and Stellar Coronae*, Kluwer Academic Publishers, Dordrecht, Holland, p. 151

Peres, G., Rosner, R., and Serio, S.: 1987, *Il Nuovo Cimento* **99B**, 29.
Peres, G., Rosner, R., Serio, S., and Vaiana, G. S.: 1982, *Astrophys. J.* **252**, 791.
Peres, G., Reale, F., Serio, S., and Pallavicini, R.: 1987, *Astrophys. J.* **312**, 895.
Priest, E. R., Foley, C. R., Heyvaerts, J., Arber, T. D., Culhane, J. L., and Acton, L. W.: 1998, *Nature* **393**, 545.
Raymond, J. C. and Smith, B. W.: 1977, *Astrophys. J. Suppl. Ser.* **35**, 419.
Reale, F., Peres, G., Serio, S., Rosner, R., and Schmitt, J. H. M. M.: 1988, *Astrophys. J.* **328**, 256.
Reale, F., Peres, G., Serio, S., De Luca, E., and Golub, L.: 2000a, *Astrophys. J.*, in press.
Reale, F., Peres, G., Serio, S., Betta, R., De Luca, E., and Golub, L.: 2000b, *Astrophys. J.*, in press
Rosner, R., Tucker, W. H., and Vaiana, G. S.: 1978, *Astrophys. J.* **220**, 643.
Serio, S., Peres, G., Vaiana, G. S., Golub, L., and Rosner, R.: 1981, *Astrophys. J.* **243**, 288.
Spitzer, L.: 1962, *Physics of Fully Ionized Gases*, Interscience, New York.
Steele, C. D. C and Priest, E. R.: 1994, *Astron. Astrophys.* **292**, 291.
Vaiana, G. S., Krieger, A. S., and Timothy, A. F.: 1973, *Solar Phys.* **32**, 81.
Vernazza, J. E., Avrett, E. H., and Loeser, R.: 1981, *Astrophys. J. Suppl. Ser.* **45**, 635.
Vesecky, J. F., Antiochos, S. K., and Underwood, J. H.: 1979, *Astrophys. J.* **233**, 987.
Walsh, R. W., Bell, G. E., and Hood, A. W.: 1997, *Solar Phys.* **171**, 81.

CROSS-SECTIONAL PROPERTIES OF CORONAL LOOPS

J. A. KLIMCHUK
U.S. Naval Research Laboratory, Washington DC 20375-5352, U.S.A.

(Received 3 October 1999; accepted 3 December 1999)

Abstract. Careful examination of 43 soft X-ray loops observed by *Yohkoh* has revealed a number of interesting properties of the loop cross section. First, the loops tend to be only slightly ($\approx 30\%$) wider at their midpoints than at their footpoints, implying less-than-expected expansion of the magnetic field. Second, the variation of width along each loop tends to be modest, implying that the cross section has an approximately circular shape. And third, cross-axis intensity profiles tend to be singly-peaked and simple, implying that the cross section is approximately uniformly filled on resolvable scales. We conclude that the energy which heats the plasma is either dissipated axially symmetrically on a scale equal to a loop diameter ($\approx 11\,000$ km) or else is dissipated with any spatial structure, but on a scale much smaller than a loop diameter, and then transported laterally in an axisymmetric fashion (perhaps via conduction along chaotic field lines). In their present form, none of the theoretical ideas concerning the magnetic structure and heating of loops are obviously capable of explaining all of the observed properties.

1. Introduction

It has long been recognized that intensity features observed in coronal images are a useful diagnostic of the coronal magnetic field (e.g., Orrall, 1981). This is very important, since the coronal field cannot be directly measured with a sufficient accuracy and since the field is thought to be essential in many important solar phenomena, including coronal heating, solar flares, and coronal mass ejections. By far the most common intensity features are loops. It is reasonable to expect that these loops trace the local magnetic field, since mass and thermal energy can flow easily along magnetic field lines but not across them. This idea is further supported by comparisons of coronal images with extrapolated photospheric magnetic fields, which show a general correspondence between observed loops and extrapolated field lines (e.g., Poletto *et al.*, 1975; Jiao, McClymont, and Mikić, 1997).

It is often instructive to think of the coronal magnetic field as a collection of rigid, thermally insulated conduits, each containing its own 'mini atmosphere'. Because parallel thermal conduction is so efficient and because the gravitational scale height is so large, plasma parameters such as temperature and density tend to be uniform along field lines. Cross-field gradients can be very large, on the other hand. The values of temperature and density on a given field line are closely related to the rate of coronal heating. Greater heating produces higher temperatures and, through chromospheric evaporation, higher densities. The fact that we observe

The physics of the solar corona and transition region. Part II.
Reprinted from Solar Physics **193,** 2000.

distinct coronal loops tells us that the heating rate has substantial spatial variation. Although the entire coronal volume is filled with magnetic field, only certain flux tubes are highlighted because of their locally enhanced heating.

One of the most fundamental properties of a loop is its cross section. Variations in cross-sectional area play an important role in determining the thermodynamic structure of loops (e.g., Vesecky, Antiochos, and Underwood, 1979; Schrijver, Lemen and Mewe, 1989; Dowdy, Emslie, and Moore, 1987), the presence and magnitude of steady flows (Mariska, 1984), the cooling of flare plasmas (Antiochos and Sturrock, 1976), and the propagation of energetic flare particles and the resulting emission of hard X-ray and gamma radiation (e.g., McTiernan and Petrosian, 1990). More importantly, the cross section reveals detailed information about the structure of the magnetic field and the spatial distribution of coronal heating. Understanding the nature of magnetic energy release in the solar corona remains one of the great challenges in solar physics. By studying the cross-sectional properties of loops, we can hope to make important progress toward solving the fundamental problem of coronal heating.

One very curious aspect of loops is their apparent tendency to have uniform thickness. The coronal magnetic field must diverge with height, on average, since the source of the field is localized in the photosphere, and the field will expand to fill the available volume in the approximately force-free environment above. One might expect a majority of coronal loops to also diverge with height and so appear thicker at their tops than at their footpoints. It is therefore surprising that soft X-ray, EUV, and visible-light images all give the impression that coronal loops have a nearly uniform thickness (e.g., Golub *et al.*, 1990; Bray *et al.*, 1991).

We have previously attempted to quantify this fascinating result by measuring thickness variations along several loops observed by the Soft X-ray Telescope (SXT) on *Yohkoh*. We found that the loops are typically only about 13% wider at their tops than at their footpoints (Klimchuk *et al.*, 1992). This is far less than one would expect based on our present understanding of coronal magnetic fields. Our initial study suffered from two limitations, however: first, it involved a rather small sample of only 10 loops; and second, all of the observations were made in half resolution mode in which 2×2 on-chip summing produces an effective pixel size of 4.9 arc sec. The need for a more extensive follow up study was clear, especially given the significance of the results. We have now completed such an investigation using many more loops and using higher resolution data. The purpose of this paper is to report our findings and to discuss their implications for magnetic structure and heating.

2. Observations

Our improved study includes 43 loops taken from the same dataset we used previously to investigate coronal heating scaling laws (Klimchuk and Porter, 1995;

Porter and Klimchuk, 1995). They were observed by SXT during the first year of the *Yohkoh* mission, beginning in late 1991, at a time when the Sun was starting its decline from an extended maximum of the solar cycle. The loops are all associated with active regions, though they do not necessarily span the dominant neutral line of a region. They were chosen because they are slowly evolving (having timescales long compared with the cooling time) and because they are relatively well isolated from overlapping bright structures that might contaminate the measurements.

An effort was made to find a truly representative sample. No preference was given to loops of any particular length or width, and the final set includes a factor of 8 range in length and a factor of 4 range in width. We can nonetheless not rule out the possibility of a selection bias. For example, the desire to have loops with an uncomplicated background meant that isolated loops were more likely to be chosen than those grouped together in a tight cluster.

Because the three-dimensional geometry of the loops is not known, we attempted to eliminate possible projection effects by selecting loops from a variety of different positions on the disk and at the limb. The hope was that any such effects would then average out statistically and not not influence the results. We note that width measurements are much less dependent on the observation geometry than are, for example, pressure measurements (Alexander and Katsev, 1996).

All of the loops in our study were observed through pairs of X-ray filters and have a median filter-ratio temperature of 5.7 MK (Porter and Klimchuk, 1995). They are hotter than the loops observed by most other coronal instruments. Width measurements were made exclusively from summed thin aluminum (Al.1) images formed by adding together individual coaligned images from a single spacecraft orbit ($\approx$ 60 min of daylight). Twenty-three of the loops were observed in full resolution mode with 2.45 arc sec pixels, and the remaining 20 were observed in half resolution mode. Several examples are shown in Figures 1–5.

The stars in the figures indicate the locations where the loop intensity falls to e^{-1} of its maximum value (after background subtraction). We take these to be the footpoints of the loops. In a few cases we instead use a subjective identification – for example, when the intensity falls just below the e^{-1} cutoff, but remains at this level for a considerable distance before ultimately dropping to zero. It is important to recognize that the photospheric footpoints of the loops could be considerably removed from the soft X-ray footpoints. Since SXT is preferentially sensitive to plasmas hotter than 2 MK (Tsuneta *et al.*, 1991), the lower portion of a loop leg fades from visibility before the flux tube reaches the photosphere. We have simulating SXT observations of static equilibrium loop models and find that the e^{-1} intensity position is separated from the photosphere by only about 5% of the full loop length (Dixon *et al.*, 1991). Nonetheless, evidence points toward a considerably greater separation in some cases. For example, Porter, Falconer, and Moore (1998) suggest that the left footpoint of the loop in Figure 1 is associated with the small bright feature located a quarter of the way down from the top of the image.

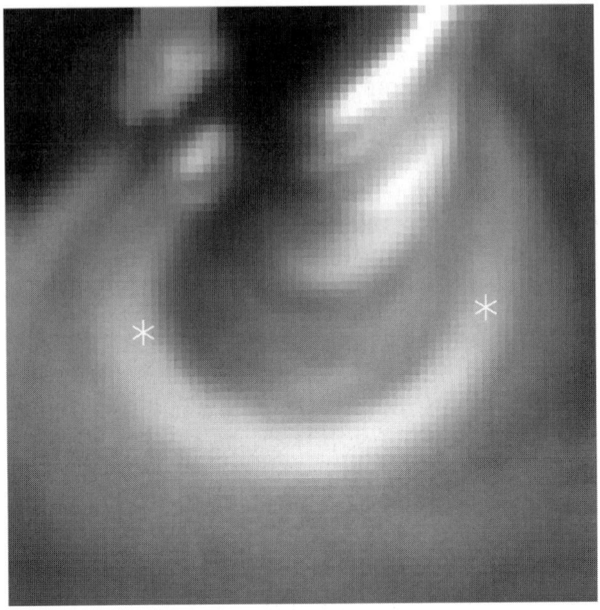

Figure 1. Full-resolution SXT image from the 26 December 1991 orbit beginning 17:58 UT. Dimensions are 189″ × 186″. Loop footpoints are marked with stars.

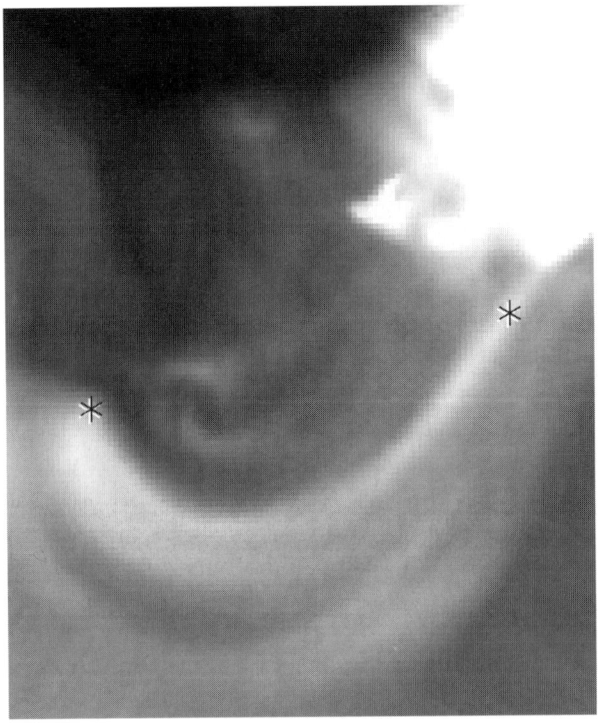

Figure 2. Full-resolution SXT image from the 18 November 1991 orbit beginning 11:34 UT. Dimensions are 270″ × 316″. Loop footpoints are marked with stars.

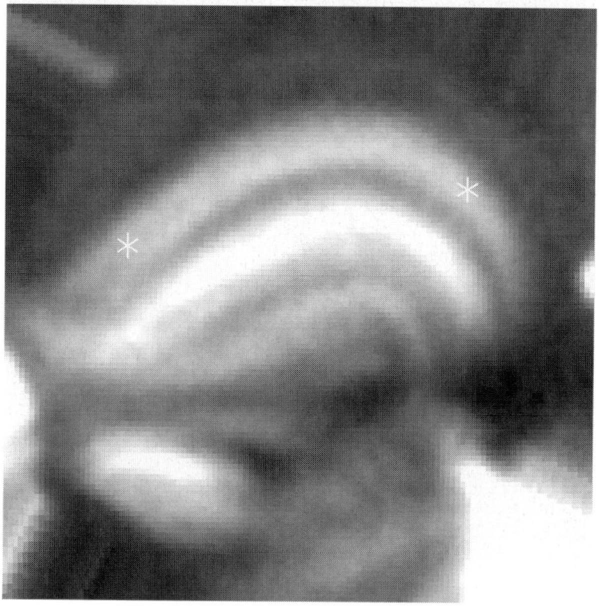

Figure 3. Half-resolution SXT image from the 7 May 1992 orbit beginning 7:13 UT. Dimensions are $282'' \times 277''$. Loop footpoints are marked with stars.

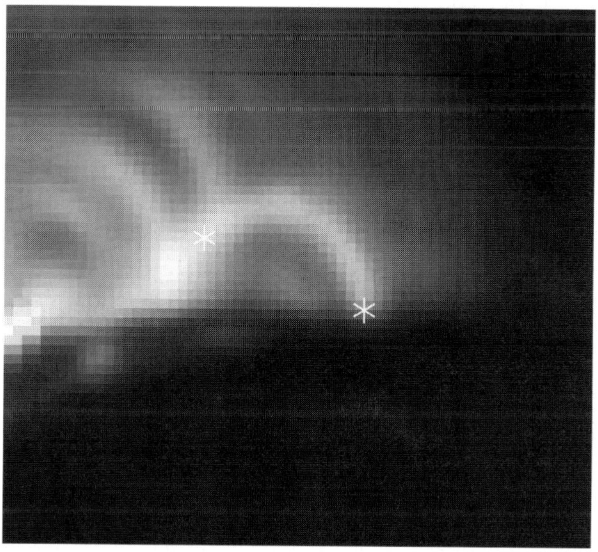

Figure 4. Full-resolution SXT image from the 11 August 1992 orbit beginning 6:59 UT. Dimensions are $145'' \times 140''$. Loop footpoints are marked with stars.

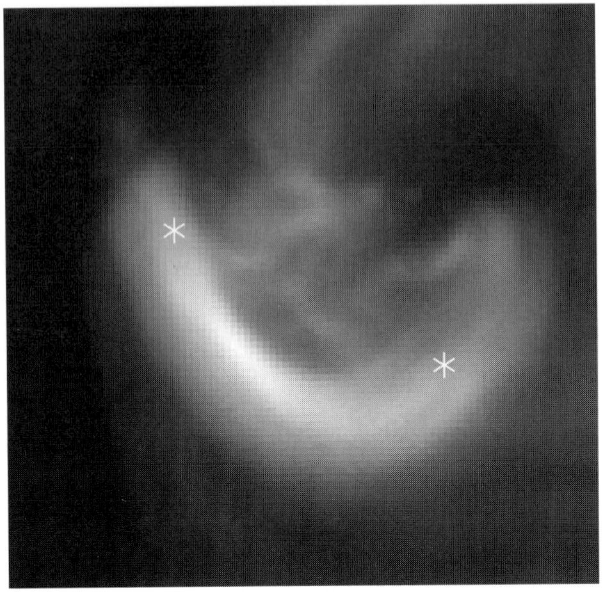

Figure 5. Full-resolution SXT image from the 17 October 1991 orbit beginning 5:42 UT. Dimensions are $225'' \times 216''$. Loop footpoints are marked with stars.

3. Data Analysis

We measured the widths of the loops using the same methodology developed for our original study. First, the intensity data are mapped onto a 'straightened loop' coordinate system in which one coordinate measures distance along the curved loop axis and the other measures distance away from it. The axis is defined by fitting a least squares polynomial to positions that are selected visually at the computer monitor. Each point in the straightened loop coordinate grid is assigned an intensity equal to the weighted average of the intensities at the four nearest neighbors in the original grid. Weighting is according to inverse distance.

Next, we subtract a solar background, recognizing that only some of the emission recorded at a given pixel comes from the physical loop structure we are attempting to isolate. To estimate this background, we visually identify the pixels in the loop and extract them from the image. We then perform a least-squares polynomial surface fit to the remaining non-loop pixels and subtract this surface (the background) from the complete image, setting any negative values to zero. Although the identification of the loop is subjective, it is usually rather obvious.

We believe that errors in the background subtraction are the greatest source of uncertainty in our width measurements. Figure 2 includes an example of the problems that can be encountered. Most of the broad swath of emission near the left footpoint is likely *not* to come from the primary loop, but it is difficult to establish this with confidence. We estimate the uncertainty associated with background subtraction by repeating our analysis while extracting a loop region that is

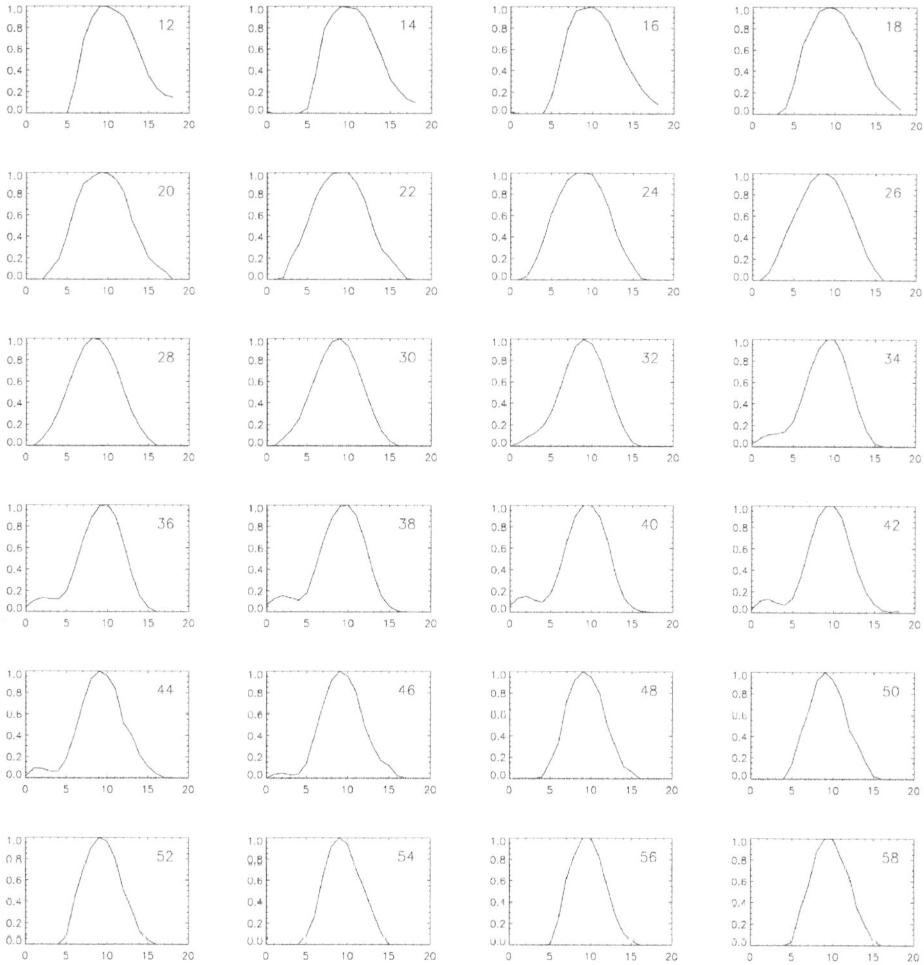

Figure 6. Cross-axis intensity profiles from every other pixel along most of the 26 December 1991 loop shown in Figure 1. Pixel numbers are given in the upper-right-hand corner (footpoints are located at pixels No.6 and No.65). Intensities are normalized to unity, and the abscissa has a range of 20 pixels.

2 pixels wider than originally defined. The results are qualitatively unchanged, as we discuss later.

Loop widths are determined from the straightened, background-subtracted images. A horizontal row in these images corresponds to an intensity profile orthogonal to the loop axis. Figure 6 displays the profiles from every other pixel along most of the loop in Figure 1. Note how simple the profiles appear. They are singly peaked and have minimal detailed structure. This is generally the case for all of the loops we have examined, although profile complexities such as enhanced shoulders and multiple peaks are not rare (e.g., Figure 7). The tendency for profiles to be simple has important implications for coronal heating, as discussed in Section 5.2.

The definition of loop width is not unique. We have chosen to adopt the second moment of the intensity profile as a characteristic measure. This moment is equivalent to the standard deviation of photon positions for all of the photons that make up the profile:

$$\sigma = \left[\frac{\sum (x_i - \mu)^2 I_i}{\sum I_i} \right]^{1/2} , \tag{1}$$

where the summation is taken over the x_i positions in the profile, and μ is the mean position:

$$\mu = \frac{\sum x_i I_i}{\sum I_i} . \tag{2}$$

To understand how σ relates to a physical dimension, consider a simple model loop with a circular cross section that is uniformly filled (i.e., that emits uniformly). These properties will be justified later. For a perfectly resolved observation, the σ for this loop would be exactly one-half of the loop radius. Of course, SXT has finite resolution, and the smearing caused by the optics will produce a broadening of the intensity profile. This is simulated in Figure 8, where the solid line indicates the observed profile for a model loop with a radius of 3 pixels. We have here assumed that the SXT point spread function is a Moffat function with parameters $a = 0.68$ and $b = 1.45$ (Martens, Acton, and Lemen, 1995). Further degradation in resolution is produced by the finite pixel size of the CCD detector. This is represented by the stars in the figure which show the intensities averaged over the individual pixels. From these 'pixelated' intensities we obtain a σ of 1.78 pix. This is only modestly greater than half of the radius, indicating that the loop is reasonably well resolved.

Intrinsically narrow loops are much more affected by the finite spatial resolution. Consider the extreme example of a loop with a radius of only 0.1 pix, as shown in Figure 9. In this case, the σ of 0.89 pix is many times greater than the radius, demonstrating that the loop is completely unresolved. The loop is so narrow, in fact, that the intensity profile is essentially equivalent to the point spread function. Note how the scattering wings are greatly reduced compared to those of earlier grazing incidence telescopes, such as the S-054 instrument on *Skylab* (see Figure 4 of Tsuneta *et al.* (1991) for a comparison).

By simulating observations of many different loops with a wide range of radii, we have generated curves of r versus σ for both full and half resolution observations. These are shown in Figure 10. The units here and throughout this paper are full resolution (2.45 arc sec) pixels. The curves asymptotically approach a straight line of slope of 2 as the loops become wider and the finite spatial resolution becomes less important. For narrow loops, the curves roll over to intersect the σ-axis near 1. The standard deviation can never be smaller than these intercept values no

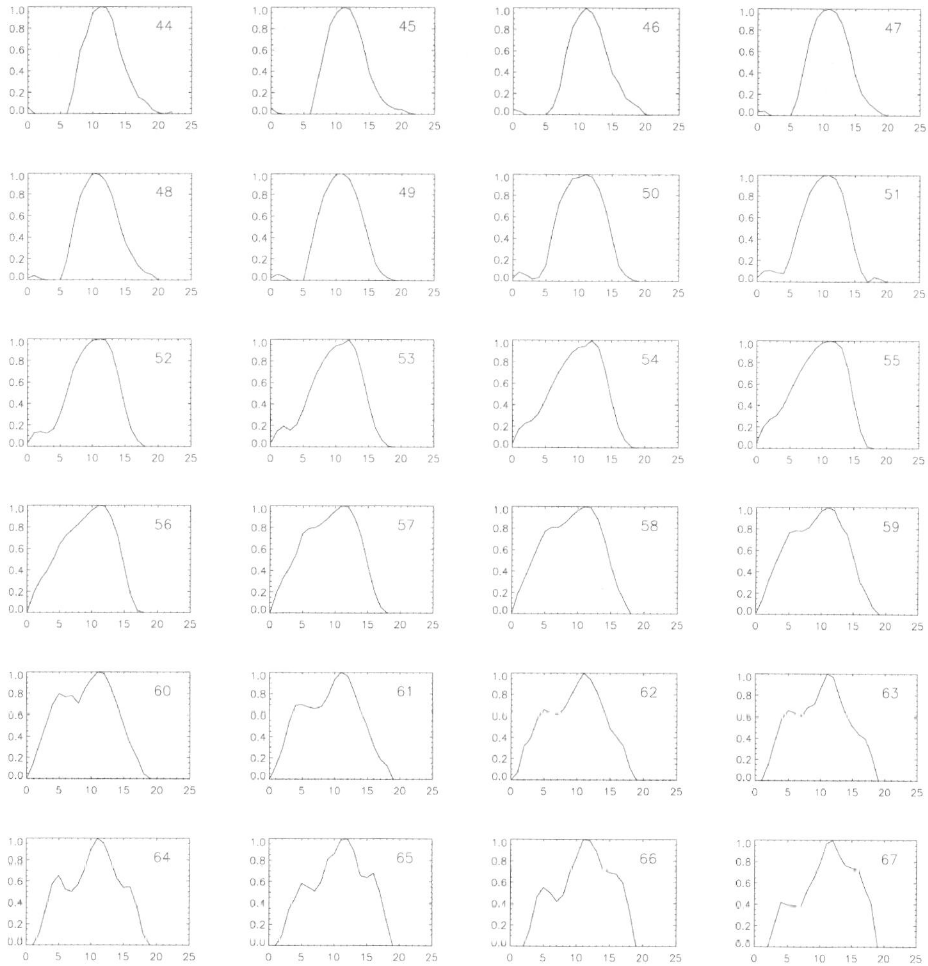

Figure 7. Cross-axis intensity profiles from every other pixel along the right half of the 17 October 1991 loop shown in Figure 5. Pixel numbers are given in the upper-right-hand corner (footpoints are located at pixels No.9 and No.67). Intensities are normalized to unity, and the abscissa has a range of 25 pixels.

matter how intrinsically narrow the loops may be. We conclude that loops with $\sigma > 1.0$ full resolution pixels and $\sigma > 0.6$ half resolution pixels are sufficiently well resolved to be reliably measured. All of the loops in our study are within these ranges. Note that the axes of the model loops in Figures 8 and 9 are taken to fall part way into a pixel from the edge. We have verified that the precise location is unimportant, and the curves in Figure 10 represent an average over all possible locations.

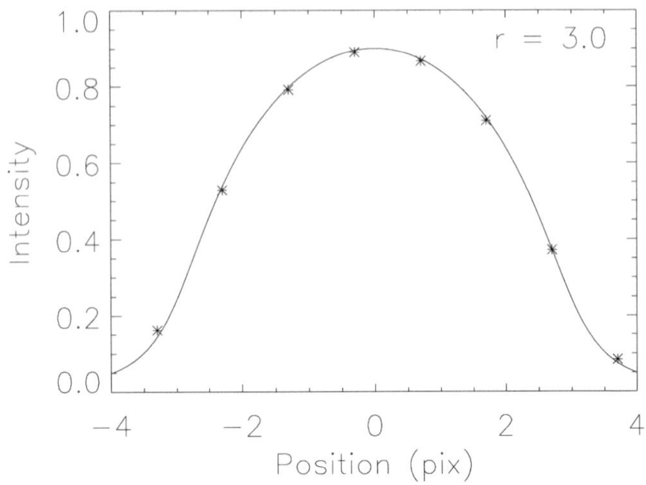

Figure 8. Cross-axis intensity profile for a simulated full-resolution observation of an idealized model loop with a radius of 3 pix (see text). Solid line includes smearing due to the point spread function, and stars indicate the pixel-integrated values.

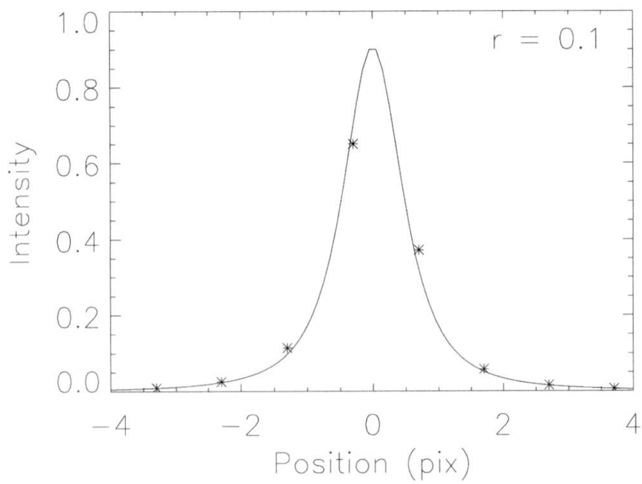

Figure 9. Cross-axis intensity profile for a simulated full-resolution observation of an idealized model loop with a radius of only 0.1 pix. Solid line includes smearing due to the point spread function, and stars indicate the pixel-integrated values. The loop is so narrow that the profile is essentially equivalent to the point spread function.

4. Results

The solid lines in Figures 11–15 show the σ measured for the loops in Figures 1–5, plotted as a function of position along the loop axis. The dashed lines give the corresponding radii determined from the curves in Figure 10, and the dotted lines give the intensities integrated over the cross-sectional profiles (normalized to unity).

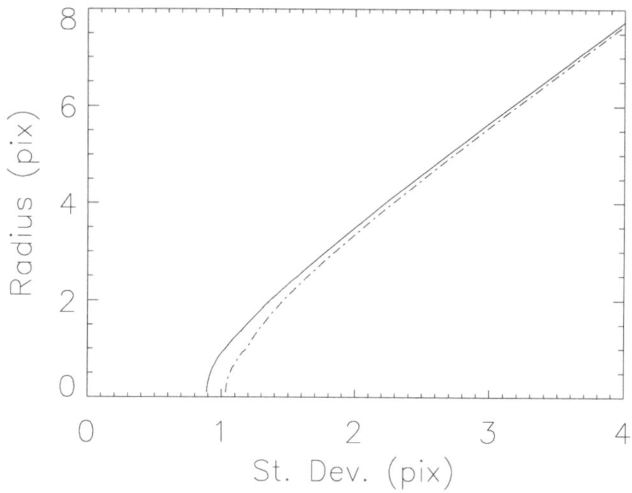

Figure 10. Radius versus standard deviation for full- (*solid*) and half- (*dot-dashed*) resolution observations of an idealized model loop having a circular uniformly-filled cross section. Units are full-resolution (2.45 arc sec) pixels in both cases.

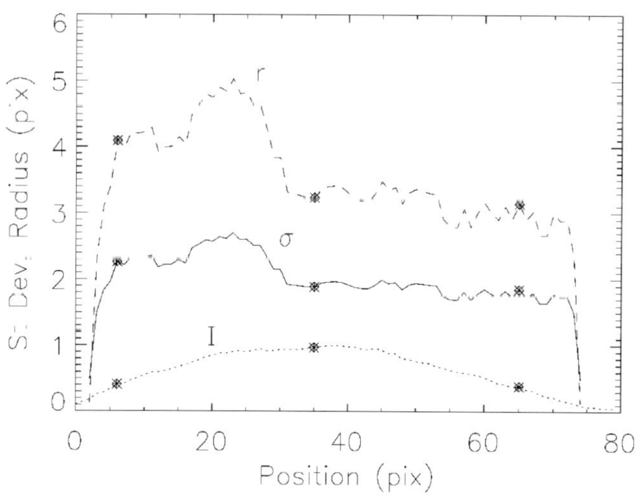

Figure 11. Standard deviation (*solid*), radius (*dashed*), and normalized intensity (*dotted*) as a function of position along the 26 December 1991 loop of Figure 1. Footpoint and midpoint positions are indicated.

The stars indicate the footpoint positions and the position midway in between. These examples were chosen because they are representative of the considerable variety that is present in the full dataset. Some loops have a nearly uniform width along their entire length (Figure 13), while others have a high degree of variability (Figure 12). Still others are uniform in one leg and variable in the other (Figure 11). There is no obvious pattern that we have been able to detect, including no clear evidence for a systematic oscillatory behavior with a preferred wavelength.

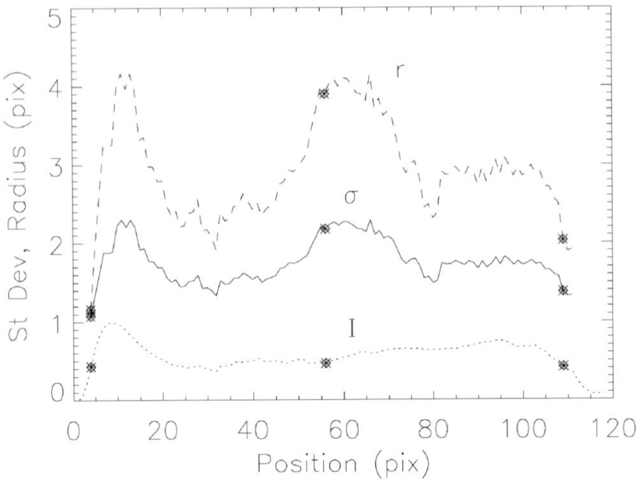

Figure 12. Standard deviation (*solid*), radius (*dashed*), and normalized intensity (*dotted*) as a function of position along the 18 November 1991 loop of Figure 2. Footpoint and midpoint positions are indicated.

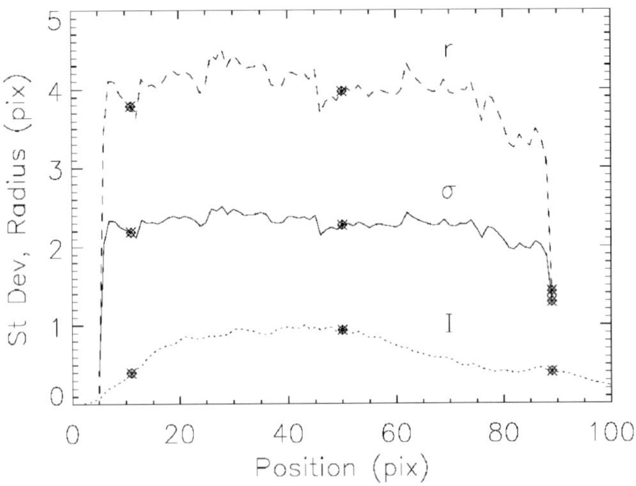

Figure 13. Standard deviation (*solid*), radius (*dashed*), and normalized intensity (*dotted*) as a function of position along the 7 May 1992 loop of Figure 3. Footpoint and midpoint positions are indicated. Units are full-resolution pixels even though these are half-resolution observations.

Figure 16 shows the mean radius for each loop – obtained by averaging the values between the two footpoints – plotted as a function of the loop length. This length has been approximately corrected for projection effects as described in Porter and Klimchuk (1995). Full-resolution observations are shown as stars and half resolution observations as diamonds. Note that the full- and half-resolution results are similar, demonstrating that the pixel size is not crucial for the width measurements. The mean radii range mostly between 1.5 and 4 pixels and have

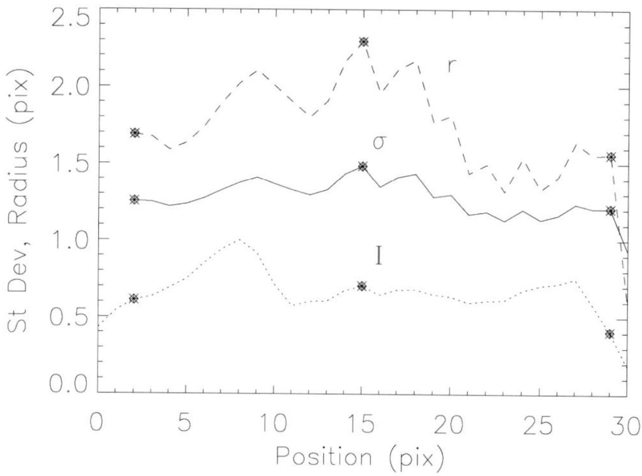

Figure 14. Standard deviation (*solid*), radius (*dashed*), and normalized intensity (*dotted*) as a function of position along the 11 August 1992 loop of Figure 4. Footpoint and midpoint positions are indicated.

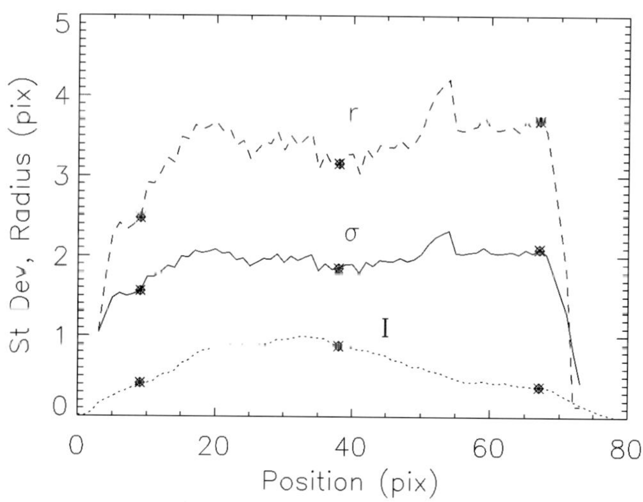

Figure 15. Standard deviation (*solid*), radius (*dashed*), and normalized intensity (*dotted*) as a function of position along the 17 October 1991 loop of Figure 5. Footpoint and midpoint positions are indicated.

median value of 2.96 pix ($\approx$ 5400 km). The mean σ have a median value of 1.84 pix. There is no strong dependence on loop length, indicating that the loops do not have a preferred aspect ratio.

To investigate whether loops tend to be wider at their tops than at their footpoints, we have determined average radii over three small loop segments: one at each of the two footpoints, and another at the loop midpoint. The length of each segment is equal to 10% of the full loop length. For each loop we then computed

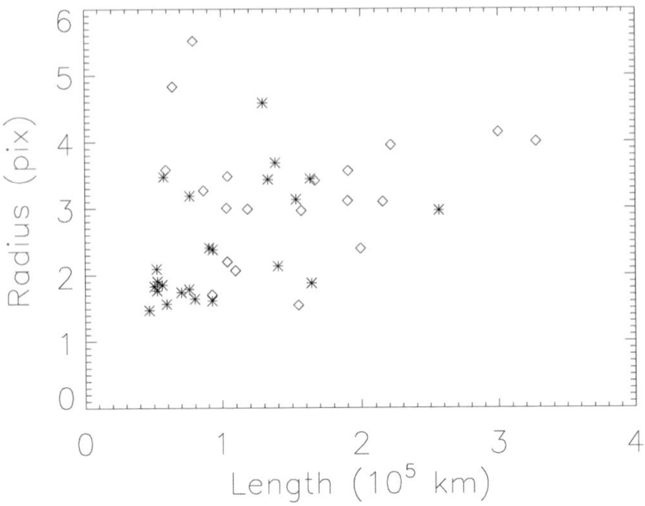

Figure 16. Mean radius for each loop as a function of loop length. Stars and diamonds are for full- and half- resolution observations, respectively. Units are full-resolution pixels in both cases.

two expansion factors, Γ_r, by dividing the average at the midpoint by the averages at the footpoints:

$$\Gamma_r \equiv \frac{r_{\text{mid}}}{r_{\text{foot}}} \,. \tag{3}$$

These are plotted against loop length in Figure 17, again with the stars and diamonds differentiating between full and half resolution observations. One outlier point is not shown. The median expansion factor for all of the loops is 1.30. Thus, a typical loop is only about 30% wider at its midpoint than at its footpoints. An expansion factor defined in terms of σ has a median value of 1.17, which is nearly identical to what we found in our original study (Klimchuk *et al.*, 1992).

As mentioned earlier, we believe the greatest source of error in our measurements is the background subtraction. To estimate the uncertainties associated with this procedure, we have repeated our analysis using a loop region that is 2 pixels wider (see above). The effect on the results is minimal, with the median values changing by the following amounts: $\Delta r = 0.64$ pix, $\Delta\sigma = 0.27$, $\Delta\Gamma_r = -0.12$, and $\Delta\Gamma_\sigma = -0.05$. The widths are slightly increased and the expansion factors are slightly decreased.

To evaluate the possible influence of projection effects, we have examined whether the loop widths and expansion factors depend on the angle between the line of sight and local vertical. They do not. Finally, to determine whether the summed images might be affected by inaccuracies in the correction for solar rotation, we have looked for a possible dependence on the proper motion due to solar rotation. Again, there is none.

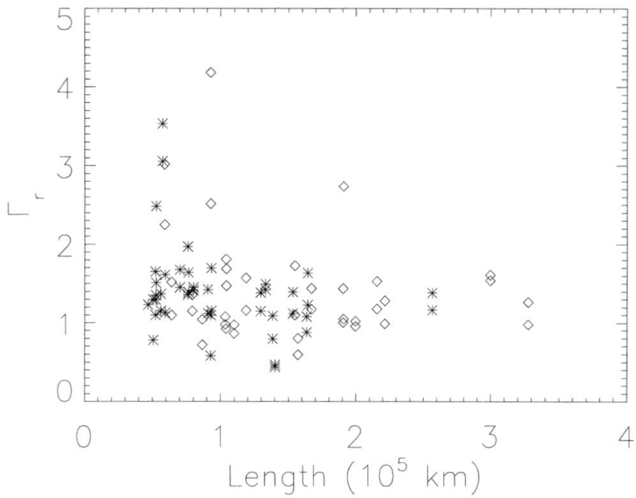

Figure 17. Expansion factor (ratio of midpoint radius to footpoint radius) as a function of loop length. Stars and diamonds are for full- and half- resolution observations, respectively.

5. Discussion

Based on these results, the conclusion seems unavoidable that *Yohkoh* soft X-ray loops exhibit very little systematic expansion with height. Yan and Sakurai (1997) also found this to be true of the *Yohkoh* loop they studied. In a parallel investigation to this one, we have recently measured the expansion of loops observed by the *Transition Region and Coronal Explorer* (TRACE). These loops, observed mostly in the 171 and 195 Å band passes, exhibit even less expansion: $\Gamma_r = 0.99$ for nonflare loops and 1.13 for post-flare loops (Watko and Klimchuk, 1999). This is important independent confirmation from a different instrument having substantially better spatial resolution (0.5 arc sec pixels). Furthermore, there is mounting evidence that TRACE loops may be physically different from *Yohkoh* loops, so it is interesting that both classes of loops have similar expansion properties. TRACE loops are considerably cooler than *Yohkoh* loops and they have fundamentally different equilibrium properties; *Yohkoh* observations are compatible with static equilibrium (Porter and Klimchuk, 1995; Kano and Tsuneta, 1995), whereas loops observed by the Extreme ultraviolet Imaging Telescope (EIT) using the same bandpasses as TRACE are necessarily far from equilibrium (Aschwanden *et al.*, 1999; Aschwanden *et al.*, 2000). In addition, *Yohkoh* images, like the hottest (284 Å) TRACE images, have a much fuzzier appearance than the cooler TRACE images, and this cannot be explained by differences in spatial resolution.

5.1. Flux tube expansion

To evaluate the importance of our results, we must compare the observed expansion of X-ray plasma loops with the expected expansion of coronal magnetic flux tubes.

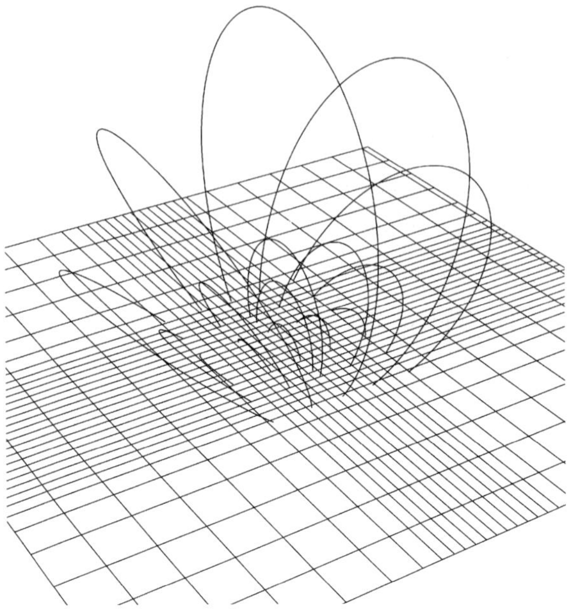

Figure 18. Selected field lines from the potential magnetic field produced by a point dipole buried below the gridded plane.

Figure 18 shows the hypothetical magnetic field of a point dipole buried below the solar surface. Considerably divergence of the field lines is apparent. This is quantified in Figure 19, which represents a view looking down onto the photosphere. The shaded contours indicate the vertical component of the photospheric magnetic field, such as would be observed at disk center by an ordinary magnetograph. A neutral line runs through the middle, separating the opposite polarity halves on the left and right sides. If we imagine that each field line is the axis of an ultra thin flux tube, then we can define an expansion factor for the tube to be

$$\begin{aligned}\Gamma_{A^{1/2}} &\equiv \frac{A_{\rm mid}^{1/2}}{A_{\rm foot}^{1/2}} \\ &= \frac{B_{\rm foot}^{1/2}}{B_{\rm mid}^{1/2}},\end{aligned} \tag{4}$$

where A is the cross-sectional area, B is the magnetic field strength, and we have made use of conservation of magnetic flux. Although this expression applies strictly to flux tubes that are infinitely thin, it is an excellent approximation for tubes of realistic size. The line contours in Figure 19 show $\Gamma_{A^{1/2}}$ plotted as a function of footpoint position. The expansion factor exceeds 1.5 over two-thirds of the field of view and reaches values in excess of 6.5 at the left and right edges. This is much greater than the observed expansion of both *Yohkoh* and TRACE loops.

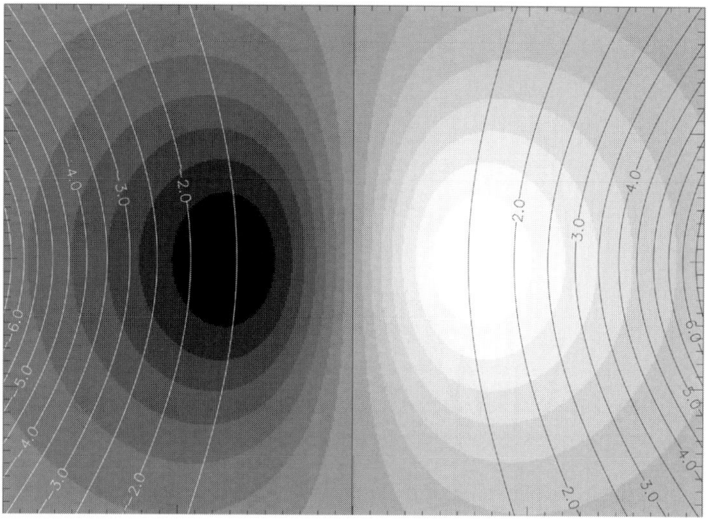

Figure 19. Field line divergence of the point dipole magnetic field of Figure 18. Shaded contours indicate the vertical component of the field in the photosphere (dark for one polarity and light for the other). Line contours indicate the $\Gamma_{A1/2}$ expansion factor plotted at the footpoint positions. The contour increment is 0.5, and the neutral line running down the middle has a value of 1.0.

Another significant difference concerns the spatial distribution of the expansion factor. As shown in Figure 17, the observed expansion is independent of loop length, and yet $\Gamma_{A1/2}$ increases strongly with distance away from the neutral line and hence length. The behavior of $\Gamma_{A1/2}$ can be easily understood in terms of the well-known inverse cube fall off of B with distance d from the subsurface dipole moment. Along short, low-lying field lines rooted close to the neutral line, d and therefore B and A are relatively constant, but along high-arching field lines with widely separated footpoints, d, B, and A all have considerable variation.

This simple dipole model is of course highly idealized, and one must ask whether the flux tubes in more realistic configurations have similar expansion properties. The answer appears to be affirmative. Both McClymont and Mikić (1994) and Gary (1997) found generally large expansion factors in their potential and nonlinear force free extrapolation models of observed active regions. There is an important issue, however, regarding the height separation between the photospheric footpoint and the footpoint seen in soft X-rays. As mentioned earlier, hydrodynamic loop modeling suggests that this separation is of order 5% of the total loop length (Dixon *et al.*, 1991). Even with an assumed separation as large as 15%, the expansion predicted by the models of McClymont and Mikić (1994) is considerably greater than that observed (although the difference is more modest in regions of large electric current). We disagree with the claim of Gary (1997) that his potential-field model is consistent with the *Yohkoh* observations. As shown in his Figure 2(a), an assumed footpoint separation of 12.5% (25% of the loop halflength) produces a median expansion factor of 1.7. Finally, we note that Aschwanden *et al.*

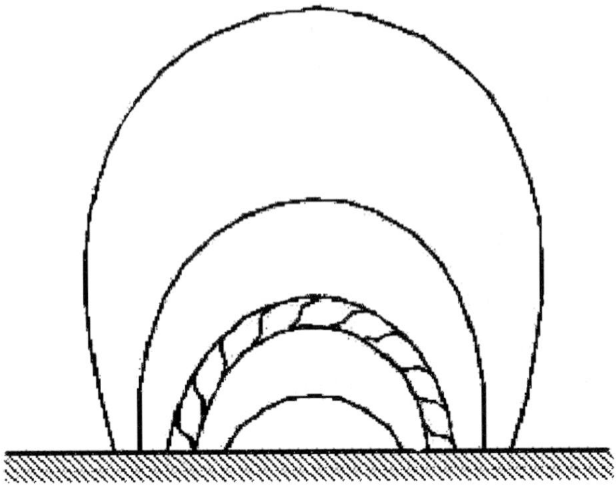

Figure 20. Twisted magnetic flux tube of uniform thickness surrounded by untwisted field that is expanding with height.

(1999) have measured minimal expansion in the lower legs of EIT loops, and these loops, like their TRACE counterparts, should extend closer to the photosphere than do the hotter *Yohkoh* loops.

How do we explain the apparently large discrepancy between the observed and expected expansion? One possibility is that observed plasma loops do not actually coincide with magnetic flux tubes (e.g., Sakurai and Yan, 1996). We do not feel that this is a plausible explanation. Another possibility is that observed loops correspond to flux tubes that are relatively highly twisted compared to the surrounding magnetic field (Klimchuk *et al.*, 1992). While there must be an overall divergence with height, a subset of flux tubes could, in principle, expand much less than the average. The idea is shown schematically in Figure 20. It was inspired by the finding that the cores of straight axisymmetric flux tubes become constricted when the tubes are twisted (e.g., Zweibel and Boozer, 1985; Robertson, Hood, and Lothian, 1992). Photospheric line-tying would prevent any constriction at the footpoints of an expanding loop, but the coronal section could constrict freely, and this would tend to promote thickness uniformity.

We have explored this idea by computing numerical models of a twisted magnetic flux tube embedded within a much larger untwisted dipole configuration (Klimchuk, Norton, and Antiochos, 2000). The models confirm that localized twist can indeed reduce the expansion factor of the tube, but only when the tube is viewed from the side, as in Figure 20. When viewed from above, the expansion factor actually increases (although line of sight integration effects tend to mitigate this). Since we expect that a wide variety of viewing angles are represented in our dataset, it appears unlikely that localized twist can explain the observation of nearly uniform thickness.

Our models have revealed an interesting property of twisted flux tubes – the tendency for the cross section to be circular in the corona. Before any twist is applied, the model flux tube has a circular cross section in the photosphere, but an oval-shaped cross section in the corona. The elongation of the oval is in the plane of the loop, vertically in this case, and is a natural part of the equilibrium structure. When the footpoints are twisted, the oval becomes progressively more circular while maintaining an approximately constant area. This is a direct consequence of the magnetic tension associated with the azimuthal field component that is introduced by the twist. Field lines trace out closed paths when viewed projected onto a cross section, and the tension in these lines will act to make the paths circular. As we discuss shortly, this tendency for circular cross sections could have important application to the observations.

Before proceeding, we note that the localized twist we envision is a form of magnetic stress. Stress is also produced by large-scale shear. McClymont and Mikić (1994) investigated the influence of such stress by comparing potential and force-free extrapolation models of the same active region. They found that, in portions of the active region, the force-free flux tubes expand less than the potential flux tubes. Similarly, Wang and Sakurai (1998) examined a family of force-free field models and found that the smallest expansion occurs in certain areas within the most highly stressed configuration. Although the interpretation of these models must be treated with some caution – for example, flux tubes in the different models cannot be directly compared because they do not have the same footpoint connectivity – they nonetheless demonstrate how large-scale magnetic stress can play a role in limiting flux tube expansion.

5.2. Coronal Heating

The previous discussion refers only to magnetic flux tubes, whereas the plasma loops that are observed are determined partly by the magnetic field and partly by the spatial distribution of coronal heating. In particular, the shape of the cross section can be strongly influenced by the heating. We suggested above how twisted magnetic flux tubes might naturally tend to have circular cross sections. We now claim that plasma loops also have circular cross sections. Our argument is based on the result that the observed variation of width along most loops is only very modest. To understand the reasoning, imagine a loop with a highly non-circular cross section. We know from vector magnetograms that virtually all field in the corona is twisted to some degree. We therefore would expect to observe the thin dimension of the cross section along some sections of the loop and the thick dimension of the cross section along others. The width would appear to vary, as indicated schematically in Figure 21 for the case of a ribbon-like loop. Since extreme variations of this kind are rare, we conclude that the actual cross sections must be approximately circular.

We can also infer that the cross sections are approximately uniformly filled with X-ray emitting plasma, at least on resolvable scales (unresolved filling factors are

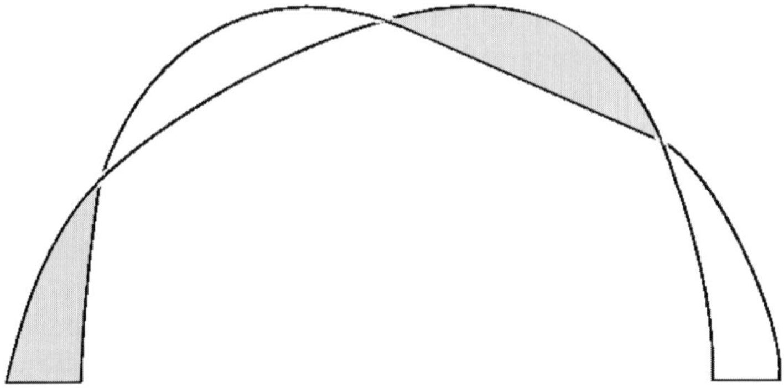

Figure 21. Hypothetical ribbon-like coronal loop that is twisted.

possible). If they were not, the intensity profiles would exhibit far more structure than is typically observed. For example, if loops were hollow tubes, the intensity profile would have a double peak instead of the single peak that is far more common.

These properties of the loop cross section place important constraints on theories of coronal heating. They imply that either: (1) the energy which heats the plasma is dissipated axially symmetrically on a scale equal to a loop diameter; or (2) the energy is dissipated with any spatial structure, but on a scale much smaller than a loop diameter, and is then transported laterally in an axisymmetric fashion. The possibilities are indicated schematically in Figure 22: (a) represents axially-symmetric dissipation, and (b) and (c) represent axially-symmetric transport. The transport can either be across field lines by waves or some diffusion process (c), or along field lines by thermal conduction if the field has a highly chaotic internal pattern (b).

Existing models of coronal heating (as reviewed, e.g., by Mandrini, Démoulin, and Klimchuk, 2000) have difficulty satisfying all the observational constraints. For example, magnetic energy dissipation in large current sheets, probably by reconnection, would produce loops that are more like ribbons than circular tubes (e.g., Galsgaard and Nordlund, 1996; Longcope and Sudan, 1994; Mikić, Schnack, and Van Horen, 1989). This is incompatible with Figure 22(a). Reconnection on much smaller scales is certainly possible, but we are left with the problem of lateral energy transport. Slow-mode MHD waves may have sufficiently short damping lengths if the frequencies are high enough (e.g., Porter, Klimchuk, and Sturrock, 1994); however, the spectrum of waves generated by reconnection is not well understood. Litwin and Rosner (1993) have examined a number of other candidates for cross-field transport and conclude that they are too slow to be important.

The picture of chaotic field lines is intriguing. Parker (1983) has long advocated the view that coronal field lines become wrapped and braided by a random shuffling of their footpoints, and Figure 22(b) is one adaptation of this idea. We are left to explain, however, why the field lines should be confined to a well-defined cylindrical

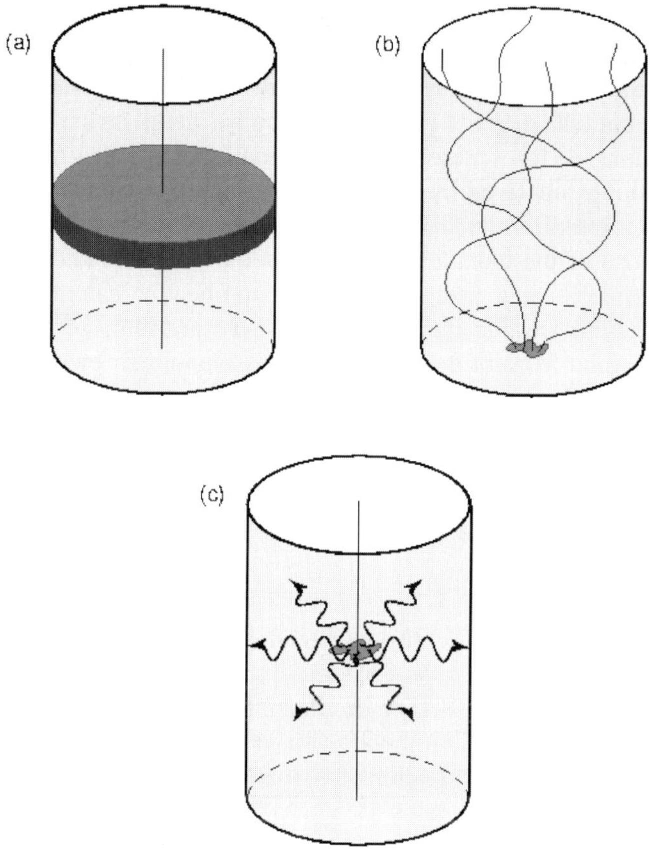

Figure 22. Schematic representation of three generic possibilities for energy release and transport in coronal loops: (a) axisymmetric release; (b) small-scale release and axisymmetric transport along internally chaotic field lines; and (c) small-scale release and axisymmetric transport across field lines.

envelope. One possibility is that the random shuffling motions are superposed on an overall rotational flow. Our force-free field modeling has shown that twisted flux tubes with a simple internal structure tend to have circular cross-sections, and it is plausible that the same may be true of flux tubes with more complicated internal structure. Confirmation is needed. We note that Galsgaard and Nordlund (1997) have found that chaotic field lines can arise not only from random footpoint motions, but also from the nonlinear development of the internal kink instability in highly twisted flux tubes.

Another popular coronal heating model involves the resonance absorption of Alfvén waves. Nearly all versions of this model predict heating confined to a thin resonance layer located a specific radial distance away from the loop axis (e.g., Davila, 1987). This would produce bright emission from a hollow tube, which we have shown is incompatible with the observations. Recently, Ofman, Klimchuk, and Davila (1998) demonstrated that the thermal and dynamic coupling between the corona and chromosphere will cause a single stationary resonance layer to break up into multiple layers that drift throughout the loop to heat the entire volume in a time average. This is in much better accord with the observations, though the reason for the circular shape of the cross section has yet to be explained.

In conclusion, we have presented observations showing that the cross sections of soft X-ray loops observed by *Yohkoh* are: (1) approximately circular in shape; (2) approximately uniformly filled on resolvable scales; and (3) *not* appreciably wider at the tops of the loops than at the footpoints. To our knowledge, none of the theoretical ideas about magnetic structure and heating can readily explain all of these properties. Despite the discrepancies, we are reluctant to reject any of the models outright. Most of them, while impressive in their own right, are highly simplified compared to the real Sun. As the above example of resonance absorption demonstrates, the inclusion of missing physical effects can greatly alter the end result. Much more work needs to be done before we can claim to have solved the coronal heating problem. Observations of the kind we have presented here can lead to valuable insights and provide important tests for the new and improved models.

Acknowledgements

This work was supported by NASA. We thank L. J. Porter for her significant role in the data analysis and S. K. Antiochos for valuable discussions. We also thank the anonymous referee for suggestions that improved the presentation.

References

Alexander, D. and Katsev, S.: 1996, *Solar Phys.* **167**, 153.
Antiochos, S. K. and Sturrock, P. A.: 1976, *Solar Phys.* **49**, 359.
Aschwanden, M. J. *et al.*: 1999, *Astrophys. J.* **515**, 842.
Aschwanden, M. J. *et al.*: 2000, *Astrophys. J.* **531**, 1129.
Bray, R. J., Cram, L. E., Durrant, C. J., and Loughhead, R. E.: 1991, *Plasma Loops in the Solar Corona*, Cambridge University Press, Cambridge.
Davila, J. M.: 1987, *Astrophys. J.* **317**, 514.
Dixon, W. W., Klimchuk, J. A., Sturrock, P. A., and Lemen, J. R.: 1991, in Y. Uchida, R. C. Canfield, T. Watanabe, and E. Hiei (eds.), *Flare Physics in Solar Activity Maximum 22*, Springer-Verlag, Berlin, p. 297.
Dowdy, J. F., Emslie, A. G., and Moore, R. L.: 1987, *Solar Phys.* **112**, 255.
Galsgaard, K. and Nordlund, A.: 1996, *J. Geophys. Res.* **101**, 13,445.
Galsgaard, K. and Nordlund, A.: 1997, *J. Geophys. Res.* **102**, 219.
Gary, G. A.: 1997, *Solar Phys.* **174**, 241.
Golub, L. *et al.*: 1990, *Nature* **344**, 842.
Jiao, L., McClymont, A. N., and Mikić, Z.: 1997, *Solar Phys.* **174**, 311.
Kano, R. and Tsuneta, S.: 1995, *Astrophys. J.* **454**, 934.
Klimchuk, J. A. and Porter, L. J.: 1995, *Nature* **377**, 131.
Klimchuk, J. A., Norton, D., and Antiochos, S. K.: 2000, *Astrophys. J.*, submitted.

Klimchuk, J. A., Lemen, J. R., Feldman, U., Tsuneta, S., and Uchida, Y.: 1992, *Publ. Astron. Soc. Japan* **44**, L181.
Litwin, C. and Rosner, R.: 1993, *Astrophys. J.* **412**, 375.
Longcope, D. W. and Sudan, R. N.: 1994, *Astrophys. J.* **437**, 491.
Mandrini, C. H., Démoulin, P., and Klimchuk, J. A.: 2000, *Astrophys. J.* **530**, 999.
Mariska, J. T.: 1984, *Astrophys. J.* **281**, 435.
Martens, P. C., Acton, L. W., and Lemen, J. R.: 1995, *Solar Phys.* **157**, 141.
McClymont, A. N. and Mikić, Z.: 1994, *Astrophys. J.* **422**, 899.
McTiernan, J. and Petrosian, V.: 1990, *Astrophys. J.* **359**, 524.
Mikić, Z., Schnack, D. D., and Van Hoven, G.: 1989, *Astrophys. J.* **338**, 1148.
Ofman, L., Klimchuk, J. A., and Davila, J. M.: 1998, *Astrophys. J.* **493**, 474.
Orrall, F. Q. (ed.): 1981, *Solar Active Regions*, Colorado Associated University Press, Boulder.
Parker, E. N.: 1983, *Astrophys. J.* **264**, 642.
Poletto, G. *et al.*: 1975, *Solar Phys.* **44**, 83.
Porter, L. J. and Klimchuk, J. A.: 1995, *Astrophys. J.* **454**, 499.
Porter, J. G., Falconer, D. A., and Moore, R. L.: 1998, in T.-D. Guyenne (ed.), *Solar Jets and Coronal Plumes*, ESA Publications, The Netherlands, p. 147.
Porter, L. J., Klimchuk, J. A., and Sturrock, P. A.: 1994, *Astrophys. J.* **435**, 482.
Robertson, J. A., Hood, A. W., and Lothian, R. M.: 1992, *Solar Phys.* **137**, 273.
Tsuneta, S. *et al.*: 1991, *Solar Phys.* **136**, 37.
Sakurai, T. and Yan, Y.: 1996, in X. S. Feng, F. S. Wei, and M. Dryer (eds.), *Proceedings of the Third SOLTIP Symposium*, Int. Academic Publishers, p. 1.
Schrijver, C. J., Lemen, J. R., and Mewe, R.: 1989, *Astrophys. J.* **341**, 484.
Vesecky, J. F., Antiochos, S. K., and Underwood, J. H.: 1979, *Astrophys. J.* **233**, 987.
Wang, H. and Sakurai, T.: 1998, *Publ. Astron. Soc. Japan* **50**, 111.
Watko, J. A. and Klimchuk, J. A.: 2000, *Solar Phys.* **193**, 77 (this issue).
Yan, Y. and Sakurai, T.: 1997, *Solar Phys.* **174**, 65.
Zweibel, E. G. and Boozer, A. H.: 1985, *Astrophys. J.* **295**, 642.

WIDTH VARIATIONS ALONG CORONAL LOOPS OBSERVED BY TRACE

J. A. WATKO[1] and J. A. KLIMCHUK[2]

[1] *The Johns Hopkins University, Praxis, Inc., U.S. Naval Research Laboratory, U.S.A.*

[2] *U.S. Naval Research Laboratory, Washington, DC 20375-5352, U.S.A.*

(Received 5 October 1999; accepted 3 December 1999)

Abstract. We have measured width variations along coronal loops observed by TRACE in the 171, 195, and 284 Å bandpasses. The loops are not significantly thicker in the middle compared to near the footpoints, and there is no correlation between the footpoint-to-midpoint expansion and the loop length. This applies to both post-flare and non-flare loops. The observations conflict with our present understanding of active region magnetic fields, and they have important implications for the structure and heating of the corona.

1. Introduction

Coronal loops are often referred to as the fundamental building blocks of the corona. In order to fully understand the corona, we must therefore understand the structure and heating of these loops. Because of the large electrical and thermal conductivities in the corona, plasma structures are coaligned with magnetic field lines. The corona is completely filled with field, however, and distinct loops are visible only because of spatial inhomogeneities in the heating rate. If the heating were uniform, the corona would appear diffuse and nearly featureless, in stark contrast to its actual appearance. By studying the cross-sectional properties of loops, we can learn important information about both the structure of the magnetic field and the spatial distribution of coronal heating (Klimchuk, 2000).

Magnetic fields diverge with distance away from their current source, and so we expect most coronal loops to expand with height above the photosphere. Klimchuk *et al.* (1992) examined 10 loops observed in half-resolution mode (4.9 arc sec pixels) by the Soft X-ray Telescope (SXT) on *Yohkoh*, and found that they are indeed wider at their midpoints than at their footpoints, but only by an average of 13%, which is far less than standard models would predict. This surprising result was recently confirmed in a follow up study involving 43 loops observed mostly in full-resolution mode (2.45 arc sec pixels) (Klimchuk, 2000).

We here report on a parallel investigation made with data collected by the *Transition Region and Coronal Explorer* (TRACE). These data have two major advantages over the *Yohkoh* observations. First, the spatial resolution is approximately three times better. The point spread function of TRACE has a full width at half maximum of 1.25 arc sec (Golub *et al.*, 1999) compared with 3.2 arc sec

The physics of the solar corona and transition region. Part II.
Reprinted from Solar Physics **193,** 2000.

for *Yohkoh* (Martens, Acton, and Lemen, 1995). Second, the loops observed by TRACE are considerably cooler than those observed by *Yohkoh*. The 171, 195, and 284 Å bandpasses on TRACE have temperature sensitivities peaking at 1.0, 1.5, and 2.0 MK, respectively (Handy *et al.*, 1999), whereas the thin filters on *Yohkoh* are most sensitive to plasmas at 5 MK (Tsuneta *et al.*, 1991). Because of the cooler temperatures, TRACE loops extend closer to the photosphere, making comparisons with the predictions of magnetic field models more straightforward.

Another important reason for studying TRACE loops is that they may be physically very different from *Yohkoh* loops. Not only are they cooler, but they also have completely different equilibrium properties. Most *Yohkoh* observations are compatible with static equilibrium (Porter and Klimchuk, 1995; Kano and Tsuneta, 1995), whereas equilibrium can be ruled out in loops observed by the Extreme ultraviolet Imaging Telescope (EIT) using the same bandpasses as TRACE (Aschwanden *et al.*, 1999, 2000). It is therefore important to determine whether TRACE loops expand similarly to *Yohkoh* loops. Any differences could provide important clues about differences in the heating mechanism, for example.

It is also potentially revealing to examine differences between post-flare loops and loops that are formed independently of flares. Magnetic reconnection is almost certainly responsible for the energy release in post-flares loops, but it is not known whether it also plays a role in ordinary loops, or whether the geometry of the reconnection is somehow different. We therefore distinguish between these two types of loops in our analysis.

2. Observations

TRACE exposures at 171, 195, and 284 Å were examined for loops isolated from other features, with two distinct footpoints and relatively quiet backgrounds. We selected nine post-flare loops from 25 August 1998 and fifteen non-flare loops from a variety of days. Details of the observations are given in Table I. All of the post-flare loops appear to be associated with a single X1.0 flare that peaked at 22:12 UT on 24 August.

To improve the signal-to-noise ratio, summed images were created by adding multiple exposures within a 10 min window (25 min in one case). Each exposure was first despiked to remove cosmic ray hits, then the pedestal was subtracted. The resulting frames were cross-correlated to determine alignment to within 0.1 pixel. The frames were examined for obvious evolution of individual loops before being co-added. If a series of frames included a loop whose position was changing, then that particular loop was not measured.

TABLE I
Summary of observations

Date	First time (UT)	Last time (UT)	Number of exposures	Number of loops	Band
25 Aug. 1998	00:13:13		1	3	171
25 Aug. 1998	10:20:53	10:27:37	8	6	195
19 May 1998	22:17:17	22:27:15	23	1	171
23 June 1998	00:47:04	00:52:39	2	1	171
25 June 1998	00:07:54	00:18:05	6	1	171
26 June 1998	00:33:46	00:56:16	8	1	171
26 June 1998	00:41:16	00:51:21	5	1	195
04 Aug. 1998	12:41:09	12:42:38	3	1	171
04 Aug. 1998	12:43:05	12:48:04	10	1	195
04 Aug. 1998	12:49:07	12:57:12	8	1	284
23 Aug. 1998	19:47:54	19:53:31	2	1	171
25 Aug. 1998	12:19:04		1	1	171
25 Aug. 1998	15:32:11	15:36:16	5	1	171
05 Sept.1998	00:11:45		1	1	171
12 Sept.1998	00:50:28	00:56:08	6	1	171
18 Sept.1998	00:19:51	00:29:56	11	2	171

3. Procedure

Our analysis followed essentially the same procedure described in Klimchuk *et al.* (1992) and Klimchuk (2000), with small modification. It is an interactive procedure performed at the computer console with software developed under the Interactive Data Language (IDL). With a mouse, the cursor was placed over several points along the loop axis from one footpoint to the other. An effort was made to define the entire visible axis, even where neighboring emission features (e.g., other loops) would interfere with width measurements. A low order polynomial fit was calculated to the axis points as shown for an example loop in Figure 1.

The leftmost and rightmost points identified on the axis are taken as the footpoints, which were defined in the following manner. In the example loop, the footpoints appear to fall along a ribbon of footpoints for an arcade of postflare loops. In this case, we chose the footpoints to be the intersections of the loop axis with the ribbon centers. In cases where a loop appeared to fade out into a rounded end, we defined the center of the radius of curvature of that end as the footpoint. In one case, for a loop off the limb, we defined the footpoints as the intersection of the loop axis with the limb.

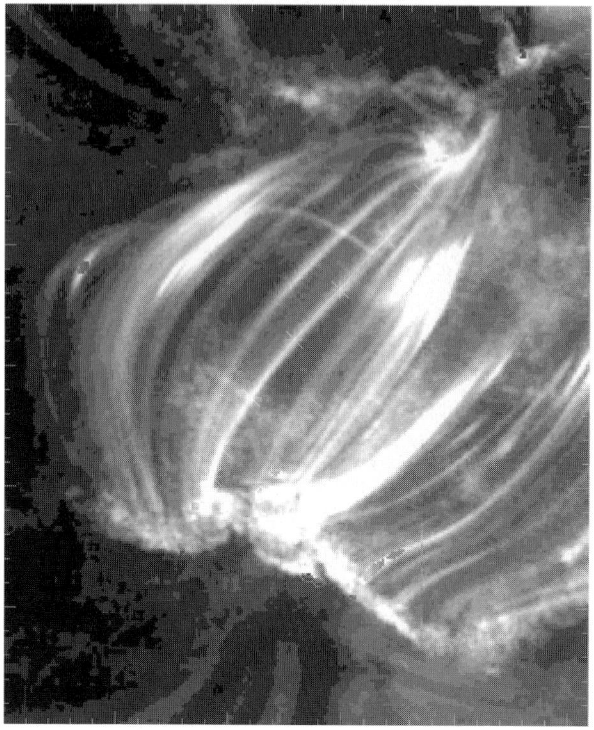

Figure 1. An example post-flare loop from 25 August 1998 is displayed with its defined axis. Each major tick mark represents 100 TRACE pixels, or 50 arc sec. The image is a single 171 Å exposure from 00:13:13 UT. The axis is a third-order polynomial fit to the points we defined via user interface.

At intervals separated by 1 pixel (0.5 arc sec) along the axis, each image was sampled in the direction perpendicular to the axis. The perpendicular sampling interval was also 1 pixel, and the values were interpolated from the four nearest neighbors on the image grid. A stack of these sampled rows is essentially a straightened image of the loop, as shown in Figure 2(a). Although the image is locally stretched or compressed parallel to the axis in positions of loop curvature, the perpendicular scale of the width is preserved.

The left and right 'edges' of the loop were identified in order to determine the background. The edges were defined subjectively to be away from the visible wings of the loop, but not so far away as to be in the wings of a neighboring loop. We performed this judgment visually, based upon the appearance of the straightened loop image. The defined loop was removed from the image (Figure 2(b)) and replaced by a linear interpolation of the background (Figure 2(c)). The interpolated background was subtracted from the original straightened loop image to leave just the loop signal (Figure 2(d)).

Each row of the straightened, background-subtracted image represents an intensity profile across the loop. A representative profile for the example loop is shown in Figure 3. Note that this is one of the narrowest loops in our dataset. To avoid

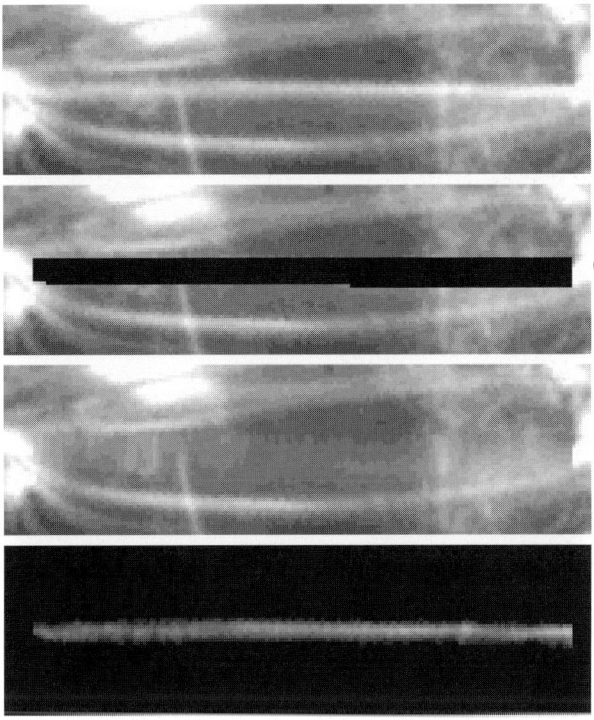

Figure 2. The straightened loop image for the example loop, formed by sampling Figure 1 perpendicular to the loop axis, is displayed in (a). (b) shows the edge we defined for the loop. The background computed by linear interpolation between the edges is shown in (c). The difference between (a) and (c) is (d), the straightened, background-subtracted loop.

imposing a functional form for the cross-loop intensity profile, we define the width as the second moment of intensity as a function of position perpendicular to the axis. This quantity is equal to the standard deviation of photon position for the slice across the loop:

$$\sigma = \left[\frac{\Sigma(x_i - \mu)^2 I_i}{\Sigma I_i}\right]^{1/2} , \tag{1}$$

where the summation is taken over the x_i positions in the profile, and μ is the mean position:

$$\mu = \frac{\Sigma x_i I_i}{\Sigma I_i} . \tag{2}$$

For a profile that is Gaussian, the e^{-1} full width is $2^{3/2} \approx 2.8$ times as large as the standard deviation.

Because noise in the data can substantially influence the width measurement in regions of faint signal, we established an objective criterion for rejecting profiles whose peak intensity was less than five times the expected noise level. This level is taken to be the quadratic sum of the shot and read noise:

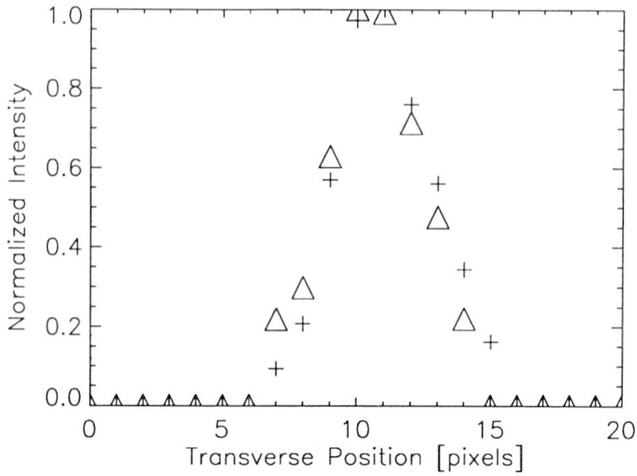

Figure 3. A typical intensity profile for one point along the axis of the example loop. Triangles show the profile based upon the edge definition shown in Figure 2(b). Pluses indicate the error-pass profile, measured by repeating the background subtraction for a loop definition that is 2 pixels wider than that shown in Figure 2(b).

$$\Delta I = 0.29\,(I + 33.3)^{1/2} , \tag{3}$$

where ΔI and I are in units of data numbers (DN), the read noise is 20 el pix^{-1}, and the amplifier gain is 12 el DN^{-1} (Handy *et al.*, 1999). To make certain that the background subtraction did not accidentally cut off the wings of the profile, measurements were also rejected if the mean photon position was too close to the defined loop edge. Although parts of some loops were not measurable by these criteria, we could nonetheless compare the width near the midpoint to the width near the footpoints as long as there were at least two valid measurements near these points (see below). Width measurements for the points that passed the two selection criteria are shown for the example loop as triangles in Figure 4.

We believe that the largest source of error in the width measurement is introduced in the background subtraction. In particular, the subjectivity of the loop edge definition makes it suspect as a source of error in the background subtraction and in the width measurement itself. To quantify this error, the analysis was repeated on the same loop data, with a definition of the background that is two pixels wider. These error-pass measurements are shown as pluses in Figure 4. The difference between the original and error-pass widths is our estimate of the uncertainty.

The error-pass width can be either greater than or less than the original width depending on the spatial variation of the background. In places where the background increases with distance from the loop axis, the interpolation overestimates the background level under the loop, and the resulting background-subtracted loop is artificially narrow. Similarly, the corrected loop is artificially wide during the error pass in places where the background decreases with distance from the axis. Artificial widening can also be produced by intensity fluctuations. Negative values

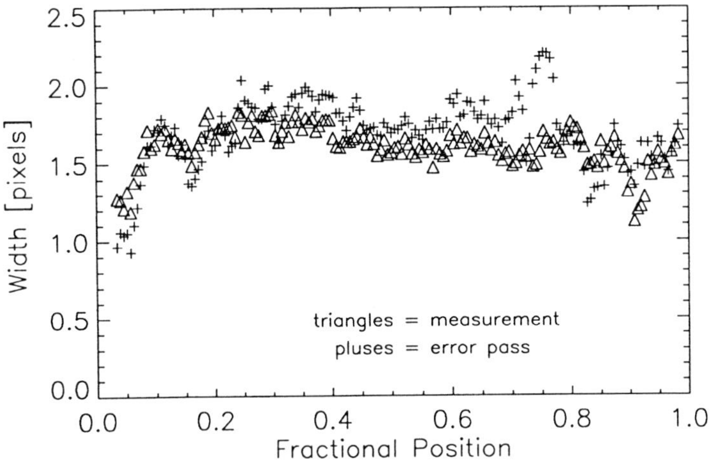

Figure 4. Loop width as a function of position along the example loop. The width shown here is the standard deviation of photon position for the intensity profile associated with the axis point. Measured widths are shown as triangles, and error-pass widths are shown as pluses. Position along the loop axis is normalized to 1 to facilitate comparison of loops with different lengths.

are set to zero in the background-subtracted images, but positive values are not, and any residual positives out in the wings can have a significant contribution to the second moment used to characterize the width (Equation (1)). For this reason, error-pass measurements or more often wider than the original measurements. The differences between the two measurements tend to increase as the separation between the defined loop edges increases (i.e., as more pixels are added to the interpolation region). However, our estimates of footpoint to midpoint expansion should not be affected as long as the background is statistically independent of loop position.

At the beginning of our investigation, we were concerned about on-board data compression as another possible source of error in the width measurements. TRACE uses the JPEG compression scheme. Most of our images have a JPEG Q-value of 6 (factor of 10 compression), but some have a value of 4 (factor of 7.5) or 0 (factor of 2.5), the latter being essentially lossless. By analyzing a single exposure compressed by different amounts, we estimate that the errors introduced into the width measurements are typically less than 0.1 pixel. This is small compared to background subtraction error of order 0.3 pixel. Also, instrumental drift and the solar rotation result in phase shifts in the discrete cosine transforms of the JPEG compression. When multiple exposures are aligned and added, these phase variations should result in a lessening of the confusion caused by JPEG residue.

TRACE's finite resolution affects the measurements of the narrower loops. The intensity profile of a TRACE observation of an infinitely thin loop, its point spread function (psf), is roughly Gaussian (Golub, 1999), with a full width at half maximum of about 2.5 pixels (Golub *et al.*, 1999). If a loop were unresolved near the

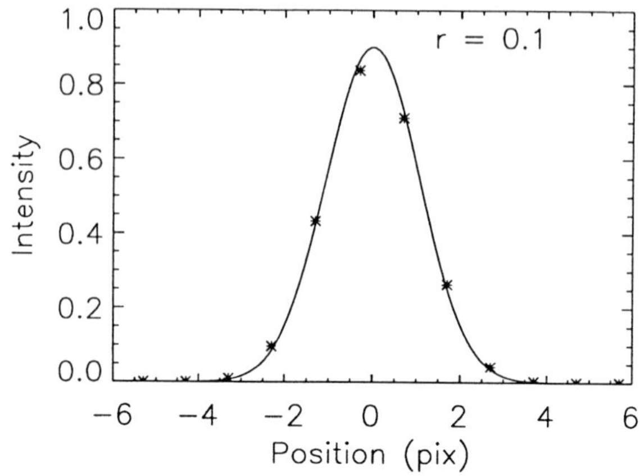

Figure 5. Intensity profile from a simulated observation of a model loop with a uniformly filled circular cross-section of radius 0.1 pix. For such an extremely narrow loop, the profile is essentially equivalent to the point spread function. Stars indicate the pixel integrated values.

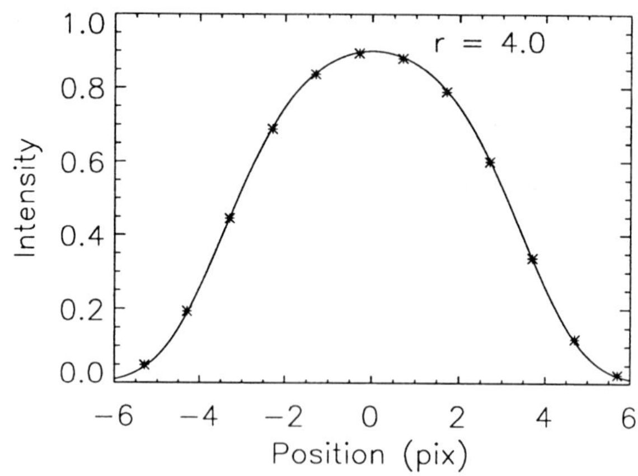

Figure 6. Intensity profile from a simulated observation of a model loop with a uniformly filled circular cross-section of radius 4.0 pix. This radius is typical of what we find in the data. Stars indicate the pixel integrated values.

footpoints, the TRACE psf might exaggerate the width of the legs and cause us to underestimate the expansion from footpoint to center. We avoid this by converting second moment measurements to loop radius values before comparing the midpoint and footpoint widths. As a model for the conversion, we use a uniformly filled loop with a circular cross-section, as justified in Klimchuk (2000). Observing such a loop would result in a TRACE intensity profile that is the convolution of the actual intensity profile with the psf. The wider the input loop, the wider TRACE's response would be. Figure 5 shows an example of an extremely narrow loop with a radius of only 0.1 pix. The observed profile is in this case nearly identical to the psf. We have found profiles this narrow in the TRACE data, confirming that the psf

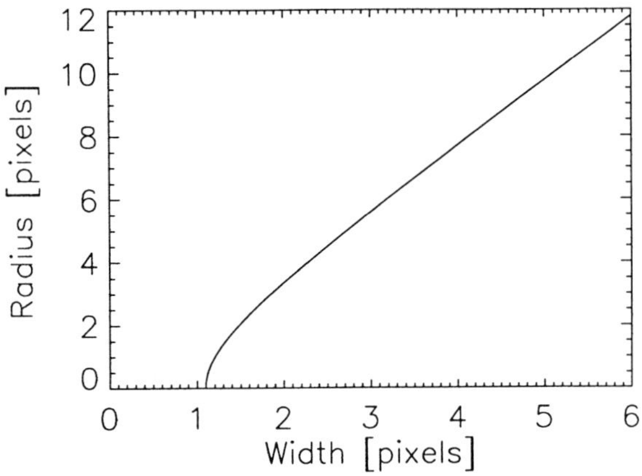

Figure 7. Our conversion curve for model loops of uniformly filled circular cross section. The ordinate is the radius of loop cross section that, when observed by TRACE, would produce the measured second moment on the abscissa. For large radii, the relation is nearly linear. For small radii, the point spread function dominates the profile. Using this curve, we remove the effects of the point spread function and finite pixel size, which might otherwise exaggerate the width of a loop's narrow regions.

is no wider than the quoted value. Figure 6 shows an example of a more typical loop with a radius of 4.0 pix. It is well resolved.

The stars in the figures indicate the intensities integrated over individual pixels, where it has been assumed that the loop axis falls part way into a pixel from the edge. Second moments are computed from these 'pixelated' intensities and plotted against radius in Figure 7 for a wide range of loop sizes. Based upon this conversion curve, for each measured loop profile width we inferred the corresponding radius of a circular cross-section loop.

4. Results

Figures 8–11 show the inferred radii as a function of position along the loop axis for four loops with a variety of width properties. In Figure 8, our example 171 Å postflare loop is plotted. This loop exhibits weak expansion; its width is nearly constant over most of the loop. Figure 9 shows a 284 Å loop from 4 August 1998. This particular loop connects a smaller footpoint and a larger footpoint. Its width increases from one end to the other. A 195 Å loop from the same day (a completely different structure) actually appears thinner in the middle than near the footpoints as shown in Figure 10. The data points placed at radius of 0 indicate that the measured loop width was so close to the width of the psf that only an upper limit can be placed upon the loop radius. The loop is unresolved at these points. The 171 Å loop from 5 September 1998, shown in Figure 11, is among those with the greatest expansion.

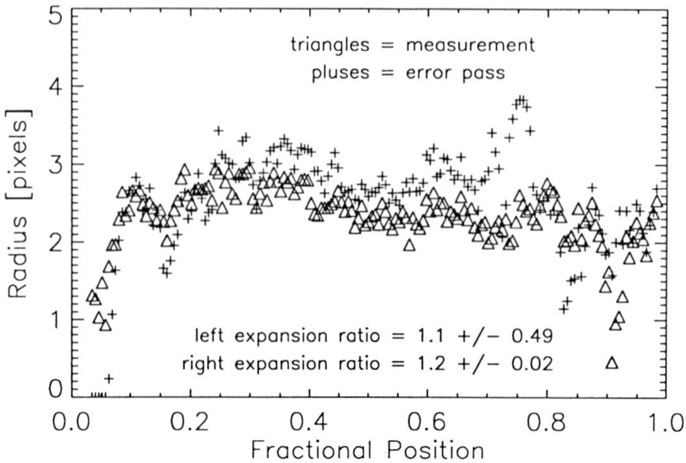

Figure 8. Loop radius plotted as a function of fractional position along the loop axis for the example post-flare loop observed in the 171 Å bandpass beginning 00:13:13 UT on 25 August 1998. Triangles indicate original measurements, and pluses indicate error-pass measurements with modified background subtraction. Two expansion factors are calculated for each loop (see text for details). The uncertainty for each expansion factor is given as the difference between the expansion factor calculated from the original radii and the expansion factor calculated from the error-pass radii.

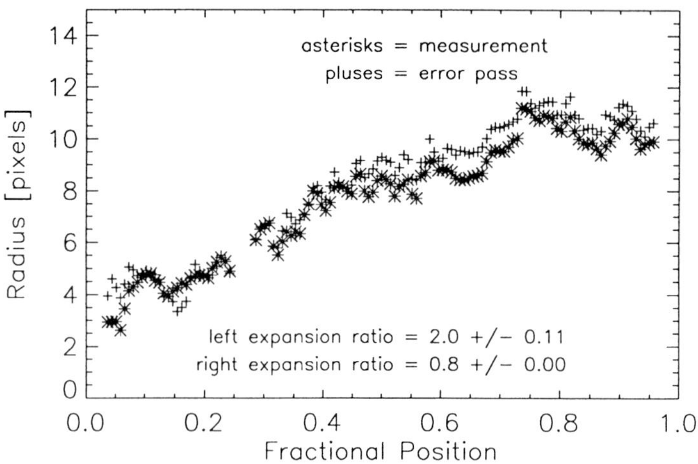

Figure 9. Loop radius plotted as a function of fractional position along the loop axis for a non-flare loop observed in the 284 Å bandpass beginning 12:49:07 UT on 4 August 1998. Asterisks indicate original measurements, and pluses indicate error-pass measurements.

To use a measure of footpoint-to-midpoint expansion that would distinguish amongst the various shapes of loops described above, we chose to define two expansion factors for each loop: the ratio of the mean width of the middle 15% of the loop to the mean width of the 15% nearest the left footpoint, and the ratio of the mean width of the middle 15% of the loop to the mean width of the 15% nearest the right footpoint. In some cases, only a portion of the 15% segment contains

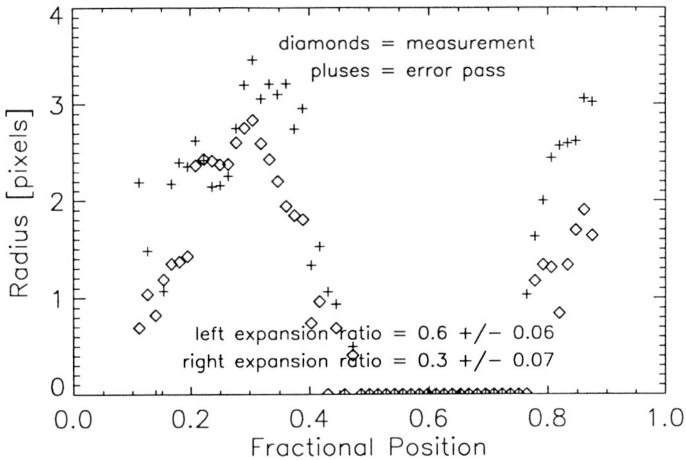

Figure 10. Loop radius plotted as a function of fractional position along the loop axis for a non-flare loop observed in the 195 Å bandpass beginning 12:43:05 UT on 4 August 1998. This is not the same structure as shown in Figure 9. Diamonds indicate original measurements, and pluses indicate error-pass measurements. Points plotted on the radius = 0 axis indicate that the measured standard deviation is too small to infer a loop radius.

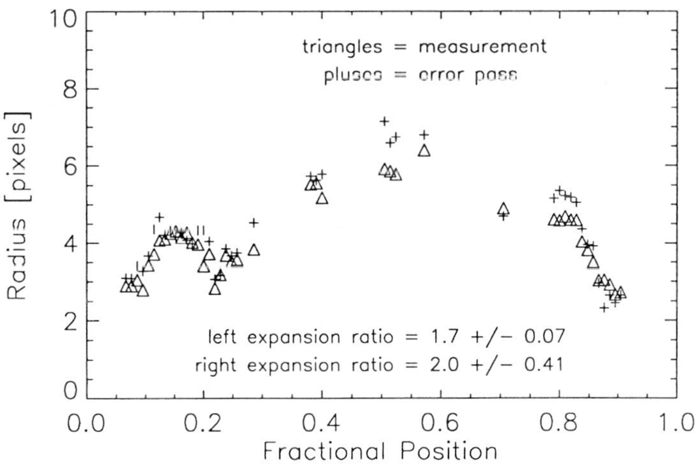

Figure 11. Loop radius plotted as a function of fractional position along the loop axis for a non-flare loop observed in the 171 Å bandpass beginning 00:11:45 UT on 5 September 1998. Triangles indicate original measurements, and pluses indicate error-pass measurements.

measurable widths due to the low intensity or wing cutoff criteria. We require at least two reliable measurements in each segment in order for an expansion factor to be computed. If one half of the loop shows expansion from footpoint to midpoint, then the expansion factor for that half would be greater than one. If one footpoint were as wide as the midpoint, then that expansion factor would equal one. If the footpoint were wider than the midpoint, then the expansion factor would be less

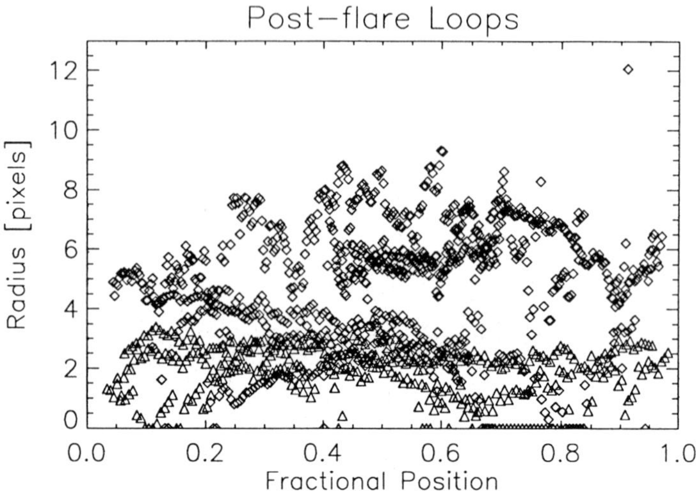

Figure 12. Loop radius is plotted versus fractional position along the loop axis for all 9 post-fare loops. Triangles and diamonds indicate 171 and 195 Å measurements, respectively. The post-flare sample does not include any 284 Å observations. Error bars have been omitted for clarity, but the errors shown for the example loop in Figure 8 are typical.

than one. Left and right expansion factors are noted in Figures 8–11. Each uncertainty given is the difference between the expansion factor calculated from the initial width measurement and the expansion factor calculated from the error-pass width measurement.

Widths for the post-flare and non-flare samples of loops are plotted in Figures 12 and 13, respectively. Error bars have been omitted to avoid crowding the plot, but the errors are similar to those shown in the four examples. Note that the majority of loops are well resolved. There is no obvious hump in the plots, as would be expected if the loops were systematically wider at their midpoints than at their footpoints.

The expansion factors for the post-flare and non-flare samples are plotted versus loop length in Figures 14 and 15. The points generally hover near unity, and in only very few cases do they approach or exceed a value of 2. The median expansion factors in the post-flare and pre-flare samples are 0.99 and 1.13, respectively. These are given in Table II. Also given are the mean values obtained from the original and error-pass measurements, and the root-mean-square (rms) difference between them. The relatively large rms difference for the post-flare loops is heavily influenced by a single case were the difference is unusually large.

A simple visual inspection of Figures 14 and 15 suggests that there is no correlation between the expansion factor and loop length. We have confirmed this with non-parametric statistical analysis using the D-statistic (Press *et al.*, 1992) and t-statistic (Efron and Petrosian, 1992).

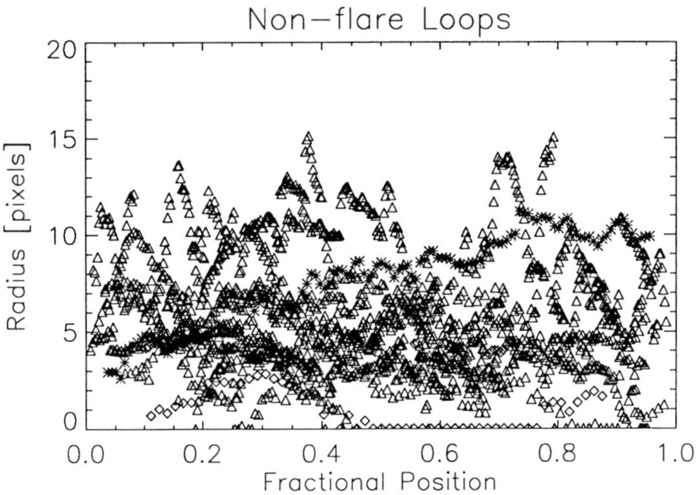

Figure 13. Loop radius is plotted versus fractional position along the loop axis for all 15 non-flare loops. Triangles, diamonds, and asterisks indicate 171, 195, and 284 Å measurements, respectively. Error bars have been omitted for clarity, but the errors shown for the example loops in Figures 9–11 are typical.

TABLE II
Expansion factors

Sample	Median	Mean	Mean (error pass)	RMS difference
Post-flare	1.13	1.20	1.54	0.59
Non-flare	0.99	1.16	1.20	0.11

5. Discussion

Our results agree remarkably well with those obtained from *Yohkoh* SXT data (Klimchuk *et al.*, 1992; Klimchuk, 2000), despite the differences in spatial resolution and loop temperature. For both datasets, the expansion factors are small and do not depend on loop length. This is completely contrary to expectations based on our present understanding of active region magnetic fields, as discussed in depth in Klimchuk (2000). It is clear that we must revise our concept of the magnetic structure of coronal loops. The field is more complicated than implied by simple potential field models or models involving only large-scale stresses. Smaller-scale complexities are implicated as playing a fundamental role. See Klimchuk, Antiochos, and Norton (2000) for a discussion of how localized twist affects the structure of coronal loops.

Our results also have important implications for coronal heating. They indicate that loops are more like tubes with circular cross-sections than like ribbons. This

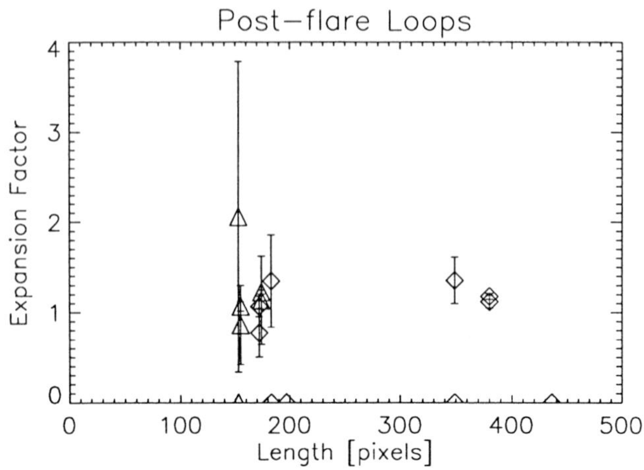

Figure 14. Loop radius expansion factors are plotted versus loop length for all 9 post-flare loops. Triangles and diamonds indicate 171 and 195 Å measurements, respectively. The post-flare sample does not include any 284 Å observations. The symmetric error bars are established based upon the difference between the expansion factor calculated from the original measurements and the expansion factor calculated from the error-pass measurements for each loop. Points plotted on the expansion factor = 0 axis indicate that fewer than 2 width measurements were selected as valid within a region of the loop required for the expansion factor calculation (see text for details).

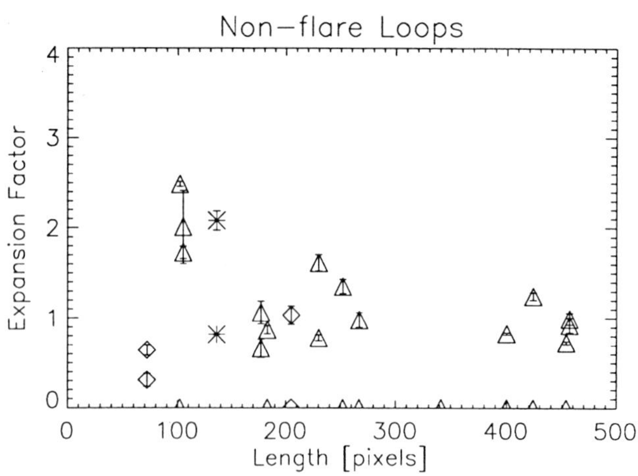

Figure 15. Loop radius expansion factors are plotted versus loop length for all 15 non-flare loops. Triangles, diamonds, and asterisks indicate 171, 195, and 284 Å measurements, respectively.

requires axisymmetric heating, which is not easy to explain. For example, magnetic reconnection should tend to produce ribbon-like structures (e.g., Galsgaard and Nordlund, 1996). In the presence of twist, these ribbons would appear much wider along some portions of their length than along others, and while variable loop widths are sometimes observed, the observations are generally inconsistent with such a picture.

It is interesting and perhaps surprising that post-flare and non-flare loops are so similar. One might imagine that the magnetic environment for these two types of loops is quite different. Post-flare loops are believed to arise from reconnection within large current sheets that are formed by the opening of the field before or during the flare. The reconnected loops should be in a magnetically relaxed state (although see Klimchuk, 1996). Reconnection in non-flare loops, if it occurs at all, is more likely to be associated with much smaller-scale current structures, and the loops may remain significantly stressed throughout the energy release (see the review of coronal heating models in Mandrini, Démoulin, and Klimchuk, 1999). It is also interesting that the post-flare loops observed in 195 Å in our sample are systematically wider than those observed in 171 Å (Figure 12). It is unclear, however, whether this reflects a temperature difference or a time difference, since the 195 Å loops were observed 10 hours later.

There are three reasons for caution when comparing our observations with the predictions of magnetic field models. First, because we searched for loops that are isolated from complicated features, and because longer loops are more likely to cross other features, we cannot rule out the possibility that our sample is biased toward shorter loops. Since longer loops are expected theoretically to have greater expansion than shorter loops (e.g., Klimchuk, 2000), this could mean that our study underestimates the average amount of expansion. On the other hand, long and short loops seem to be equally represented in our sample, and there is no evidence for an expansion/length correlation (Figure 15).

Second, TRACE footpoint positions do not represent the true intersection of magnetic flux tubes with the photosphere. Matter low in the legs of a loop is cooler than at the top, and because of the narrow TRACE passband, a loop will appear to terminate somewhat above the photosphere. This effect is even greater for the hotter loops observed by *Yohkoh*. It is very significant therefore that the expansion factors measured by TRACE and *Yohkoh* are so similar. We note that Aschwanden *et al.* (1999) found little expansion in the lower portions of loops observed by EIT (within one or two scale heights of the photosphere). Since their study did not require that complete loops be visible, there is no issue of a possible bias toward shorter loops.

Lastly, there is the question of how the geometry of the observations might influence the measurements. Width measurements are much less susceptible to these effects than, say, pressure measurements (e.g., Alexander and Katsev, 1996). Nonetheless, geometry does play a role. Our approach to this problem was to select loops from a variety of positions across the disk and at the limb and to assume that the geometrical effects average out in a statistically large sample. To verify this, we have begun to model a uniformly filled loop with circular cross-section. We plan to rotate the model to different viewing angles and generate synthetic images by integrating the square of the density along the line of sight and adding a background and random noise comparable to those of the real TRACE observations. We will then perform our width measurements using the same procedure we employed with the real solar data.

These issues not withstanding, it seems clear that the observation of limited expansion in coronal loops presents us with a very real and interesting challenge, the pursuit of which is likely to provide breakthroughs in our understanding of how the corona is structured and heated.

Acknowledgements

This work was supported by the NASA Sun-Earth Connections Guest Investigator Program. We thank the TRACE team, especially H. Warren, C. Schrijver, R. Shine, and T. Tarbell, for their valuable assistance with the data, and S. Antiochos and J. Karpen for helpful discussions and comments on the manuscript.

References

Alexander, D., and Katsev, S.: 1996, *Solar Phys.* **167**, 153.
Aschwanden, M. J. *et al.*: 1999, *Astrophys. J.* **515**, 842.
Aschwanden, M. J. *et al.*: 2000, *Astrophys. J.* **531**, 1129.
Efron, B. and Petrosian, V.: 1992, *Astrophys. J.* **399**, 345.
Galsgaard, K. and Nordlund, A.: 1996, *J. Geophys. Res.* **101**, 13445.
Golub, L.: 1999, private communication.
Golub, L. *et al.*: 1999, *Phys. Plasmas* **6**, 2205.
Handy, B. N. *et al.*: 1999, *Solar Phys.* **187**, 229.
Kano, R. and Tsuneta, S.: 1995, *Astrophys. J.* **454**, 934.
Klimchuk, J. A.: 1996, in R. D. Bentley and J. T. Mariska (eds.), *Magnetic Reconnection in the Solar Atmosphere*, Pub. Astron. Soc. Pacific, San Francisco, p. 319.
Klimchuk, J. A.: 2000, *Solar Phys.* **193**, 53 (this issue).
Klimchuk *et al.*: 1992, *Publ. Astron. Soc. Japan* **44**, L181.
Klimchuk, J. A., Antiochos, S. K., and Norton, S.: 2000, *Astrophys. J.*, submitted.
Mandrini, C. H., Démoulin, P., and Klimchuk, J. A.: 2000, *Astrophys. J.* **530**, 999.
Martens, P. C., Acton, L. W., and Lemen, J. R.: 1995, *Solar Phys.* **157**, 141.
Porter, L. J. and Klimchuk, J. A.: 1995, *Astrophys. J.* **454**, 499.
Press, W. H., Teukolsky, S. A., Vetterling, W. T., and Flannery, B. P.: 1992, *Numerical Recipes*, Cambridge University Press, Cambridge.
Tsuneta, S. *et al.*: 1991, *Solar Phys.* **136**, 37.

HOW ACCURATELY CAN WE DETERMINE THE CORONAL HEATING MECHANISM IN THE LARGE-SCALE SOLAR CORONA?

D. H. MACKAY[1], K. GALSGAARD[1], E. R. PRIEST[1] and C. R. FOLEY[2]
[1] *School of Mathematical Sciences, University of St. Andrews, St.Andrews, Fife, Scotland, KY16 9SS*
[2] *MSSL, University College London, Holmbury St. Mary, Dorking, Surrey, RH5 6NT, U.K.*

(Received 20 September 1999; accepted 3 November 1999)

Abstract. In recent papers by Priest *et al.*, the nature of the coronal heating mechanism in the large-scale solar corona was considered. The authors compared observations of the temperature profile along large coronal loops with simple theoretical models and found that uniform heating along the loop gave the best fit to the observed data. This then led them to speculate that turbulent reconnection is a likely method to heat the large-scale solar corona. Here we reconsider their data and their suggestion about the nature of the coronal heating mechanism. Two distinct models are compared with the observations of temperature profiles. This is done to determine the most likely form of heating under different theoretical constraints. From this, more accurate judgments on the nature of the coronal heating mechanism are made. It is found that, due to the size of the error estimates in the observed temperatures, it is extremely difficult to distinguish between some of the different heat forms. In the initial comparison the limited range of observed temperatures ($T > 1.5$ MK) in the data sets suggests that heat deposited in the upper portions of the loop, fits the data more accurately than heat deposited in the lower portions. However if a fuller model temperature range ($T < 1.0$ MK) is used results in contridiction to this are found. In light of this several improvements are required from the observations in order to produce theoretically meaningful results. This gives serious bounds on the accuracy of the observations of the large-scale solar corona in future satellite missions such a *Solar-B* or *Stereo*.

1. Introduction

One of the main challenges facing solar physics is the nature of the coronal heating mechanism. It is still a mystery why the Sun should have a hot corona ($T > 1$ MK) in comparison with a much cooler photosphere ($T \sim 6000$ K). Many structures are found in the solar corona, including X-ray bright points (Golub *et al.*, 1974, 1977), coronal holes (Hara *et al.*, 1994) and coronal loops (Uchida, 1993). Satellite missions (NIXT, *Yohkoh*, SOHO, TRACE) have shown that the solar corona is an evolving system with much internal structure. Coronal loops, the structures studied in this paper are observed to have a wide range of temperatures, sizes and lifetimes. They may however be broadly classified into two main types, short lived active region loops ($T > 3$ MK, Yoshida and Tsuneta, 1996; Kano, 1997) and the cooler large-scale loops of the quiet solar corona ($T = 1$–2 MK, Sturrock, Wheatland, and Acton, 1996; Culhane, 1997). The mechanisms that produce the high temperatures in these two distinctive types of loop have yet to be identified.

The physics of the solar corona and transition region. Part II.
Reprinted from Solar Physics **193,** 2000.

However, many ideas have been put forward (Ulmschneider, Priest, and Rosner, 1991) including magnetic flux braiding (Galsgaard and Nordland, 1996), phase mixing (Hood, Ireland, and Priest, 1997) and resonant absorption (Erdelyi, 1998) and in each case references therein. As well as revealing different types of loop, the Soft X-ray Telescope on *Yohkoh* showed that there are two distinct parts to the coronal heating mechanism. First there is a local time-dependent, impulsive part which occurs on short time scales and secondly a global steadier part which acts on a much larger scales (Acton, 1996; Tsuneta, 1996). This may indicate that different mechanisms are working together to heat different parts of the corona in different ways. Although many theories on coronal heating have been produced over the years, few have tried to compare theory with observations (Jordan, 1992). Recently in two papers (Priest *et al.*, 1998, 1999) a new method for determining the coronal heating mechanism from observations has been put forward. The method suggests that the coronal heating problem may be tackled in two distinct steps. First, observed temperature profiles along coronal loops are compared with temperature profiles from theoretical models. From this, the form of the heating profile along the loop that best fits the observations is determined. Once the form of heating has been found, the second step is to determine what mechanism (or mechanisms) can produce that form of heating (Galsgaard *et al.*, 1999).

In this paper the nature of the coronal heating mechanism in the large-scale solar corona is considered and from this the form of the heating that best fits the observations is determined. The data used in this paper was considered for a limited range of theoretical parameter space in Priest *et al.* (1998, 1999, hereafter P1 and P2). In this paper a full range of the parameter space is considered to check the validity of the previous results. Also, two separate models are compared with the observations and the statistical significance of each fit is determined to give an indication of the validity of each model and to see how the results of P1 and P2 depend on the model. In Section 2 the statistical method for determining the best fit between the observed temperatures and model temperatures is briefly discussed. In Section 3 the basic model which includes a static energy balance between thermal conduction and heating is described. From this, the heat form for this model that gives the best fit to the observational data is determined along with its statistical significance. In Section 4 the basic model is extended to include the effects of radiation and it is found that extra information that is not included in the present data sets, namely the temperature profile down to at least 1 MK, is required in order to determine the best fit for the more advanced model beyond any doubt. Finally, in Section 5 the conclusions are given and improvements that are required from the observations of the large-scale solar corona so that theoretically meaningful results can be produced are discussed.

2. Chi-Squared Statistic

To try and determine the coronal heating mechanism, a model is fitted to data of the observed temperature profile along a coronal loop. In each case the agreement between the model and data is tested using the chi-squared method (Bevington and Robinson, 1994). In the chi-squared method the function

$$\chi^2 = \sum_{i=1}^{N} \left(\frac{T_i - T_{\text{model}}(s_i)}{\sigma_i} \right)^2 \tag{1}$$

has to be minimised, where T_i is the observed temperature at a distance s_i along the loop, $T_{\text{model}}(s_i)$ is the model temperature at the same point and σ_i is the standard deviation in the observed temperature. The smaller the value of the χ^2 the better the fit and model parameters are adjusted to find the minima of the χ^2. If the system has N data points and M parameters to be varied, the number of degrees of freedom is $v = N - M$ and a 'good fit' is said to be found when $\chi^2/v \approx 1$. In general to obtain a good fit the model temperature must pass between all of the error bars in the data points. However, since the data is not exact (i.e., subject to measurement errors), even if the correct model is compared with the data, the model will not give an exact fit. Therefore for the χ^2 to be useful one must not only deduce the parameters for the minimum value of the χ^2 but must also give a statistical measure of the 'goodness' of the fit. It is not sufficient just to say that the fit 'looks good' or that a low enough value of the χ^2 has been obtained.

From Equation (1) it can be seen that to determine the χ^2 the standard deviation in the observed temperature is required at each data point. Now, for the observational data in P1 and P2 only three values are available at each location, the observed temperature (T_{obs}), an upper limit (T_{upp}) and a lower limit (T_{low}). From three data points it is impossible to determine the standard deviation exactly but an approximation is possible. First, assume that the temperature measurements form a normal distribution. Now for a normal distribution, measurements should fall within $\pm\sigma$ of the true value 68% of the time and $\pm 3\sigma$, 99.7% of the time. Therefore, in a normal distribution there is very little chance of the measurements lying outside $\pm 3\sigma$. If the normal distribution is centred at the observed temperature (the 'true value') then the tails of the distribution would lie close to the error bars. Therefore the probability of having a temperature outside the error bars would be very small and so the half width of the error bars can be said to lie around $\pm 3\sigma$. Therefore the complete width of the error bar is 6σ ($= T_{\text{upp}} - T_{\text{low}}$). In P1 and P2 the full width of the error bar ($T_{\text{upp}} - T_{\text{low}}$) was used as one standard deviation (1σ) and so an over weighting of the χ^2/v was produced and this gave a misleading result (the actual value of χ^2/v is approximately 36 times larger than quoted in P1 and P2 since the true value of σ is $(T_{\text{upp}} - T_{\text{low}})/6$). This is an important point since a χ^2/v value of approximately unity should be sought. As well as taking the half width of the error bar to be $\pm 3\sigma$, results are also calculated here for a half width of

$\pm 2\sigma$ and $\pm\sigma$ and the results for this are shown in the Appendix. However, for the main part of the paper the most strict use of $\pm 3\sigma$ will be used.

Now, once the best fit has been calculated, it must be shown that the results are statistically significant. This is done by comparing the computed χ^2 values with those from statistical tables (Table 7, Kmietowicz and Yannoulis, 1988). For the fit to be statistically significant the obtained χ^2 value should be less than some critical value χ^2_{crit} which gives an area of say 0.05 in the chi-squared distribution. If $\chi^2 < \chi^2_{\text{crit}}$ for a given v then a 'good fit' in absolute terms has been obtained. As well as comparing with the critical value of χ^2, the significance level of the fit can be calculated as a percentage or p-value (expressed as a fraction between 0 and 1). The closer to 1 the more significant the fit. A p-value of 0.5 is strong evidence in favor of the model, while a p-value of 0.1 is weak evidence for the model. On the other hand if a significance level of 0.001 is obtained then the model is clearly incorrect. We now proceed to carry out the comparison between observations and models.

3. Basic Model

For the basic model the same one that was used in P2 will be taken where an energy balance between thermal conduction and heating is considered. For the present data sets there are only observations at coronal temperatures (Sturrock, Wheatland, and Acton, 1996) and so only the coronal part of the loop will be considered. Also, a balance between thermal conduction and heating that does not include the effect of radiation is only valid in the hot coronal part of a loop that is not too dense. The equation

$$\frac{1}{A(s)}\frac{\mathrm{d}}{\mathrm{d}s}\left(A(s)\kappa_0 T^{5/2}\frac{\mathrm{d}T}{\mathrm{d}s}\right) = -H(s) \tag{2}$$

is solved for the temperature $T(s)$, where $A(s)$ is the cross-sectional area of the loop, s is the distance along the loop from the northern coronal footpoint and $\kappa_0 T^{5/2}$ is the coefficient of thermal conduction ($\kappa_0 = 9.24 \times 10^{-12}$ Jm^{-1} s^{-1}). The area of the loop is assumed to be constant (Klimchuk *et al.*, 1992) and $H(s)$ is the heating rate per unit volume which may take any of the following forms:

(1) Uniform heating,

$$H(s) = H_0. \tag{3}$$

(2) Uniform heating over distance lh from base,

$$H(s) = \begin{cases} H_o & s \leq lh \\ 0 & s > lh. \end{cases} \tag{4}$$

(3) Uniform heating over distance lh from top,

$$H(s) = \begin{cases} 0 & 0 \leq s \leq L - lh \\ H_0 & L - lh < s \leq L . \end{cases} \tag{5}$$

(4) Exponential decay of heating from base,

$$H(s) = H_0 \exp^{-(s/(lh))} . \tag{6}$$

(5) Exponential decay of heating from top,

$$H(s) = H_0 \exp^{((s-L)/(lh))} . \tag{7}$$

Each of the above heat forms represents energy dumped in different ways and for them Equation (2) may be solved analytically with the boundary conditions, $T = T_0$ at $s = 0$ and $dT/ds = 0$ at $s = L$, where the observed loop length is $2L$ (P2). The loop is assumed to be symmetrical about the summit ($s = L$) and there is a maximum of three parameters that can be varied in the comparison with the observational data, namely, H_0 the maximum heating rate, T_0 the base temperature and lh the heating length scale.

The observational data sets of the large-scale solar corona obtained from Yohkoh and the models are now compared. In the present study three data sets are compared with the models. One of these sets gives the temperature profile along the loop considered by P1 and has 10 data points ($L = 348.5$ Mm). The second set is for the same loop but uses 74 data points ($L = 348.5$ Mm). The errors in the temperatures are much larger in the second data set due to the fact that there are more measurements, but this has to be balanced with the fact that the more data points used the more accurately the fit can be determined. The third data set represents the temperature profile along a different loop given with 33 data points ($L = 505$ Mm). For the third data set the temperature profile along only one half of the loop is compared with the models, whereas the first and second sets have the whole temperature profile from one footpoint to the other. This is because the other half of the third loop showed strong contamination in the optically thin radiation due to an active region that lay at its southern end. The second and third data sets are taken from P2 where they are compared with models for a heating length scale of $lh = 0.1L$. Other data sets were studied for $lh = 0.1L$ in P2 but they are not included in this paper since they produce very similar results.

For the three data sets the minimum χ^2 value for each heat form is determined as a function of lh. To obtain the minimum values of the χ^2 as a function of lh the parameters T_0 and H_0 are varied ($M = 2$). In Figure 1 the graphs of the reduced chi-squared (χ^2/v) versus the heating length scale (lh) can be seen for three data sets ((a) data set 1, (b) data set 2 and (c) data set 3). In each of the graphs the different forms of heating are shown as the same line types. For each data set the same heat forms follow a similar pattern as the heating length scale is increased.

From the graphs it can be seen that in each case the lowest values of χ^2/v are obtained for the largest heating length scales. However, in each case the χ^2/v are always much greater than 1. For each of the data sets uniform heating gives one

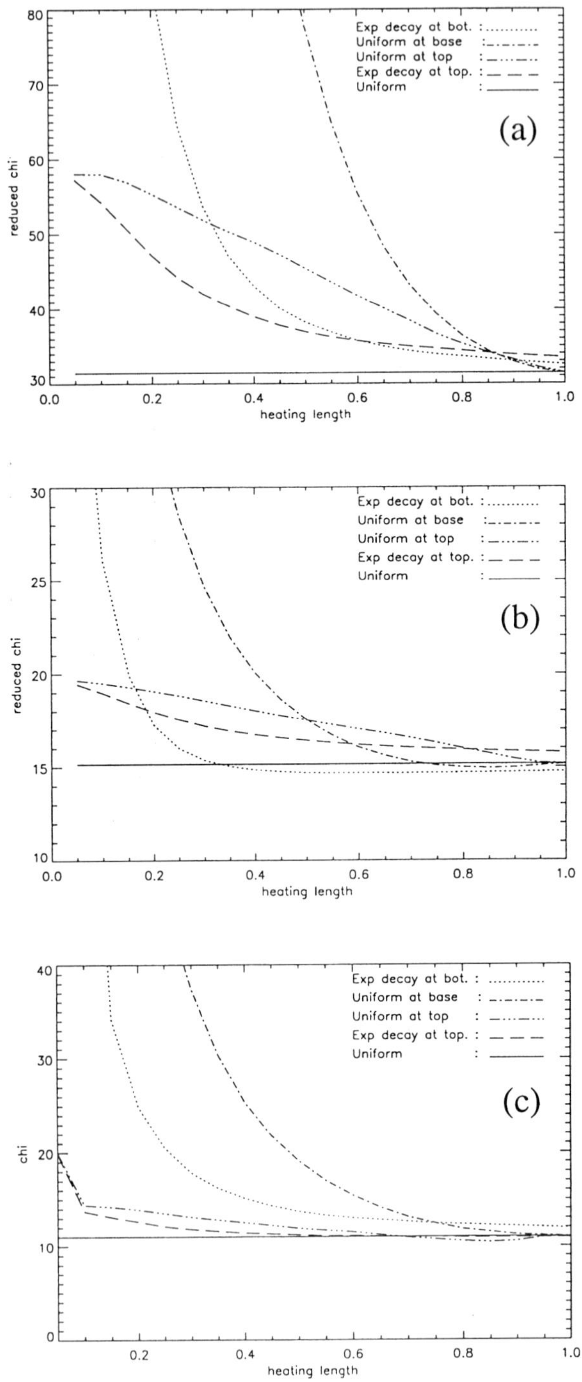

Figure 1. Graphs of reduced chi-squared (χ^2/v) versus length scale of heating (lh) for various heat forms for (a) data set 1, (b) data set 2 and (c) data set 3 and the basic model.

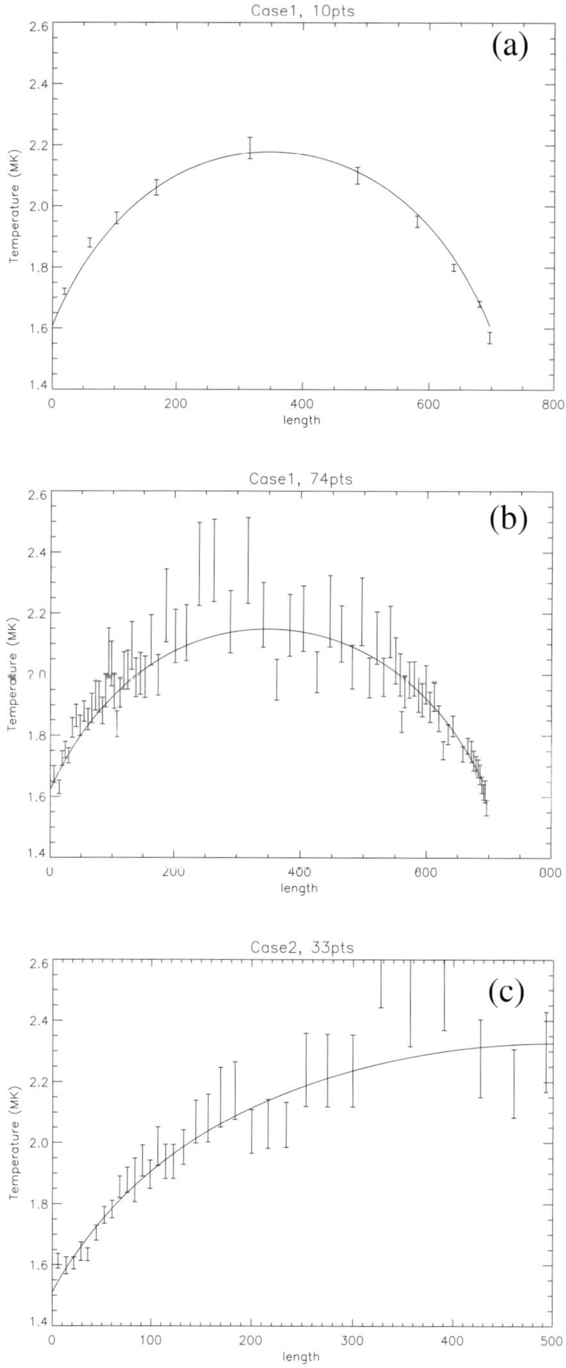

Figure 2. Observational data and fitted temperature profile for uniform heating and the basic model for (a) data set 1, (b) data set 2 and (c) data set 3.

of the best fits. For small length scales the other forms do not give as good a fit as uniform heating, but, as their length scale increases, their fit becomes comparable to that of uniform heating.

From the graphs it can be seen that in general uniform heating over a length scale near the top (Equation (5)) gives a reasonably good fit for all values of lh (compared to that of uniform heating) and gives a much better fit than uniform heating near the base (Equation (4)). Uniform heating at the base gives a very bad fit in each data set until the heating length scale reaches around $lh = 0.7L$, where it becomes similar to uniform heating at the top. Exponential decay from the base (Equation (6)) gives nearly as good a fit as uniform heating when the heating length scale becomes around $lh = 0.5L$ or greater, but for small values of lh it gives a very poor fit. Exponential decay from the top (Equation (7)) gives a relatively good fit for all length scales (again compared to that of uniform heating) with a slightly better fit being obtained as the length scale increases.

From this it is clear that when heat is deposited in the upper part of a loop (Equations (5) and (7)) there is very little change in the χ^2/v value as the length scale is increased, although increasing the length scale does give a slightly better fit. On the other hand, when heat is deposited at the footpoints (Equations (4) and (6)), there is a large change in the χ^2/v value as the length scale of the heating is increased, with larger length scales giving a much better fit. This indicates that the three observed temperature profiles of the large-scale solar corona compared here agree better with heat deposited at the top or uniformly along the loop than heat deposited at the footpoints for this model. Heat deposited at the footpoints only gives as good a fit as uniform heating for large length scales. For this model depositing heat in the upper part of the loop gives better results than depositing heat in the lower part. This would suggest that from the comparison with this model, coronal heating mechanisms that tend to dump their energy in the lower portions of the loop seem less likely to explain the temperature profile seen here for the large-scale solar corona than mechanisms with uniform heating or heating near the summit. The other data sets considered in P2 give very similar results to those obtained above.

From this comparison it is clear that heat deposited in the upper part of the loop gives a better fit than heat deposited at the footpoints. However, it is difficult to make strong judgments on which exact form of heating best fits the data. The reason for this is that all of the heating profiles converge as the heating length scale increases. Although they converge to the same chi-squared (χ^2) value there is still a significant difference in each of the heating profiles compared to that of uniform heating. One reason for this fast convergence is because the error bars are so large and the data points so scattered that all of the temperature profiles are eventually able to pass through a large number of the data points to give a relatively low value of χ^2. To be able to distinguish between the different heat forms data with much smaller error bars (especially at the top of the loop) would be required. This would allow the form of the heating to be determined more accurately. In Section

TABLE I
Statistical results for Model 1

Data set	v	χ^2/v	χ^2	χ^2_{crit} (area 0.05)	p-value
1	8	31.5	241.5	15.5	$\ll 0.0001$
2	72	15.15	1090.8	92.8	$\ll 0.0001$
3	31	11.1	344.1	47.36	$\ll 0.0001$
4	72	10.0	720.0	92.8	$\ll 0.0001$

5 different temperature profiles are compared to determine how small the error bars would have to become in the observations of the large-scale coronal loops to distinguish between the different models in the fitting process.

An important feature to consider now is the magnitude of the χ^2/v values that are obtained in the fitting procedure. In all of the cases the χ^2/v values do not lie close to one even though a 'good looking' fit is produced (Figure 2). For each data set the plot is shown for uniform heating. For the first data set of 10 points $\chi^2/v = 31.5$ (Figure 2(a), $T_0 = 1.61$ MK, $H_0 = 1.39 \times 10^{-6}$ Jm^{-3} s^{-1}) and for the second data set $\chi^2/v = 15.15$ (Figure 2(b), $T_0 = 1.62$ MK, $H_0 = 1.28 \times 10^{-6}$ Jm^{-3} s^{-1}). For the third data set $\chi^2/v = 11.1$ (Figure 2(c), $T_0 = 1.51$ MK, $H_0 = 1.04 \times 10^{-6}$ Jm^{-3} s^{-1}). Since these values do not lie close to unity the next task is to work out the statistical significance of the results. The statistical significance of uniform heating is calculated for each data set since the results converge to this value.

In Table I the results can be seen for data sets 1-3 and the uniform heating profile. From the table it can be seen that all of the chi-squared (χ^2) values obtained are much larger than the χ^2_{crit} values. So in absolute terms a 'good' fit has not been obtained, even though the fit to the data looks reasonable. The χ^2 values are in fact so large that a p-value of $\ll 0.0001$ is obtained in each case, effectively giving a significance level of 0. Thus, on a statistical basis the results obtained are very insignificant and we cannot be sure that the models fit the data. The main reason for such a poor significance level is that many of the data points lie above or below the temperature curve without the curve passing through them (Figure 2). The results here show that the heating profiles considered could not give such a strong deviation in temperature along the loop as the observations suggest. If the outlying data points do represent correct temperature values then this shows that the models are completely wrong, but, if the deviations in the observations are due to scatter in the data, the scatter is strongly affecting the result.

Above it has been shown that it is possible to obtain a reasonably good fit to the observed data in a 'chi-by-eye' but the results are not significant statistically.

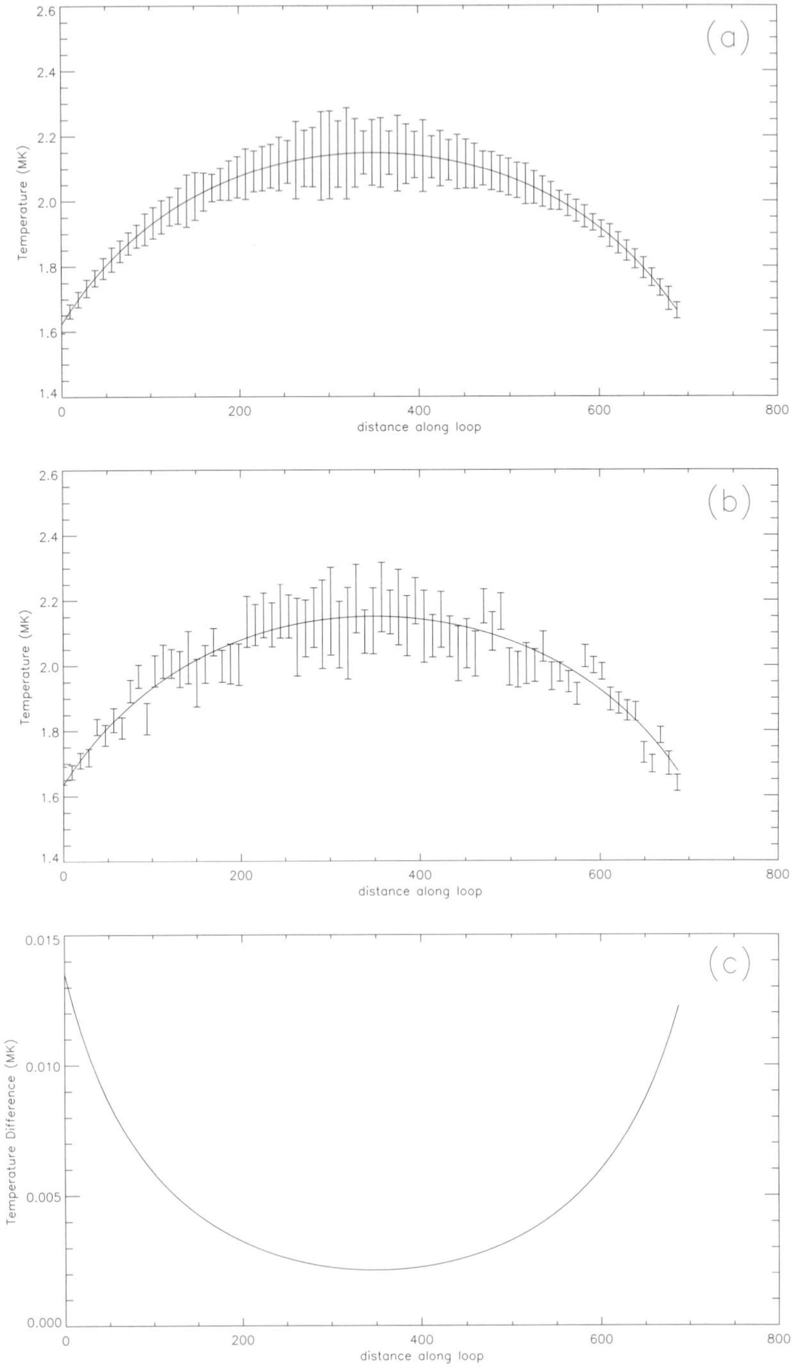

Figure 3. (a) Theoretical temperature profile with error bars added. (b) Data set after average 2% scatter is added along with the deduced temperature profile. (c) Difference between theoretical temperature profile and fitted temperature profile with distance along loop.

This means that there is a very low certainty that the models compared could in fact produce the observed temperature profiles. The following question can now be asked, 'How much of a scatter does it take to produce statistically insignificant results?' To try and answer this question a theoretical data set is produced of a uniform heating temperature profile in the same temperature range as observed. It is calculated for a base temperature $T_0 = 1.62$ MK and a base heating rate $H_0 = 1.29 \times 10^{-6}$ Jm^{-3} s^{-1} and can be seen in Figure 3(a). Error bars are then added to the temperature profile at 74 points along the loop in a manner that resembles the temperature profile for data set 2. If a chi-squared test were performed on this artificial data set, it would produce a chi-squared value of zero. A random scatter of the data points is now carried out, such that each data point along with its error bar is moved up or down by a small amount. In Figure 3(b) the scattering of the data points can be seen for a mean scattering of 2% of the observed temperature (maximum shift 4% up or down). The locations of the centres of the temperatures and error bars resemble to a high degree the type of observed distribution seen in Figure 2(b). When the chi-squared method is applied to the scattered data, the best fit is found for $T_0 = 1.64$ MK and $H_0 = 1.27 \times 10^{-6}$ Jm^{-3} s^{-1}, which is very similar to the exact values and the fitted temperature profile can be seen in Figure 3(b). The fit gives a χ^2/v of 10 which gives statistically insignificant results (data set 4, Table I) . The difference in the initial temperature profile and the fitted profile can be seen with distance along the loop in Figure 3(c) and is very small. Only an average 2% scattering in the theoretical data set is enough to produce a chi-squared value so large that the data becomes statistically insignificant, even though the difference in temperature between the exact profile and fitted profile is very small. Such a scatter in the observational data could be due to the effect of optically thin radiation with loops lying slightly in front of or behind the observed loop affecting the photon count. From this it can be seen that the χ^2 method can be critically affected by a small scatter in the data. Therefore, the results found here may not be as bad as they seem. So, even if the true model were compared with the observations it would be extremely hard to obtain a statistically significant result. Therefore it is unclear at the moment whether the results for the heating profile are significant or not. The only way to determine this is to obtain observations with much less scatter. One obvious method of improving the fit is not just to compare one observation of a loop with the models but to compare an average of many observations of a single loop with the models since this would reduce the scattering in the data. However, in the present observations the time scale to make the observations of the large-scale coronal loops is very long, and so making several observations of the same loop may not be feasible. However, shortening the time scale for making the observations is something that should be considered for future space missions such as *Solar-B* or *Stereo*.

4. Inclusion of Radiation

In the previous section with the basic model it was found that for each of the three data sets considered the observed temperature profile in the large-scale coronal loops was more consistent with energy deposited in the upper half of the loop than energy deposited in the lower half. The data sets are now reconsidered, but this time the basic model is changed to include the effect of radiation (Rosner, Tucker, and Vaiana, 1978; Hood and Priest, 1979; Roberts and Frankenthal, 1980; Hood and Anzer, 1988 and Walsh, Bell, and Hood, 1997). This is carried out to see whether the same conclusion ensues when new features are included in the model. The static energy equation is again solved, but this time a balance between thermal conduction, heating and radiation is assumed, so that the energy equation becomes

$$\frac{\mathrm{d}}{\mathrm{d}s}\left(k_0 T^{5/2}\frac{\mathrm{d}T}{\mathrm{d}s}\right) = \frac{p^2}{k_b^2}\chi T^{\alpha-2} - H(s), \tag{8}$$

and each of the parameters have the same definition as before. The first term on the right hand side represents the radiative loss for optically thin radiation. The parameters χ and α depend on the temperature range (Hildner, 1974). The pressure (p) is given by $p = n_e k_B T$, where n_e is the electron number density and $k_B = 1.38 \times 10^{-23}$ J/deg. In solving Equation (8) it will be assumed that the pressure is held constant along the loop. All of the variables are now non-dimensionalised such that

$$T = T_c\bar{T},\ s = L\bar{s},\ p = p_c\bar{p},\ H = H_c\bar{H},\ \chi = \bar{\chi}\chi_c\frac{T_c^{\alpha_c}}{T_c^{\alpha}},$$

where $T_c = 10^6$ K, L as before is the loop half length, $H_c = (p_c^2/k_b^2)\chi_c T_c^{\alpha_c-2}$ is the typical radiative loss for the corona, $p_c = 2 \times 10^{-3}$ Pa, $\alpha_c = -1.0$ and $\chi_c = 5.51 \times 10^{-30}$. From this it can be seen that the heating rate is scaled to multiples of the typical radiative loss in the corona and the dimensionless form of the temperature dependent parameters is given in Table II. On reducing Equation (8) to two dimensionless first-order equations, where the bars have been removed for convenience the problem reduces to solving

$$\frac{\mathrm{d}T}{\mathrm{d}s} = VT^{-5/2}, \tag{9}$$

$$\frac{\mathrm{d}V}{\mathrm{d}s} = b\left(p^2\chi T^{\alpha-2} - H(s)\right), \tag{10}$$

where

$$b = \frac{p_c^2\chi_c T_c^{\alpha_c-2}L^2}{k_b^2 k_o T_c^{7/2}} \tag{11}$$

is the ratio of conduction and radiation time scales and these equations are solved subject to the boundary conditions $T = T_0$ at $s = 0$ and $V = 0$ at $s = L$, so that the conductive flux is zero at the top of the loop which is symmetrical. The loop pressure is fixed such that $p = 1$ and the temperature at the top of the loop and the conductive flux at the base are the two free boundary conditions.

The model described above is very simple and much more realistic models which include features such as gravity could be constructed (Galsgaard *et al.*, 1999). However, the aim of the present paper is to find out which form of heat produces the temperature profile that best fits the data. For the above model the temperature profiles produced for each form of heat are not significantly different from those produced by more complicated models. More complicated models will usually give slightly higher or lower temperature values at each point along the loop but the overall shape of the profile will be similar for a given form of heat. As long as the temperature profiles from the other models are similar to the ones used above, a similar fit and χ^2 value will be obtained for a given heat form. This occurs since the minimisation tends to produce the same temperature profile as the above model but with slightly different values of T_0 and H_0. Thus, the comparison carried out there can only deduce which form of heat gives the best fitting temperature profile to the observational data. It cannot tell us the exact value of the physical parameters that give the best fit as they will vary slightly from one model to the next when the models produce similar profiles. Also Galsgaard *et al.* (1999) showed that the difference in temperature profile along a loop for a model which includes gravity or not is much smaller than the error bars in the present observations. Along with this, the difference in the temperature profiles obtained by using different boundary conditions (such as $V = 0$ at $s = 0, L$) is also much smaller than the width of the error bars. Therefore, the present observations cannot be used to distinguish between more sophisticated models or boundary conditions. The simple model also assumes a static situation at the lower boundary in the transition region since it does not include dynamic or flow effects. Recent observations from SOHO and TRACE show that on the Sun these effects may be important, however, models which include them have not commonly been constructed. Due to this and the uncertainty of what type of flows to add it is beyond the scope of the present comparison to consider these effects and that is also why the simple static model is adopted.

The new model is now compared with the observational data sets for the heat forms described in Section 3. As in the previous section, the minimum χ^2/v value is determined as a function of the heating length scale (lh). To find the minimum value, the base temperature (T_0) and maximum heating rate (H_0) are again allowed to vary. When this is carried out, it is found that for each heat form and data set, exactly the same graph of χ^2/v is produced as in Figure 1. Although exactly the same graph is obtained there are small differences in the T_0 and H_0 values obtained in the minimisation. In all cases slightly higher H_0 values are required due to the losses in the optically thin radiation. For the uniform heating profile and

TABLE II

α and χ values for optically thin radiation obtained from Hildner (1974)

T	α	χ
$T < 0.015$	7.4	2.24×10^{13}
$0.015 \leq T < 0.08$	1.8	1.37×10^{3}
$0.08 \leq T < 0.3$	0	1.45×10^{1}
$0.3 \leq T < 0.8$	−2.5	7.15×10^{-1}
$0.8 \leq T < 10$	−1.0	1

data set 2 the best fit occurs for $T = 1.62$ MK, $H_0 = 1.31 \times 10^{-6}$ Jm^{-3} s^{-1}, which is very similar to the values obtained with the basic model ($T_0 = 1.62$ MK, $H_0 = 1.28 \times 10^{-6}$ Jm^{-3} s^{-1}). The slightly higher values effectively reproduce the same temperature profile as seen in Figure 2(b) but with radiation included. It is important to note that in all of the cases considered the ratio of the heating term and radiative term was in general greater than 10, showing that for the range of temperatures given by the observations the radiative term is much less important than the heating term in the model. Therefore, the basic model in Section 3 is reasonable for the temperature range considered. Since exactly the same results are obtained as in Figure 1, it again shows that in the temperature range considered the observed temperature profiles are in fact more consistent with heat deposited at the top or uniform along the loop compared to heat deposited at the footpoints.

In all of the data sets considered so far, the temperature profile only goes down to 1.6 MK and we are unsure of the temperature profile below this. Since the arcades considered here are anchored in the photosphere the temperature must fall down to 1 MK at some point along their length and then down through transition region values. This means that in the previous comparisons of the observations the full temperature range of the loops has not been taken into consideration. For the basic model the temperature range considered was confined to coronal values since it was only in this range that the observations were obtained and the model valid. However, for the new model a greater temperature range can be considered since by adding the radiative term into Equation (8) the energy balance is now valid down to transition region values. It would therefore be interesting to calculate the best fit when the model temperature profile extends down to 2×10^4 K.

The fitting procedure is now repeated, but this time the base temperature of the loop is fixed at 2×10^4 K. Since the present loop length only deals with the temperature above 1.6 MK in order for the temperature to fall lower than this the loop length must be allowed to increase. By allowing the loop length to increase, it suggests that the loop is embedded slightly behind the Sun out of the field of view

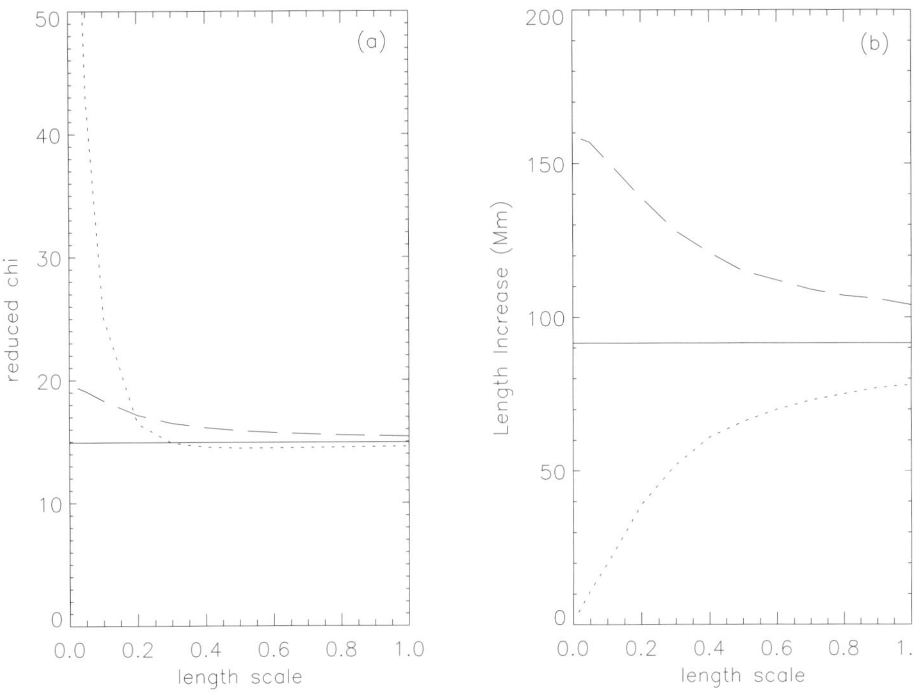

Figure 4. (a) Graph of reduced chi-squared for a model that includes radiation when the length of the loop is allowed to increase. Uniform heating (*solid line*), exponential decay from base (*dotted line*) and exponential decay from the summit (*dashed line*). (b) Length increase required to find the best fit in region of observed data.

of the present observations. The fitting procedure is now repeated for each data set but this time the length of the loop along with the heating rate is allowed to vary to determine the minimum χ^2/v for each form of the heating profile as a function of the heating length scale (again there are two parameters to be varied, L and H_0). For this comparison only three of the five previous forms of heat will be considered, namely uniform heating, exponential decay from the base and exponential decay from the top. Only these forms are chosen since they give the three characteristic temperature profiles that can be compared with the observations. Also, for the other two forms, solutions are numerically hard to find when energy is not deposited everywhere along the loop. It should be noted that in the comparison the lower bound for the temperature lies well outside the temperature range given by the observations. Although the solution does extend far outside the observed temperature range the fit is only tested at the observed points and the rest of the solution without the observations should only be regarded as the necessary form of the temperature lower down in the theoretical loop to give the profile in the observed region for a given form of heat.

In Figure 4(a) the minimum value of the χ^2/v can be seen as a function of the heating length scale (lh) and in Figure 4(b) the half length increase (Mm) that is required to obtain the best fit is shown. The first thing to note is that for each form of heating the graph of χ^2/v is again exactly the same as that obtained from the analytical model in Section 3. This tells us that the minimisation increases the length of the loop until the temperature solution for each form of heat matches directly onto the previous solutions found in Sections 3 and 4. This therefore has serious consequences for the conclusions of the previous comparisons, since it gives an estimate of the total length of the loop that is required for a specific form of heating so that the solution can fall down to transition region values. For the uniform heating profile, which in general gives one of the best fits, a half length increase of 91.5 Mm is required. This is an extremely large increase, about 26% of the half length of the loop and for this to be true the loop would have to be deeply embedded behind the Sun. The majority of the increase in length of the loop is required for the temperature to fall to 1 MK and therefore it lies in a temperature range where the physics is reasonably well known and where adding extra features such as gravity does not strongly affect the result. The transition region part of the loop, where the physics is not so well known, only corresponds to a small part of the length increase and so adding dynamic effects at the lower boundary should not change the results significantly. Thus, for uniform heating to fit the data and for the temperature to match down to chromospheric values, the loop must be significantly longer but this seems unlikely as the loops here were chosen by P1 and P2 to be nearly perpendicular off the limb to avoid projection effects. From Figure 4(b) it can be seen that exponential decay from the top requires the largest increase in length of the loop for each length scale, while exponential decay from the base requires the smallest increase in length. Since a smaller increase in length of the loop is more reasonable as the observational data is taken down to the limb (Figure 1(a), Priest *et al.*, 1998) and the loop is chosen to be nearly perpendicular off of the limb this seems to suggest that heat decaying from the base gives better results which, is in contradiction to the previous results of Section 3 and 4. In Figure 5 the plots of the temperature profile along with the observed data can be seen where the dashed line in each plot represents the outer bounds for the observational data. In Figure 5(a) the uniform heating case is shown (half length increase 91.5 Mm, $H_0 = 1.31 \times 10^{-6}$ Jm^{-3} s^{-1}), (b) exponential decay from the base ($lh = 0.1L$, half length increase 21.5 Mm, $H_0 = 42.1 \times 10^{-6}$ Jm^{-3} s^{-1})) and in (c) exponential decay from the top ($lh = 0.1L$, half length increase 150.5 Mm, $H_0 = 9.45 \times 10^{-6}$ Jm^{-3} s^{-1})). As can be seen in Figures 5(a) and (c) a large increase in the length of the loop is required for the temperature to fall to 1 MK. For exponential decay from the base a slightly worse fit is obtained, but only a 5% increase in loop length is required for the temperature to fall below 1 MK. Although the fit is slightly worse, it has a χ^2/v of 23.5 compared to 15.1 for uniform heating, which is not that much higher, since they are both greater than 1. The higher value of the reduced chi-squared is mainly due to a bad fit in the upper half of the loop, but this is where the errors are the largest and the scatter in the data the most. The comparison above has also been carried out with the feature of gravity included (Hood and Anzer, 1987).

When this is done exactly the same results as those shown in Figure 4 were found. Again this is mainly a result of the fact that the majority of the increase in loop length is required in the coronal part of the loop ($T > 1$ MK) where the effects of gravity are negligible.

By considering a larger temperature range (but the full temperature range of the solar atmosphere) it has been found that the data could possibly be more consistent with heat decaying from the base. This form of heat produces a sharper temperature rise which flattens out nearer the top and requires the shortest increase in length of the loop, which seems the most feasible. The increase in length of the loop can in this case be justified by having the loop anchored behind the limb of the sun. Thus, the opposite result to that of Section 3 is found when a fuller temperature range is used (but the full temperature range of the sun). In the data there are no temperature values below 1.6 MK and so we are unsure of the temperature profile lower down. The temperature profile in the range 1 MK to 1.6 MK is very important as it helps determine the coronal heating mechanism more readily than the temperature structure higher up and so it would be useful to included it in future modeling. Also, this shows that possibly misleading results may be obtained if a full range of temperatures from 1 MK is not considered. Without comparing the observations down to 1 MK misleading or contradictory results may be obtained. However, for small coronal loops such as those observed by Lenz *et al.* (1999) the whole loop can be seen so the length is known, so even if there are no observations down to 1 MK the temperature variation with length is 'known'. However, making observations down to 1 MK would help in distinguishing the different mechanisms from one another. Such measurements would be difficult for the type of large-scale coronal loop seen here as there is a problem with contamination of the optically thin radiation as we get closer to the photosphere. It would therefore be useful to have simultaneous observations of the large-scale corona in X-rays as seen here as well as at other wavelengths such as those observed by TRACE that could provide the link from low coronal values to chromospheric values. This is a necessary requirement for the theory to produce meaniful results. The results above which deal with a fuller temperature range seem to indicate that the mechanism that heats the large-scale corona may dump its energy low down in the loop. One such mechanism could be the 'magnetic carpet' where reconnection between small magnetic fragments may occur low down in the solar atmosphere (Schrijver *et al.*, 1998). However better observational data sets and models which include dynamic effects will be required to show this.

5. Discussion

In this paper observational data of the temperature profile along large-scale coronal loops have been compared with theoretical models. Two different models have been considered and for theoretical reasons both can be considered over different

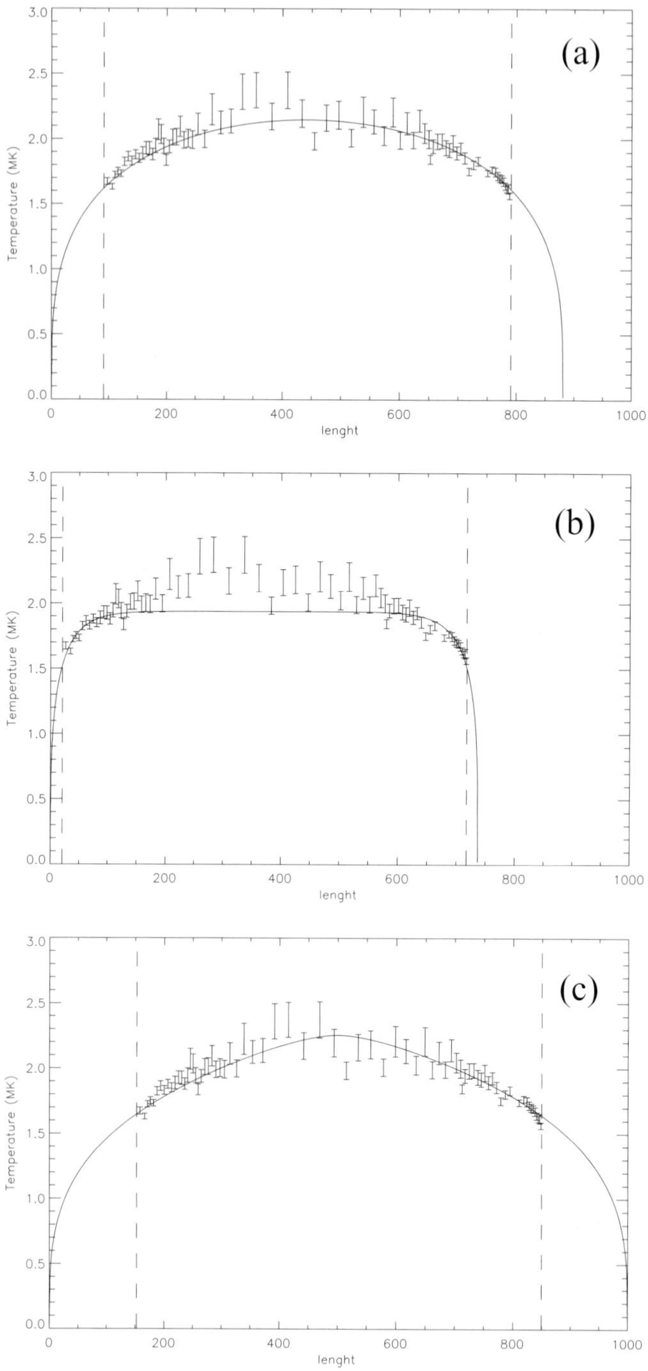

Figure 5. Deduced temperature profiles for data set 2 when the length of the loop is allowed to increase for (a) uniform heating, (b) exponential decay from the base ($lh = 0.1$) and (c) exponential decay from the summit ($lh = 0.1$) for the model that includes radiation.

temperature ranges. Unfortunately both of the models give conflicting results when considered over different temperature ranges but this shows the inadequacies of the present data sets. The first model uses a balance between heat conduction and heating and therefore is only valid in the hot corona. It indicates that heat deposited uniformly or in the upper part of the loop fits the data better than heat deposited in the lower part. This seems to indicate that turbulent reconnection could be the heating mechanism for the large-scale solar corona. However, no clear distinction between which specific form of heating gave the best fit could be made as, when the length scale of heating becomes long enough, all of the heat forms give comparable results even though there are significant differences between each of the heat forms. This is mainly due to the size of the error bars which allow all the solutions to lie within the data points once (lh) becomes large enough. The second model which includes the above effects along with radiation again found that heat deposited in the upper half of the loop or uniformly along the loop gives the best fit when considered over the same temperature range as the basic model.

One problem with the observational data sets is that they only consider temperature values above 1.6 MK. This means that the observations along the loop do not consider the full temperature range of the solar atmosphere. Since the energy balance for the second model is valid down to the transition region a fuller temperature range through the solar atmosphere can be considered. In order to consider a fuller temperature range through the solar atmosphere so that the temperature can fall to below 1 MK it was hypothesised that the loop would have to be slightly longer than that given by the observations. In view of this the base temperature of the loop was fixed at 2×10^4 K and the length of the loop was then allowed to vary to see if results in agreement with the previous cases could be obtained. Uniform heating or heat deposited at the top only gave the same fit as in the previous cases when there is a 26% increase in the loop length, which seems too large to be feasible. On the other hand, heat decaying from the base only needs a 5% increase in length to reproduce the previous results. Thus by considering a fuller temperature range with the second model it was found that heat decaying from the base gave the more reasonable fit as it represented the most feasible increase in loop length for the temperature to fall below 1 MK. This was in contradiction to the results above and suggested that magnetic reconnection of small fragments on the solar surface, the so called ‘magnetic carpet’ may be the method that heats the large-scale solar corona. Thus from the data sets conflicting results are obtained but these results show the inadequacy of the present data sets when compared to theory and how only considering a certain temperature range may be misleading. For meaningful theoretical results to be produced the following improvements in the observations of the large-scale solar corona have to be made.

First, the width of the error bars must be substantially reduced. This would allow the various heat forms to be distinguished over a much larger range of length scales without the fast convergence in the reduced chi-squared value. A simple estimate of how small the error bars would have to become to distinguish the heat forms over

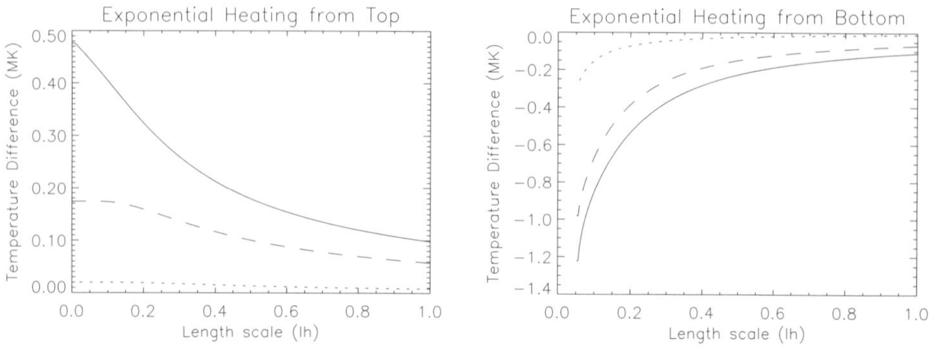

Figure 6. Estimate of how small the error bars must become to distinguish each model from uniform heating as a function of the heating length scale. The solid line gives the difference in summit temperature, the dashed line the difference at $0.5L$ and the dotted line the difference at $0.1L$.

each heating length scale can be seen in Figure 6. The graph shows the difference in summit temperature (solid line), temperature at $0.5L$ (dashed line) and temperature at $0.1L$ (dotted line) as a function of the heating length scale for the two exponential forms of heating compared to uniform heating. The difference is computed for the model which includes radiation and a base temperature of 2×10^4 K and the same total amount of energy is deposited in the loop by each heat form. The loop has a half length of $L = 348.5$ Mm such as that of data sets 1 and 2 and for the uniform heating profile $H_0 = 2.22 \times 10^6$ Jm^{-3} s^{-1}. The uniform heating temperature profile has a summit temperature of 2.2 MK which is around the observed value seen in the data sets. Now for any two forms of heating to be distinguished by the chi-squared method the width of the error bar at any specific point would have to be less than the difference in temperatures between the two models at any point. This provides an upper estimate on the width of the error bars along the loop and in general the width of the error bar would have to be at least less than half of the difference in temperature for a significantly different χ^2 value to be obtained. From the graph it can be seen that the difference in temperature between any two heat forms is very small and the width of the error bar would in general have to become less than 0.1 MK to distinguish the heat forms over all length scales. In the present data sets where there are many data points along the loop the error bars at the top of the loop are on the order of 0.22 MK. From Figure 6 it can be seen that with the present data sets at the summit of the loop, exponential decay from the base and top can only be distinguished up to $0.5L$. These simple results agree in principle with those found in Figure 1 which shows the convergence of the χ^2. This shows in principle how small the error bars have to become to distinguish the models and it remains to be seen if observations of the large-scale corona can be produced with such small errors. Initial results from TRACE indicate that TRACE may be able to distinguish the various heat forms for the type of loops that it observes (Lenz *et al.*, 1999). However, TRACE cannot go up to a high enough temperature to observe the large-scale solar corona.

TABLE III

Statistical results for Model 1 when the half width of the error bar is $\pm 2\sigma$

Data set	v	χ^2/v	χ^2	χ^2_{crit} (area 0.05)	p-value
1	8	13.4	107.2	15.5	$\ll 0.0001$
2	72	6.73	484.4	92.8	$\ll 0.0001$
3	31	4.93	152.9	47.36	$\ll 0.0001$
4	72	4.44	320	92.8	$\ll 0.0001$

Secondly, the amount of scatter in the data must be reduced if statistically significant results are to be produced and we are not just to rely on a 'chi-by-eye' approach . This is a result of the fact that the chi-squared method treats all points along the loop equally and outlying points strongly affect the fit. Results show that only a 2% average scatter is enough to make the results insignificant. One method of reducing the scatter is to compare a number of observations of the same loop taken simultaneously or on a short time scale and average out the scatter. This would need both an improvement in the cadence of observations for the large-scale solar corona and making simultaneous observations would be ideal for a mission such as *Stereo*.

Next, a reasonably large number (> 16) of observations is required along the loop in general to obtain a good fit. However, in the present data sets the more observations that are made the greater the error bars and the harder it is to distinguish the different forms of heat. Many observations with small error bars are required. Also, the observations should be made equally spaced along the loop since for the data here the fit is weighted towards the footpoint compared to the top of the loop because there are many more measurements at the footpoints. At the present time, it is easier to make observations at the footpoints compared to the summit since the photon count is much greater at the footpoints.

Finally, the temperature profile from at least 1 MK is required as only considering a finite temperature range above this may possibly lead to conflicting results especially when the footpoints of the arcades are not seen. This is not so important for small coronal loops such as those considered by Lenz *et al.* (1999) as the entire length of the loop is known. However taking the temperature down to at least 1 MK would help distinguish the models. For the large-scale loops of the solar corona this may mean that more than one instrument has to be used to give the full temperature range along the loop.

In this paper a comparison of theory and observations has shown that the temperature profiles observed along large-scale coronal loops could in fact be more

TABLE IV

Statistical results for Model 1 when the half width of the error bar is $\pm\sigma$

Data set	υ	χ^2/υ	χ^2	χ^2_{crit} (area 0.05)	p-value
1	8	3.35	26.8	15.5	0.0003
2	72	1.69	121.1	92.8	0.0001
3	31	1.23	38.2	47.36	0.0285
4	72	1.11	80.0	92.8	0.0243

consistent with footpoint heating compared to that of uniform heating and heat decaying from the summit. Footpoint heating gave the more reasonable result when the full temperature range of the solar atmosphere was considered, however, to be entirely sure of the result improvements in the observations are required. For future studies, theoretical requirements have been suggested that must be fulfilled by the observations if the nature of the coronal heating mechanism in the large-scale corona is to be determined from temperature profiles along coronal loops. These requirements may not be easy to fulfill, however they are theoretically required. It must now be seen if observational instruments on TRACE or future satellite missions such as *Solar-B* and *Stereo* can deliver these requirements.

Appendix: Results for $\pm 2\sigma$ and $\pm\sigma$

The results of Section 3 are now reconsidered but this time the half-width of the error bars are taken to be of size $\pm 2\sigma$ and $\pm\sigma$. This means that the weight factor in Equation (1) will be slightly larger and lower chi-squared values will be obtained. This represents a relaxation of the previous results found in the paper to see if the results become more significant. Since only the weighting factor is different exactly the same type of solution is obtained as in Figure 1 and the only difference is the absolute value of the chi-squared. Due to this only the uniform heating results are quoted in Table III for the $\pm 2\sigma$ case and Table IV for the $\pm\sigma$. From Table III it can be seen that even with the less strict value of $\pm 2\sigma$ statistically insignificant results are still found. For $\pm\sigma$ there are now some chi-squared values that are less than the critical values giving a good fit in absolute terms, however, the best p-value obtained is at most 0.0285. This indicates that if the model is correct then such deviations in the data could only occur 3% of the time which is not a good result. This shows that if the coronal heating mechanism for the large-scale corona is to be determined with some statistical significance then the scatter in the data must be reduced. For the theoretical data set number 4 only a 2% average scatter in the data is enough to give a p-value as low as 0.0243 which is a relatively insignificant result. Also from the results above it can be seen that in order to get some significant results a strong relaxation on the definition of σ from the observed data is required.

Acknowledgements

The authors would like to thank Prof. A. Hood, Dr S. Wood and Prof. R. Cormack for helpful discussions. The work was funded by the UK Particle Physics and Astronomy Research Council.

References

Acton, L. W.: 1996, *ASP Conf. Ser.* **109**, 45.
Bevington, P. R. and Robinson, D. K.: 1994, McGraw-Hill, New York.
Culhane, J. L.: 1997, *Adv. Space Res.* **19**, 1839.
Erdelyi, R.: 1998, *Solar Phys.* **180**, 213.
Galsgaard, K. and Nordlund, Å.: 1996, *J. Geophys. Res.* **101**, 13445.
Galsgaard, K., Mackay, D. H., Priest, E. R. and Nordlund, Å.: 1999, *Solar Phys.* **189**, 95.
Golub, L., Krieger, A., Silk, J., Timothy, A. and Vaiana, G.: 1974, *Astrophys J.* **189**, L93.
Golub, L., Krieger, A., Harvey, J. and Vaiana, G.: 1977, *Solar Phys.* **53**, 111.
Hara, L., Tsuneta, S., Saku, Acton, L. W., Bruner, M., Lemen, J. R., and Ogawara, Y.: 1994, *Publ. Astron. Soc. Japan* **46**, 493.
Hildner, E.: 1972, *Solar Phys.* **35**, 123.
Hood, A. W. and Anzer, U.: 1988, *Solar Phys.* **115**, 61.
Hood, A. W. and Priest, E. R.: 1979, *Astron. Astrophys.* **77**, 233.
Hood, A. W., Ireland, J. and Priest, E. R.: 1997, *Astron. Astrophys.* **318**, 957.
Jordan, C.: 1992, *Journal of the Italian Astronomical Society* V63.
Kano, R.: 1997, *Proceedings of the Fifth SOHO Workshop*, ESA SP-404, p. 445.
Klimchuk, J. A., Lemen, J. R., Feldman, U., Tsuneta, S. and Uchida, Y.: 1992, *Publ. Astron. Soc. Japan* **44**, L181.
Kmietowicz, Z. W. and Yannoulis, Y.: 1976, *Statistical Table Longman Science and Technical*, p. 33.
Lenz, D. D., DeLuca, E. E, Golub, L., Rosner, R. and Bookbinder, J. A.: 1999, *Astrophys. J. Letters*, accepted.
Priest, E. R., Foley, C. R., Heyvaerts, J., Arber, T. D., Culhane, J. L., and Acton, L. W.: 1998, *Nature* **393**, 545.
Priest, E. R., Foley, C. R., Heyvaerts, J., Arber, T. D., Culhane, J. L., Acton, L. W., and Mackay, D. H.: 1999, *Astrophys. J.*, submitted.
Roberts, B. and Frankenthal, S: 1980, *Solar Phys.* **68**, 103.
Rosner, R., Tucker, W. H., and Vaiana, G. S.: 1978, *Astrophys. J.* **220**, 643.
Schrijver, C. J., Title, A. M., Harvey, K. L., Sheeley, N. R., Wang, Y. M., van den Oord, G. H. J., Shine, R. A., Tarbell, T. D., and Hurlburt, N. E.: 1998, *Nature* **394**, 152.
Sturrock, P. A., Wheatland, M. S., and Acton, L. W.: 1996, *Astrophys. J.* **461**, L115.
Tsuneta, S.: 1996, *Solar and Astrophysical MHD Flows*, Kluwer Academic, Publishers, Dordrecht, p. 85.
Uchida, Y.: 1993, *Physics of the Solar and Stellar Coronae*, Kluwer Academic Publishers, Dordrecht, p. 97.

the lower transition region lines and extended measurements of the phenomenon to higher temperatures (Dere, 1982; Dere, Bartoe, and Brueckner, 1984; Hassler, Rottman, and Orrall, 1991; Brekke 1993; Brekke, Hassler, and Wilhelm, 1997; Chae, Yun, and Poland, 1998; Brosius, Thomas, and Davila, 1999; Brynildsen *et al.*, 1999; Thompson and Brekke, 1999).

Dere (1982) used active region plage and sunspot observations obtained with the NRL High-Resolution Telescope and Spectrograph (HRTS) sounding rocket experiment to measure typical downflow velocities $\sim$ 10 km s^{-1} at log T $\sim$ 5.0, and to extend the observed red-shift phenomenon to O V (log T $\sim$ 5.37). Dere, Bartoe, and Brueckner (1984) found an average red shift corresponding to 5 km s^{-1} in the quiet Sun using the two strong C IV resonance lines at 1548.2 and 1550.7 Å (log T = 5.0), while Brekke (1993) used HRTS spectra to measure average quiet-Sun red shifts $\sim$ 8 km s^{-1} in O V lines. Hassler, Rottman, and Orrall (1991) conducted an in-flight wavelength calibration of the LASP EUV Coronal Spectrometer during a 1987 sounding rocket experiment, allowing them to directly measure wavelengths of observed solar lines against platinum lines of known wavelength generated on-board. The C IV line at 1548 Å and Fe II line at 1563 Å yielded net red shifts corresponding to Doppler velocities of 7.5 $\pm$ 1.0 km s^{-1} and 2.7$\pm$1.5 km s^{-1}, respectively, in the quiet Sun. The Si II line at 1533 Å and Ne VIII line at 770 Å yielded no net red shifts. Brekke, Hassler, and Wilhelm (1997) and Chae, Yun, and Poland (1998) used the SUMER (Solar Ultraviolet Measurements of Emitted Radiation) spectrometer aboard the SOHO (Solar and Heliospheric Observatory) spacecraft to find systematic red shifts in transition region as well as low coronal lines, extending the temperature range of the phenomenon to O VI (log T $\sim$ 5.46), Ne VIII (log T $\sim$ 5.80), and Mg X (log T $\sim$ 6.05). Some of these results must be viewed with caution, since, as Brekke, Hassler, and Wilhelm (1997) point out, conflicting reference wavelengths have been reported for O VI and Mg X. (The Mg X reference wavelength that they adopted, 624.95 Å from Kelly (1987), is given to only two decimal places rather than three or four, indicating the relative uncertainty in its value.) Finally, Chae, Yun, and Poland (1998) report that the Mg X line appears to be blended in the SUMER spectrum with an unknown first- or second-order component that can affect its derived centroid wavelength.

Unfortunately, the absolute rest wavelengths for coronal EUV lines are not well known. For example, the Kelly (1987) finding list and the Fuhr, Martin, and Wiese (1988) atomic parameter compilation both adopt wavelengths for almost all of their Fe XI–XV EUV lines from the solar observations of Behring *et al.* (1976) or Behring, Cohen, and Feldman (1972). Behring *et al.* (1976) obtained a full-Sun solar spectrum from 160 to 770 Å on photographic plates during a sounding rocket flight in 1973. Their wavelength scale is based entirely on the wavelengths of 10 strong reference lines of He I, He II, O III, O IV, O V, and Ne VII compiled in Kelly and Palumbo (1973). However, the O IV and O V ions used by Behring *et al.* (1976) to derive their wavelength calibration were later found to exhibit persistent red shifts in a variety of solar features (Doschek, Feldman, and Bohlin 1976;

Feldman, Cohen, and Doschek, 1982; Dere, 1982; Brekke, 1993; Brekke, Hassler, and Wilhelm, 1997). Further, Brekke, Hassler, and Wilhelm (1997) also found red shifts in lines of Ne VIII, formed at a similar temperature (log $T \sim 5.80$) as the Ne VII line (log $T \sim 5.71$) used as a wavelength standard by Behring *et al.* (1976). Thus the EUV wavelengths compiled by Kelly (1987) for lines of Fe XI, XII, XIII, XIV, and XV (as well as several other ions), many of which are adopted from the two Behring references, are not laboratory rest values but solar values likely affected by whatever mechanism produces the observed solar wavelength shifts. It is, therefore, problematic to use those wavelengths as references for measuring coronal Doppler velocities. To measure *absolute* wavelength shifts on the Sun it is essential to compare observed solar wavelengths with accurate (to, say, three decimal places) laboratory rest wavelengths. An absence of the latter necessitates the measurement of *relative* wavelength shifts in the present work. Improved laboratory and/or theoretical values for EUV wavelengths from coronal ions would be extremely valuable.

Brosius, Thomas, and Davila (1999) used 170–340 Å spectra of NOAA region 7870 and its environs (its 'edge') obtained with SERTS-95 to measure relative wavelength shifts of coronal emission lines from various ion species, including He II, Si XI, and Fe X–XVI. (Achour *et al.* (1995) applied a similar approach to measure transition region differential red shifts between an active region and its quieter surroundings using HRTS spectra.) The average 'edge' spectrum was used to provide reference wavelengths against which to measure relative shifts in the active region. The lines of He II at 303.78 Å, Fe XII at 193.51 Å, Fe XIII at 202.05 Å, Fe XIV at 211.33 Å, Fe XV at 284.15 Å, and Fe XVI at 335.41 Å were used for measuring spatial variations of the relative wavelength shifts across the region because all of these lines are free from known blends in the SERTS-95 spectra and are either intrinsically strong or near the SERTS-95 peak sensitivity. The iron ions are the hottest ions ever used for this type of analysis. All six lines reveal statistically significant spatial variations in their measured relative wavelength shifts in the active region core, including mixtures of blue shifts and red shifts (each with maximum values corresponding to relative Doppler velocities $\sim$ 15 km s^{-1}), indicating a dynamic, turbulent corona. For each of these lines we calculated intensity-weighted relative Doppler velocities from wavelength shifts in the spatially averaged core spectrum, obtaining values of $+5.8 \pm 0.6$ km s^{-1} for He II, $+5.7 \pm 0.9$ km s^{-1} for Fe XII, $+0.4 \pm 0.7$ km s^{-1} for Fe XIII, -2.1 ± 1.0 km s^{-1} for Fe XIV, $+0.8 \pm 0.8$ km s^{-1} for Fe XV, and -1.1 ± 0.5 km s^{-1} for Fe XVI. Combining the above six lines with several additional ones that are strong enough in both the edge and average core spectra to provide reliable centroid measurements, we found statistically significant net relative red shifts for lines of He II, Fe X, Fe XI, and Fe XII; lines of Fe XIII and Fe XV show no significant shift while lines of Si XI, Fe XIV, and Fe XVI show a small net relative blue shift.

In the present work we apply a similar technique to that used by Brosius, Thomas, and Davila (1999) to measure *relative* Doppler velocities between an active region and its quiet surroundings. Section 2 describes the SERTS instrument flown in 1997 (SERTS-97), Section 3 presents the solar observations, Section 4 provides wavelength shift measurements, Section 5 discusses implications of the measured shifts, and Section 6 summarizes our conclusions.

2. The SERTS-97 Instrument

The version of SERTS flown on 18 November 1997 recorded spectrographic data on a CCD-intensified detector over the wavelength range 297.5–353.5 Å with a spectral resolution (instrumental FWHM) of 115 mÅ. It incorporated the same multi-layer-coated toroidal diffraction grating that was flown in 1991 and 1993 (Brosius *et al.*, 1996), which enhanced the instrumental sensitivity in this wavelength range over that of a standard gold-coated grating by factors up to nine. The spatial resolution was about 7 arc sec. SERTS-97 simultaneously recorded high resolution spectra along a 6 arc min slit, and spectroheliograms of two 3 arc min × 8.5 arc min areas on each end of the slit. See Figure 1.

An end-to-end radiometric calibration of the rocket instrument was carried out at Rutherford-Appleton Laboratory in the same facility used to characterize the CDS experiment on SOHO, and using the same EUV light source recently recalibrated by PTB against the BESSY synchrotron. SERTS-97 has been used to carry out an intensity cross-calibration with SOHO/CDS, one of the major objectives of this flight (Thomas *et al.*, 1999).

Pre- and post-flight wavelength calibrations were carried out at GSFC using EUV spectra of a hollow-cathode gas-discharge lamp emitting He II and Ne II lines with rest wavelengths known to better than 0.3 mÅ (Kaufman and Edlen, 1974). Although the wavelength scale itself remained constant for the pre- and post-flight laboratory wavelength calibrations, the zero point of this scale shifted slightly, indicating a small, constant offset between the in-flight wavelength scale and those measured in the laboratory. This translates into a systematic, constant error on every wavelength measured from each area observed on the Sun. Fortunately, systematic wavelength shifts introduced by this offset are eliminated (directly canceled) by treating *relative* wavelength shifts between two different physical areas. In other words, because wavelengths measured simultaneously in any two different areas on the Sun include the same systematic shift, this shift is eliminated by subtracting the two wavelengths. Because the He II line at 303.78 Å is strong in both the laboratory and the solar spectra, we use it as the reference wavelength on which the solar spectrum is fixed to the laboratory wavelength scale. Thus we set the He II wavelength in our solar reference spectrum equal to its laboratory rest wavelength of 303.7822 Å, and base all other wavelengths in the reference spectrum as well as in the spatially resolved active region spectra on this choice. Of course, this solar line could in fact be significantly Doppler-shifted relative to its rest value, which would impose an additional systematic offset on all measured

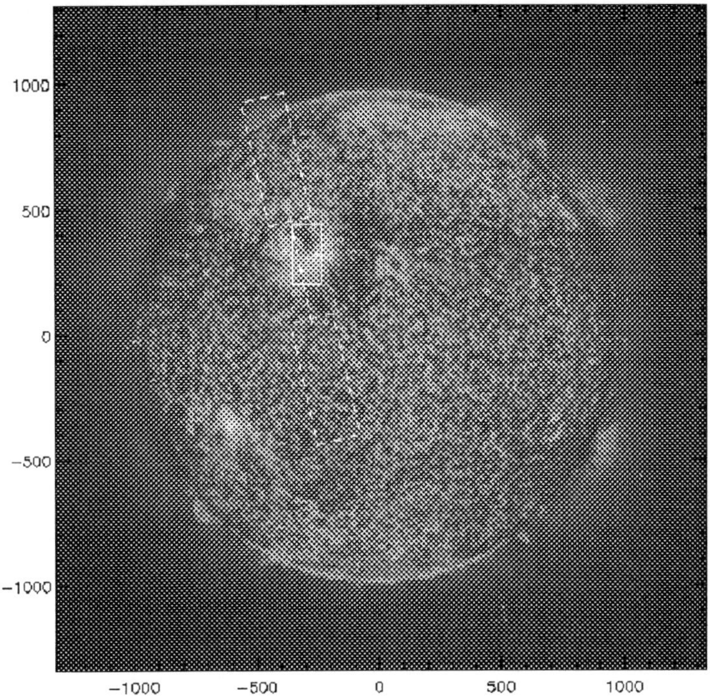

Figure 1. SOHO/EIT full-disk He II 304 Å image obtained during the SERTS rocket flight on 18 November 1997. The dashed rectangles outline the 3 arc min × 8.5 arc min areas within which SERTS spectroheliograms were obtained, and the dashed line connecting them shows the location of the 7 arc sec × 6 arc min SERTS slit. The solid rectangle shows the 2 arc min × 4 arc min CDS field of view, centered on NOAA region 8108. Solar north is at the top.

wavelengths. However, this offset too is eliminated from the *relative* wavelength shifts by subtraction.

3. Observations

SERTS was launched on a Terrier-boosted Black Brant rocket from White Sands, NM, at 19:35 UT on 18 November 1997. It reached a maximum altitude ~ 330 km, and recorded spectrographic data for nearly 400 s. Spectroheliograms of NOAA active region 8108 (N21 E18) and slit spectra of quiet areas south and slightly west of this region were obtained in the first pointing position. Slit spectra of region 8108 and spectroheliograms of quiet areas to the northeast and southwest were obtained in the second. The absolute pointing of the instrument was derived by co-aligned the 3 arc min × 8.5 arc min He II 303.78 Å SERTS spectroheliograms with coordinated, full-disk He II 304 Å images from SOHO/EIT.

The 4 arc sec × 4 arc min CDS slit was used to raster spectral data over a 2 arc min × 4 arc min field of view. Spectral data were obtained in 16 wave bins,

Figure 2. SOHO/CDS Fe XVI 335 Å image of NOAA region 8108 showing the location of the SERTS-97 slit. The field of view is 2 arc min × 4 arc min, and solar north is at the top.

each covering 25 spectral pixels, for each 4 arc sec × 1.68 arc sec CDS spatial pixel. Fifteen of the wave bins are from NIS1, and one (covering the Si XI 303.32 and the He II 303.78 Å lines, seen in second order) is from NIS2. All wave bins include one or more lines observed by SERTS. We determined which CDS pixels mapped onto the SERTS slit, and used the coordinated spectral data to cross-calibrate CDS with SERTS (Thomas *et al.*, 1999) and to intercompare measured relative wavelength shifts.

SERTS spectra were recorded in 222 1.59-arc sec spatial pixels along the narrow slit, with each spatial pixel comprised of 2830 spectral elements. The 42 spatial pixels in the southernmost 67 arc sec (19%) of the slit in Figure 1 were used to obtain the average spectrum from the 'quiet surroundings'; this is the reference spectrum against which relative Doppler velocities in the active region are measured. For the active region, we treat only spectra from within the 221 arc sec segment (63% of the SERTS slit) within the CDS field of view. Figure 2 shows the Fe XVI 335.4 Å image of the active region from SOHO/CDS, with the SERTS slit superimposed for reference.

We examined the spatial variation of the He II 303.78 Å line centroid wavelength in laboratory spectra as well as in the quiet-Sun spectra from the first SERTS pointing position, and found that the wavelength varies systematically from one end of the slit to the other. The effect is somewhat more pronounced in the laboratory spectra than it is in the quiet-Sun spectra. We fit the systematic wavelength shift with a second order polynomial (see Figure 3), and used this polynomial to remove the systematic trend from the active region spectra in the second SERTS pointing position.

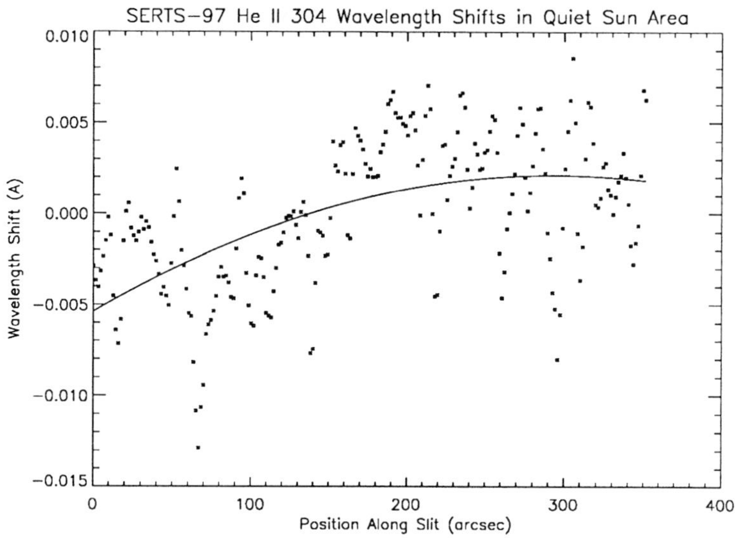

Figure 3. Quiet-Sun He II systematic wavelength shift along the SERTS-97 slit in pointing position 1, and the quadratic fit used to remove it.

4. Relative Wavelength Shifts

Gaussian fits to emission lines in the SERTS-97 spectra were carried out using a procedure similar to that described in Brosius *et al.* (1996) and Brosius, Davila, and Thomas (1998). Examples are shown in Figure 4. These fits yield the line centroid wavelength in Å, the total (integrated) intensity in erg cm^{-2} s^{-1} sr^{-1}, and the full width at half maximum intensity (FWHM) in mÅ, along with the 1-sigma standard deviation for each. Values of these quantities for five lines from the 'quiet surroundings' spectrum are given in Table I. All of these lines are relatively strong and free from known blends in the SERTS-97 spectra. T_{max} is the temperature that maximizes the fractional ion abundance in the available ionization equilibrium calculations. For iron, log T_{max} is derived from quadratic fits to the ionization equilibrium calculations of Arnaud and Raymond (1992); for other elements, it is derived from fits to similar calculations by Monsignori-Fossi (1992). The FWHM includes the instrumental width as well as both thermal and non-thermal contributions. The uncertainties listed for each wavelength are statistical only.

We fit Gaussian profiles to the lines listed in Table I for each of the 139 SERTS slit spatial pixels within the CDS field of view. Doppler shifts correspond to differences between fitted line profile centroids; Figure 5 shows examples of typical shifts observed. Figures 6–8 show the relative Doppler velocities derived between the 'quiet surroundings' and NOAA region 8108 as functions of position along the SERTS slit for He II 303.78 Å, Fe XIV 334.17 Å, and Fe XVI 335.40 Å. For reference we also display the total line intensity (in erg cm^{-2} s^{-1} sr^{-1}) in the top

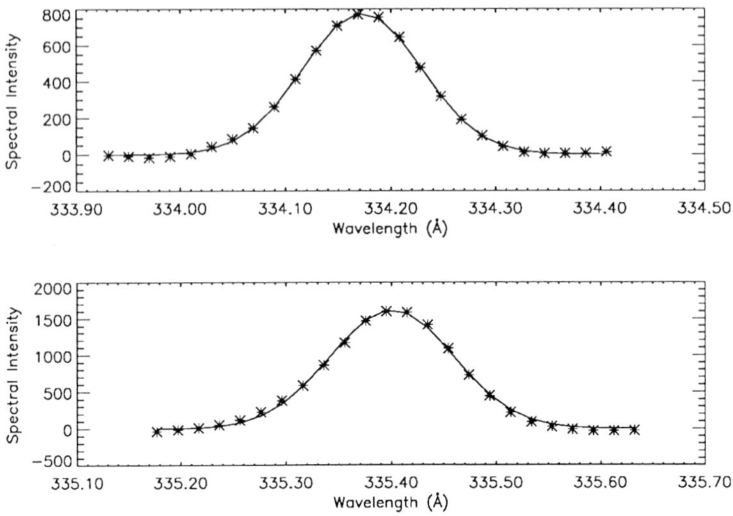

Figure 4. Gaussian fits to the Fe XIV 334.17 Å (*top*) and the Fe XVI 335.40 Å (*bottom*) lines from the SERTS-97 'quiet surroundings' spectrum. SERTS spectral data are shown as asterisks and the Gaussian fits are shown as solid lines. Spectral intensity is in erg cm^{-2} s^{-1} sr^{-1} Å^{-1}.

TABLE I

Spectral line data for the 'quiet surroundings' spectrum

Ion	log $T_{\max}$	Wavelength	Intensity	FWHM
He II	4.67	303.7822 ± 0.0005	8270 ± 929	144 ± 7
Si X	6.14	347.3984 ± 0.0011	102 ± 12	148 ± 8
Si XI	6.20	303.3188 ± 0.0012	643 ± 84	145 ± 8
Fe XIV	6.27	334.1728 ± 0.0004	110 ± 12	133 ± 7
Fe XVI	6.43	335.4006 ± 0.0007	239 ± 27	138 ± 7

frame of each figure. The first pixel in each figure corresponds to the southernmost SERTS slit pixel within the CDS field of view (see Figures 1 and 2). (The top frame in Figure 5 corresponds to $x \approx 75$ arc sec in Figure 7, and the bottom frame corresponds to $x \approx 190$ arc sec.) For He II the maximum measured relative red shift corresponds to a Doppler velocity $\sim +13$ km s^{-1}, and the maximum relative blue shift corresponds to a Doppler velocity ~ -3 km s^{-1}. For Fe XIV and Fe XVI the maximum relative red shifts correspond to Doppler velocities $\sim +10$ and $\sim +13$ km s^{-1}, respectively, while the maximum relative blue shifts correspond to Doppler velocities of ~ -10 and ~ -5 km s^{-1}, respectively. For Si X and Si XI the maximum relative red shifts correspond to Doppler velocities $\sim +19$ and $\sim +23$ km s^{-1}, and the maximum relative blue shifts correspond to ~ -14 and ~ -7.

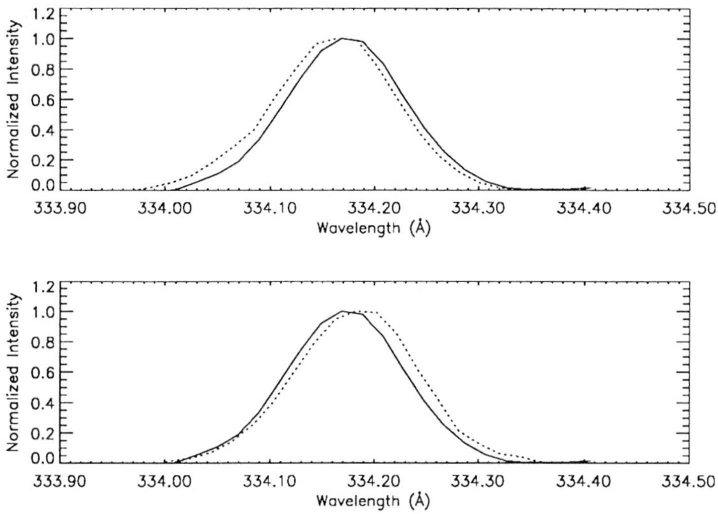

Figure 5. Fe XIV 334.17 Å line profile shifts between the 'quiet surroundings' (*solid*) and two different locations along the SERTS slit (*dotted*). We show normalized measured spectral data rather than Gaussian profile fits. The profile in the top frame is blue-shifted by a Doppler velocity corresponding to $\Delta V = -9.4 \pm 0.7$ km s^{-1}, and the profile in the bottom frame is red-shifted by a Doppler velocity corresponding to $\Delta V = +9.4 \pm 0.6$ km s^{-1}. Note that the entire profile moves symmetrically either blueward or redward of the reference profile.

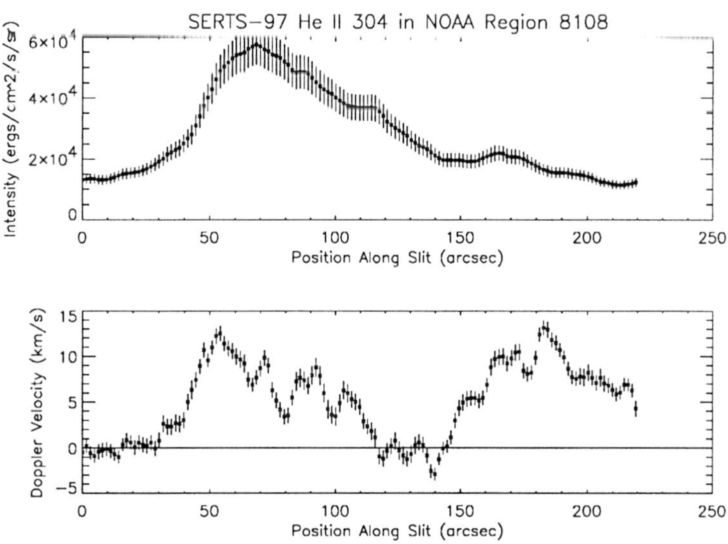

Figure 6. He II 303.78 Å line intensity (*top frame*) and relative Doppler velocity (*bottom frame*) as a function of position across region 8108.

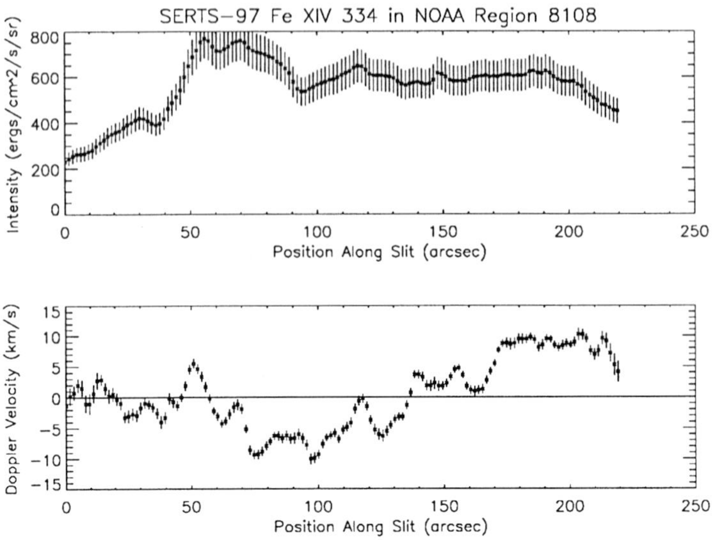

Figure 7. Fe XIV 334.17 Å line intensity (*top frame*) and relative Doppler velocity (*bottom frame*) as a function of position across region 8108.

Figure 9 shows relative wavelength shifts measured for He II 303.78 Å and Fe XVI 335.40 Å from spectra in the SOHO/CDS pixels that mapped onto the SERTS slit. Here, wavelength shifts are measured relative to (almost) average wavelengths rather than relative to the 'quiet surroundings' values. CDS wavelengths were corrected for spectral line tilt as described in Thompson and Brekke (1999). Note the qualitative similarity between these and the wavelength-shift variations seen in Figures 6 and 8.

5. Discussion

Brekke *et al.* (1997) argue that the Fe XVI resonance line at 335.41 Å, originally thought to be ideal for studying coronal mass motions with SOHO/CDS, is contaminated with a Mg VIII line at 335.23 Å and should not be used for coronal dynamical studies. Fortunately this is not a problem for SERTS since these two lines are well separated in SERTS spectra. The Mg VIII line is weak and not evident in either the active region or the 'quiet surroundings' spectra; however, it is easily visible in the quiet-Sun spectrum from the first pointing position.

Region 8108 appears to be divided into two different flow areas. The He II velocity pattern seen in Figure 6 suggests a southern half (from about 0 to 111 arc sec) wherein the maximum red shift corresponds to a Doppler velocity $\sim +12$ km s^{-1}, and a northern half (from about 111 to 221 arc sec) wherein the maximum red shift corresponds to a Doppler velocity $\sim +13$ km s^{-1}. Table II gives the (intensity-weighted) relative Doppler velocities for He II and five coronal ions in spatially

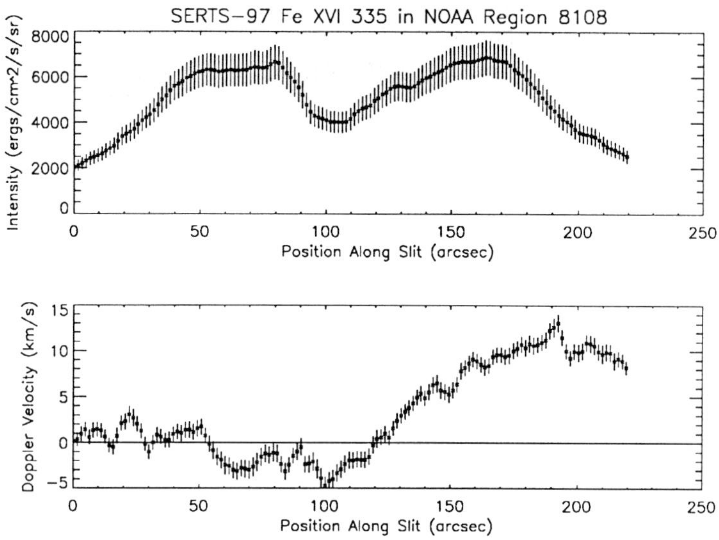

Figure 8. Fe XVI 335.40 Å line intensity (*top frame*) and relative Doppler velocity (*bottom frame*) as a function of position across region 8108.

TABLE II

Average relative Doppler velocities (km s^{-1}) in the southern and northern halves of region 8108

Ion	log T_{max}	ΔV_{south}	ΔV_{north}
He II	4.67	$+6.61 \pm 0.85$	$+4.34 \pm 0.77$
Si X	6.14	-2.24 ± 1.08	$+8.46 \pm 1.04$
Si XI	6.20	$+1.98 \pm 1.68$	$+7.31 \pm 1.37$
Fe XIII	6.20	-2.32 ± 1.05	$+5.25 \pm 1.00$
Fe XIV	6.27	-3.23 ± 0.57	$+3.50 \pm 0.51$
Fe XVI	6.43	-0.72 ± 0.88	$+6.88 \pm 0.82$

averaged spectra from the southern half and the northern half of the region. The hot coronal lines are predominantly red-shifted in the northern half and either blue-shifted or nearly unshifted in the southern half. This may be evidence that material flows up from the southern part of the region, and down into the northern part. An alternative explanation may be different velocities in adjacent loops, reflecting different heating and/or cooling histories. Teriaca, Erdelyi, and Doyle (1999) have interpreted red shifts of low to mid transition region lines and blue shifts of low coronal lines ($T > 6 \times 10^5$ K) observed in quiet-Sun spectra with SOHO/SUMER as signatures of nanoflares in the solar atmosphere.

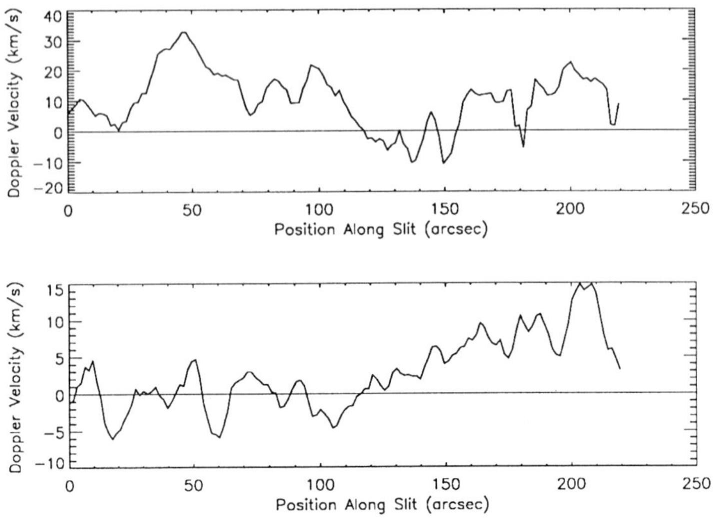

Figure 9. Relative Doppler velocities for He II 303.78 Å (*top*) and Fe XVI 335.40 Å (*bottom*) measured with SOHO/CDS, as a function of position across region 8108. Note that these exhibit the same general trends observed for the same lines with SERTS.

Although the CDS wavelength shifts are not measured relative to the same 'quiet surroundings' as those used for SERTS, both He II and Fe XVI show very similar trends to those seen with SERTS. Differences may be due to the different spatial resolutions. Also, although the CDS raster scanning was centered on the time of the SERTS flight, the scanning began before the flight and ended after it, so that the SERTS and CDS spectra are not everywhere exactly simultaneous.

6. Summary

We used SERTS-97 spectra to measure wavelength shifts of coronal emission lines in a solar active region *relative to* its quiet surroundings. This method removes any systematic uncertainties in the SERTS-97 wavelength calibration by cancellation, and circumvents the unavailability of reliable laboratory wavelengths for coronal emission lines by using lines from the 'quiet surroundings' spectrum as references. This means that we measure wavelength shifts of active region coronal emission lines not relative to a frame at rest (such as a laboratory frame), but relative to an area on the Sun which may itself be in motion. Statistically significant wavelength shifts between the spatially resolved active region slit spectra and the 'quiet surroundings' reference spectrum are found. For He II 303.78 Å the maximum measured relative red shift corresponds to a Doppler velocity $\sim +13$ km s^{-1}, and the maximum relative blue shift corresponds to a Doppler velocity ~ -3 km s^{-1}. For Fe XIV 334.17 Å the corresponding values are $\sim +10$ and ~ -10 km s^{-1}. For Fe XVI 335.40 Å the corresponding values are $\sim +13$ and ~ -5 km s^{-1}.

The active region appears to be divided into two different flow areas; hot coronal lines are predominantly red-shifted in the northern half and either blue-shifted or nearly un-shifted in the southern half. This may be evidence for material flows up from the southern part of the region, and down into the northern part. Qualitatively similar relative wavelength shifts (and flow patterns) are obtained with SOHO/CDS spectra that are found with SERTS spectra.

Acknowledgements

J. W. B. acknowledges NASA support through contracts W-91618 and NAS5-99145. J. M. D. and R. J. T. acknowledge NASA support for the SERTS program by RTOP grants from the Solar Physics Office of NASA's Space Physics Division. W. T. T. acknowledges NASA support through contract NAS5-32350.

References

Achour, H., Brekke, P., Kjeldseth-Moe, O., and Maltby, P.: 1995, *Astrophys. J.* **453**, 945.
Arnaud, M. and Raymond, J.: 1992, *Astrophys. J.* **398**, 394.
Behring, W. E., Cohen, L., and Feldman, U.: 1972, *Astrophys. J.* **175**, 493.
Behring, W. E., Cohen, L., Feldman, U., and Doschek, G. A.: 1976, *Astrophys. J.* **203**, 521.
Brekke, P.: 1993, *Astrophys. J.* **408**, 735.
Brekke, P., Hassler, D. M., and Wilhelm, K.: 1997, *Solar Phys.* **175**, 349.
Brekke, P., Kjeldseth-Moe, O., and Harrison, R. A.: 1997, *Solar Phys.* **175**, 511.
Brekke, P., Kjeldseth-Moe, O., Brynildsen, N., Maltby, P., Haugan, S. V. H., Harrison, R. A., Thompson, W. T., and Pike, C. D.: 1997, *Solar Phys.* **170**, 163.
Brosius, J. W., Davila, J. M., and Thomas, R. J.: 1998, *Astrophys. J. Suppl.* **119**, 255.
Brosius, J. W., Thomas, R. J., and Davila, J. M.: 1999, *Astrophys. J.* **526**, 494.
Brosius, J. W., Davila, J. M., Thomas, R. J., and Monsignori-Fossi, B. C.: 1996, *Astrophys. J. Suppl.* **106**, 143.
Brynildsen, N., Maltby, P., Brekke, P., Haugan, S. V. H., and Kjeldseth-Moe, O.: 1999, *Solar Phys.* **186**, 141.
Chae, J., Yun, H. S., and Poland, A. I.: 1998, *Astrophys. J. Suppl.* **114**, 151.
Dere, K. P.: 1982, *Solar Phys.* **77**, 77.
Dere, K. P., Bartoe, J. -D. F., and Brueckner, G. E.: 1984, *Astrophys. J.* **281**, 870.
Doschek, G. A., Feldman, U., and Bohlin, J. D.: 1976, *Astrophys. J.* **205**, L177.
Feldman, U., Cohen, L., and Doschek, G. A.: 1982, *Astrophys. J.* **255**, 325.
Fuhr, J. R., Martin, G. A., and Wiese, W. L.: 1988, *J. Phys. Chem. Ref. Data* **17**, Supp. No. 4.
Hassler, D. M., Rottman, G. J., and Orrall, F. Q.: 1991, *Astrophys. J.* **372**, 710.
Kaufman, V., and Edlen, B.: 1974, *J. Phys. Chem. Ref. Data* **3**, 825.
Kelly, R. L.: 1987, *J. Phys. Chem. Ref. Data* **16**, Supp. No. 1 (Finding List).
Kelly, R. L., and Palumbo, L. J.: 1973, *NRL Report 7599*, (Naval Research Laboratory, Washington, D.C.
Monsignori-Fossi, B. C.: 1992, unpublished.
Teriaca, L., Erdelyi, R., and Doyle, J. G.: 1999, presented at the Monterey Workshop.
Thomas, R. J., Davila, J. M., Thompson, W. T., Kent, B. J., and Hollandt, J.: 1999, *Bull. Am. Astron. Soc.* **31**, 850.
Thompson, W. T. and Brekke, P.: 1999, presented at the Monterey Workshop.

MODELING CORONAL LOOP ABUNDANCES: EFFECTS OF PARAMETER VARIATIONS ON OBSERVABLES

DAWN D. LENZ
Department of Astronomy and Astrophysics, University of Chicago, 5640 South Ellis Avenue, Chicago, IL 60637, U.S.A. (e-mail: lenz@oddjob.uchicago.edu)

(Received 13 December 1999; accepted 11 February 2000)

Abstract. Coronal abundance variations arise from and affect solar atmospheric processes such as coronal heating and structural dynamics. Lenz (1999) presented initial results of a numerical theoretical study of abundances and ion heating rates in static, steady-state coronal loops. We present here a closer investigation of two fundamental aspects of the physics of coronal abundances, relating modeling parameters to observables: (1) the effect of varying the abundances on the electron temperature and (2) the effect of varying the ion heating rate on abundances.

1. Introduction

Heavy ions, while minor by number in the corona, play a major role in coronal energetics. They are the dominant radiative coolants in the corona. They can also act as tracers of physical processes in the solar corona and wind, providing clues to heating and acceleration.

Observations of solar atmospheric abundances indicate species-dependent abundance variations. The FIP (First Ionization Potential) effect, in which higher-FIP elements show more severe depletion relative to photospheric abundances than lower-FIP elements, has been consistently observed in the solar corona and wind (Feldman, 1992, and references cited therein; Raymond *et al.*, 1997a), though the extent of the FIP effect may differ in different coronal structures and may in fact be absent in loops (Saba, 1995). The physical processes responsible for the FIP effect have not been conclusively identified.

Recent observations further indicate that the temperature and density of the corona are species-dependent (Raymond *et al.*, 1997a, b, 1998; Seely *et al.*, 1997); temperature and velocity are also known to vary with species in the solar wind (McKenzie, Banaszkiewicz, and Axford, 1995). Thus, the coronal heating problem may conceivably involve species-dependent heating.

A multi-species numerical model (Lenz, 1999; hereafter L99) for momentum and energy balance in coronal loops enables us to study abundance structuring in loops and to make observational predictions. Initial results obtained with this model were presented in L99; here, we report on further study of two fundamental issues, relating the model parameters to observables, that were not considered in

The physics of the solar corona and transition region. Part II.
Reprinted from Solar Physics **193,** 2000.

detail by L99: (1) the effect of varying the abundances on the electron temperature and (2) the effect of varying the ion heating rate on abundances.

2. The Model Problem

Our study of coronal abundances is based on a multi-species model of a static, steady-state coronal loop (L99). Here we summarize the model; a full description of the equations and assumptions is given in L99. The loop geometry is approximated by a thin semicircular cylindrical tube in which transverse variations in plasma properties are negligible; we thus confine attention to variations along the loop. The model allows for separate momentum and energy equations for a combined electron/proton/helium component and for each heavy ion species. The ion heating rate is a free parameter of the solution for each species heavier than He.

For reference in the discussion of our results, it is useful to note here the force balance equation,

$$\frac{\mathrm{d}p_s(s)}{\mathrm{d}l} + n_s m_s g - n_s Z_s e E = F_{\mathrm{therm},s} \tag{1}$$

(Burgers, 1969; Schunk, 1975, 1977), where s denotes the plasma species ($s = e$ denotes electrons, $s = p$ denotes protons, $s = i$ denotes any of the ions, $s = j$ denotes any ion heavier than He), n is the number density, m is the particle mass, p is the gas pressure, g is the component of gravity along the loop, e is the electron charge, Z_s is the number of ionized electrons ($Z_e \equiv -1$), E is the polarization electric field, and $F_{\mathrm{therm},s}$ is the thermal diffusive force. We also note the energy equation for ions heavier than He that is used in the model,

$$E_{C,j} = E_{H,j} \, , \tag{2}$$

where $E_{H,j}$ denotes the volumetric heating rate for the species and $E_{C,j}$ denotes the collisional energy exchange between the ion species and the rest of the plasma:

$$E_{C,j} \equiv \sum_{k=e,p,\mathrm{He}} \left\{ \frac{n_j m_j \nu_{jk}}{m_j + m_k} [3k(T_j - T_e)] \right\} \, , \tag{3}$$

where ν is the collision frequency, k is Boltzmann's constant, and T is the temperature. We take $E_{\mathrm{H},j}/n_j \equiv E^*_{\mathrm{H},j}$ to be uniform for each species.

3. Results

3.1. Abundance Variations

Abundance variations resulting from model calculations using a set of representative parameters are discussed in L99; here we summarize the main features of the

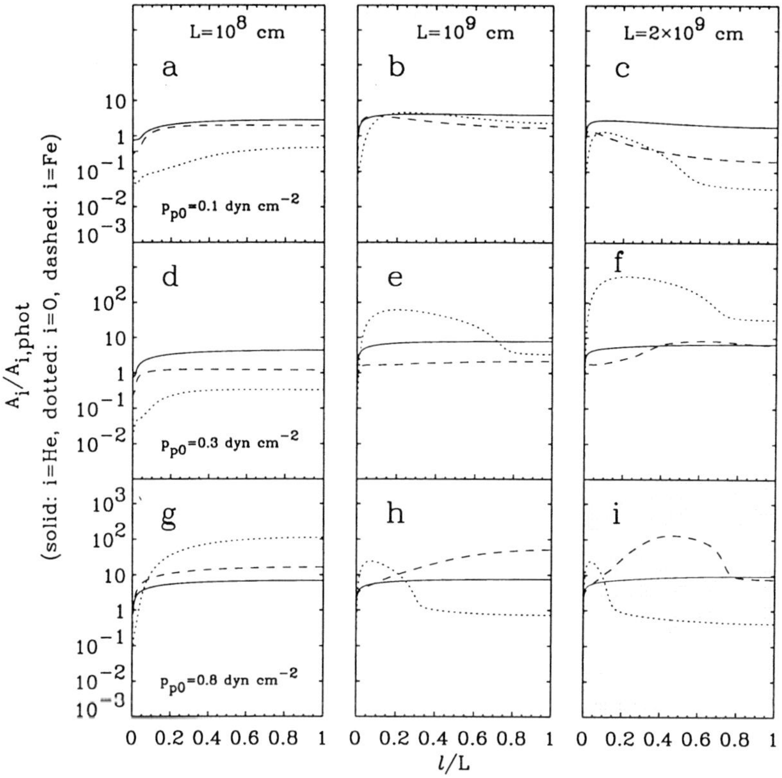

Figure 1. Abundances ($A_i \equiv n_i/n_p$) relative to photospheric abundances for He, O, and Fe for the indicated values of L and p_{p0}. Solution i is outside the applicable range of our model. See Section 3.1 for discussion.

results. Figure 1 shows the abundances relative to photospheric for i = He, O, and Fe for three loop semi-lengths and three base pressures. The abundance of a given ion depends on the relative importance of the thermal force (which is proportional to the square of the charges on the interacting particles) and the gravitational force. The thermal force causes abundance *enhancements* in all cases. The calculated models are sensitive to the footpoint densities.

The FIP effect is present in the assumed base abundances, but is not maintained in all the loops shown in Figure 1. Thus, if the FIP effect determines relative abundances low in the loop, the physics of our model can modify the FIP effect for some combinations of loop parameters.

3.2. Dependence of Electron Temperature on Abundances

The electron temperature is one of the most important coronal observables. We investigate the effects of variations in base abundances on the electron temperature using two different models: a model with uniform loop abundances (Serio *et al.*, 1981) and our variable-abundances model. Figure 2 shows the effects of

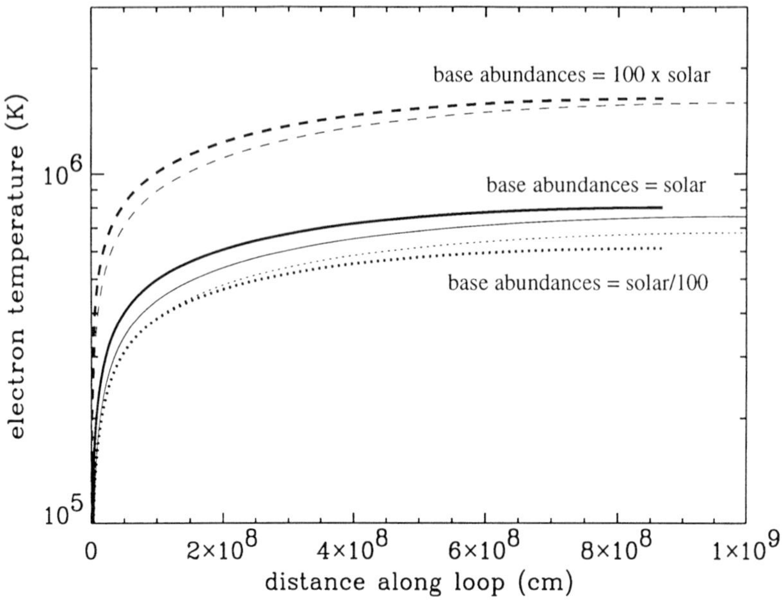

Figure 2. Electron temperature profiles as a function of base abundances: solar (*solid lines*), 100 × solar (*dashed lines*), and solar/100 (*dotted lines*). *Thick lines* are profiles in which abundances are uniform throughout the loop; *thin lines* are profiles in which abundances are allowed to vary with height in the loop. The loop semi-length is 10^9 cm and the base proton pressure is 0.1 dyn cm^{-2}. See Section 3.2 for discussion.

variations in base abundances on the electron temperature, for both uniform loop abundances and variable loop abundances. The electron temperature changes little with relatively large changes in the abundances, in spite of the strong dependence of the radiative loss function on the abundances. This is because there are no free parameters in the fundamental single-fluid scaling laws (Rosner, Tucker, and Vaiana, 1978); specifying the loop length and base pressure determines the loop-top temperature, independent of the details of the radiative loss and heat input. Energy balance of radiative losses with heating and conductive flux results in higher temperatures for higher abundances.

The electron temperature profiles are more sensitive to the base (boundary condition) abundances than to abundance variations along the loop; the temperature profile of a loop with uniform abundances does not differ much from that of a loop with varying abundances, as long as their base abundances are the same. This is because the impact of abundance variations is strongest near the base, where the temperature gradients are steepest and the abundance- and temperature-sensitive radiative losses thus vary the most; once the background structure flattens out, which happens a relatively short distance from the base compared to the scales on which the abundances can vary, abundance variations have less impact.

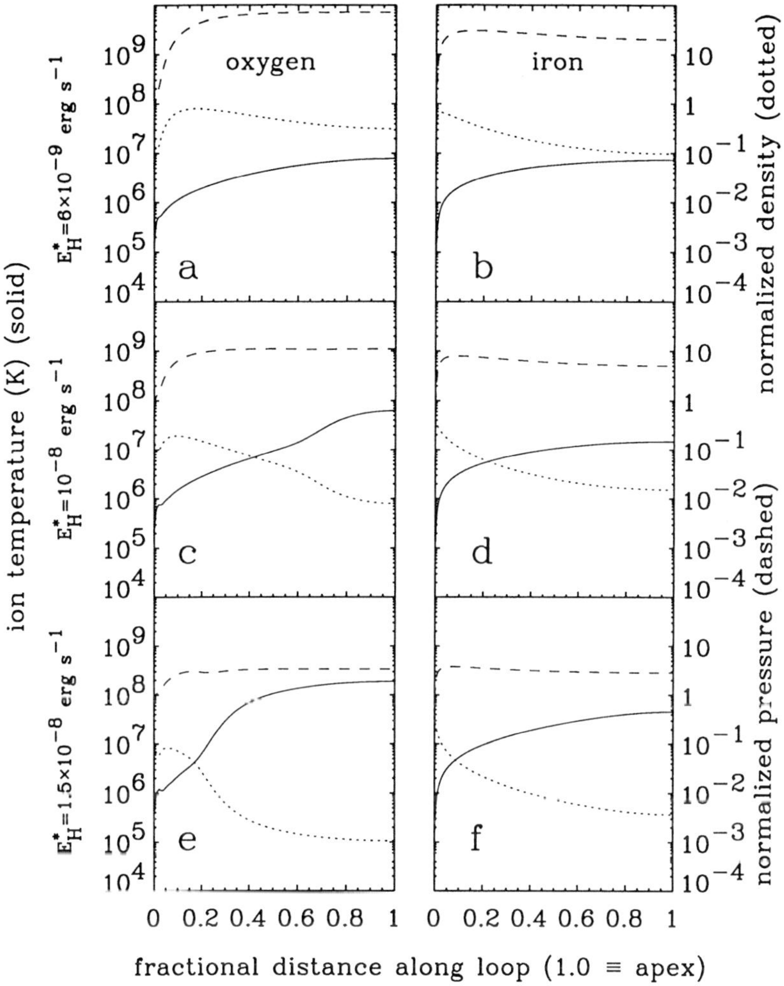

Figure 3. Temperature (*solid*), pressure (*dashed*), and density (*dotted*) profiles for oxygen (*left column*) and iron (*right column*) as a function of ion heating rate in a loop with semi-length 10^9 cm and base proton pressure 0.1 dyn cm^{-2}. See Section 3.3 for discussion.

3.3. Effect of Varying the Ion Heating Parameter on Abundances

To examine the effect of varying the ion heating rate on the ion structure, we have calculated ion profiles in which all loop parameters except the ion heating rate are held fixed. Figure 3 shows the effect of varying the ion heating rate on the ion temperature, pressure, and density structure for a loop with a semi-length of 10^9 cm and a base proton pressure of 0.1 dyn cm^{-2}. The systematic decrease of ion pressure and density with increasing ion temperature can be understood by considering the force balance equation (1) and the ion energy equation (2). In a region of shallow temperature gradient, i.e., the coronal part of a loop, the gravitational and electric forces nearly balance, so the plasma is nearly isobaric. Since the collision rates responsible for cooling the minor ions scale as the square of the ion charge, higher ion temperatures lead to nonlinearly lower ion pressures and correspondingly lower ion densities.

The ion temperature is bounded below by the electron temperature in the limit of no ion heating. This is because in our formulation of the problem, the ions can only cool by colliding with the protons, and hence cannot be cooler than the background fluid. Because of the strong dependence of the minor-ion temperature/density structure on the heating rate, there is a very narrow range of heating rates that result in physically reasonable minor-ion temperatures. Depending on the amount of gravitational settling of the minor ions, very high ion temperatures ($T \approx 10^9$–10^{10} K) can result because of the decreased (density-dependent) cooling of the ions by collisions with the major component (Figure 1(i)). These cases illustrate the limit of applicability of our formulation, since our assumptions (i.e., classical conductivity, negligible electron–ion collisional energy exchange) are invalid for such high minor-ion temperatures.

4. Summary

Building on the initial results reported in L99, we have used a numerical model of multi-species solar coronal loops to perform a closer study of two fundamental aspects of coronal abundance variation that relate modeling parameters to observables: (1) the effect of varying the abundances on the electron temperature and (2) the effect of varying the ion heating rate on abundances. We find that variation of abundances along the loop has little effect on the electron temperature; variation of the base abundances has a noticeable, but relatively weak, effect on the electron temperature. Higher ion heating rates lead to higher ion temperatures, lower ion pressures, and lower ion densities. There is a relatively narrow range of ion heating rates that yields physically reasonable minor ion temperature profiles. Further study will benefit from inclusion of effects such as ionization non-equilibrium and momentum deposition. The results also require rigorous testing with observations of coronal abundances.

Acknowledgements

This work is further development of part of Ph.D. thesis research conducted at the University of Chicago with Professor Robert Rosner, and was supported in part by NASA/SOHO UVCS and TRACE subcontracts to the University of Chicago and by the NSF/ATM Solar Terrestrial Research Program.

References

Burgers, J. M.: 1969, *Flow Equations for Composite Gases*, Academic Press, New York.

Feldman. U.: 1992, *Phys. Scr.* **46**, 202.

Lenz, D. D.: 1999, *Astrophys. J.* **517**, 497 (L99).

McKenzie, J. F., Banaszkiewicz, M., and Axford, W. I.: 1995, *Astron. Astrophys.* **303**, L45.

Raymond, J. C., *et al.*: 1997a, *Solar Phys.* **175**, 645.

Raymond, J., Suleiman, R., van Ballegooijen, A., and Kohl, J.: 1997b, in A. Wilson (ed.), *Correlated Phenomena at the Sun, in the Heliosphere and in Geospace, ESA SP-415, Proceedings of the 31st ESLAB Symposium*, ESA Publications, The Netherlands, p. 383.

Raymond, J. C., Suleiman, R., Kohl, J. L., and Noci, G.: 1998, *Space Sci. Rev.* **85**, 283.

Rosner, R., Tucker, W. H., and Vaiana, G. S.: 1978, *Astrophys. J.* **220**, 643.

Saba, J. L. R.: 1995, *Adv. Space Res.* **15**, 7, 13.

Schunk, R. W.: 1975, *Planetary Space Sci.* **23**, 437.

Schunk, R. W.: 1977, *Rev. Geophys. Space Phys.* **15**, 429.

Seely, J. F., Feldman, U., Schühle, U., Wilhelm, K., Curdt, W., and Lemaire, P.: 1997, *Astrophys. J.* **484**, L87.

Serio, S., Peres, G., Vaiana, G. S., Golub, L., and Rosner, R.: 1981, *Astrophys. J.* **243**, 288.

WAVES AND OSCILLATIONS IN THE CORONA

(Invited Review)

B. ROBERTS
Mathematical Sciences, University of St. Andrews, St. Andrews, Fife KY16 9SS, Scotland
(e-mail: bernie@mcs.st-and.ac.uk)

(Received 25 October 1999; accepted 2 February 2000)

Abstract. It has long been suggested on theoretical grounds that MHD waves must occur in the solar corona, and have important implications for coronal physics. An unequivocal identification of such waves has however proved elusive, though a number of events were consistent with an interpretation in terms of MHD waves. Recent detailed observations of waves in events observed by SOHO and TRACE removes that uncertainty, and raises the importance of MHD waves in the corona to a higher level. Here we review theoretical aspects of how MHD waves and oscillations may occur in a coronal medium. Detailed observations of waves and oscillations in coronal loops, plumes and prominences make feasible the development of *coronal seismology*, whereby parameters of the coronal plasma (notably the Alfvén speed and through this the magnetic field strength) may be determined from properties of the oscillations. MHD fast waves are refracted by regions of low Alfvén speed and slow waves are closely field-guided, making regions of dense coronal plasma (such as coronal loops and plumes) natural wave guides for MHD waves. There are analogies with sound waves in ocean layers and with elastic waves in the Earth's crust. Recent observations also indicate that coronal oscillations are damped. We consider the various ways this may be brought about, and its implications for coronal heating.

1. Introduction

The recent discovery by the Transition Region and Coronal Explorer (TRACE) spacecraft of oscillations in coronal loops (Aschwanden *et al.*, 1999; Nakariakov *et al.*, 1999) raises the twin prospects of a greater insight into the process of coronal heating and the use of such oscillations as a diagnostic tool for coronal seismology. The theoretical basis for coronal seismology has been known for some time, with the demonstration that magnetohydrodynamic (MHD) waves in coronal structures may be refracted by regions of low Alfvén speed (Uchida, 1968, 1970; Habbal, Leer, and Holzer, 1979) and thus wave guided by such regions and so able to form distinctive wave packets (Edwin and Roberts, 1983; Roberts, Edwin, and Benz, 1983, 1984). In fact the evidence for MHD waves in the corona has increased dramatically over the last few decades. Their existence in coronal loops had been inferred from modulated radio emission (see Roberts, Edwin, and Benz (1984) and the review by Aschwanden (1987)). Oscillations have been detected in prominences (e.g., Tsubaki, 1988) and given a theoretical description in terms of MHD waves (Joarder and Roberts, 1992, 1993; Oliver *et al.*, 1993; Roberts and

The physics of the solar corona and transition region. Part II.
Reprinted from Solar Physics **193,** 2000.

Joarder, 1994; Joarder, Nakariakov, and Roberts, 1997). Sunspots support a variety of wave phenomena that may be interpreted in terms of MHD waves (see Thomas and Weiss, 1992, and articles therein). Recently, coronal plumes have been shown to harbour waves, and these may be interpreted as slow MHD waves (DeForest *et al.*, 1997; DeForest and Gurman, 1998; Ofman, Nakariakov, and DeForest, 1999; Ofman, Nakariakov, and Sehgal, 1999). Finally, there are global disturbances in the form of EIT waves (Thompson *et al.*, 1999, 2000) and quakes (Kosovichev and Zharkova, 1999).

The fact that the corona is highly inhomogeneous complicates the theoretical description of MHD waves, giving rise to a variety of phenomena including phase mixing, resonant absorption and dispersive ducting. But the ability to measure oscillations in closed structures (prominences, coronal loops) and waves in open structures (plumes) establishes the basis for coronal seismology, suggested earlier on theoretical grounds (Uchida, 1970; Roberts, Edwin, and Benz, 1984; Roberts, 1986). Of course, waves are interesting in their own right, simply because they are there, but their use in coronal seismology, in accelerating the solar wind, and in coronal heating adds greatly to that interest.

Any description of MHD waves in coronal structures modelled as magnetic flux tubes leads to the occurrence of two particular speeds, in addition to the familiar sound speed c_s and Alfvén speed c_{A}. The slow magnetoacoustic speed c_t and the kink speed c_k are defined by

$$\frac{1}{c_t} = \left(\frac{1}{c_s^2} + \frac{1}{c_{\mathrm{A}}^2}\right)^{1/2}, \qquad c_k = \left(\frac{\rho_0 c_{\mathrm{A}}^2 + \rho_e c_{\mathrm{A}e}^2}{\rho_0 + \rho_e}\right)^{1/2}, \tag{1}$$

where ρ_0 and c_{A} are the plasma density and Alfvén speed inside the tube, and ρ_e and $c_{\mathrm{A}e}$ are their values in the tube's environment. The speed c_t is sub-sonic and sub-Alfvénic; the speed c_k is the mean Alfvén speed of the medium, intermediate between the Alfvén speed inside the tube and the Alfvén speed in the environment. In the low β coronal plasma, where sound speeds are much smaller than Alfvén speeds, the slow speed c_t is close to the sound speed inside the flux tube. If the magnetic field strength B_0 inside a coronal flux tube (loop) is comparable to the field strength B_e in the environment, so $B_0 \approx B_e$, but the plasma densities are very different, so that the inside of the tube is much denser than its surroundings (i.e., $\rho_0 \gg \rho_e$), then $c_k \approx \sqrt{2} c_{\mathrm{A}}$. Thus the kink speed may be some 41% larger than the Alfvén speed inside the tube. The speeds c_t and c_k are important also in studies of isolated photospheric flux tubes (e.g., Roberts, 1985, 1990a, b, 1992; Hollweg, 1990a; Ryutova, 1990).

2. Wave Equations

The observed complexity of the coronal plasma places severe demands on any theory to model its MHD behaviour. In wave studies it is common to consider a magnetically structured atmosphere in which the effects of gravity and flow are ignored (e.g., Roberts, 1981). A flux tube geometry is of particular interest. Ignoring the effect of twisted magnetic fields (both twisted fields and steady flows can in principle also be included, with interesting and important results; see Bennett, Roberts and Narain (1999) for a recent discussion of twisted tubes), we consider the equations of linear MHD describing a small amplitude perturbative flow $\mathbf{v}$ about an equilibrium magnetic field $\mathbf{B}_0$. The equilibrium field is taken to be unidirectional and aligned with the z-axis of a cylindrical coordinate system (r, θ, z), though it may vary in strength across the field; specifically $\mathbf{B_0} = B_0(r)\hat{\mathbf{z}}$ (for unit vector $\hat{\mathbf{z}}$ aligned with the z-axis).

This model aims to describe waves in a magnetically structured coronal atmosphere but it should be noticed that it ignores the effects of field-line curvature – coronal flux tubes are curved loops – and gravitational stratification. Curvature is likely to be of some importance in describing waves in loops; similarly, stratification may be of some significance, especially in any description of the coupling of a coronal loop to the dense photosphere–chromosphere. Ultimately, both these effects need to be incorporated into a description of the MHD modes, though they are not likely to be first-order effects.

The equilibrium plasma pressure $p_0(r)$ and density $\rho_0(r)$ of a radially structured magnetic atmosphere are such as to maintain total pressure balance: the sum of the plasma pressure p_0 and the magnetic pressure $B_0^2/2\mu$ is a constant,

$$\frac{\mathrm{d}}{\mathrm{d}r}\left(p_0(r) + \frac{B_0^2(r)}{2\mu}\right) = 0\,. \tag{2}$$

Where the field is strong, the plasma pressure is correspondingly reduced.

Small amplitude motions $\mathbf{v} = (v_r, v_\theta, v_z)$ about the equilibrium (2) satisfy the coupled wave equations (e.g., Roberts, 1981, 1991b)

$$\rho_0(r)\left(\frac{\partial^2}{\partial t^2} - c_{\mathrm{A}}^2(r)\frac{\partial^2}{\partial z^2}\right) v_r + \frac{\partial^2 p_T}{\partial r\,\partial t} = 0\,, \tag{3}$$

$$\rho_0(r)\left(\frac{\partial^2}{\partial t^2} - c_{\mathrm{A}}^2(r)\frac{\partial^2}{\partial z^2}\right) v_\theta + \frac{1}{r}\frac{\partial^2 p_T}{\partial \theta\,\partial t} = 0\,, \tag{4}$$

$$\left(\frac{\partial^2}{\partial t^2} - c_t^2(r)\frac{\partial^2}{\partial z^2}\right) v_z = -\left(\frac{c_s^2(r)}{c_{\mathrm{A}}^2(r) + c_s^2(r)}\right)\frac{1}{\rho_0(r)}\frac{\partial^2 p_T}{\partial z\,\partial t}\,. \tag{5}$$

The evolution of the total perturbed pressure $p_T(r, \theta, z)(= p + B_0 B_z/\mu$, for perturbed plasma pressure p and magnetic field component B_z in the direction of the applied field) is described by

$$\frac{\partial p_T}{\partial t} = \rho_0(r)c_{\mathrm{A}}^2(r)\frac{\partial v_z}{\partial z} - \rho_0(r)(c_s^2(r) + c_{\mathrm{A}}^2(r))\ \mathrm{div}\,\mathbf{v}\,. \tag{6}$$

Equation (6) results from a combination of the isentropic equation and the induction equation of ideal MHD. Equations (3)–(5) come from the components of the momentum equation in which the magnetic force has been expressed in terms of p_T.

One solution of the above system of equations is $p_T = 0$ with $v_r = v_z = 0$ and $\partial/\partial\theta = 0$; this describes a torsional Alfvén wave $\mathbf{v} = (0, v_\theta, 0)$ which satisfies the wave equation with Alfvén speed $c_{\rm A}(r)$. The torsional Alfvén wave exhibits phase mixing (Heyvaerts and Priest, 1983). We return to this topic later.

More generally, when $p_T \neq 0$, it is usual to describe pressure variations (and the associated motions) in terms of a Fourier representation, writing

$$p_T(r, \theta, z, t) = p_T(r) \exp i(\omega t + n\theta - k_z z) \tag{7}$$

for frequency ω, azimuthal number $n = 0, 1, 2, \ldots$, and longitudinal wavenumber k_z. The resulting equations may then be manipulated to yield the ordinary differential equation (e.g., Edwin and Roberts, 1983)

$$\rho_0(r)(k_z^2 c_{\rm A}^2(r) - \omega^2)\frac{1}{r}\frac{{\rm d}}{{\rm d}r}\left\{\frac{1}{\rho_0(r)(k_z^2 c_{\rm A}^2(r) - \omega^2)} r \frac{{\rm d}p_T}{{\rm d}r}\right\} = \left(m^2(r) + \frac{n^2}{r^2}\right) p_T \tag{8}$$

with m^2 given by

$$m^2(r) = \frac{(k_z^2 c_s^2 - \omega^2)(k_z^2 c_{\rm A}^2 - \omega^2)}{(c_s^2 + c_{\rm A}^2)(k_z^2 c_t^2 - \omega^2)} .$$

Equation (8) possesses singularities at $\omega^2 = k_z^2 c_{\rm A}^2$ and $\omega^2 = k_z^2 c_t^2$; these singularities generate the Alfvén and slow continua, respectively. The presence of these singularities is an indication of a number of interesting effects connected with the phenomenon of resonant absorption (Goedbloed, 1971, 1983; Goossens, 1991; Goossens and Ruderman, 1995; Tirry and Goossens, 1996), of particular interest in the question of coronal heating.

We consider the specific case of a flux tube of radius a, field strength B_0 and plasma density ρ_0 embedded in a magnetic environment with field strength B_e and plasma density ρ_e:

$$B_0(r) = \begin{cases} B_0, & r < a, \\ B_e, & r > a, \end{cases} \qquad \rho_0(r) = \begin{cases} \rho_0, & r < a, \\ \rho_e, & r > a. \end{cases} \tag{9}$$

The Alfvén, sound and tube speeds within the tube are $c_{\rm A}$, c_s and c_t, and their values in the external medium are $c_{{\rm A}e}$, c_{se} and c_{te}.

For the equilibrium (9) the differential equation (8) may be solved for p_T in terms of Bessel functions $J_n(n_0 r)$ in $r < a$ and $K_n(m_e r)$ in $r > a$, with the result (see McKenzie, 1970; Wentzel, 1979; Wilson, 1980; Spruit, 1982; Edwin and Roberts, 1983; Cally, 1986; Evans and Roberts, 1990)

$$\frac{n_0}{\rho_0(k_z^2 c_{\rm A}^2 - \omega^2)} \frac{J_n'(n_0 a)}{J_n(n_0 a)} = \frac{m_e}{\rho_e(k_z^2 c_{{\rm A}e}^2 - \omega^2)} \frac{K_n'(m_e a)}{K_n(m_e a)} , \tag{10}$$

where

$$n_0^2 = \frac{(\omega^2 - k_z^2 c_A^2)(\omega^2 - k_z^2 c_s^2)}{(c_s^2 + c_A^2)(\omega^2 - k_z^2 c_t^2)}, \qquad m_e^2 = \frac{(k_z^2 c_{se}^2 - \omega^2)(k_z^2 c_{Ae}^2 - \omega^2)}{(c_{se}^2 + c_{Ae}^2)(k_z^2 c_{te}^2 - \omega^2)}.$$

Equation (10) is the dispersion relation describing waves in a magnetic flux tube that is embedded in a magnetic environment; it applies when $m_e > 0$, corresponding to waves that are confined to the tube. The integer n that arises in the description of tube waves defines the geometry of the vibrating tube. The case $n = 0$ corresponds to the sausage wave (a symmetric pulsation of the tube, with the central axis of the tube remaining undisturbed). The case $n = 1$ describes the kink mode (involving lateral displacements of the tube, maintaining a circular cross-section, with the axis of the tube resembling a wriggling snake). Finally, there are fluting modes ($n \geq 2$), which ripple the boundary of the tube. Only the kink mode displaces the central axis of the vibrating tube.

The restriction $m_e > 0$ imposed on the flux tube dispersion relation means that the amplitude of a wave declines exponentially with radius $r (> a)$, so that far from the tube there is no appreciable disturbance. Inside the tube (for $r < a$) the disturbance is oscillatory if $n_0^2 > 0$ or non-oscillatory (evanescent) if $n_0^2 < 0$. Modes that inside the tube are oscillatory in r are called body waves. In the strongly magnetized coronal plasma, the modes are body waves. The waves are dispersive, the phase speed $c (= \omega / k_z)$ of a tube wave depending upon its wavelength, $2\pi / k_z$, through the combination $k_z a$.

The dispersion relation (10) possesses two sets of modes, namely fast and slow body waves. (There are no surface ($n_0^2 < 0$) waves.) The fast waves are strongly dispersive, and arise only if $c_{Ae} > c_A$; if $c_{Ae} < c_A$, then the fast waves are leaky and propagate energy away from the region of high Alfvén speed. Fast body waves, then, are trapped in regions of *low* Alfvén speed, typically corresponding to regions of *high* plasma density. Regions of low Alfvén speed in a strongly magnetized plasma provide wave guides for fast magneto-acoustic waves (Uchida, 1974; Habbal, Leer, and Holzer, 1979; Edwin and Roberts, 1983; Roberts, Edwin, and Benz, 1984).

There are close analogies between the behaviour of fast magnetoacoustic body modes in a strongly magnetized plasma and Love waves in the Earth's crust and Pekeris sound waves in an internal ocean layer; the fast sausage waves are sometimes referred to as magnetic Pekeris waves and the fast kink waves as magnetic Love waves (Edwin and Roberts, 1983; Roberts, Edwin, and Benz, 1984).

The slow waves (with phase speed between the slow speed c_t and the tube's sound speed) are only weakly dispersive. In a strongly magnetized plasma, their speed is close to the sound speed in the tube. The coronal loop provides an almost rigid tube for the one dimensional ducting of sound waves.

Following the treatment in Roberts, Edwin, and Benz (1984), we may discuss both standing and propagating modes. Standing waves occur in closed structures such as loops, provided the wave has had time to travel the entire length of the loop and back again; for shorter times, the wave has not reached the ends of the loop, where line-tying in the dense lower atmosphere causes reflection, and so the wave propagates freely as if the structure were open. Setting $k_z = N\pi/L$ in a tube (coronal loop) of length L gives the allowed wavelengths $2L/N$ $(= 2\pi/k_z)$ that fit along the closed tube. The integer $(N - 1)$ describes the number of nodes of the vibration along the axis of the tube, with $N = 1$ being the principal mode (and having no nodes within the tube). The period τ $(= 2\pi/\omega)$ of a *slow* wave $(c \approx c_t)$ standing in a loop is

$$\tau = \frac{2L}{Nc_t} \approx \frac{2L}{Nc_s} \,. \tag{11}$$

This generally produces long periods and requires quite some time to set up. For example, in a loop of length $L = 100$ Mm $(= 10^5$ km) with Alfvén speed $c_A = 10^3$ km s^{-1} and sound speed $c_s = 200$ km s^{-1} it requires some 10^3 s for a slow wave to traverse the extent of the loop and so set up a standing wave with this period (for the principal mode $N = 1$). This is so long that it may be difficult to observe in general, except perhaps in short loops where the period can be reduced to the order of 100 s.

Turning to fast waves, the *kink* mode has speed $c \approx c_k$ and so produces a standing wave of period τ given by

$$\tau = \frac{2L}{Nc_k} \,. \tag{12}$$

We may view this as a global mode of oscillation of a coronal flux tube, moving the whole tube in its vibration. The mode exists for all wavelengths and frequencies as a trapped oscillation of the tube. In the extreme of a coronal tube with internal field strength B_0 comparably with the field strength B_e in the environment but plasma density ρ_0 much greater than the density ρ_e in the environment, so that $B_0 \approx B_e$ and $\rho_0 \gg \rho_e$, the kink speed is $c_k \approx \sqrt{2}c_A$ and the period (12) of a standing kink mode becomes

$$\tau = \frac{\sqrt{2}L}{Nc_A} \,. \tag{13}$$

This is much shorter than the period of a slow wave; for the above illustration, we obtain $\tau = 140$ s for the principal mode $N = 1$, and about 100 s must elapse before the wave has had time to become aware of the loop ends.

When the mode number N is of low or moderate value (say, $N = 1$–10) and $a \ll L$ – the usual situation in a coronal loop – then the slow modes and the fast kink mode are the only magneto-acoustic waves to be ducted within a loop. The fast sausage mode *leaks* from the loop. This suggests that it is the kink mode that is most likely to be seen as a *standing* wave in a coronal loop.

The conclusion that for standing waves it is the fast kink mode or the slow modes that most likely arise (the sausage wave leaking for low $k_z a$) does not apply when waves are impulsively excited; then all modes are equally possible. The case of impulsively excited waves is in fact of interest. Impulsive excitation of waves may be due to a flare or reconnection or indeed any process that imparts a sudden motion or pressure variation into a loop. Our discussion considers the case of an open magnetic structure, which includes the case of waves in a loop provided the wave has not had time to reach the ends of the loop (and so feel the effect of line-tying). Of particular interest is the fast sausage wave, since this mode is leaky for small $k_z a$ and so unable to exist as a trapped standing wave (unless the mode number N is moderately large, of order L/a). However, when impulsively excited the wave is able to propagate freely along the tube and does so in accordance with its strong dispersive nature. The result is that the sausage wave produces a distinctive wave signature. The distinctive nature of this wave packet makes it potentially useful as a means of coronal diagnostics with high temporal resolution (to be contrasted with the use of standing waves for more moderate temporal resolution). The impulsively generated fast wave produces a signature which consists of three parts: a low-amplitude periodic phase, followed by a larger amplitude quasi-periodic phase, and finally a decay phase. The longest time scales in the motion are those in the periodic phase; the quasi-periodic phase produces high frequency oscillations. An estimate of the time scale τ^{pulse} in the periodic phase is provided by (Roberts, Edwin, and Benz, 1984)

$$\tau^{\text{pulse}} = \frac{2\pi a}{j_0 c_{\text{A}}} \left(1 - \frac{\rho_e}{\rho_0}\right)^{1/2}, \tag{14}$$

where j_0 (≈ 2.40) denotes the first zero of the Bessel function J_0. In this estimate we have taken the Alfvén speed to be much larger than the sound speed. For a tube of radius $a = 10^3$ km and Alfvén speed $c_{\text{A}} = 10^3$ km s^{-1}, with plasma density much higher than that in the environment (so $\rho_0 \gg \rho_e$), this produces a time scale of 2.6 s; such short time scales make for high temporal resolution seismology. In addition to the time scale τ^{pulse}, the decay time of a propagating disturbance carries seismic information too. The diagnostic use of the decay time of a propagating pulse is considered in Roberts, Edwin, and Benz (1984).

We end our discussion with a specific illustration, considering the loop oscillations observed by TRACE. Aschwanden *et al.* (1999) report the detection of oscillations in five loops, with lengths L extending from 90 to 160 Mm and periods τ in the band 258–320 s; oscillation amplitudes are in the range 2.0–5.6 Mm. The oscillations occurred following a flare that began at 14 July 1998, 12.55 UT, their onset apparently being due to a disturbance that propagated from the central flare site with a radial speed of about 700 km s^{-1}. Aschwanden *et al.* conclude that the loop oscillations are most likely standing kink modes. The same event is considered by Nakariakov *et al.* (1999) who report the oscillations of a loop, again interpreted as a standing kink mode, and pay particular attention to the damping of the oscillation. The oscillation has a displacement amplitude of 2 Mm (speed 47 km s^{-1}), with a cyclic frequency ν ($= \omega/2\pi$) of 3.9 mHz, corresponding to a period of 256 s. The loop is 130 Mm in length. The oscillation lasts for over 20 min; Nakariakov *et al.* estimate the decay time scale to be about 14.5 min (i.e., the oscillation amplitude falls by a factor of e in a time of about 14.5 min).

Consider, then, the application of the period formula (12) to the TRACE oscillations, interpreted as standing kink waves. The shortest loop observed by Aschwanden *et al.* is of length $L = 90$ Mm and has a period of 258 s; for the principal mode, Equation (12) produces a kink speed of almost 700 km s^{-1}. On the other hand, the longest loop observed is of length 160 Mm and has a period of 278 s, producing a kink speed of close to 1000 km s^{-1}. This difference in the kink speed may perhaps be understood as a consequence of differing density contrasts between a loop and its environment. For the loop oscillation observed by Nakariakov *et al.*, an interpretation in terms of a standing kink mode in its principal oscillation ($N = 1$) produces a kink speed of $c_k = 1014$ km s^{-1}. Assuming that $\rho_0 = 10\rho_e$ (the actual density ratio is not known) then produces a coronal loop Alfvén speed of $c_A = 752$ km s^{-1}.

3. Damping

The observed extraordinary damping reported in the coronal loop oscillations detected by TRACE raises a major puzzle. Why do the oscillations decay so rapidly, in only a relatively few periods? To answer our question is not easy because as yet there is no theory of coronal oscillations that takes full account of the many complexities of the corona. For example, the theory described above makes no allowance for loop curvature, for stratification or for continuous (as opposed to discrete) field or plasma inhomogeneity, nor is there any realistic modelling of the coupling of the coronal plasma to the denser chromospheric–photospheric plasma. Moreover, the medium has been treated as ideal. So a number of effects crowd in and complicate our picture with some of these effects leading to damping. By damping we mean that the oscillation decays in time; this may indicate a local physical damping with heating resulting or it may indicate a transfer of energy from the visible oscillations to some other agency (with no heating resulting). Both effects are quite possibly occurring. Here we consider briefly some of the effects that require evaluation if we are to explain the observed damping. A fuller discussion of this problem will ultimately require a detailed and probably numerical treatment of a realistic coronal loop. In the meantime, it behoves us to enquire as to what effects are most likely, and where possible to estimate the time scales they produce.

Non-ideal effects, such as viscous and ohmic damping, optically thin radiation, and thermal conduction, act to damp wave motion. Damping due to optically thin (or thick) radiation and thermal conduction have been assessed for an unbounded uniform medium (e.g., Bogdan and Knölker, 1989; Ibañez and Escalona, 1993) or for a slab geometry (Van der Linden and Goossens, 1991; Laing and Edwin, 1994), who generally find that fast waves are not readily damped, requiring at least 20 (and maybe several hundred) damping periods before suffering any appreciable decay. This effect therefore is too slow to explain the observed decay.

An effect that leads to the appearance of damping, but is in fact operating in an ideal medium, is wave leakage. Wave leakage is likely to occur when loop curvature is taken into account in any description of the oscillations or when there is leakage from the ends of the loop (as energy is ducted from the coronal part of the loop into the dense footpoints of the loop). Leakage of fast waves from loop footpoints has been considered by Berghmans and De Bruyne (1995), who conclude that only a very small amount of wave energy is transmitted through to the lower and denser medium from the corona, leading to a decay of order 100 times the transit time of a wave along the loop. This would amount to some 10^4 s, too long to explain the observed decay.

Consider then another contribution to wave leakage: loop curvature. In a coronal loop modelled as a straight magnetic flux tube, we have described a division of modes into sausage and kink oscillations and the global kink mode is trapped within the flux tube for all frequencies. In contrast, the fast sausage modes are trapped at high frequencies but leak at low frequencies. The trapping of the kink mode in a straight flux tube led Nakariakov *et al.* to presume that wave leakage does not occur. But in a *curved* loop, it seems likely that the precise distinction between kink and sausage modes in a straight flux tube is lost, since now a kink-like oscillation will tend to compress the plasma – as the oscillating loop tries to fit into the curved geometry – and so couple with a sausage-like mode. If this is indeed the case (a detailed calculation of the effect is not available), then the fact that the sausage mode leaks at low frequencies suggests that there would be a corresponding leakage in the kink mode, as energy in the kink oscillation is partly transferred into sausage-like oscillations which then leak away by radiating a wave out to infinity. In the absence of a detailed study it is difficult to assess the efficiency of this process, but it would seem that whatever the leakage in the kink mode, due to curvature coupling with the sausage mode, it would not be larger than the leakage experienced by the sausage mode in a straight tube. The leakage of the sausage mode in a straight tube may be estimated as follows (see Spruit, 1982; Cally, 1986). Taking (for simplicity) the plasma β to be zero (corresponding to zero sound speed), leakage of the sausage mode leads to a decay time τ_d which is of order $(L/a)^2$ times the period of the wave. For $a/L = 0.01$, this produces a decay time that is about 10^4 wave periods, and so much longer than the observed decay. This suggests that such leakage is probably not important, but until a proper study of curvature effects becomes available (so far only Cartesian arcade geometries have been examined; see Smith, Roberts, and Oliver, 1997; Terradas, Oliver, and Ballester, 1999) this conclusion must remain tentative.

Viscous and ohmic damping are generally considered to be very small in the coronal plasma, but the presence of density inhomogeneity on a small scale is expected to strongly enhance dissipation through the process of resonant absorption (see, for example, reviews by Hollweg, 1990b; Goossens, 1991; see also Mok and Einaudi, 1985; Ruderman, Tirry, and Goossens, 1995). We can most easily *illustrate* the effect by examining the process of *phase mixing* of an Alfvén wave (Heyvaerts and Priest, 1983). This is not to imply that phase mixing is the actual cause of the damping; but resonant absorption processes are operating in the system, and do so in a manner similar to phase mixing and it is phase mixing that is easier to describe mathematically. The Alfvén wave in an inhomogeneous plasma with Alfvén speed $c_A(x)$ structured in x (the Cartesian coordinate perpendicular to the applied magnetic field) propagates transverse motions $v(x, z, t)\hat{\mathbf{y}}$ according to the wave equation (Heyvaerts and Priest, 1983)

$$\frac{\partial^2 v}{\partial t^2} = c_A^2(x)\frac{\partial^2 v}{\partial z^2} + \nu\left(\frac{\partial^2}{\partial x^2} + \frac{\partial^2}{\partial z^2}\right)\frac{\partial v}{\partial t}, \tag{15}$$

where ν is the coefficient of kinematic viscosity (considered dominant over ohmic diffusivity). An approximate solution of Equation (15) is (cf., Heyvaerts and Priest, 1983; Roberts, 1988; Hood, Ireland, and Priest, 1997; Ruderman, 1999)

$$v = u(t)\sin(k_z z)\cos(k_z c_A(x) t), \tag{16}$$

where the amplitude $u(t)$ is given by

$$u(t) = u_0 \exp\{-[\tfrac{1}{2}k_z^2 \nu(t + \tfrac{1}{3}c_A'^2 t^3)]\}, \tag{17}$$

the prime ($'$) denoting the derivative of the Alfvén speed $c_A(x)$.

In a uniform medium (for which c_A is constant) the decay in amplitude $u(t)$ is exponential in time t, producing a decay time τ_d given by

$$\tau_d = \frac{2}{\nu k_z^2}. \tag{19}$$

For a standing wave with $k_z = N\pi/L$ this produces a decay time

$$\tau_d = \frac{2L^2}{\nu N^2 \pi^2}. \tag{20}$$

This time is very long; with a coronal viscosity of $\nu = 4 \times 10^9$ m^2 s^{-1} and a loop of length $L = 10^5$ km, the principal ($N = 1$) mode decays by a factor of e in 5×10^5 s (some 139 hr). However, in a magnetically structured plasma the simple exponential decay of a uniform medium persists only for a time of order $\sqrt{3}/c_A'$ after which the t^3-dependence in Equation (17) dominates, and phase mixing is established. In the corona, the transition time is likely to be short, about one second. So phase mixing effects are expected to dominate. Damping then proceeds on a time scale τ_d given by

$$\tau_d = \left[\frac{6}{\nu k_z^2 c_A'^2}\right]^{1/3}. \tag{21}$$

Write $c'_{\mathrm{A}} \approx c_{\mathrm{A}}/l$, corresponding to an Alfvén speed that varies spatially on a scale of order l. Then phase mixing produces damping (of the $N = 1$ mode) on a scale

$$\tau_d = \left[\frac{6L^2 l^2}{\nu \pi^2 c_{\mathrm{A}}^2} \right]^{1/3} . \tag{22}$$

Assuming a scale l of variation in the Alfvén speed c_{A} that is one-tenth of the loop length ($l = L/10$), then (22) produces a decay time scale in a loop of length $L = 10^5$ km and Alfvén speed $c_{\mathrm{A}} = 10^3$ km s^{-1} of 530 s (for $\nu = 4 \times 10^9$ m^2 s^{-1}), dramatically shorter than in a uniform medium and comparable with that observed.

We note that the damping time τ_d that operates when phase mixing is established may be rewritten in terms of the viscous Reynolds number $R = lc_{\mathrm{A}}/\nu$, based upon the Alfvén speed, c_{A}, and the spatial scale of inhomogeneity, l. Writing also $\tau_{\mathrm{A}} = l/c_{\mathrm{A}}$ for the transit time across the scale l of a wave moving with speed c_{A}, Equation (22) becomes

$$\tau_d = \left(\frac{6L^2}{\pi^2 l^2} \right)^{1/3} R^{1/3} \tau_{\mathrm{A}} \approx 0.85 \left(\frac{L^2}{l^2} \right)^{1/3} R^{1/3} \tau_{\mathrm{A}} . \tag{23}$$

Thus, for $l = L/10$ we obtain $\tau_d \approx 3.9 R^{1/3} \tau_{\mathrm{A}}$, and for $l = L/100$ (comparable with the radius of a loop) we obtain $\tau_d \approx 18.2 R^{1/3} \tau_{\mathrm{A}}$. These results of phase mixing compare reasonably well with scalings deduced from numerical studies of resonant absorption; for example, in a slab geometry (see Ofman, Davila, and Steinolfson, 1994, 1995; Erdélyi and Goossens, 1995), Nakariakov *et al.* (1999) quote $\tau_d \approx 32.6 R^{0.22} \tau_{\mathrm{A}}$.

However, it is in fact more appropriate to use the *shear* viscosity coefficient of a plasma than the compressional viscosity we have used, reflecting the strong anisotropy of the coronal plasma, and this has the effect of reducing the effective viscosity coefficient by a factor $(\omega_i \tau_i)^{-2}$ where ω_i is the ion gyro-frequency and τ_i is the ion collision time (see Braginskii, 1965; Hollweg, 1985, 1987; see also Ofman, Davila, and Steinolfson, 1994; Erdélyi and Goossens, 1995). Under coronal conditions, this results in a severe reduction in the effective value of ν. Consequently, the theoretically estimated damping time by this process is much longer than the observed damping time. If we turn the result around, and set τ_d to be the observed damping time, then an effective Reynolds number R of order 10^5 or 10^6 results (Nakariakov *et al.*, 1999), much smaller than the usual estimate of 10^{14} or so. Similar considerations apply if we consider ohmic damping instead.

Finally, following a suggestion by M. S. Ruderman (1999, private communication), we consider the time scale for which a global oscillation decays as a result of the development of localized oscillations (Alfvén waves) in regions where plasma inhomogeneity is strong. The importance of the effect (resonant absorption) in coronal physics was pointed out by Ionson (1978), following a related development for cold plasma oscillations by Sedlacek (1971). In an *incompressible* medium with plasma inhomogeneity on a spatial scale l, Lee and Roberts (1986; see also Rae and Roberts, 1981) have shown that global disturbances of a magnetic structured medium decay on a time scale τ_d given by

$$\tau_d = \frac{16c_k}{\pi l k_z^2 (c_{\mathrm{A}e}^2 - c_{\mathrm{A}}^2} . \tag{24}$$

Applied to a standing wave of period $2L/c_k$ in a loop length $L = 130$ M, the observed decay time is produced by a plasma inhomogeneity of scale $l \approx 15 \times 10^3$ km if we assume $c_{\mathrm{A}e}^2$ or $l \approx 6 \times 10^3$ km for $c_{\mathrm{A}e}^2 = 10c_{\mathrm{A}}^2$. These scales are broadly consistent with the cross-sectional size of a loop. Thus, the process seems capable of producing an oscillation decay that is comparable with that observed by Nakariakov *et al.* The resonance decay time (24) is not an indicaton of plasma heating but rather of a mode conversion process, as global oscillations transfer energy into localized Alfvén waves (Lee and Roberts, 1986). Of course, heating may also be involved, since small-scale oscillations are produced and so readily damped by non-adiabatic processes (e.g., viscosity) operating efficiently in the resonant absorption layers.

4. Concluding Remarks

The discovery of damped oscillations in coronal loops (Aschwanden *et al.*, 1999; Nakariakov *et al.*, 1999), interpreted as the global kink mode, raises the prospect of a proper development of coronal seismology in which both the period of the oscillation and its damping time provide important (and largely independent) sources of information about the corona. This seismic information offers the means of determining local conditions (e.g., magnetic field strength) within coronal structures (loops and plumes), a task that has not hitherto proved possible. Such progress is made possible by the use of high resolution observations combined with reasonable time resolution. The addition of high time resolution makes possible a finer scale of seismology, using rapid pulsations within a loop as well as its global mode of oscillation. Such a programme of development requires detailed observational studies combined with more sophisticated theoretical models of oscillations. Our current theoretical understanding is limited to simple models only but there is now a pressing need to extend our understanding to more realistic configurations with an appropriate incorporation of certain physical effects, including field line curvature, non-adiabaticity, plasma inhomogeneity, the presence of flows, and an appropriate representation of the coupling of a coronal structure to the dense medium at its base. Such a programme has in view the twin goals of coronal seismology and a greater understanding of coronal heating. The prospects for such theoretical developments, spurred by observations, are bright.

Acknowledgements

I am grateful for discussions with Drs Leon Ofman, Valery Nakariakov, and Michael Ruderman which helped clarify my thoughts on this topic. The invitation from Drs Carl Schrijver and Neal Hurlburt to attend the Monterey Workshop was much appreciated.

References

Aschwanden, M. J.: 1987, *Solar Phys.* **111**, 113.
Aschwanden, M. J., Fletcher, L., Schrijver, C. J., and Alexander, D.: 1999, *Astrophys. J.* **520**, 880.
Berghmans, D. and De Bruyne, P.: 1995, *Astrophys. J.* **453**, 495.
Bennett, K., Roberts, B., and Narain, U.: 1999, *Solar Phys.* **185**, 41.
Bogdan, T. J. and Knölker, M.: 1989, *Astrophys. J.* **339**, 579.
Braginskii, S. I.: 1986, in M. A. Leontovich (ed.), *Review of Plasma Physics*, Consultant Bureau, New York, **1**, 205.
Cally, P. S.: 1986, *Solar Phys.* **103**, 27.
DeForest, C. E. and Gurman, J. B.: 1998, *Astrophys. J.* **501**, L217.
DeForest, C. E. *et al.*: 1997, *Solar Phys.* **175**, 393.
Edwin, P. M. and Roberts, B.: 1983, *Solar Phys.* **88**, 179.
Erdélyi, R. and Goossens, M.: 1995, *Astron. Astrophys.* **297**, 597.
Evans, D. J. and Roberts, B.: 1990, *Astrophys. J.* **348**, 346.
Goedbloed, J. P.: 1971, *Physica* **53**, 412.
Goedbloed, J. P.: 1983, *Lecture Notes in Magnetohydrodynamics*, Rijnhuizen Report 83-145, Assoc. Euratom - FOM, pp. 289.
Goossens, M.: 1991, in E. R. Priest and A. W. Hood (eds.), *Advances in Solar System Magnetohydrodynamics*, Cambridge University Press, Cambridge, p. 137.
Goossens, M. and Ruderman, M. S.: 1995, *Physica Scripta* **T60**, 171.
Habbal, S. R., Leer, E., and Holzer, T. E.: 1979, *Solar Phys.* **64**, 287.
Heyvaerts, J. and Priest, E. R.: 1983, *Astron. Astrophys.* **117**, 220.
Hollweg, J. V.: 1985, *J. Geophys. Res.* **90**, 7620.
Hollweg, J. V.: 1987, *Astrophys. J.* **320**, 875.
Hollweg, J. V.: 1990a, in C. T. Russell, E. R. Priest, and L. C. Lee (eds.), *Physics of Magnetic Flux Ropes*, Washington, American Geophysical Union Mono. **58**, 23.
Hollweg, J. V.: 1990b, *Computer Phys. Rep.* **12**, 205.
Hood, A. W., Ireland, J., and Priest, E. R.: 1997, *Astron. Astrophys.* **318**, 957.
Ibañez, S. M. H. and Escalona, T.: : 1993, *Astrophys. J.* **415**, 335.
Joarder, P. S. and Roberts, B.: 1992, *Astron. Astrophys.* **256**, 264.
Ionson, J. A.: 1978, *Astrophys. J.* **226**, 650.
Joarder, P. S. and Roberts, B.: 1993, *Astron. Astrophys.* **277**, 225.
Joarder, P. S., Nakariakov, V. M., and Roberts, B.: 1997, *Solar Phys.* **176**, 285.
Kosovichev, A. G. and Zharkova, V. V.: 1999, *Nature* **393**, 317.
Laing, G. B. and Edwin, P. M.: 1994, *Solar Phys.* **151**, 191.
Lee, M. A. and Roberts, B.: 1986, *Astrophys. J.* **301**, 430.
McKenzie, J. F.: 1970, *J. Geophys. Res., Space Phys.* **75**, 5331.
Mok, Y. and Einaudi, G.: 1985, *J. Plasma Phys.* **33**, 199.
Nakariakov, V. M., Ofman, L., DeLuca, E. E., Roberts, B., and Davila, J. M.: 1999, *Science* **285**, 862.
Ofman, L., Davila, J. M., and Steinolfson, R. S.: 1994, *Astrophys. J.* **421**, 360.

Ofman, L., Nakariakov, V. M., and DeForest, C. E.: 1999, *Astrophys. J.* **514**, 441.
Ofman, L., Nakariakov, V. M., and Sehgal, N.: 1999, preprint.
Oliver, R., Ballester, J. L., Hood, A. W., and Priest, E. R.: 1993, *Astrophys. J.* **409**, 809.
Rae, I. C. and Roberts, B. 1981, *Geophys. Astrophys. Fluid Dyn.* **18**, 197.
Roberts, B.: 1981, *Solar Phys.* **69**, 27.
Roberts, B.: 1985, in E. R. Priest (ed.), *Solar System Magnetic Fields*, D. Reidel Publ. Co., Dordrecht, Holland, p. 37.
Roberts, B.: 1986, in A. Poland (ed.), *Coronal and Prominence Plasmas*, NASA Conf. Publ. 2442, p. 325.
Roberts, B.: 1988, in B. Buti (ed.), *Cometary and Solar Plasma Physics*, World Scientific, Singapore, p. 1.
Roberts, B.: 1990a, in E. R. Priest and V. Krishan (eds.), *Basic Processes on the Sun*, Kluwer Academic Publishers, Dordrecht, Holland, p. 159.
Roberts, B.: 1990b, in C. T. Russell, E. R. Priest, and L. C. Lee (eds.), *Physics of Magnetic Flux Ropes*, Washington, American Geophysical Union Mono. **58**, 113.
Roberts, B.: 1991a, *Geophys. Astrophys. Fluid Dynamics* **62**, 83.
Roberts, B.: 1991b, in E. R. Priest and A. W. Hood (eds.), *Advances in Solar System Magnetohydrodynamics*, Cambridge University Press, Cambridge, p. 105.
Roberts, B.: 1992, in J. H. Thomas and N. O. Weiss (eds.), *Sunspots: Theory and Observations*, Kluwer Academic Publishers, Dordrecht, Holland, p. 303.
Roberts, B. and Joarder, P. S.: 1994, in G. Belvedère, M. Rodono and G. M. Simnett (eds.), *Advances in Solar Physics*, Springer-Verlag, Berlin, p. 173.
Roberts, B., Edwin, P. M., and Benz, A. O.: 1983, *Nature* **305**, 688.
Roberts, B., Edwin, P. M., and Benz, A. O.: 1984, *Astrophys. J.* **279**, 857.
Ruderman, M. S.: 1999, *Astrophys. J.* **521**, 851.
Ruderman, M. S., Tirry, W., and Goossens, M.: 1995, *J. Plasma Phys.* **54**, 129.
Ryutova, M. P.: 1990, in J. O. Stenflo (ed.), 'Solar Photosphere: Structure, Convection and Magnetic Fields', *IAU Symp.* **138**, 229.
Sedlacek, Z.: 1971, *J. Plasma Phys.* **5**, 239.
Smith, J. M., Roberts, B., and Oliver, R.: 1997, *Astron. Astrophys.* **317**, 752.
Spruit, H. C.: 1982, *Solar Phys.* **75**, 3.
Terradas, J., Oliver, R., and Ballester, J. L.: 1999, *Astrophys. J.* **517**, 488.
Thomas, J. H. and Weiss, N. O.: 1992, *Sunspots: Theory and Observations*, Kluwer Academic Publishers, Dordrecht, Holland, 428.
Thompson, B. J. *et al.*: 1999, *Astrophys. J.* **517**, L151.
Thompson, B. J. *et al.*: 2000, *Solar Phys.*, this issue.
Tirry, W. J., and Goossens, M.: 1996, *Astrophys. J.* **471**, 501.
Tsubaki, T.: 1988, in R. C. Altrock (ed.), *Solar and Stellar Coronal Structures and Dynamics*, Sunspot: National Solar Observatory, p. 140.
Uchida, Y.: 1968, *Solar Phys.* **4**, 30.
Uchida, Y.: 1970, *Publ. Astron. Soc. Japan* **22**, 341.
Uchida, Y.: 1974, *Solar Phys.* **39**, 431.
Van der Linden, R. A. M. and Goossens, M.: 1991, *Solar Phys.* **131**, 79.
Wentzel, D. G.: 1979, *Astron. Astrophys.* **76**, 20.
Wilson, P. R.: 1980, *Astron. Astrophys.* **87**, 121.

CAN GRAVITATIONAL EFFECTS DAMP ALFVÉN WAVES?

J. F. MCKENZIE[1,2] and W. I. AXFORD[1]
[1] *Max-Planck-Institute für Aeronomie, D-37191 Katlenburg-Lindau, Germany*
[2] *School of Pure and Applied Physics, University of Natal, Durban, South Africa*

(Received: 27 October 1999; accepted: 18 February 2000)

Abstract. We show that Alfvén-gravity waves propagating in a gravitationally stratified atmosphere do not suffer damping as a result of the rate of working of the gravity drift current on the electric field of the waves. A self-consistent treatment involving conservation of total energy, Poynting's theorem, and the rate of working of the various drift currents on the electric field demonstrates that dissipation only arises from 'real' dissipative processes such as Ohmic heating or viscous effects, otherwise the system is adiabatic.

1. Introduction

It has been suggested (Khabibrakhmanov and Mullan, 1994, 1997, hereafter KM) that Alfvén waves propagating in an atmosphere stratified under gravity can be damped as a result of the rate of working of the gravity drift current (Heikilla and Axford, 1965) on the electric field of the waves, thus giving rise to some form of Joule-like heating. These authors have proposed this as a mechanism for heating the solar corona and it has been used in calculations of solar wind models (Cuseri *et al.*, 1999).

The purpose of this paper is to point out that Alfvén waves propagating under gravity do not in fact suffer damping and that no dissipative heating can arise from gravity drift currents in general. The confusion arises from an inconsistent approach to the overall energy budget of the system in which Poynting's theorem is not invoked. If the latter is used, together with the conservation of total energy of the system, the various drift currents from gravity, pressure gradient and inertia and their rate of working on the electric field, it follows that the system is adiabatic and that dissipation effects only arise from either Ohmic heating, if the plasma is not perfectly conducting, or viscous effects. In the collisionless case one merely retrieves the double adiabatic CGL theory and in the isotropic case conservation of entropy.

2. Outline of the Problem

KM claim that gravitational effects can give rise to the dissipation of Alfvén waves propagating in a stellar atmosphere. The basis of their argument goes as follows.

The physics of the solar corona and transition region. Part II.
Reprinted from Solar Physics **193,** 2000.

Consider an Alfvén wave propagating vertically upwards into a plasma atmosphere stratified hydrostatically according to

$$\frac{\mathrm{d}p}{\mathrm{d}z} = -\rho g \,. \tag{1}$$

The ambient magnetic field $\mathbf{B}_0$ is vertical (anti-parallel to gravity $\mathbf{g}$) and the wave magnetic field $\mathbf{B}_w$ is horizontal (in the x-direction, say) with an associated electric field $E_w = -V_A B_w$ in the y-direction in which V_A is the Alfvén speed. KM point out that in such a configuration the wave induces a gravity drift current $\mathbf{J}_\mathrm{g}$ given by

$$\mathbf{J}_\mathrm{g} = \frac{-\mathbf{B} \times (\rho \mathbf{g})}{B^2} = \frac{-B_w \hat{x} \times (-\rho g)\hat{z}}{B_0^2} = -\rho g \frac{B_w}{B_0^2} \hat{y} \,. \tag{2}$$

According to KM's reasoning there is therefore a 'Joule heating' associated with the wave given by

$$\mathbf{J}_\mathrm{g} \cdot \mathbf{E}_\mathrm{w} = \rho g V_\mathrm{A} \frac{B_w^2}{B_0^2} \,, \tag{3}$$

which is to be attributed to a rate of increase of the plasma energy, $\mathrm{d}W_p/\mathrm{d}t$, say. This (self-consistent) result is similar to that obtained by KM in considering the behaviour of a single particle influenced by an Alfvén wave travelling parallel to a uniform magnetic field in the presence of gravity. However, their approach is not self-consistently applicable to a plasma with a finite ion temperature and scale height.

How are we to interpret this result? After all we know that there is no dissipative heating associated with an Alfvén wave in the absence of viscous or resistive effects. Let us recall that a vertically propagating Alfvén wave is governed by the horizontal components of the momentum equation and Faraday's law, namely

$$\rho \frac{\partial u_w}{\partial t} = J_y B_0 = \frac{B_0}{\mu_0} \frac{\partial B_w}{\partial z} \,, \tag{4a}$$

$$\frac{\partial B_w}{\partial t} = \frac{\partial E_w}{\partial z} = B_0 \frac{\partial u_w}{\partial z} \,, \tag{4b}$$

in which we have used ampère in Equation (4a) and perfect conductivity ($\mathbf{E} = -\mathbf{u} \times \mathbf{B}$) in Equation (4b). These equations immediately yield the full wave Equations for the wave velocity u_w and magnetic field B_w,

$$\frac{\partial^2 u_w}{\partial t^2} = V_\mathrm{A}^2 \frac{\partial^2 u_w}{\partial z^2} \,, \tag{5a}$$

$$\frac{\partial^2 B_w}{\partial t^2} = \frac{\partial}{\partial z} \left(V_\mathrm{A}^2 \frac{\partial B_w}{\partial z} \right) \,, \tag{5b}$$

in which V_A is the Alfvén speed given by

$$V_A^2 = B_0^2/\mu_0\rho \ . \tag{5c}$$

Note that the phase speed of the wave is the local Alfvén speed which varies with height. The full wave equations can be solved exactly for an exponential atmosphere (Ferraro and Plumpton, 1966), or in general by using the WKB method. The time-averaged $\mathbf{J} \cdot \mathbf{E}$ associated with this wave is given by

$$\begin{aligned} \langle J_y E_y \rangle &= \left\langle \frac{1}{\mu} \frac{\partial B_w}{\partial z} u_w B_0 \right\rangle \\ &= \left\langle \rho u_w \frac{\partial u_w}{\partial t} \right\rangle \\ &= 0 \end{aligned} \tag{6}$$

for any time harmonic wave. J_y and E_y are in quadrature! Thus there is no 'Joule heating' associated with an Alfvén wave. This would appear to contradict the KM result (3).

The paradox is resolved by observing that there are drift currents associated with the pressure gradient ($\mathbf{J_p}$) and inertia ($\mathbf{J_i}$) as well as with gravity. These are given by

$$\begin{aligned} \mathbf{J_p} &= \frac{\mathbf{B} \times \nabla p}{B^2} = \frac{B_w \hat{x} \times (\mathrm{d}p/\mathrm{d}z)\hat{z}}{B_0^2} \\ &= \frac{-\mathrm{d}p}{\mathrm{d}z} \frac{B_w}{B_0^2} \hat{y} = \rho g \frac{B_w}{B_0^2} \hat{y} \ , \end{aligned} \tag{7a}$$

$$\begin{aligned} \mathbf{J_i} &= \frac{\mathbf{B} \times \rho(D\mathbf{u}/Dt)}{B^2} = \frac{B_0 \hat{z} \times \rho(\partial u_w/\partial t)\hat{x}}{B_0^2} \\ &= \frac{\rho}{B_0} \frac{\partial u_w}{\partial t} \hat{x} \ . \end{aligned} \tag{7b}$$

Clearly the 'Joule cooling' $\mathbf{J_p} \cdot \mathbf{E}$ associated with the pressure gradient drift current exactly compensates the 'Joule heating' $\mathbf{J_g} \cdot \mathbf{E}$ of the gravity drift current by virtue of hydrostatic pressure balance (Equation (1)) leaving only the inertial contribution

$$\langle \mathbf{J_i} \cdot \mathbf{E} \rangle = \left\langle \rho u_w \frac{\partial u_w}{\partial t} \right\rangle = 0 \tag{8}$$

in accordance with the classical result (6) above. Thus the KM mechanism for the dissipation of Alfvén waves is an illusion stemming, as it does, from not placing all the drift currents (pressure and inertial as well as gravity) on the same footing. It is straightforward to extend this result to an Alfvén wave propagating in a stellar wind in the presence of a gravitating mass. However, in the next section we provide a general proof that gravity and the other drift currents do not give rise to dissipation in any circumstances.

3. Total Energy Conservation

Conservation of total energy in a (electron-proton) plasma may be expressed in the usual divergence form (Braginskii, 1965; Kulsrud, 1989)

$$\frac{\partial W}{\partial t} + \operatorname{div} \mathbf{F} = 0 \,. \tag{9}$$

This equation holds irrespective of whether the plasma is collisional or collisionless, or high- or low-β. The energy density W and energy flux $\mathbf{F}$ may be written, in the non-relativistic approximation, as

$$W = \tfrac{1}{2}\rho u^2 + e + \rho\phi + B^2/2\mu_0 \,, \tag{10a}$$

$$\mathbf{F} = \rho\mathbf{u}\left(\frac{1}{2}u^2 + \phi\right) + e\mathbf{u} + \underline{\Pi}\cdot\mathbf{u} + \mathbf{q} + \mathbf{E}\times\mathbf{B}/\mu_0 \,, \tag{10b}$$

where ρ is the mass density and $\mathbf{u}$ the bulk velocity (defined in the usual way by the zeroth and first moments of the proton distribution function $f_\phi(t, \mathbf{r}, \mathbf{v})$), ϕ is the gravitational potential ($\mathbf{g} = -\nabla\phi$) and $\mathbf{E} \times \mathbf{B}/\mu_0$ is the Poynting vector. The internal energy e, the stress tensor $\underline{\Pi}$ (including pressure and viscous terms) and the heat flux $\mathbf{q}$ are related in the standard way to moments of the particle distribution functions. The particular forms they take play no role in this development but see below for the cases of a collisional isotropic plasma and a collisionless anisotropic pressure tensor. The distribution functions obey the Boltzman equation or, if one wishes, the guiding centre Vlasov equation (see Kulsrud, 1989; KM).

The associated continuity (zeroth moment) and momentum (first moment of Vlasov) equations are, respectively

$$\frac{\mathrm{D}\rho}{\mathrm{D}t} + \rho \operatorname{div} \mathbf{u} = 0 \,, \tag{11}$$

$$\rho\frac{\mathrm{D}\mathbf{u}}{\mathrm{D}t} = -\operatorname{div}\underline{\Pi} + \mathbf{J}\times\mathbf{B} + \rho\mathbf{g} \,. \tag{12}$$

Electron inertia has been neglected and the plasma is regarded as strictly charge neutral, both of which assumptions adequately cover phenomena which change on time scales greater than either the lower hybrid gyroperiod or the electron plasma period.

By using Poynting's theorem (in the non-relativistic approximation)

$$\frac{\partial}{\partial t}(B^2/2\mu_0) + \operatorname{div}(\mathbf{E}\times\mathbf{B}/\mu_0) = -\mathbf{J}\cdot\mathbf{E} \,, \tag{13}$$

the energy equation may be cast in the form

$$\frac{De}{Dt} + e \operatorname{div}\mathbf{u} + \operatorname{div}\mathbf{q} + (\operatorname{div}(\underline{\Pi}\cdot\mathbf{u}) - \mathbf{u}\cdot\operatorname{div}\underline{\Pi}) =$$

$$= \mathbf{u} \cdot \left(\rho \mathbf{g} - \operatorname{div} \underline{\Pi} - \rho \frac{D\mathbf{u}}{Dt} \right) + \mathbf{J} \cdot \mathbf{E} \,. \tag{14}$$

The first set of terms on the right-hand side of (14) may be thought of as the rate of working of the various forces (gravity, stress and inertia), while the second term $\mathbf{J} \cdot \mathbf{E}$ is the power expended by the current on the 'impressed' electric field. How do we calculate this term? Firstly, we note that in a perfectly conducting plasma the electric field is frozen into the electrons, that is to say

$$\mathbf{E} = -\mathbf{u_e} \times \mathbf{B} \,, \tag{15}$$

where $\mathbf{u_e}$ is the electron (bulk) velocity. One can of course allow for finite conductivity by including an Ohmic term $\mathbf{J}/\sigma$ on the right-hand side of (15). Secondly, the current $\mathbf{J}$ may be expressed as the sum of the various drift currents (see, e.g., Chen, 1977) by taking the vector product of the momentum equation with $\mathbf{B}$ to yield

$$\mathbf{J}_{\perp} = -\frac{\mathbf{B} \times (\rho \mathbf{g})}{B^2} + \frac{\mathbf{B} \times \rho (\mathrm{D}\mathbf{u}/\mathrm{D}t)}{B^2} + \frac{\mathbf{B} \times \operatorname{div} \underline{\Pi}}{B^2} \,. \tag{16}$$

Thus the rate of working of the gravity drift current (first term on the right-hand side of (16)) on $\mathbf{E}$ is

$$\mathbf{J_g} \cdot \mathbf{E} = -\rho \mathbf{u_e} \cdot \mathbf{g} \,. \tag{17a}$$

Similarly the inertial and stress drift currents yield

$$\mathbf{J_i} \cdot \mathbf{E} = \rho \mathbf{u_e} \cdot \frac{\mathrm{D}\mathbf{u}}{\mathrm{D}t} \,, \tag{17b}$$

$$\mathbf{J_p} \cdot \mathbf{E} = \mathbf{u_e} \cdot \operatorname{div} \underline{\Pi} \,. \tag{17c}$$

Hence the right-hand side of the re-cast energy equation (14) becomes

$$(\mathbf{u} - \mathbf{u_e}) \cdot \left(\rho \mathbf{g} - \rho \frac{\mathrm{D}\mathbf{u}}{\mathrm{D}t} - \operatorname{div} \underline{\Pi} \right) = -\frac{\mathbf{J}}{en} \cdot (\mathbf{J} \times \mathbf{B}) = 0 \,, \tag{18}$$

which demonstrates that the $\mathbf{J} \cdot \mathbf{E}$ is compensated exactly by the first set of terms on the right-hand side of (14). The only dissipative processes are those associated with viscous stresses (through $\underline{\Pi}$), heat conduction ($\mathbf{q}$), and Ohmic terms (J^2/σ). The energy equation then takes the form appropriate to the transport of internal energy, namely

$$\frac{De}{Dt} + e \operatorname{div} \mathbf{u} + \operatorname{div} \mathbf{q} + \left(\operatorname{div} (\underline{\Pi} \cdot \mathbf{u}) - \mathbf{u} \cdot \operatorname{div} \underline{\Pi} \right) = J^2/\sigma \,. \tag{19}$$

It should be stressed that dissipation arising from the stress tensor can be either of the classical fluid type or of the kinetic type in which Landau damping plays a role (see Stubbe, 1990, on kinetic transport theory in which kinetic expressions for the transport coefficients are developed). If the collision time is very much greater than the characteristic fluid time, which in turn is much greater than the ion gyroperiod, one can make contact with the classical collisionless formulation (e.g., Kulsrud, 1989), in which we put $\mathbf{q} = 0$, $\sigma = \infty$, and the stress tensor and internal energy become

$$\underline{\Pi} = p_\perp \mathbf{I} + (p_\parallel - p_\perp)\hat{\mathbf{n}}\hat{\mathbf{n}} , \tag{20a}$$

$$e = p_\parallel/2 + p_\perp , \tag{20b}$$

where $\mathbf{I}$ is the unit dyadic, $\hat{\mathbf{n}}$ is the unit vector along $\mathbf{B}$. The perpendicular ($p_\perp$) and parallel ($p_\parallel$) pressures are given by

$$p_\perp = \sum_s m_s \int f_s \frac{v_\perp^2}{2} \, \mathrm{d}^3 v , \tag{21a}$$

$$p_\parallel = \sum_s m_s \int f_s (v_\parallel - \mathbf{u} \cdot \hat{\mathbf{n}})^2 \, \mathrm{d}^3 v , \tag{21b}$$

where $f_s(t, \overline{r}, \mathbf{v})$ is governed by the guiding centre Vlasov equation as described by Kulsrud or KM. The energy equation (19) then assumes its adiabatic (dissipationless) form

$$\frac{\mathrm{D}e}{\mathrm{D}t} = \frac{(e + p_\parallel)}{\rho} \frac{D\rho}{Dt} - \frac{(p_\parallel - p_\perp)}{B} \frac{\mathrm{d}B}{\mathrm{d}t} , \tag{22}$$

in which we have used

$$\mathrm{div}\,(\underline{\Pi} \cdot \mathbf{u}) - \mathbf{u} \cdot \mathrm{div}\,\underline{\Pi} = \Pi_{ij} \frac{\partial u_j}{\partial x_i} = p_\perp \mathrm{div}\,\mathbf{u} + (p_\parallel - p_\perp) \frac{\partial u_\parallel}{\partial x_\parallel} \tag{23}$$

and

$$\frac{1}{B} \frac{\mathrm{d}B}{\mathrm{d}t} = \frac{\partial u_\parallel}{\partial x_\parallel} + \frac{1}{\rho} \frac{\mathrm{D}\rho}{\mathrm{D}t} \tag{24}$$

with the latter following from Faraday. The above 'adiabatic' result along with conservation of magnetic moment

$$\frac{\mathrm{D}}{\mathrm{D}t} \left(\frac{p_\perp}{\rho B} \right) = 0 \tag{25}$$

constitute the well-known double adiabatic theory of CGL (see, e.g., Kulsrud, 1989). This is a full blown non-linear result which is quite independent of the solution of the Vlasov equation for the distribution function or any microscopic/kinetic processes associated with its distortion or evolution in the presence or absence of waves. Thus the KM mechanism for damping Alfvén waves in a gravitationally stratified atmosphere (which is merely a special case of the foregoing general development) is an illusion arising from not placing all the drift currents on an equal footing and an inconsistent approach to total energy conservation.

Finally, we note that the isotropic case ($p_\parallel = p_\perp$) may be cast in the familiar form of conservation of entropy,

$$\frac{\mathrm{D}S}{\mathrm{D}t} = 0 , \tag{26}$$

where $S = \ln(p/\rho^{\gamma})$, $\gamma = \frac{5}{3}$.

4. Conclusions

The idea put forward by KM that Joule-like heating arising from the rate of working of the current induced by gravity drift on a wave electric field can produce non-adiabatic heating is shown to be erroneous, essentially because it is implicitly based on an inconsistent account of the energy budget of the system as a whole and a failure to recognize that all the drift currents, including the stress and inertial currents, should be placed on the same footing.

If one wishes to search for possible Landau damping effects a proper dispersion equation for propagation in a non-uniform medium should first be sought: this was not attempted by KM. It is not correct to simply assume that a transformation to the frame of the wave is inertial. The magnitude of the vertical bulk speed of the plasma must be determined by the full MHD equations: in fact in the case under consideration the vertical speed should be zero (a hydrostatic atmosphere, e.g., Axford and McKenzie, 1997, Figure 2). One should expect that a proper treatment along such lines will give the same result as that for Landau damping of Alfvén waves travelling parallel to a magnetic field in a uniform medium, namely that there is no damping because in the frame of the wave there is no electric field (Oraevsky, 1989).

References

Axford, W. I. and McKenzie, J. F.: 1997, *Cosmic Winds*, University of Arizona Press, Arizona, p. 31.

Braginskii, S. I.: 1965, in M. A. Leontovich (ed.), *Reviews of Plasma Physics*, Vol. 1, Consultants Bureau, New York, p. 605.

Chen, F. F.: 1977, *Introduction to Plasma Physics*, Plenum Press, New York, p. 60.

Cuseri, I., Mullan, D., Noci, G., and Poletto, G.: 1999, *Astrophys. J.*, (in press).

Ferraro, V. C. A. and Plumpton, C.: 1966, *An Introduction to Magneto Fluid Dynamics*, Clarendon Press, New York.

Heikilla, W. J., and Axford, W. I.: 1965, in C. O. Hines, I. Paghis, T. R. Hartz, and J. A. Fejer (eds.), *Physics of the Earth's Upper Atmosphere*, Chapter 5, Prentice Hall, New York, p. 117.

Hollweg, J. V.: 1997, *Astrophys. J.* **488**, 895.

Khabibrakhmanov, I. K. and Mullan, D. J.: 1994, *Astrophys. J.* **430**, 814.

Khabibrakhmanov, I. K. and Mullan, D. J.: 1997, *Astrophys. J.* **488**, 898.

Kulsrud, R. M.: 1989, in A. A. Galeev and R. N. Sudan (eds.), *Basic Plasma Physics*, Ch. 1.4, North-Holland Personal Library, Amsterdam, p. 14.

Oraevsky, V. N.: 1989, in A. A. Galeev and R. N. Sudan (eds.), *Basic Plasma Physics*, chapter 2.2, North-Holland Personal Library, Amsterdam, p. 103.

Stubbe, P.: 1990, *Phys. Fluids* **B2 (1)**, 22.

OBSERVATIONS OF THE 24 SEPTEMBER 1997 CORONAL FLARE WAVES

B. J. THOMPSON[1], B. REYNOLDS[2], H. AURASS[3], N. GOPALSWAMY[4], J. B. GURMAN[5], H. S. HUDSON[6], S. F. MARTIN[7] and O. C. ST. CYR[8]

[1]*NASA Goddard Space Flight Center, Greenbelt, MD, U.S.A.*

[2]*Perth, Australia*

[3]*Astrophysikalisches Institut Potsdam, Potsdam, Germany*

[4]*Catholic University of America, Washington, DC, U.S.A.*

[5]*NASA Goddard Space Flight Center, Greenbelt, MD, U.S.A.*

[6]*Solar Physics Research Corp., Tucson, AZ, U.S.A.*

[7]*Helio Research, La Crescenta, CA, U.S.A.*

[8]*Computational Physics Inc., Fairfax, VA, U.S.A.*

(Received 20 September 1999; accepted 8 November 1999)

Abstract. We report coincident observations of coronal and chromospheric 'flare wave' transients in association with a flare, large-scale coronal dimming, metric radio activity and a coronal mass ejection. The two separate eruptions occurring on 24 September 1997 originate in the same active region and display similar morphological features. The first wave transient was observed in EUV and Hα data, corresponding to a wave disturbance in both the chromosphere and the solar corona, ranging from 250 to approaching 1000 km s^{-1} at different times and locations along the wavefront. The sharp wavefront had a similar extent and location in both the EUV and Hα data. The data did not show clear evidence of a driver, however. Both events display a coronal EUV dimming which is typically used as an indicator of a coronal mass ejection in the inner corona. White-light coronagraph observations indicate that the first event was accompanied by an observable coronal mass ejection while the second event did not have clear evidence of a CME. Both eruptions were accompanied by metric type II radio bursts propagating at speeds in the range of 500–750 km s^{-1}, and neither had accompanying interplanetary type II activity. The timing and location of the flare waves appear to indicate an origin with the flaring region, but several signatures associated with coronal mass ejections indicate that the development of the CME may occur in concert with the development of the flare wave.

1. Introduction

Metric type II radio bursts have normally been taken to be the manifestation of a shock wave propagating in the solar corona (Uchida, 1960; Wild, Smerd, and Weiss, 1963; Weiss, 1965). An MHD shock wave excites plasma radiation, by accelerating electrons and creating an energized population which serves as the source of the radio bursts (e.g., Wild, Sheridan, and Neylan, 1959; Smith, 1972; review by Mann, 1995). Their association with flares was quickly established, and several authors, including Maxwell and Thompson (1962), demonstrated that the initiation time of type II bursts corresponded extremely well with the timing of the

The physics of the solar corona and transition region. Part II.
Reprinted from Solar Physics **193,** 2000.

associated flare. However, Dodson, Hedeman, and Chamberlain (1953) performed a study which led them to conclude that metric radio bursts were more clearly associated with high-velocity ejections than solar flares. Roberts (1959) followed this up by stating that the majority of flares of any magnitude range are not accompanied by type II activity, implying that an additional condition must be present for the radio bursts to form. Compelling arguments for and against the possible candidates for this special condition have been developed.

The problem with the 'high-velocity ejecta' or 'piston-driven' shock idea was that the estimated speeds of the shocks exceeded most of the candidate drivers. To resolve this discrepancy, Wagner and MacQueen (1983) proposed that for a flare to be accompanied by type II activity, the flare blast wave must overtake a density enhancement or CME. Uchida and co-authors (Uchida, Altschuler, and Newkirk, 1973; Uchida, 1974) noted that an MHD wave produced by a flare or other exciting impulse could steepen into a shock, which could excite Langmuir turbulence and high-frequency radio emission. Uchida's mechanism allowed for a weak MHD fast-mode wave to produce metric type II bursts, without requiring the continued presence of a piston to drive the wave. The debate over the 'piston-driven' vs 'blast wave' origin of shock waves continues today, as detailed in a review by Cliver, Webb, and Howard (1999, and references therein).

The development of several of the models of type II radio burst production occurred in conjunction with studies of rapidly moving phenomena on the Sun which were presumed to be solar shock waves or the manifestations of solar shock waves. Studies prior to 1961 focused on 'action at a distance' by a flare, such as sympathetic flaring (e.g., Richardson, 1936, 1951) and the report of the initiation of filament oscillations during a flare by Dodson (1949). The report of Moreton (1960), Moreton and Ramsey (1960) and Athay and Moreton (1961) of arc-shaped transients quickly (>1000 km s^{-1}) sweeping through the corona observed in Hα images was swiftly incorporated into the study of radio burst production. Authors such as Wild, Smerd, and Weiss (1963) proposed that these 'flare waves' or 'Moreton waves' could have the same origin as the radio type II bursts, and Kai (1970) demonstrated a strong relationship between the waves and the radio bursts. Studies such as those in Smith and Harvey (1971) demonstrated that the timing of the flare wave initiation corresponded well with the flare, and that the associated radio type II activity commenced in parallel with the wave observations.

Based on the morphology of the waves and estimates of coronal magnetic field and density, Meyer (1968) deduced that the waves probably were fast MHD disturbances in nature, and endeavored to explicitly determine the coronal magnetic field using these considerations. Uchida (1968) further developed this notion by considering the refraction of such a wave, and performed simulations which reproduced the location of the chromospheric disturbance through the initiation of a three-dimensional coronal disturbance. The correspondence was more compelling as the radio observations were incorporated into the model.

Pinter (1977) reported that 70% of the flare waves reported in his study were accompanied by type II radio bursts, and argued that the correlation could be much higher, as the association appeared to improve as the number and sensitivity of ground-based radio observations improved. Pinter found that the estimated speed of the radio observations exceeded, but were typically proportional to, the associated flare wave velocities.

Thompson *et al.* (1998, 1999) detailed the observation of coronal wave transients which were candidates for the coronal counterpart of the chromospheric observations. Because there were no associated Hα observations of Moreton waves, and because of several differences (most notably their slow speed, quasi-isotropic propagation, and diffuse morphology) from the sharp, arc-shaped Hα waves, the phenomena were labeled 'EIT waves' (after SOHO/EIT, the observing instrument) in anticipation of more definitive observations.

Klassen *et al.* (2000) extended the results of previous flare wave and radio correlative studies by performing a similar study with the 'EIT wave' transients. They found a strong association between Potsdam type II radio bursts and the EUV activity as observed by SOHO/EIT. The survey found at least 90% of the radio type II bursts had associated EUV activity indicative of an 'EIT wave' transient. Individual studies by Klassen *et al.* (1999) and Pick *et al.* (1999) have added to our understanding of the correspondence between radio and coronal observations. Other authors, such as Torsti *et al.* (1999) and Krucker *et al.* (1999) investigated correlations between these waves and interplanetary particle events.

Despite these studies, which indicated strong associations between 'EIT wave' observations and phenomena which had previously been linked to flare waves as well, the timing and location of the 'EIT waves' indicated that it was very possible that they could be driven by coronal mass ejections. Although the observations were not thorough enough to exclude flares as a driver of the EUV waves, the development of CME-associated features in conjunction with the waves introduced additional complexity into the study.

Cliver, Webb, and Howard (1999) performed a study of metric type II bursts and CMEs observed during the Solar Maximum Mission, and found that the correlation between the two phenomena increased dramatically as the origin of the radio activity approached the solar limb. They went on to argue that since the correlation appeared to improve with the visibility of the CME (i.e., as the CME launch angle approached the solar limb), the type II bursts had a more clear association with CMEs. Furthermore, this association improved remarkably if the speed of the CME exceeded 400 km s^{-1}. Because of the prior connection that was made between metric type II bursts and flare waves, Cliver, Webb, and Howard further favored a strong flare wave/CME correspondence. It became clear that the coronal observations were essential to unravel the mystery of what drives coronal shock waves.

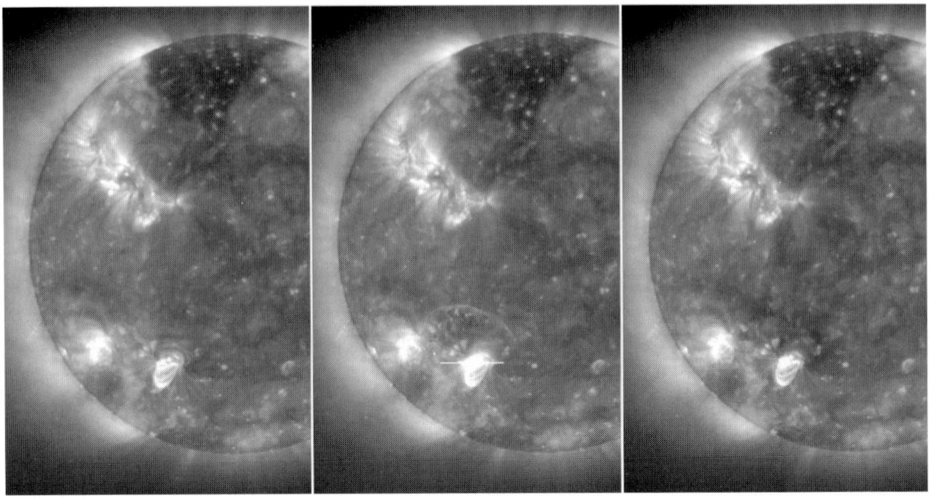

Figure 1. Subfields of SOHO/EIT 195 Å images at 02:33, 02:49 and 03:03 UT on 24 September 1997, showing the eastern portion of the Sun during the development of the first event.

2. Coronal Observations

In this study we examine two EIT waves associated with two M-class flares with GOES peak times at 02:48 UT (M5.9) and 11:06 UT (M3.0) respectively. Neither event had complete hard X-ray coverage either from the Compton Gamma Ray Observatory or from the *Yohkoh* spacecraft, but for the first event Hα observations are available as well, and for the second event coronal metric radio burst data are discussed.

The Extreme ultraviolet Imaging Telescope (EIT) (Delaboudinière *et al.*, 1995) images the solar disk and inner corona of the Sun in four pass bands consisting of strong EUV spectral lines. The observations in this paper consist of a wavelength range which is centered on 195 Å, as the three other wavelength regimes are obtained at a much lower cadence and were not available for the times we discuss. The primary wavelengths observed with the 195 Å pass band are Fe XII 192.3, 193.5 and 195.1 Å. The temperature at which these emissions peak at coronal densities is approximately 1.5–1.6 MK. Images are recorded on a 1024×1024 pixel (44.2×44.2 arc min) EUV-sensitive CCD camera, with a pixel size of 2.59 arc sec.

EIT was running a campaign consisting of full-disk 195 Å images consisting of a single half-resolution image every 10–23 min. A 'pre-event' image is shown in the first frame of Figure 1, restricted to the eastern portion of the image, where the activity is observed. The first frame of Figure 1 was recorded at 02:33 UT, while the second and third frames were recorded at 02:49 and 03:03 UT. NOAA active region No. 8088 is visible in the southeast at a heliographic location of S28, E20, while NOAA region No. 8087 was positioned to the east at S25, E45. The complex containing active region No. 8089 was located to the north ranging N10 to N45

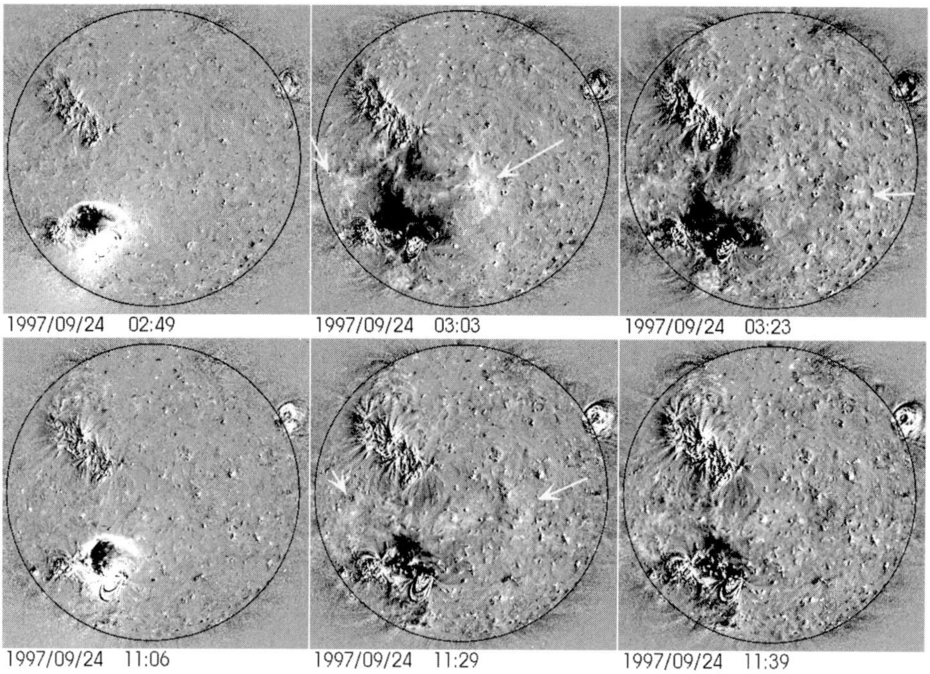

Figure 2. Differenced SOHO/EIT images for the two events on 24 September 1997. The first row of images, recorded at 02:49, 03:03 and 03:23 UT, have a pre-event image at 02:33 UT digitally subtracted from them. The second set of images (11:06, 11:29 and 11:39 UT) had an image at 10:46 UT subtracted from each. The arrows indicate the progression of the weak wavefront.

and from E20 to near the observable limb. Two features of the sequence are worth noting. First, a comparison of the first and third images reveals a large degree of decreased emission to the north of AR No. 8088, to the northeast of AR No. 8087, and to the south of AR No. 8089. Also, the same image pair displays a relatively low degree of evolution observed to the south of the flaring region No. 8088. Most of the evidence of this eruption in the EUV are confined to regions north of the flaring region and south of AR No. 8089. An exception is the center image, near the time of the flare, when a broad increase in emission is observed over an area roughly five arc min in diameter centered on the flare; this is assumed to be due primarily to scattered light in the instrument as a result of the flare (F. Auchere, personal communication). Otherwise, the most dramatic changes occur to the north of the flaring region.

To illustrate this point more clearly, a series of differenced images are shown in Figure 2 for the eruptions beginning at 02:49 UT and at 11:06 UT. Each image has a 'pre-event' image subtracted from it; for the first set of images (02:49, 03:03 and 03:23 UT) the image at 02:33 UT was digitally subtracted from each one to enhance changes occurring in the corona. The second set of images (11:06, 11:29 and 11:39 UT) had an image at 10:46 UT subtracted from each. These 'pre-event' images were the last images recorded before clear evidence of the eruptions was observed. Therefore they also serve as a lower limit to the times in which various aspects of the eruption developed.

The series of images shows a strong correspondence between the two separate eruptions. Both start off with a sharp, compact bright front propagating northward from region No. 8088 and extending over an angle of approximately 120 degrees. The bright front in the first (02:49 UT) panel of Figure 2 is also visible in the middle EIT image in Figure 1. The horizontal line in this frame corresponds to an instrumental artifact of the flare resulting from a local overfilling of CCD pixels. The bright front exhibits an average increase of more than 100 %, with small (<20 arc sec) areas exceeding a 200 % increase.

For context, Figure 3 shows the local evolution of the two eruptions. The middle row of two figures are subfields of EIT images at 02:49 and 03:03 UT, and the lower two figures are subfields from 11:06 and 11:29 UT. The top row shows larger subfields of images from the *Yohkoh* Soft X-ray Telescope (SXT) obtained at 02:35 and 03:51 UT. The first image was obtained at a quarter of the SXT resolution and the second at a half, both with the Aluminum Magnesium (AlMg) filter. A single SXT pixel spans 2.46 arc sec. The portion of the brightening extending to the north of the flaring region in the 03:51 UT SXT image is an instrumental artifact due to the brightness of the flaring region. The GOES soft X-ray photometer observed a flare of magnitude M5.9 beginning at 02:43 UT, peaking at 02:48 UT, and ending at 02:52 UT. The later event was associated with an M3.0 flare from 10:57–11:10 UT, peaking at 11:06 UT. Although the BATSE instrument on the Compton Gamma Ray Observatory was not observing at the time of the first flare, BATSE recorded a hard X-ray flare beginning at 1107:17 UT and peaking at 1107:33 UT in association with the later event.

The SXT images indicate that additional loop structures either brightened or became visible following the flare and eruption. The loops, marked by arrows in Figure 3, formed primarily to the immediate north of the flaring region, while there was some evidence of a new soft X-ray loop spanning the equator extending from region No. 8088 to No. 8089. The SXT images were not obtained at a sufficient cadence to show the development of these features. These features are typically identified as signatures of a large-scale magnetic reconfiguration usually associated with a coronal mass ejection (Rust, 1983; Hudson, Acton, and Freeland, 1996; Hudson and Webb, 1997; Nitta and Akiyama, 1999). A 'cusp-shaped' feature, extending to the south from the same region, is formed before the event takes place and shows little sign of evolution during the event, indicating that it was perhaps independent of it.

The bright front and the EUV flare both appear in the same image for both events (02:49 UT and 11:06 UT); based on the EUV observations alone, the low cadence of the images make it impossible to determine which originated first. Another phenomenon first appears in these images as well, a dark region between the flaring region and the bright front. This dark 'dimming' region remains for a

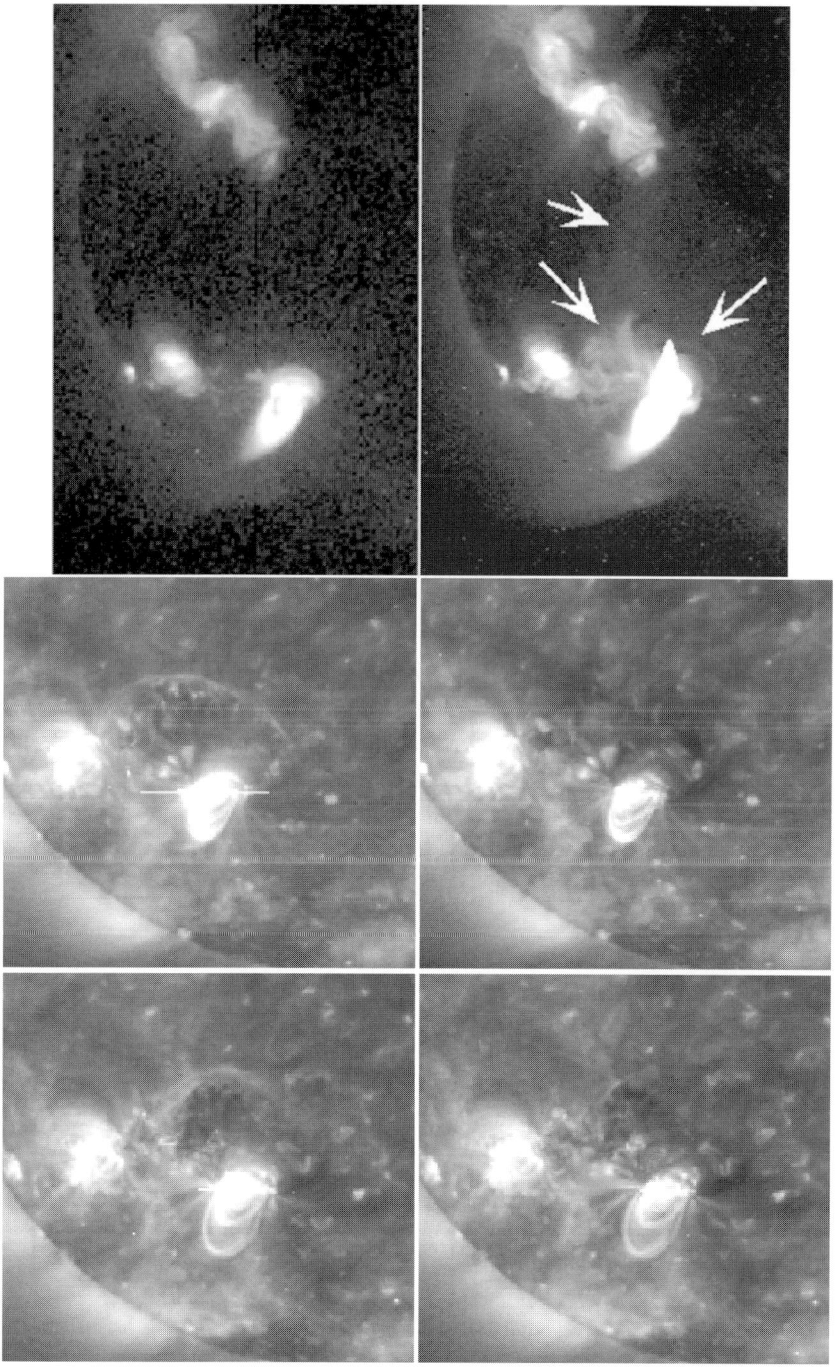

Figure 3. The top row shows larger subfields of images from the *Yohkoh* Soft X-ray Telescope (SXT) obtained at 02:35 and 03:51 UT, detailing the development of the first eruption. The middle row are subfields of EIT images at 02:49 and 03:03 UT, and the lower two figures are subfields from 11:06 and 11:29 UT. Each pair shows the development of the eruption close to the flaring site; a sharp bright arc is visible in the 02:49 and 11:06 EIT images.

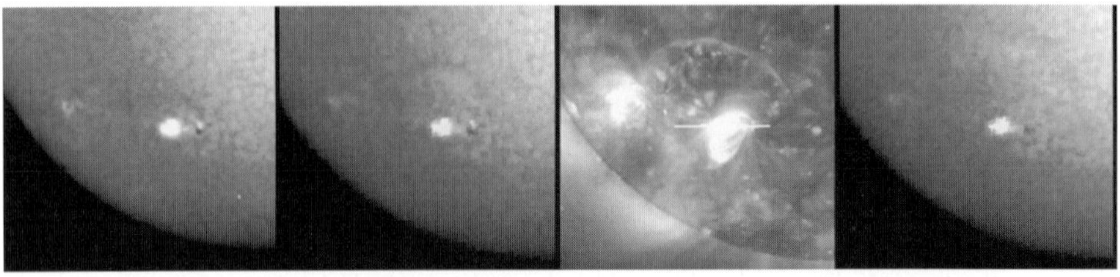

Figure 4. Hα images of the first eruption obtained at 02:45, 02:47 and 02:50 UT. The corresponding EIT sub-frame at 02:49:21 UT is shown in the third frame of the figure.

fairly long duration (>8 hours) and consists of a decrease of 30–40% on average, exceeding 50% in small (<20 arc sec) regions.

The dimming region develops gradually, appearing only up to the sharp wavefront in the images at 02:49 and 11:06 UT, but extending over greater distances towards active region No. 8089 in later images. The relative lack of evolution observed north of the wavefront in the images at 02:49 and 11:06 UT, compared to the drastic changes following, indicates that the dimming region and wavefront developed in conjunction.

The difference images in Figure 2 show evidence of the later development of a more diffuse wavefront, most notable in the image at 03:03 UT. A faint bright front (marked by arrows) appears to the east and west of the dimming region, and the total emission increase of less than 20% and the broad expanse are more characteristic of the 'EIT wave' transients described by Thompson *et al.* (1998, 1999). Unlike the sharp propagation fronts, the weaker fronts tend to propagate through the corona and leave very little evidence of their transit which can be observed with the EIT image cadence. Wills-Davey and Thompson (1999) describe the transit of a similar diffuse wave in TRACE EUV data, and discuss the observable local effects. Running-difference images (images with the immediately previous image subtracted from them) and movies show later evidence of the wave (marked by an arrow in Figure 2) at 03:23 near the west limb of the Sun.

3. Hα and Metric Radio Observations

Figure 4 shows Hα observations obtained in conjunction with the EIT observations of the first eruption. The Hα images were obtained with a Vixen 102ED refractor stopped down to 63 mm and a Daystar 0.5 heated T scanner with a 2× Barlow lens. Exposures of $\frac{1}{125}$ s were recorded on Tech Pan 2415 film. The times of the first, second and fourth frames of Figure 4 were approximately 02:45, 02:47, and 02:50 UT. The corresponding EIT sub-frame at 02:49:21 UT is shown in the third frame of the figure.

The development of the wavefront in the Hα images appears to indicate that the EUV wave and Hα waves are relatively cospatial. These data show a dark, roughly concentric wavefront propagating outwards from the flare site, similar to early Moreton wave observations. A more diffuse, slightly brightened area follows the dark Hα wavefront, typical of the Doppler shifting into the red wings of the Hα passband as the chromosphere relaxes following the passage of the wave. Another Hα image obtained at 02:55 UT showed no further evidence of the wavefront. Unfortunately, neither the timing of the Hα observations nor that of the EIT observations are accurate to the degree required to properly determine speed estimates; the Hα times are accurate to approximately 30 s, while the EIT times are slightly more accurate. For this reason, it is not possible to determine whether the location of the EUV front leads that of the Hα wavefront, or if they are cospatial. Additionally, it is impossible to determine the altitude at which the EUV emission is being produced, based on the assumption that the EUV emission could be coronal in origin while the Hα emission would be primarily chromospheric. However, the images indicate that the EUV and Hα observations are approximately cospatial.

Figure 5 illustrates the locations of the Hα and EUV wavefronts, assuming that the reported observation times are accurate and that the wavefronts observed in the EUV and Hα are cospatial. The active regions are outlined and the shaded region corresponds to the area which exhibited EUV dimming. The four arcs drawn within the shaded region correspond to the Hα and EUV observations of the sharp wavefront. The line running north-south to the left of the dimming region indicates the location of the farthest edge of the diffuse EUV brightening at 03:03 UT, and the lines running north-south to the right of the dimming region correspond to the location of the diffuse front at 03:03 and 03:23 UT. Five lines are drawn radially from the flare site, as a means of sampling the wave speed at various locations. These lines we will identify as Track 1 through Track 5, numbered from the left to the right.

Assuming the reported observation times are accurate, we compute the speeds along the five tracks. The three data points along Track 1 yield speeds (beginning with the pair closest to the flaring region) of 620 and 630 km s^{-1}, where the speeds correspond to the distance traveled along a great circle on a sphere of one solar radius. The speeds along Track 2 are 430 and 720 km s^{-1}, along Track 3 are 360, 330, and 1000 km s^{-1}, along Track 4 are 380, 280, 1220 and 460 km s^{-1}, and along Track 5 are 310, 410, 1000, 390, and 260 km s^{-1}. The speeds of 1000 and 1220 km s^{-1} correspond to measurements made between the EIT observation at 02:49:21 UT and the Hα observation at 02:50 UT. The large inhomogeneity represented by these speeds indicates two possible difficulties: either the EUV observations are not cospatial with the Hα observations, or the inaccuracy of the observation times is grossly influencing the speed estimates. The speed estimates along Tracks 2-5 based solely on the location of the Hα images at 02:47 and 02:50 UT are 510, 424, 570 and 530 km s^{-1}. This implies that either the reported time of the EIT image at 02:49:21 UT is later than the true observation time, or that the EUV wavefront location leads the location of the Hα wavefronts. The error in the speed estimates is primarily due to the difficulty in determining the

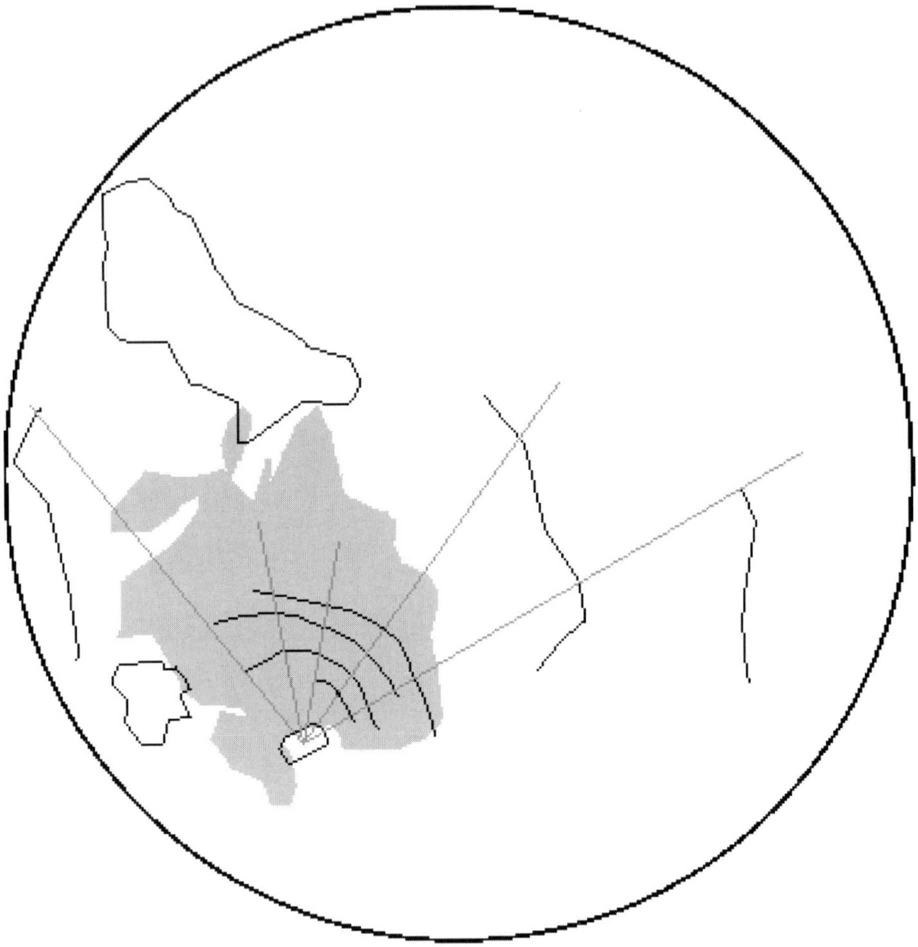

Figure 5. Drawing of the locations of the Hα and EUV wavefronts, corresponding to the wave progression at 02:45, 02:47, 02:49, 02:50, 03:03 and 03:23 UT. The active regions correspond to the black outlines and the shaded region corresponds to the area which exhibited EUV dimming. Five lines are drawn emanating from the flare site, identified from left to right as Track 1 through Track 5.

exact location of the wavefronts and due to timing errors; most of the estimates are probably not accurate to within 25%.

The speed and locations of the wavefront appear to indicate that a plausible origin intersects the flaring site at close to 02:43 UT. The GOES flux was observed to be increasing at this time, and peaked later at 02:48 UT.

Metric radio observations of the eruptions were obtained with the radio spectral polarimeter at the Potsdam Astrophysical Institute (Mann *et al.*, 1992) and with the Hiraiso Radio Spectrograph (HiRAS) in Japan. The Hiraiso observations indicate that a metric type II radio burst began around 02:49 UT (although a type III burst beginning around 02:47 UT would have obscured any earlier type II activity) and ended at approximately 02:59 UT. Based on the model of Mann *et al.* (1999), the speed of the type II burst was estimated to be approximately 620 km s^{-1}.

The second event was accompanied by type II bursts observed by the Potsdam Astrophysical Institut. This event is reported in a survey by Klassen *et al.* (2000) of the association between Potsdam type II radio bursts and EUV activity as observed by SOHO/EIT. The survey found at least 90% of the radio type II bursts had associated EUV activity indicative of an 'EIT wave' transient. Figure 6 shows the radio observations developing in conjunction with the second EUV event. The EIT observations indicated that activity developed between 10:46 and 11:06 UT. The type II activity associated with this event had several stages. Based on the Mann *et al.* (1999) model, the frequency of the bursts between 11:03:30 and 11:05:30 was relatively constant, indicating that the direction of propagation was transverse to the density gradient. The frequency drift of the 11:07:30 to 11:09:00 bursts was also slight. From 11:05:30 to 11:08:00 and 11:09:00 to 11:13:00 the estimated speed of the burst was approximately 510 km s^{-1}.

Alternative models can cause the estimated speeds to vary; Klassen *et al.* (2000) used the model described by Newkirk (1961) to estimate the speed at 740 km s^{-1} later in the event. Both bursts display several type II lanes and split bands, indicating the disturbance is meeting with many sites of comparatively low magnetic field. The estimated speed of the EUV disturbance, based on the 11:06 and 11:29 UT EIT images, was approximately 200 km s^{-1} to the east and 300 km s^{-1} to the west. If the earlier wave transient at 02:45 UT can be compared to the later event, it is understandable that the speeds obtained with the 11:06 and 11:29 images will be quite slow, as the fastest speeds generally were observed early in the wave's motion, when it had a sharper profile. Klassen *et al.* found a similar result in their catalog: when the wave speeds were determined exclusively using EIT data, the average speed of the type II bursts was nearly three times that of the EIT speeds. However, as seen in the earlier event at 02:45 UT, the speed of the wave based on the Hα observations averaged more than 500 km s^{-1}, which is closer to the estimated type II speed of 620 km s^{-1}. Additionally, the speed determined by the radio bursts represents the speed along a density gradient, while the Hα (and apparently the early EUV) wavefront remains at low altitudes.

4. The Correspondence with a Coronal Mass Ejection

Figure 7 illustrates a series of SOHO/LASCO C2 (Brueckner *et al.*, 1995) images of the coronal mass ejection associated with the first eruption on 24 September 1997, first appearing in the C2 field of view at 03:38 UT. The mass ejection is indicated with arrows in the southeast in the 03:38 UT image and over a 180-deg expanse in a later image at 04:20 UT. Its speed, based on the brightest portion in the southeast in the 03:38 and 04:20 UT images, is estimated to be between

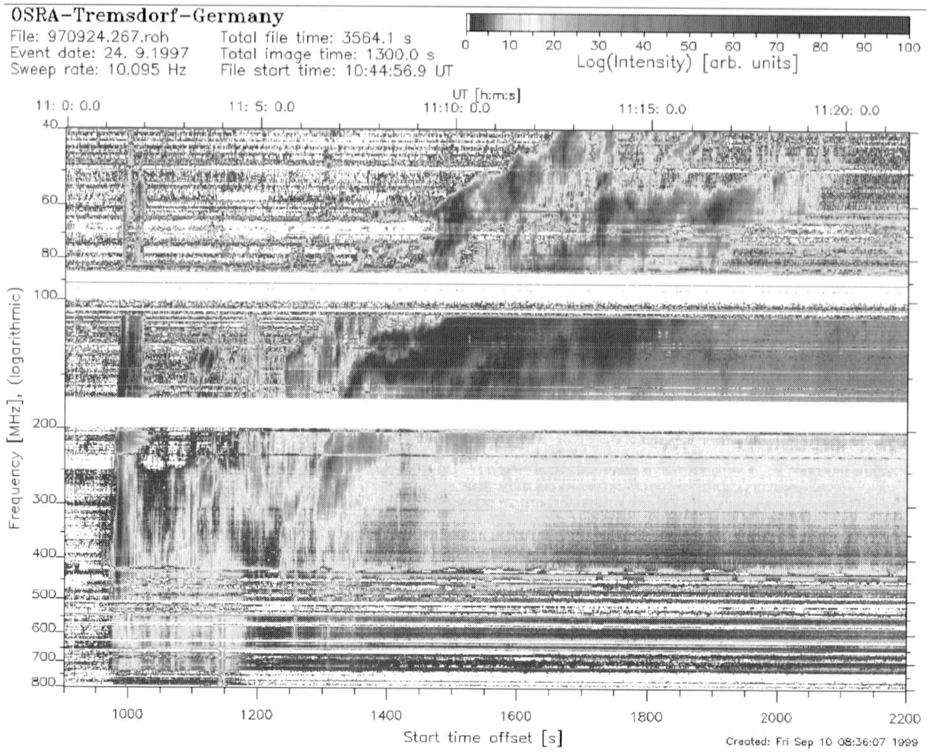

Figure 6. Metric radio data from the second event observed by the radio spectral polarimeter at the Potsdam Astrophysical Institut. Based on the frequency drift of the radio bursts, the speed in the corona was estimated to be between 500 and 750 km s^{-1}.

280 and 350 km s^{-1}. The CME was not visible in the 02:55 UT image, indicating that the leading portion of the CME in the 03:38 UT image was traveling at least 300 km s^{-1}.

The eruption is difficult to observe for several possible reasons. The southeast limb was the site of continuous coronal mass ejections throughout 24 September 1997. At least six mass ejections were observed to the south and southeast, including the loop eruption which was in progress in the southeast in all frames of Figure 7. The mass ejection appearing at 03:38 UT is only seen as a superposition over the existing loop eruption, and is therefore difficult to track.

Another reason the coronal mass ejection was difficult to observe was because of its launch angle. As mentioned previously, the coronal dimming observed in the EUV by SOHO/EIT occurred primarily on the solar disk. As demonstrated by Thompson *et al.* (2000), the white-light extent of a coronal mass ejections corresponds most strongly with the large-scale dimming observed in the EUV images (if such a large-scale dimming is observed). Therefore, most of the CME material would presumably have been launched at an angle of greater than 50 deg from the plane of the sky. The Thomson scattering angle of the material associated with this eruption would not allow the CME to be as visible as an eruption occurring near the solar limb, as was the case for the earlier loop CME in the southeast.

The brightness of the CME associated with the first eruption in our study is not unusually weak for a 'halo' coronal mass ejection. The 6 January 1997 (Webb *et al.*, 1998) halo coronal mass ejection had roughly the same brightness as the one discussed here. However, the second event commencing around 11:00 UT was accompanied by little or no white light activity (another loop eruption to the southeast was in progress during this event as well). The extent of the EUV dimming was much less than the first eruption, but no systematic study has been done investigating any possible correlation between the EUV or soft X-ray coronal dimming and the white-light brightness of a mass ejection. Studies of halo coronal mass ejections observed by SOHO/LASCO (St. Cyr *et al.*, 1999, and references therein) indicate that the majority of CMEs occurring along the Earth–Sun axis are observed by SOHO/LASCO, inferring that very few mass ejections are 'missed' by LASCO.

The WAVES instrument on board the WIND spacecraft (Bougeret *et al.*, 1995) did not observe interplanetary type II activity with either event (M.L. Kaiser, personal communication). It is not rare that a series of metric radio type II bursts not be accompanied by interplanetary type II bursts (e.g., Robinson, Stewart, and Cane, 1984) but it is still difficult to determine the similarities and differences between their origins (e.g., Gopalswamy *et al.*, 1998; Cliver, 1999). It is generally accepted that the presence of interplanetary type II activity is a strong indicator of a coronal mass ejection (e.g., Cane, 1983).

Additionally, it is important to note that the SOHO and WIND spacecraft, which were both obtaining *in-situ* observations of the solar wind for the days following the eruptions, observed no clear evidence of a disturbance of solar origin (N. Fox, personal communication).

5. Discussion

The Hα observations of the 'flare wave' or 'Moreton wave' beginning at 02:45 on 24 September 1997 correspond well with previous observations of these disturbances, as first reported by Moreton (1960) and Moreton and Ramsey (1960), and as summarized in Smith and Harvey (1971) and Pinter (1977). The wave develops into a discrete arc spanning approximately 120 degrees, and propagates at speeds ranging from a few hundred to over a thousand km s^{-1} (though the highest speed estimates are suspicious due to the unavailability of exact image times). The Hα observations consisted of a dark front propagating from a flare site, with evidence of a brightening following closely behind it. The event is strongly reminiscent of the images taken in Hα + 0.5 Å on 20 September 1963 (Moreton, 1965; Figure 4 of Smith and Harvey, 1971). The 24 September 1997 Hα observations may have contained contributions from the red and blue wings of Hα, commensurate with

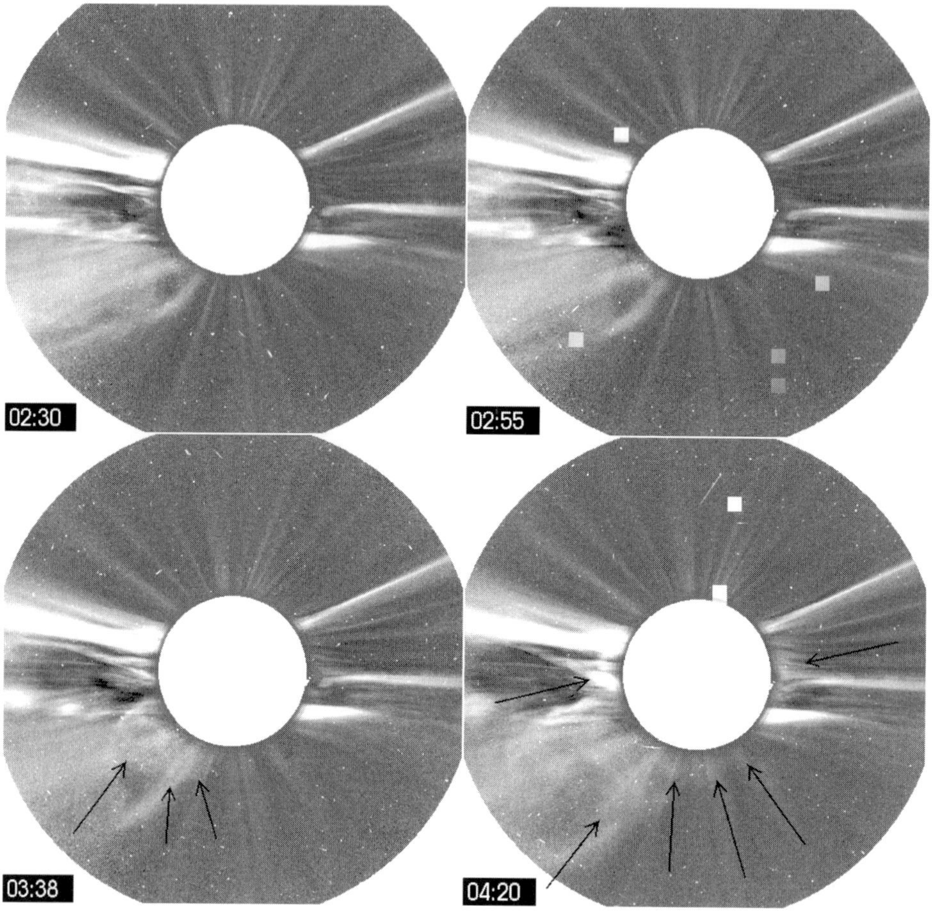

Figure 7. SOHO/LASCO C2 images of coronal mass ejections on 24 September 1997. A CME is in progress in the southeast (lower left) portion of the image, and a second disturbance (indicated by arrows) begins at 03:38 UT.

the interpretation that the wavefront consists of a chromospheric compression followed by relaxation. However, Athay and Moreton (1961) discuss the visibility of these waves in both on-band and in the wings of Hα. Additionally, Smith and Harvey (1971) indicate that some of the wave observations consist of moving material and not just compressional motions; the resolution of the data prohibited a determination of the relative contributions of material motion and chromospheric depression.

The range of frequencies of the observed type II bursts is similar to those reported in the summaries mentioned above, and the type II bursts and wave transient originate at close to the same time. The estimated type II burst velocity is slightly greater than the measured velocity of the chromospheric/coronal disturbance, also concurring with observations of prior flare wave transients. It is fortunate that these observations are not unusual for a flare wave, as the incorporation of the coronal observations into this phenomenon may be valid for most of these transients.

This study was initiated for two reasons: to examine the coronal counterpart of flare wave transients, and to utilize the broader variety of available data as a means of differentiating between initiators and drivers of these waves. Sadly, only the first of these two goals was accomplished. Both the CME and the flare seem integral to our observations. Additionally, the data are not sufficient to differentiate between different models proposed to explain the early Hα and metric radio data; even with the increased range of data available, we are still left with many of the same questions.

Fortunately, the EUV and Hα observations fairly overlap. Although it was not clear whether the EUV and Hα waves were exactly cospatial, it was clear that they appeared roughly in the same location and had very similar morphologies. The EIT observation of a sharp wavefront at 02:49:21 UT showed a simple correspondence with the Hα data, confirming suspicions that the chromospheric observations of the waves were a manifestation of a phenomenon that was coronal in origin.

Two additional aspects are brought into the study when the coronal observations are incorporated. First, a strong 'coronal dimming' is observed in association with the wave's development, and this, combined with the *Yohkoh* SXT and SOHO/LASCO observations, indicate the possible presence of a coronal mass ejection. Additionally, in later images, a diffuse transient appears which is more typical of the 'EIT waves' discussed earlier. It is impossible to determine from the data available whether the later appearance of a weak 'EIT wave' corresponded with the disappearance of the sharp EUV emission front. This would enable observers to determine whether the two had an associative correspondence, and whether the two were two separate phenomena or are manifestations of the same. It is possible that the sharp coronal fronts, which appear to correspond to potential observability in chromospheric Hα, represent the high-amplitude limit of the weak 'EIT wave' observations. The waves could be more strongly driven or could be steeper during the early stages, and become freely propagating or weaker in its subsequent development.

Nearly all of the EIT wave transients reported in the Klassen *et al.* (2000) study were weak and diffuse and did not have the morphology of the Hα waves. The timing of the Klassen *et al.* events indicated that the radio bursts occurred early in the production of the EIT wave transients. This may imply that the EIT wave transients occur in two stages: an earlier, possibly driven stage, and a later, freely propagating weaker stage. It is conceivable that the weak, quasi-isotropic waves observed in EIT could take on a sharper and more rapidly propagating form in the early stages of propagation, or in the strongly driven limit (thereby resolving the discrepancy in morphology), but there is still a question of the initiator or driver.

The first event showed evidence of an eruption in the LASCO C2 white-light coronagraph data, while the second event was lacking a white light counterpart. Both the EIT and SXT observations had what is usually interpreted as evidence of a mass ejection in both events, but the studies establishing these signatures were developed in conjunction with white light observations of CMEs. The CME may have been directed far from the plane of the sky, and therefore was not easily observable against a background of steadily evolving corona. If that is not the case, the reliability of the coronal disk signatures comes into question.

The timing and location of the flare relative to the Hα wave transient are similar to early observations; the initiation of the wave appears to be centered on the flare. However, it is possible to propose the CME as the driver of the wave, with the flare serving as an associated (but not initiating) phenomenon. The co-temporal appearance of the coronal dimming may support this explanation. Indeed, the coincident appearance and parallel development of the dimming region and the bright front imply an association, but causality is difficult to determine. The gradual formation of the dark depleted region could represent an expanding driver of the bright front, with the dimming region (CME) causing the bright front to form. Conversely, the bright front could sweep through the ambient corona, causing a drastic change in its wake. It is difficult to believe that a disturbance as strong as the one shown in the wavefront would result in no change in coronal emission; one would expect that heating or cooling out of the EUV 195 Å band pass could easily result. However, the long duration ($>$3 hours in both cases) and the fairly stable nature of the emission depletion exceeds most estimates of the time it would take for the coronal plasma to return to its original temperature. Additionally, the *Yohkoh* SXT observations indicate that some change of magnetic topology occurred in the regions which exhibited emission depletion, which gives additional evidence that the change in EUV emission was not solely due to a shock front propagating through the region.

The observations are not sufficient to distinguish between possible dimming mechanisms. Certainly, the 'plowed up' material in the moving front could consist of evacuated material from the dimming region. The relative changes in emission could support this theory, if we make the poor assumption that the emission recorded in the plasma does not vary with its density or rapid evolution.

In addition to the evidence possibly linking the wave to the coronal mass ejection, the 'traditional' aspects of the Hα and metric type II data (timing, location) still implicate the flare as a possible initiation mechanism. Additionally, the formation of the dimming also appears to emanate from the flaring site. The location of the flare seems central to all of the associated phenomena.

The metric radio type II bursts began at 02:47 UT and continued until approximately 02:59 UT. As mentioned previously, an Hα image obtained at approximately 02:55 UT showed no evidence of the wavefront. By that time, the type II bursts had decreased significantly in frequency, implying that they were occurring at a higher altitude. An Hα image obtained at 02:45 UT indicated that the wave transient had begun developing by that time (though it had not reached its full angular expanse). Smith and Harvey (1971) demonstrated that for the wave observed on September 20, 1963, the type II radio bursts began within 1 minute of the appearance of the wave transient. Smith and Harvey used these timing arguments to link a number of flare wave transients to a flare origin.

The model developed by Uchida (1968, 1974; Uchida, Altschuler, and Newkirk, 1973) was able to explain both the location and speed of both flare waves and type II bursts. The Uchida model consisted of tracing a fast-mode MHD impulse through a varying coronal density and magnetic field profile. It was shown that the three-dimensional disturbance was focused by areas of low Alfvén speed. The regions where the front intersected the chromosphere reproduced the observed wavefronts quite convincingly.

Based on the Uchida model, however, one would expect the coronal observations to appear spread out over a large area, with an increase in intensity and finer structure where the wavefront was steepening towards the chromosphere. Instead, the coronal observations look very similar to those of the chromosphere, as the location and extent of the EUV sharp front maps quite well with the Hα front. This is not sufficient to exclude the Uchida model, however; a flare occurring close to the edge of an active region (as is this case for this flare) would have an impulse which is refracted towards higher altitudes (and therefore out of EUV visibility) on one side of the flare. Additionally, the model did not rule out the possibility of a directional impulse being delivered by the flare.

The Uchida model is readily applied to recent coronal observations of weak impulses. Typical 'EIT wave' transients appear to be freely propagating and can be observed moving across the entire solar disk. Using high-resolution and high-cadence TRACE EUV data, Wills-Davey and Thompson (1999) show strong evidence of the coronal magnetic field motion as the wave propagates through it, and the guiding of the wave by the magnetic field. Wu *et al.* (1999) have utilized a technique similar to that of Uchida, and have been able to reproduce the propagation and morphology of an 'EIT wave' transient to a stunning degree of accuracy. Neither the Uchida nor the Wu *et al.* model necessarily require that the impulse have a specific initiator; only the existence of the impulse is central to the model.

However, the weak 'EIT wave' transients may be easier to simulate and to reconcile with a weak MHD wave/shock model. Again, the high-amplitude limit of the weak waves may have an appearance similar to those observed in the 24 September 1997 data. One possible explanation for the appearance of coronal dimming regions is rapid heating by the shock transit, and many of the features of the observations can be incorporated by slight modifications of the MHD models.

An entirely different approach also shows promise. The model described by Delannée and Aulanier (1999) provides a mechanism by which a coronal mass ejection which is triggered by a flare reconnection can produce a bright arc-shaped EUV 'front.' They report the observation of a bright arc in the EIT data which apparently remains stationary for more than 10 min; this arc also appears in conjunction with coronal dimming regions, but clearly cannot be wave-like in nature because it does not propagate. Furthermore, the identification of a transient as a 'wave' implies that its propagation involves the transfer of information through a medium; simple material motion, or a series of brightenings which are caused by a non-propagating disturbance should be excluded from classification as a wave.

The Delannée and Aulanier model predicts both the location of EUV brightenings as the result of opening magnetic fields lines in a CME. The dimming regions are regions predicted by their model to be depleted by the eruption, and consist of areas both close to the flaring region and other areas (including near other active regions) which are magnetically connected by loops overlaying the flaring region. The events described have three pieces of evidence supporting this theory: first, dimmings occur near the flaring region and also additional dimmings are observed north of the equator near active region No. 8089. Second, both of the observed EUV sharp wavefronts on 24 September 1997 occur roughly at the edge of the dimming region closest to the flaring active region, suggesting the location of a separatrix in the model. Finally, the *Yohkoh* SXT data show evidence of post-eruption arcades also extending over this area, possibly indicating a change in magnetic topology. The model proposes that an observed propagation of the wavefront (or arc-shape) may come from the consecutive opening of field lines and the resulting compression of material.

Unfortunately, both the MHD approach and the opening magnetic field line approach are both able to adequately explain many features of the 24 September 1997 observations, and we are unable to eliminate either model from consideration, and are similarly left with the same question of whether coronal shock waves are piston-driven or blast waves. However, the presence of both sharp and diffuse wavefronts, as well as evidence that both the CME and the flare are essential aspects of the observations, may cause us to return to an old lesson eventually learned by anyone studying the Sun: posing a question by attempting to differentiate between one thing *or* another can drastically impede your ability to assess the relative contribution of *both*.

Acknowledgements

BJT would like to gratefully acknowledge discussions with Y. Uchida, C. Delannée, G. Aulanier, M. Wills-Davey, and E. Cliver. NG is an NAS/NRC Senior Research Associate.

References

Athay, R. G. and Moreton, G. E.: 1961, *Astrophys. J.* **133**, 935.
Bougeret, J.-L. *et al.*: 1995, *Space Sci. Rev.* **71**, 231.
Brueckner, G.E. *et al.*: 1995, *Solar Phys.* **162**, 357.
Cane, H. V.: 1983, in M. Neugebauer (ed.), *Solar Wind Five*, p. 703.
Cliver, E. W.: 1999, *J. Geophys. Res.* **104, A3**, 4743.
Cliver, E. W., Webb D. F., and Howard, R. A.: 1999, *Solar Phys.* **187**, 89.

Delaboudinière, J.-P. *et al.*: 1995, *Solar Phys.* **162**, 291.
Delannée, C. and Aulanier, G.: 1999, *Solar. Phys.* **190**, 107.
Dodson, H. W.: 1949, *Astrophys. J.* **110**, 382.
Dodson, H. W., Hedeman, E. R., and Chamberlain, J.: 1953, *Astrophys. J.* **117**, 66.
Gopalswamy, N., Kaiser, M. L., Lepping, R. P., Kahler, S. W., Ogilvie, K., Berdichevsky, D., Kondo, T., Isobe, T., and Akioka, M.: 1998, *J. Geophys. Res.* **10**, 307.
Hudson, H. S., Acton L. W., and Freeland, S. L.: 1996, *Astrophys. J.* **470**, 629.
Hudson, H. S. and Webb, D.: 1997, in Crooker, Joselyn, and Feynman (eds), *Coronal Mass Ejections*, American Geophysical Union Geophysical Monograph 99.
Kai, K.: 1970, *Solar. Phys.* **11**, 310.
Klassen, A., Aurass, H., Klein, K.-L., Hofmann, A., and G. Mann, G.: 1999, *Astron. and Astrophys.* **343**, 287.
Klassen, A., Aurass, H., Mann G., and Thompson, B. J.: 2000, *Astron. Astrophys.*, in press.
Krucker, S., Larson, D., Lin R., and Thompson, B. J.: 1999, *Astrophys. J.* **519**, 864.
Mann, G.: 1995, in A. O. Benz and A. Krüger (eds), *Lecture Notes in Physics 44*, *Coronal Magnetic Energy Releases*, Springer-Verlag, Berlin, p. 183.
Mann, G., Aurass, H., Voigt, W., and Paschke, J.: 1992, ESA-Journal, SP-348, 129.
Mann, G., Jansen, F., MacDowall, R. J., Kaiser, M. L., Stone, R. G.: 1999, *Astron. Astrophys.* **348**, 614.
Maxwell, A. and Thompson, A. R.: 1962, *Astrophys. J.* **135**, 138.
Meyer, F.: 1968, in K. O. Kiepenheuer (ed.), 'Structure and Development of Solar Active Regions', *IAU Symp.* **35**, 485.
Moreton, G. E.: 1960, *Astron. J.* **65**, 494.
Moreton, G. E. and Ramsey, H. E.: 1960, *Publ. Astron. Soc. Pacific* **72**, 357.
Moreton, G. E.: 1965, in 'Stellar and Solar Magnetic Fields', *IAU Symp.* **22** 371.
Newkirk, G.: 1961, *Astrophys. J.* **133**, 983.
Nitta, N. and Akiyama, S.: 1999, *Astrophys. J.* **525**, L57.
Pick, M., Maia, D., Vourlidas, A., Benz, A. O., Howard, R. A., and Thompson, B. J.: 1999, in S. R. Habbal, R. Esser, J. V. Hollweg, and P. A. Isenberg (eds), *Solar Wind Nine: Proceedings of the Ninth International Solar Wind Conference*, American Institute of Physics, 649.
Pinter, S.: 1977, Spec. Rep. AFGL-SR-209, Hanscom Air Force Base, 35.
Ramsey, H. E. and Smith, S. F.: 1966, *Astron. J.* **71**, 197.
Richardson, R. S.: 1936, Annual Report of the Director, Mt. Wilson Obs., Vol. **35**, 871.
Richardson, R. S.: 1951, *Astrophys. J.* **114**, 356.
Roberts, J. A.: 1959, *Australian J. Phys.* **12**, 327.
Robinson, R. D., Stewart R. T., and Cane, H. V.: 1984, *Solar. Phys.* **91**, 159.
Rust, D. M.: 1983, *Space Sci. Rev.* **34**, 21.
Smith, D. F.: 1972, *Astrophys. J.* **174** 163.
Smith, S. F. and Harvey, K. L.: 1971, in C. J. Macris (ed.), *Physics of the Solar Corona*, Reidel Publ. Co., Dordrecht, p. 156.
St. Cyr, O. C. and Plunkett, S. P.: 1999, on-line LASCO CME List.
St. Cyr, O. C. *et al.*: 1999, *J. Geophys. Res.*, in preparation.
Sterling, A. C. and Hudson, H. S.: 1997 *Astrophys. J.* **491**, L55.
Thompson, B. J., Plunkett, S. P., Gurman, J. B., Newmark, J. S., St. Cyr, O. C., Michels, D. J., and Delaboudinière, J.-P.: 1998, *Geophys. Res. Lett.* **25**, 246.
Thompson, B. J., Newmark, J. S., Gurman, J. B., Neupert, W., Delaboudinière, J.-P., St. Cyr, O. C., Stezelberger, S., Dere, K. P., Howard, R. A., and Michels, D. J.: 1999, *Astrophys. J.* **517**, L151.
Thompson, B. J., Cliver, E. W., Nitta, N., Delannée, C., and Delaboudinière, J.-P.: 2000, *Geophys. Res. Lett.*, submitted.
Torsti, J., Kocharov, L., Teittinen, M., and Thompson, B.: 1999, *Astrophys. J.* **510**, L460.
Uchida, Y.: 1968, *Solar Phys.* **4**, 30.

Uchida, Y.: 1970, *Publ. Astron. Soc. Japan* **22**, 341.
Uchida, Y.: 1974, *Solar Phys.* **39**, 431.
Uchida, Y., Altschuler, M. D., and Newkirk G., Jr.: 1973, *Solar Phys.* **28**, 495.
Wagner, W. J. and MacQueen, R. M.: 1983, *Astron. Astrophys.* **120**, 136.
Webb, D. F., Cliver, E. W., Gopalswamy, N., Hudson, H. S., and St. Cyr, O. C.: 1998, *Geophys. Res. Lett.* **25**(14), 2469.
Weiss, A. A.: 1965, *Australian J. Phys.* **18**, 167.
Wild, J. P., Sheridan, K. V., and Neylan, A. A.: 1959, *Australian J. Phys.* **12**, 369.
Wild, J. P., Smerd, S. F., and Weiss, A. A.: 1963, *Ann. Rev. Astron. Astrophys.* **1**.
Wild, J. P.: 1970, *Proc. Astron. Soc. Australia* **1**, 365.
Wills-Davey, M. and Thompson, B. J.: 1999, *Solar. Phys.* **190**, 467.
Wu, S. T. *et al.*: 1999, *Astrophys. J.*, in preparation. 1999.

STRUCTURE OF A LARGE LOW-LATITUDE CORONAL HOLE

B. J. J. BROMAGE[1], D. ALEXANDER[2], A. BREEN[3], J. R. CLEGG[1], G. DEL ZANNA[1], C. DEFOREST[4], D. DOBRZYCKA[5], N. GOPALSWAMY[6], B. THOMPSON[7] and P. K. BROWNING[8]

[1]*Centre for Astophysics, University of Central Lancashire, Preston, U.K.*
[2]*Lockheed Martin Solar and Astrophysics Lab., Palo Alto, CA, U.S.A.*
[3]*Max-Planck Institut für Aeronomie, Katlenburg-Lindau, Germany*
[4]*SOHO SOI/MDI, Stanford University, / NASA GSFC, Greenbelt, MD, U.S.A.*
[5]*Harvard-Smithsonian Center for Astrophysics, Cambridge, MA, U.S.A.*
[6]*The Catholic University of America / NASA GSFC, Greenbelt, MA, U.S.A.*
[7]*Space Applications Corp. / NASA GSFC, Greenbelt, MD, U.S.A.*
[8]*University of Manchester Institute of Science and Technology, Manchester, U.K.*

(Received 17 September 1999; accepted 28 November 1999)

Abstract. Coronal holes on the Sun are the source of high-speed solar wind streams that produce magnetic disturbances at the Earth. A series of multi-wavelength, multi-instrument observations obtained during the 1996 'Whole Sun Month' campaign examined a large coronal hole in greater detail than ever before. It appeared on the Sun in August, and extended from the north pole to a large active region in the southern hemisphere. Its physical and magnetic structure and subsequent evolution are described.

1. Introduction

The existence of coronal holes as regions of open magnetic field lines on the Sun was predicted in the 1960s (Wilcox, 1968). They were subsequently seen from space observations as striking dark features in X-ray and EUV (extreme ultraviolet) images of the Sun, and were confirmed as sources of the recurrent high-speed solar wind streams which produce geomagnetic disturbances (Krieger, Timothy, and Roelof, 1973). The nature and distribution of coronal holes over the solar surface was found to vary in a characteristic way through the 22-year solar activity cycle, appearing to be closely linked to variations in the global magnetic field of the Sun (Fox, McIntosh and Wilson, 1998). At solar minimum, coronal holes are largely confined to the poles, with opposite magnetic polarity in the north and the south. As the solar cycle progresses, the polar holes shrink and more low-latitude holes appear. Near solar maximum, the polar holes disappear altogether before re-emerging with their polarity reversed. They then increase in size as the solar activity decreases towards minimum again (Wang, Hawley, and Sheeley, 1996).

A series of SOHO (Domingo, Fleck, and Poland, 1995) coronal hole observations was planned for August 1996, then at the minimum of the solar cycle. At about the same time, the 'Whole Sun Month Campaign' (10 August–8 September

The physics of the solar corona and transition region. Part II.
Reprinted from Solar Physics **193,** 2000.

1996) was scheduled, with the aim of characterising the Sun at solar minimum by observing it with many different instruments over one Carrington rotation. The classic eclipse picture at solar minimum consists of an axisymmetric corona with large polar holes on either side of equatorial helmet streamers, suggesting that EUV images would reveal a quiet Sun with near-circular polar coronal holes and no strong active regions on the disk. However, during August 1996, the situation was very different: on one side of the Sun the classic solar minimum conditions prevailed, but on the other there was a complex active region just south of the equator, associated with a large low-latitude coronal hole. This latter feature became known as the 'Elephant's Trunk' because of its characteristic shape.

Many characteristics of this coronal hole have been reported already. Here, some new results concerning the boundary regions are presented and discussed in relation to its magnetic structure and to some of the earlier findings.

2. Observations of the Elephant's Trunk Coronal Hole

The passage of this large coronal hole was observed by a number of coronal and in-situ particle instruments, both in space and at ground-based observatories (see Table I). Some examples of the different data obtained are shown in Figure 1.

2.1. SOHO EIT DATA

The EIT image (top left) was formed from the Fe XII line at around 195 Å, emitted by coronal plasma at a temperature of about 1.6×10^6 K. The large active region appears as a bright feature in the south-east quadrant of the Sun (east is to the left on the Sun). The Elephant's Trunk coronal hole is the long dark feature extending from a broad, triangular-shaped cap at the north pole, vertically down along the meridian to just below the equator, where the tip of the 'trunk' curves round to the east and connects to the western side of the active region. To its east, running almost parallel to the boundary of the hole there is clearly another dark feature. This is a prominence channel (Mackay, 1997), formed in the magnetic arcade structure which abuts the open field region of the coronal hole. They are frequently observed running parallel to the polar hole boundaries. The prominence channel appears dark in this wavelength because it contains much cooler material (at about 10^4 K) and so does not emit in Fe XII. The darkness of the coronal hole, on the other hand, is due to lower density as well as somewhat reduced temperature (about 8×10^5 K) of the plasma, which is here free to expand into space along the open field lines. Otherwise, apart from a small active region just south of the hole and a small southern polar hole, the rest of the solar disk consists of 'quiet sun', where the coronal plasma is constrained by small magnetic loop structures. The boundary of the coronal hole seen here is quite irregular, reflecting the structure of the adjacent closed-field loops.

TABLE I

Instruments observing the coronal hole in August 1996.

Instrument		Type of Observation
EUV Imaging Telescope (Delaboudinière *et al.*, 1995)	EIT	Images in 4 EUV wavelengths, emitted from levels in the solar atmosphere from the upper chromosphere to the corona
Coronal Diagnostic Spectrometer (Harrison *et al.*, 1995)	CDS	EUV spectra in the range 151–785 Å
Solar Ultraviolet Measurements of Emitted Radiation (Wilhelm *et al.*, 1995)	SUMER	EUV spectra in the range 500–1610 Å
Michelson Doppler Imager (Scherrer *et al.*, 1995)	MDI	The line-of-sight component of the photospheric magnetic field
UltraViolet Coronagraph Spectrometer (Kohl *et al.*, 1995)	UVCS	Coronal emission in the H I Lyman-α (1216 Å) and O VI (1032, 1037 Å) UV lines, at between 1.5 and 2.5 $R_\odot$ (solar radii) from Sun centre
Yohkoh Soft X-ray Telescope (Tsuneta *et al.*, 1991)	SXT	X-ray images of the Sun
Nobeyama radioheliograph (Nakajima *et al.*, 1994)	Nobeyama	Ground-based radio images of the Sun at 17 GHz
Kitt Peak National Observatory solar telescope	KPNO	Solar images in the 10830 Å infrared spectral line and chromospheric magnetograms
Ulysses spacecraft (Balogh *et al.*, 1994)	*Ulysses*	*In-situ* solar wind measurements at several AU
WIND satellite (Ogilvie *et al.*, 1995)	WIND	*In-situ* solar wind measurements at 1 AU
European Incoherent SCATter radar (Breen *et al.*, 1996)	EISCAT	Solar wind measurements between the Earth and the Sun using the interplanetary scintillation technique

2.2. *Yohkoh* X-RAY DATA

The *Yohkoh* (SXT) image is formed by coronal X-ray emission from a somewhat broader range of plasma temperatures above about a million degrees, but with a sensitivity biased towards hotter emission in the 3 to 6 million degree range. For this reason the Sun looks a little different here, with the coronal hole appearing slightly wider, particularly at high latitudes where the region to the west of the hole appears darker in X-rays than in the EUV. This suggests that there is less high-temperature material to the west of the hole than on the eastern side (see below). The smoother appearance of the boundary in X-rays also reflects the physical structure at higher temperatures, and possibly indicates the increased scale and a

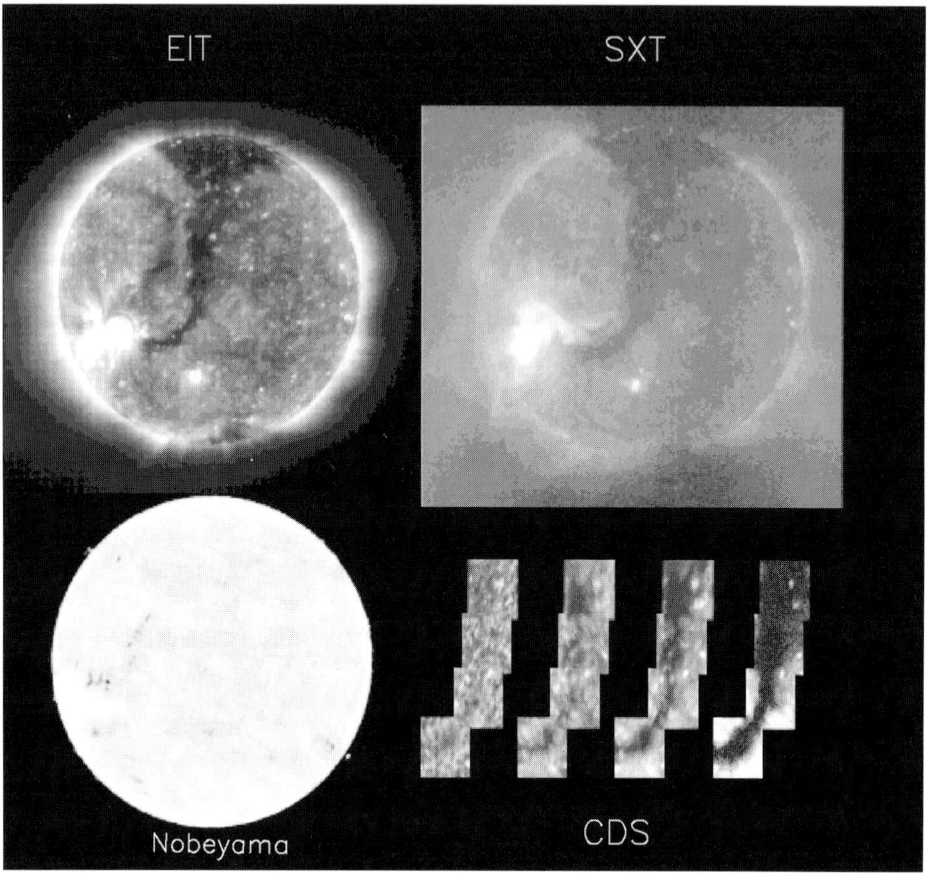

Figure 1. Images of the Elephant's Trunk obtained on 26 August 1996 by four instruments: EIT on SOHO (EUV wavelength); SXT on the Japanese satellite *Yohkoh* (X-rays); radioheliograph from Nobeyama, Japan (17GHz); CDS on SOHO (four separate EUV wavelengths emitted from the chromosphere (*left*) to the corona (*right*)).

decrease in complexity of the overlying loop structures at this height, compared to those (of typical size about 100 arc sec) in the lower corona.

2.3. Ground-based Radio Observations

The Nobeyama image reveals a totally different picture at radio (centimetre) wavelengths. The cool, dense filament material in the prominence channel appears very dark, while the strongest emission is from the active region. Short-lived (tens of minutes) bright regions are evident in both the low-latitude and polar coronal holes, together with a general diffuse enhancement relative to the adjacent quiet sun region. The bright regions are associated with the underlying unipolar magnetic flux which can be seen in photospheric magnetograms. The source of this enhanced emission in coronal holes is uncertain, but seems likely to be associated with dynamical processes in the upper chromosphere or transition region (Gopalswamy *et al.*, 1999).

2.4. Expansion of the Coronal Hole with Height in the Solar Atmosphere

At the bottom right of Figure 1 are shown four monochromatic images of the Elephant's Trunk near the meridian. These were each formed from the same mosaic of rastered observations made with SOHO CDS over a period of 2–3 hours on 26 August. Each image shows the emission at a different wavelength, representing the solar plasma at four different temperatures and, hence, emission from different heights in the solar atmosphere. (Temperature in this region of the corona typically increases with height. However, it is important to bear in mind that the images do not necessarily represent an actual slice in altitude through the atmosphere. Different structures may exhibit different temperature and density distributions with height and may contain plasma with different elemental abundances, affecting the observed intensity of the spectral lines.) The images represent emission at temperatures increasing from left to right. The first image is He I, formed at about 3×10^4 K. Here the chromospheric network structure is clearly visible, with the coronal hole appearing as a slight darkening, superimposed on the network boundaries. In Mg IX and X (1×10^6 K and 1.3×10^6 K), the hole is seen more clearly but there is still some evidence of diffuse 'network' structure, possibly associated with the magnetic field lines from the network boundaries diverging above the chromosphere. The long narrow 'trunk' of the coronal hole also appears to be broken up by regions of diffuse emission. However, in the right-most image, Si XII (about 2×10^6 K) shows the hole as a broader, more continuous feature, comparable with the EIT Fe XII image.

The expansion of the hole with height is seen more clearly in Figure 2(A). The profiles presented here show intensity versus distance across the coronal hole near its southern end. They are formed by averaging the CDS and SXT data in Figure 1 over one arcmin in the north-south direction. Figure 2(B) shows profiles of intensity observed above the solar limb in the north polar hole on 17 August. (For a more detailed description of the emission above the limb, see Fludra *et al.*, 1999). With the exception of Mg IX, the height of the peak emission increases steadily with the temperature of the line. Mg IX is found to be particularly sensitive to emission from plume structures. These extend above the limb, so the upper part of plumes originating behind the limb contribute to the line-of-sight integrated intensity, enhancing the emission and extending the peak intensity of Mg IX to about 20 arc sec above the chromospheric 'limb'. Thus, assuming the relationship of peak emission to height applies also to the *low-latitude* hole (and ignoring Mg IX), an estimate was obtained of the expansion of the coronal hole boundary with height, from the upper chromosphere to the corona, and this is displayed in Figure 2(C). From this graph it is found that the field lines at the boundary make an angle (θ) with the radial direction of 37 deg ±3 deg. This is consistent with the value of θ commonly seen in EIT images at the polar coronal hole boundaries above the limb and is also similar to the value of θ for plumes close to the boundary (DeForest *et al.*, 1997).

SOHO UVCS data obtained when the most southerly part of the Elephant's Trunk was on the eastern limb of the Sun, on 23 August, show that the hole was about 40 deg (subtended at the centre of the Sun) wide at 1.75 $R_\odot$ from sun-centre, and about 60 deg wide at 2.25 $R_\odot$. This clearly suggests that the boundaries of the hole were diverging super-radially, with the mean value of θ (the angle between the boundary field line and the radial direction) reaching about 55 deg at 2.25 $\mathrm{R}_\odot$ from Sun centre. (For more details of the UVCS observations, see Dobrzycka *et al.*, 1999). This expansion of the coronal hole is confirmed by the Mauna Loa synoptic data shown in Figure 3 where the hole can be seen to have formed a gap in the streamer belt of about 30 deg. This is at a distance of 1.7 $\mathrm{R}_\odot$ from Sun centre. The expansion of the coronal hole and its effect on the streamer belt have been modelled by Zhao, Hoeksema and Scherrer (1999).

2.5. Density and Temperature in and around the Coronal Hole

CDS spectroscopic data have been analysed to investigate the density and temperature structure of the hole and surrounding boundary region (Del Zanna and Bromage, 1999). The results for the low-latitude hole are similar to those obtained for polar coronal holes (Fludra *et al.*, 1999; Bromage *et al.*, 1997), with average coronal electron densities $\simeq 2 \times 10^8$ cm^{-3}, (about one-third the values found in the surrounding quiet-Sun regions) and temperatures typically about 8×10^5 K. A region on the western boundary observed with CDS gave an average temperature of 1.1×10^6 K. In addition, *Yohkoh* data has yielded an average quiet-Sun temperature on the western boundary of 1.37×10^6 K at high latitude, while the eastern boundary was hotter, at 1.79×10^6 K. Near the southern extent of the hole the corresponding temperatures were found to be 1.53×10^6 K to the east and 1.50×10^6 K on the western boundary. These results were obtained from SXT data using the filter ratio technique of Tsuneta *et al.* (1991), and they constitute weighted average values over the line of sight. The slightly higher value obtained with *Yohkoh* is probably due to the greater sensitivity of SXT to higher temperature material.

2.6. Relation to the Solar Wind

Breen *et al.* (1999) have studied the relationship between in situ solar wind observations and the Elephant's Trunk coronal hole. They have reported observations by the Ulysses mission and the WIND satellite, of 'fast' wind emanating from within the hole as it crossed the meridian. EISCAT (Interplanetary Scintillation) observations of fast wind have also been mapped back to the region of the coronal hole, as shown in Figure 3. Here the lines of sight of the EISCAT measurements are projected onto a Mauna Loa synoptic map of the streamer belt at 1.7 $R_\odot$. The

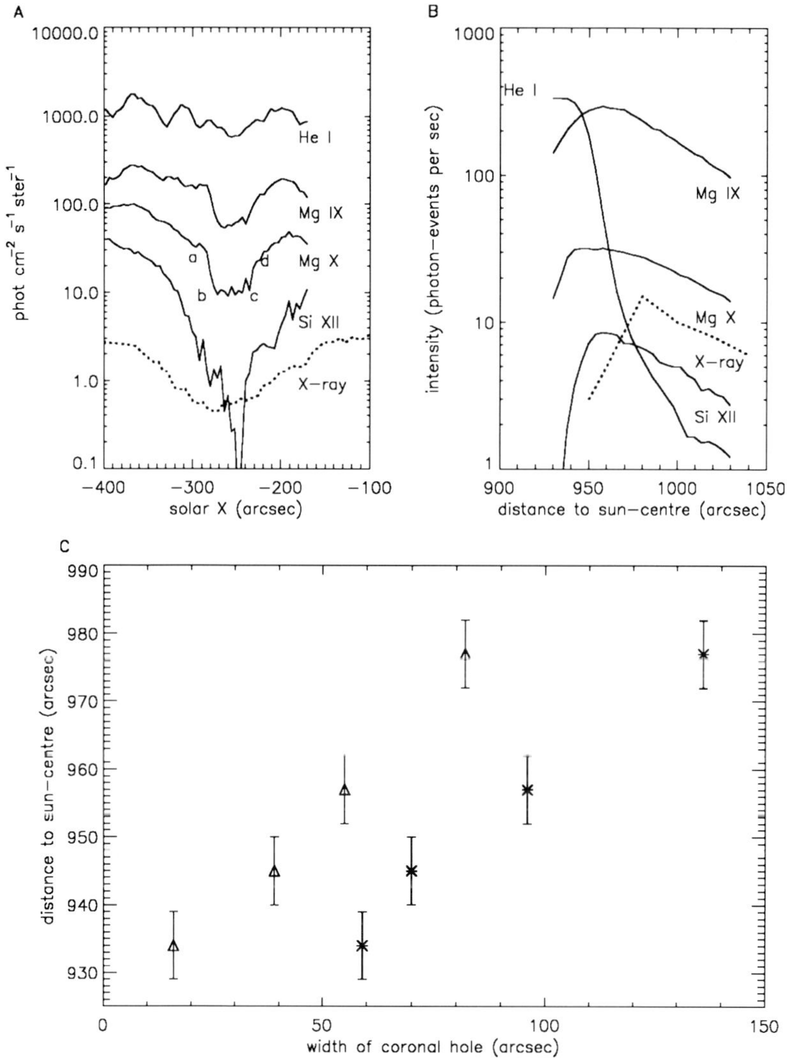

Figure 2. (A) CDS intensity profiles (solid lines) in four different wavelengths, showing the expansion of the coronal hole with height from the chromosphere (He I) to the corona (Fe XII). The points a,d shown on the Mg X profile define the 'outer boundary' of the hole and points b,c define the 'inner boundary'. The corresponding X-ray profile from *Yohkoh* is superimposed (broken line) using an arbitrary scale. (B) Average radial profiles of intensity in the same lines, obtained above the north polar limb on 17 August; the peak value in each profile shows the position of maximum emission of that line (which will be slightly higher than the peak emission from the disk because of 'limb-brightening'). The X-ray profile was obtained from *Yohkoh* observations above the west limb on 17 August. (C) The widths of the hole, a–d (triangles) and b–c (asterisks) are plotted for He I, Mg X, Si XII and the *Yohkoh* X-ray emission, versus the heights obtained from (B). All spatial units are arc secs, where the radius of the Sun ($R_\odot$) is about 960 arc sec.

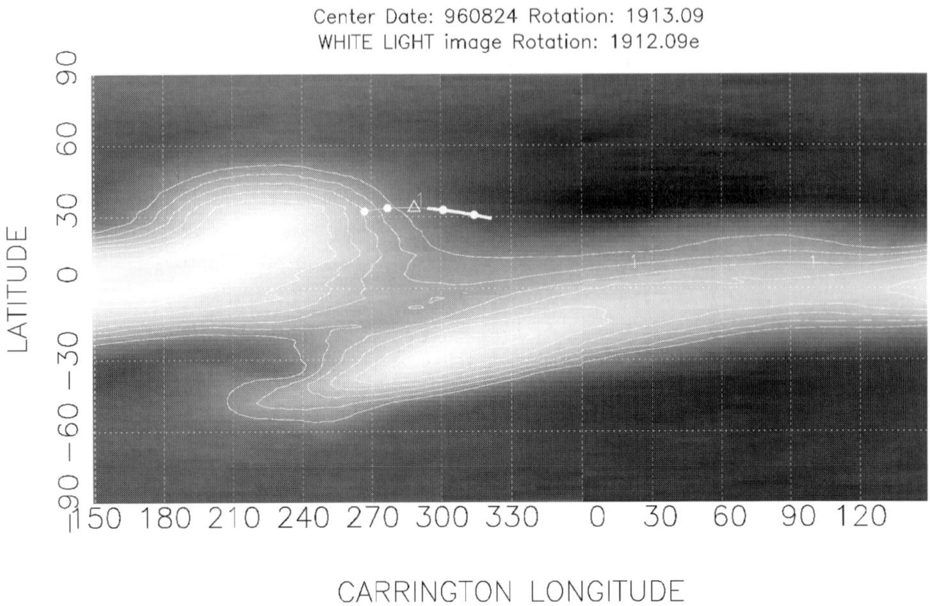

Figure 3. A synoptic image of the corona seen in white light by the coronagraph at Mauna Loa Observatory. The Mauna Loa measurements are made at a distance of 1.7 $R_\odot$ from Sun centre. This image was formed by successively plotting the daily observations of the white-light intensity of the corona on the east side of the Sun over 27 days, from Carrington rotation 1912 to 1913. It represents a cross-section through the coronal streamer belt around the Sun at a distance of 0.7 $R_\odot$ from the limb. The coronal hole appears as a gap in the streamer belt. The source of EISCAT observations of the fast solar wind is indicated by a triangle. These observations were made in rotation 1913 and map to this position in the upper part of the coronal hole.

speed of the wind coming from the *low*-latitude region of the hole was found by WIND and EISCAT to be 500–600 km s^{-1}, whereas the *Ulysses* observations which mapped back to broader, *high*-latitude regions of the coronal hole, yielded more typical 'fast' wind speeds of 700–750 km s^{-1}. This is consistent with the findings of Wang, Hawley, and Sheeley (1996) that the solar wind speed from a hole is related to its size.

EISCAT results also suggest that 'slow' solar wind (about 350 km s^{-1}) emanates predominantly from the eastern (hotter) boundary region of the hole (Breen *et al.*, 1999). This result may, with further study, help towards understanding the mechanism of slow solar wind production, which appears to be associated with the boundary regions of coronal holes.

3. Magnetic Structure of the Elephant's Trunk

A comparison between MDI photospheric magnetograms and EIT images of the Elephant's Trunk (Bromage *et al.*, 1997) reveals an area of largely unipolar magnetic field within both the hole itself and the region immediately to the east of it (i.e., between the hole and the prominence channel). This is also the dominant polarity in the connecting north polar hole. To the east of the prominence channel is a corresponding area of the opposite polarity, with a neutral region following the line of the channel.

Figure 4 (upper), modelled from Kitt Peak chromospheric field data of 28 August, clearly shows the north polar crown arcade structure turning southward and extending down to the active region just below the equator, connecting to an area of the same polarity as that in the Elephant's Trunk. This model, based on the spherical harmonic technique of Nakagawa (1973), shows the neutral radial field surface ($B_r = 0$) through the north and south polar crowns, connecting across the active region. It is useful to compare this with Figure 4 (lower) which shows the dramatically contrasting situation on the Sun in May 1997. Then there were two separate polar crowns, and two bands of arcade-like structures encircling the Sun at the latitudes of the recently emerged new-cycle magnetic flux. The Elephant's Trunk coronal hole of August 1996 is thus defined by (a) the boundary of the loop structure containing the prominence channel and (b) a similar but less well-defined region of closed field to the west of the hole. The presence of these closed field structures is clearly integral to the existence of the coronal hole and likely to be instrumental in its formation.

Further details of modelling the magnetic structure of this coronal hole using MDI photospheric field data are reported in Clegg, Bromage, and Browning (1999) and Breen *et al.* (1999).

4. Further Evolution of the Elephant's Trunk

Small-scale changes in the shape of the boundary could be seen from EIT images to occur on a time-scale of a few hours, with sections of the hole sometimes appearing to close and re-open. However, the main structure persisted for several months, appearing to rotate quasi-rigidly with the active region, with a synodic period of 27 days (Bromage *et al.*, 1997; Zhao, Hoeksema, and Scherrer, 1999). At the next passage, in September, the lower part of the hole was still firmly fixed to the active region but differential rotation seemed to be affecting the structure at higher latitudes, with the northern part of the prominence channel appearing to bend round more towards the east at this time. The hole itself had closed near the top of the 'trunk', separating it from the polar hole. Remnants persisted for a further two or three months, still rotating with the active region.

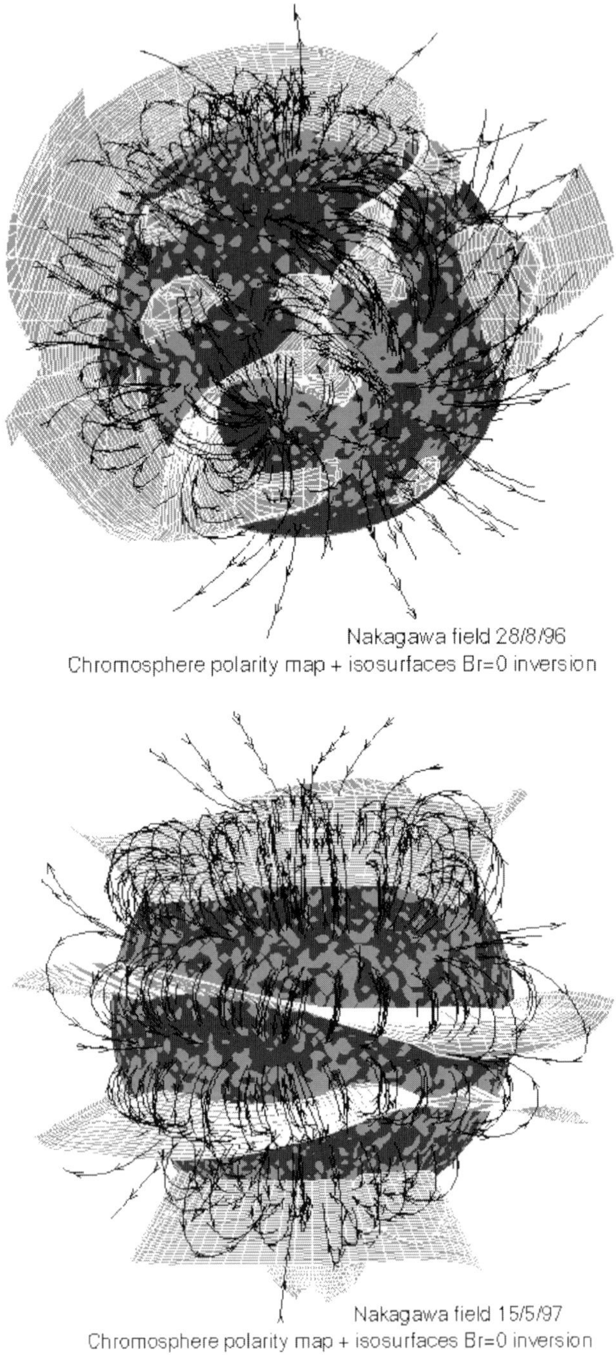

Figure 4. Models of the solar magnetic field (*above*) on 28 August 1996, showing the Elephant's Trunk and, for comparison (*below*) on 5 May 1997, when the new-cycle active regions had begun emerging. The light and dark areas on the disk indicate the chromospheric magnetic polarity seen by Kitt Peak Solar Observatory, while the lighter, hatched surfaces represent the 3-dimensional surfaces where the radial magnetic field is zero. These form the axis of magnetic arcade structures such as the 'polar crown' which normally surrounds each polar coronal hole.

5. Discussion and Conclusions

The equatorial regions of the Elephant's Trunk coronal hole have a similar density and temperature structure to the polar holes. The low-latitude section of the coronal hole was found to be expanding superradially away from the Sun, at an angle to the radial of about 37 deg, in passing from the transition region to the corona. This angle, θ, increased to about 55 deg at 2.25 $R_\odot$ from Sun centre. The value of θ obtained near the surface is similar to that observed at the limb, at the boundary of the polar holes. The superradial angle of polar plumes depends on this angle θ at the interface between open and closed field, with both angles being a result of pressure balance as the magnetic field configuration changes at the boundary of the coronal hole.

The fast solar wind appears to originate from within the narrow 'trunk' section of the hole, although with a somewhat slower speed than the typical fast solar wind. This is consistent with the findings of Wang, Hawley, and Sheeley (1996) that solar wind speed is related to the width of the coronal hole. Breen *et al.* (1999) observed slower wind appearing to emanate from the eastern boundary of the hole which was found here to be hotter than the western boundary, particularly at higher latitude. There may be a connection between the enhanced temperature here and the generation of the slow wind. The maintenance of such a rigid open-field structure against differential rotation at higher latitudes could be achieved by continuous reconnection between open and closed field lines at the boundary of the coronal hole (Wang, Hawley, and Sheeley, 1996). Plasma heating and solar wind generation could both be by-products of this reconnection, resulting in the observed difference in temperature between the eastern and western boundaries.

Whilst detailed structure in the boundary of the hole in the lower corona varied on a time-scale of hours, the overall feature rotated quasi-rigidly for several months, as was the case with the Skylab 'CH1' hole (Nolte *et al.*, 1976), the first transequatorial coronal hole to be observed. This apparently rigid rotation has also been seen in other (although not all) low-latitude holes (Insley, 1995). However, differential rotation slowly distorted the structure at high latitude, eventually causing disconnection from the polar hole in later disk passages. The development of this large polar hole extension and the subsequent disconnection from the polar hole represent a transfer of unipolar flux from the north polar hole to low latitudes – perhaps heralding the beginning of the decline of the poloidal field as the solar cycle rises towards maximum. Migration of flux poleward to eventually cancel with flux of the opposite polarity in the polar hole is thought to be the main way of reducing the poloidal field in the rise phase of the solar cycle. However, the type of event reported here might make a significant contribution to the change of the solar field from predominantly poloidal at solar minimum, to toroidal at solar maximum.

Acknowledgements

The authors are grateful for the provision of data by the SOHO CDS, MDI and EIT instrument teams. SOHO is a project of international co-operation between ESA and NASA. We would like to thank the Director and staff of EISCAT for data used in this study; EISCAT is supported by the scientific research councils of Finland, France, Germany, Japan, Norway, Sweden and the U.K. *Yohkoh* is a mission of the Japanese Institute for Space and Astronautical Science. We are grateful for provision of data by the Kitt Peak National Solar Observatory. The Nobeyama radioheliograph was built with the support of the Ministry of Education and Culture, Japan. DA was supported by NASA under contract NAS-8-37334, and NG by NASA grant NAG-5-6139. BJIB, JRC and GDZ have made use of Starlink computing support provided by PPARC. JRC is supported by a PPARC grant.

References

Balogh, A. *et al.*: 1994, *Phil. Trans. R. Soc. London A* **349**, 227.
Breen, A. R. *et al.*: 1996, *J. Atmospheric Terrest. Phys.* **58**, 507.
Breen, A. R., Mikic, Z., Linker, J. A, Lazarus, A. J., Thompson, B. J., Biesecker, D. A., Moran, P. J., Varley, C. A., Williams, P. J. S., and Lecinski, A.: 1999, *J. Geophys. Res.* **104** (A5), 9847.
Bromage, B. J. I., Del Zanna, G., Fludra, A., DeForest, C., Thompson, B.: 1997, 31st ESLAB Symp., *ESA Publication* **SP-415**, 259.
Clegg, J. R., Bromage, B. J. I., and Browning, P. : 1999, *J. Geophys. Res.* **104** (A5), 9831.
DeForest, C. E. *et al.*: 1997, *Solar Phys.* **175**, 393.
Delaboudinière, J.-P. *et al.*: 1995, *Solar Phys.* **162**, 291.
Del Zanna, G. and Bromage, B. J. I.: 1999, *J. Geophys. Res.* **104** (A5), 9753.
Dobrzycka, D., Strachan, L., Panasyuk, A., Cranmer, S. R., Kohl, J. L.: 1999, *J. Geophys. Res.* **104** (A5).
Domingo, V., Fleck, B., and Poland, A. I.: 1995, *Solar Phys.* **162**, 1.
Fludra, A., Del Zanna, G., Alexander, D. and Bromage, B. J. I.: 1999, *J. Geophys. Res.* **104** (A5), 9709.
Fox, P., McIntosh, P., and Wilson, P. R.: 1998, *Solar Phys.* **177**, 375.
Gopalswamy, N. *et al.*: 1999, *J. Geophys. Res.* **104** (A5), 9767.
Harrison, R. A. *et al.*: 1995, *Solar Phys.* **162**, 233.
Insley, J. E.: 1995, *Solar Phys* **160**, 1.
Kohl, J. L. *et al.*: 1995, *Solar Phys.* **162**, 313.
Krieger, A. S., Timothy, A. F., and Roelof, E. C.: 1973, *Solar Phys.* **29**, 505.
Mackay, D. H. *et al.*: 1997, *Astrophys. J.* **486**, 534.
Nakagawa, Y.: 1973, *Astron. Astrophys.* **27**, 95.
Nakajima, H. *et al.*: 1994, *Proc. IEEE* **82**, 705.
Nolte, J. T. *et al.*: 1976, *Solar Phys.* **46**, 303.
Ogilvie, K. W. *et al.*: 1995, *Spare Sci. Rev.* **71**, 55.
Scherrer, P. H. *et al.*: 1995, *Solar Phys.* **162**, 129.
Tsuneta, S. *et al.*: 1991, *Solar Phys* **136**, 37.
Vaiana, G. S.: 1973, *Astrophys. J.* **185**, L47.

Wang, Y.-M., Hawley, S. H., and Sheeley Jr., N. R.: 1996, *Science* **271**, 417.
Wilcox, J. M.: 1968, *Space Sci. Rev* **8**, 258.
Wilhelm, K. *et al.*: 1995, *Solar Phys.* **162**, 189.
Zhao, X. P., Hoeksema, J. T., and Scherrer, P. H.: 1999, *J. Geophys. Res.* **104** (A5), 9735.

ELECTRO-MECHANICAL COUPLING BETWEEN THE PHOTOSPHERE AND TRANSITION REGION

T. D. TARBELL[1], M. RYUTOVA[2] and R. SHINE[1]
[1]*Lockheed Martin Solar and Astrophysics Laboratory, 3251 Hanover Street, Palo Alto, CA 94304, U.S.A. (E-mail: tarbell@lmsal.com)*
[2]*Lawrence Livermore National Laboratory, Institute of Geophysics and Planetary Physics, Livermore, CA 94550, U.S.A.*

(Received 28 September 1999; accepted 24 November 1999)

Abstract. We study the response of the chromosphere and transition region to dynamic changes in the photospheric network magnetic fields. We present results from simultaneous measurements taken by TRACE in chromospheric and transition region (C IV) images, high-resolution magnetograms taken by MDI, and spectra of chromospheric (C II) and transition region lines (O VI) obtained with the SUMER instrument on SOHO. Enhanced emission in the C IV line is generally co-spatial with the magnetic pattern in the photosphere. We propose a mechanism of electro-mechanical coupling between the photosphere and upper layers of atmosphere based on hydrodynamic cumulation of energy produced by reconnecting flux tubes in the photosphere/chromosphere region (Tarbell *et al.*, 1999). We believe that a basic process causing energetic events is the cascade of shock waves produced by colliding and reconnecting flux tubes. The continuous supply of flux tubes in the 'magnetic carpet' ensures the ubiquitous nature of this process and its imprint on the upper atmosphere. The appearance of bright transients often, but not always, correlates with canceling mixed polarity magnetic elements in the photosphere. In other cases, transients occur in regions of unipolar flux tubes, suggesting reconnection of oblique components. Transients are also seen in regions with no fields detected with the MDI sensitivity; these may be reconnections of tiny features with diameters less than 100 km. Blinkers and other bright transients are often accompanied by two directional plasma jets. These may be generated by cylindrical self-focusing of shock fronts or by collision of shocks produced by neighboring reconnection processes. The observations suggest that stronger emissions correspond to lower velocity jets, and vice versa; this property is a natural consequence of the proposed mechanism. Plasma flows are always seen whenever the slit crosses strong magnetic flux tubes or vertices of converging flows in the supergranular network. The overall energy distribution between heating and plasma flows is an intrinsic feature of our mechanism.

1. Introduction

The unique opportunity to compare simultaneous time series of data of the photosphere and overlying layers of atmosphere provided by the Transition Region and Coronal Explorer (TRACE) and Solar and Heliospheric Observatory (SOHO) opens a new phase-space for exploring and understanding the energy transfer mechanisms in the solar atmosphere. Although earlier ground-based and space observations have accumulated rich information on particular classes of events, from properties of the photospheric fields to systematics of flares and background pa-

The physics of the solar corona and transition region. Part II.
Reprinted from Solar Physics **193,** 2000.

rameters of the solar wind, the links between various events have been subjects of suggestions and modeling. In this paper we present observational evidence for a clear connection between dynamic changes in the photospheric magnetic fields and energetic events in the chromosphere and transition region.

Many regularities in the observed properties of mass flows in the chromosphere and transition region have been found in the first detailed studies performed by the High-Resolution Telescope and Spectrograph (HRTS) (Dere, Bartoe, and Brueckner, 1989; Dere, 1994; Kjeldseth-Moe *et al.*, 1994, and references therein). Flow velocities observed in a wide temperature range (3×10^4–3×10^5 K) vary from a few tenths of km s^{-1} at the chromosphere up to hundreds of km s^{-1} in the transition zone. The most persistent flows in the chromosphere are associated with spicules which exhibit 20–25 km s^{-1} flows directed predominantly upward (although $\frac{1}{3}$ of spicules show downward motions). Unlike quasi-steady spicular flows, short-lived jet-like phenomena are observed with somewhat lower velocities as well as with velocities exceeding 80 km s^{-1} (Dere, 1994). Upward mass flows seen in the transition region are usually adjacent to downflows and often appear like two directional plasma jets. In some cases downflows exceed the upflows and may even appear as one directional jets. The velocity amplitudes of downflows are observed to increase with temperature, reaching a maximum at 10^5 K, and decrease at higher temperatures (Kjeldseth-Moe *et al.*, 1988). In regions of maximum downflows, complex dynamical structures with multiple velocities are observed. Explosive events are observed in the transition zone as small scale regions ($\approx$ 2 arc sec) with high velocity flows ranging from 100 to 400 km s^{-1}. The velocity field in explosive events seems to be isotropic (see, e.g., Dere, Bartoe, and Brueckner, 1989). It was also found that the hot localized regions emitting soft X-rays correspond to lower velocity regions of 40 km s^{-1} rather than the high velocity explosive events (Moses *et al.*, 1994).

These results have been confirmed and extended in the recent studies performed with Coronal Diagnostic Spectrometer (CDS) (Wikstol *et al.*, 1997; Brekke, Hassler, and Wilhelm, 1997) and the instrument for Solar Ultraviolet Measurements of Emitted Radiation (SUMER) (Innes *et al.*, 1997) on SOHO. For example, comparison of high-resolution MDI (Michelson Doppler Imager) magnetograms with simultaneous CDS images of the chromosphere and transition region (Wikstol *et al.*, 1997; Tarbell *et al.*, 1999) show that sporadic bright transient emissions, 'blinkers' (Harrison *et al.*, 1997), strongly correlate with the cancelation (total or partial) of opposite polarity network magnetic fields in the photosphere. It was also found that X-ray bright points (and other bright areas in the transition region) do not necessarily correspond to the strong flows (Kjeldseth-Moe *et al.*, 1988).

Tarbell *et al.* (1999) have shown that ubiquitous small-scale magnetic flux 'tubes' constantly emerging from subsurface layers may cause the formation of plasma jets and a sporadic excess of temperature near the solar surface. They proposed a mechanism of energy release in the solar atmosphere associated with strong hydromagnetic activity of the photospheric flux tubes. Brought together by convective motions, flux tubes collide and reconnect, creating a sling-shot effect which generates complex 3-D shock waves with curved surfaces. Self-focusing of these shocks occurs as they propagate upward in the rarefied atmosphere. Depending on the geometry of the shock collision, highly concentrated energy may be either converted entirely into heat or into strong jets, or be distributed between the two.

In the present paper we confirm the results obtained by Tarbell *et al.* (1999) and extend our studies to simultaneous measurements of the enhanced emission and plasma flows in the chromosphere and transition region and their connection to changes in the photospheric magnetic fields. We use observations collected by MDI, TRACE and SUMER in SOHO Joint Observing Program (JOP) 72, on 16 May 1998. High-resolution magnetograms taken by MDI compiled in the 2.2-hour movie show changes in small scale magnetic flux concentrations. TRACE images in the C IV lines and UV continuum are taken simultaneously with 15 s cadence and co-aligned with the MDI magnetograms and SUMER data. The field of view is 128 $\times$ 320 arc sec of very quiet Sun near disk center. The time series of C IV images shows the appearance and evolution of bright transient phenomena. The SUMER spectra allow us to study mass flows along the slit position. We use here C II and O VI lines which correspond to chromospheric ($\sim 2.5 \times 10^4$ K) and transition region ($\sim 3 \times 10^5$ K) temperatures. These observations show distinctive properties in the response of the chromosphere and transition region to dynamic changes in the photospheric network. These properties include the following elements:

(1) At any moment of time the enhanced emission in the C IV line generally mimics the magnetic pattern. In transients, the C IV emission pattern has interesting, complex shapes, and changes dramatically in tens of seconds.

(2) The appearance of bright transients often correlates with reduction of magnetic flux in the photosphere, observed in canceling mixed polarity elements. Sometimes, however, it occurs in regions of unipolar flux tubes or no strong magnetic fields (at the MDI sensitivity).

(3) Blinkers and other bright transients are often accompanied by two directional plasma jets. In bright transients, the jets in the SUMER spectra are sometimes observed several thousands of kilometers away from the center of a C IV transient that is over the canceling magnetic features identified in magnetograms.

(4) If the radiative transient is accompanied by plasma flows, the energy is distributed between the two: usually the stronger localized emissions correspond to lower velocity jets, and *vice versa*.

(5) Plasma flows are always seen whenever the slit crosses the site of magnetic flux tubes or a region of converging supergranules, a vertex.

(6) If the magnetic flux concentrations in the vicinity of a 'vertex' are weak or even below the MDI resolution, multiple flows are seen in cooler (C II) lines without much activity in the transition region.

(7) Multiple flows and explosive events occurring repeatedly are seen in the O VI line above the converging supergranules with significant magnetic flux concentrations.

(8) Explosive events may appear in both SUMER lines or in only one. Events seen in both also have transient emission in C IV, as would be expected from the temperatures of formation. Some events seen in only one SUMER line are very weak or absent in C IV.

We suggest that a coupling mechanism between the photosphere and upper layers of atmosphere is associated with hydrodynamic cumulation of energy produced by reconnecting flux tubes in the photosphere and chromosphere, as proposed by Tarbell *et al.* (1999). We show further that the evolution of generated mass flows may lead to several different phenomena depending on the onset of particular flow instabilities. For example, sub-Alfvénic downflows may evolve into explosive events due to the instability of negative energy waves (NEW) (Ostrovski, Rybak, and Tsimring, 1986; Ryutova, 1988). Super-Alfvénic upflows generate high-frequency Alfvén waves due to an instability similar to the classical Kelvin –Helmholtz instability (KH). If the generated flows are below the threshold of any instability, then their presence leads to strong enhancement of the Alfvén wave dissipation and a 'patchy' pattern in the energy deposition observed as a mosaic of unevenly distributed emission regions (Ryutova and Habbal, 1995).

2. Key Elements of the Coupling Mechanism

As shown by Tarbell *et al.* (1999), a unified mechanism of heating and jet formation associated with post-reconnection dynamics of photospheric flux tubes can explain many of the diverse high velocity and heating events observed in the solar chromosphere and transition region. The mechanism is made possible by the specific conditions near the solar surface, including the following.

(i) Photospheric magnetic fields are concentrated in well-defined flux tubes embedded in an almost non-magnetic environment; i.e., the plasma beta in the surrounding medium is very large, $\beta = 8\pi p_{\rm ext}/B_{\rm ext}^2 \gg 1$. At the same time, because of the pressure equilibrium, the ratio of external gas-dynamic pressure and magnetic pressure inside flux tubes is finite:

$$\beta^* = \frac{8\pi p_{\rm ext}}{B^2} \geq 1 \,. \tag{1}$$

(ii) Flux tubes are non-collinear.

(iii) The low atmosphere is sharply stratified.

Brought together by convective or other motions, flux tubes collide and reconnect. The reconnection of photospheric flux tubes results in essentially different effects than that in the coronal plasma. Unlike the corona where the reconnection process liberates a large amount of energy stored in the magnetic field (coronal

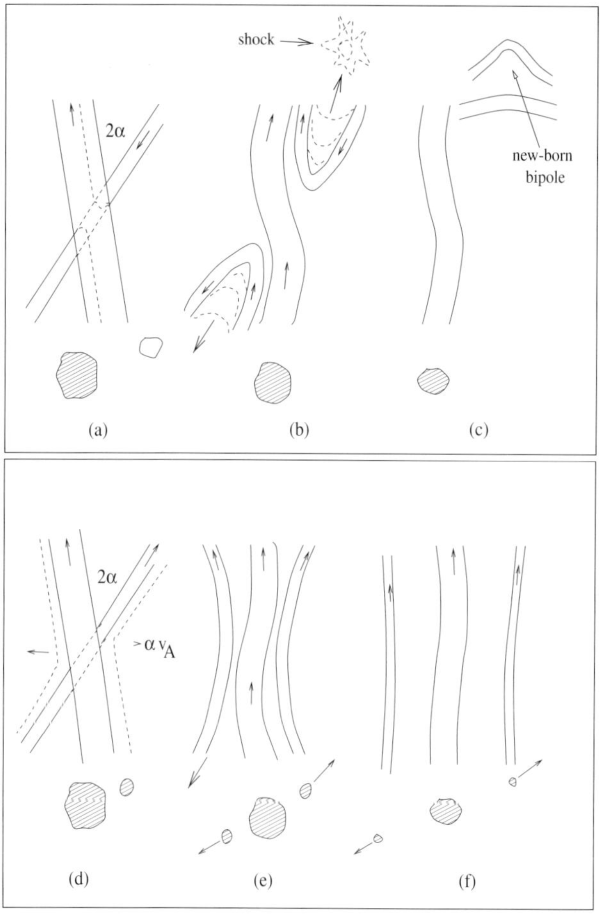

Figure 1. Reconnection of non-collinear flux tubes. The upper panel shows interaction of opposite polarity flux tubes with the apparent total cancellation of the thin flux tube and partial cancellation of the thick one: (a) just before the collision; (b) after reconnection 'sling shot' effect; (c) almost straightened segments of the reconnection product and the remnant of the flux tube. The lower panel shows a fragmentation process in a unipolar 'group' of flux tubes: (d) before the collision; (e) reconnection products moving apart with velocities $\sim \alpha v_A$; (f) the final action of the restoring magnetic force.

plasma beta $\beta = 8\pi p/B^2 \ll 1$) and deposits it 'on the spot', in the photosphere the reconnection does not give an immediate gain in the energy. It rather puts the system in an unsteady state whose further evolution is determined by the dynamics of the reconnection products. After reconnection, strongly curved photospheric flux tubes behave like elastic bands: straightening and shortening they create a sling-shot effect triggering strongly nonlinear processes in the external plasma. Note that non-collinearity of flux tubes plays an important role: first, reconnection can occur whether the colliding fluxes tubes are of the same or opposite polarities, and second, the probability of collision and reconnection is much higher than that for parallel flux tubes. Examples of the interaction of flux tubes are shown in Figure 1; a collision and reconnection of opposite polarity elements is shown in the upper panel and same polarity elements in the lower panel. In this example we show the interaction of flux tubes with different cross-sections to emphasize the fact that generally only a limited portion of magnetic flux may participate in each elemental act of reconnection.

When two flux tubes are not collinear they can overlap only in a short region. The characteristic 'radius' of this region, R, can be estimated as follows. For the flux tube to participate in the reconnection process over its whole cross-section (πR^2), the characteristic reconnection time should be less than a straightening time. We assume that the reconnection is fast. The maximum rate in the fast reconnection process (Petschek, 1964) is estimated as $w_{\max} \simeq v_{\rm A}/\ln Re_m$, where Re_m is magnetic Reynolds number. The reconnection time is, therefore $\tau_{rec} \simeq R \ln Re_m/v_{\rm A}$. The time for the flux tube straightening is $t \simeq L/2v_{\rm A}$, where L is a characteristic length of the flux tube. This gives the following estimate for the maximum radius at which the flux tube participates 'completely' in one act of reconnection:

$$R_{\max} < \frac{L}{2 \ln Re_m} \,. \tag{2}$$

For photospheric conditions the Reynolds number may be taken between 10^3–10^6. For $L = 10^3$ km, the estimate (2) gives $R_{\max} \sim$ 36–70 km. Therefore, magnetic flux concentrations having as small a radius as 150–200 km may be the sites of multiple reconnection processes. Moreover, the straightened segment of the reconnection product (Figure 1) may participate further in similar processes.

In each act of reconnection the kick produced by the sling-shot effect generates acoustic or MHD waves. Those waves that propagate upward in the stratified atmosphere quickly become evolutionary shocks. These are 3-D shocks with a complex uneven surface. As it propagates further upward, the front of a shock will become more uneven: faster elements of a shock-front will lead and slower elements will lag. At some height self-focusing and collision of the shock-fronts occurs leading to hydrodynamic cumulation; i.e., concentration of the kinetic energy of a system in a small volume. As shown by Tarbell *et al.* (1999), depending on the geometry of the shock collision, highly concentrated energy may be either converted entirely into heat or into strong jets, or be distributed between the two. If the shocks experience a head-on collision (fronts of colliding shocks are parallel), immediately after the collision the plasma behind the reflected shocks is at rest. Therefore all the initial kinetic energy is converted into thermal energy of plasma confined in a small volume. If the colliding shocks are oblique and intersect at some angle, 2θ, the reflected shocks will form a finite angle, 2ϕ. In this case, a temperature increase occurs in a larger volume; i.e., heating becomes less efficient, and the angle between the reflected shock fronts gives the plasma space to escape. Therefore, only part of the kinetic energy of the system is converted into heat, and another part is converted into directed flows of plasma or cumulative jets. With

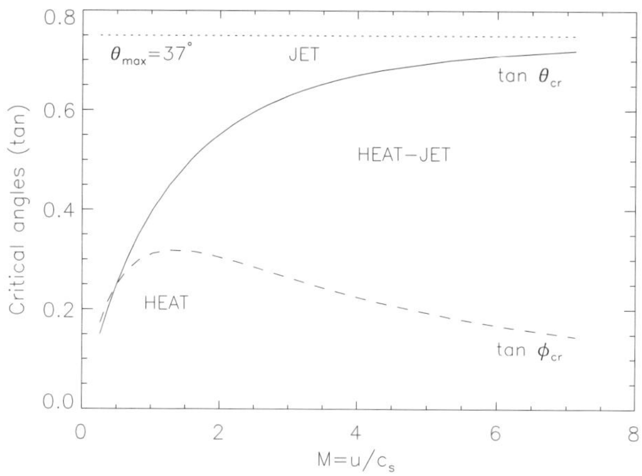

Figure 2. Critical angles of shock front convergence θ_{cr} and reflection ϕ_{cr} vs Mach number, defining regions of predominant heating, jet formation, and distribution of energy between the two. The dotted line shows the limiting angle corresponding to maximum jet velocity. (After Tarbell *et al.*, 1999.)

increasing collision angle, θ, the velocity of the jet increases at the expense of local heating. At some critical angle there is no reflected wave and most of the energy is concentrated in hot and dense plasma jets. Figure 2 shows the critical angles of shock front convergence, θ_{cr}, and corresponding angles of reflection, ϕ_{cr}, vs Mach number: regions of predominant heating (almost head-on collision), jet formation, and distribution of energy between the two are indicated. Note that the largest range of parameters corresponds to the region between the two curves where the energy is distributed between enhanced heating and cumulative jets. In the middle of this region energy may be distributed quite equally between the radiative transients and plasma flows. Closer to the solid curve the heating will be weaker and flows will be stronger, while closer to the dashed curve the stronger transients will be accompanied by weaker flows. As one can see in Figure 2, with increasing angle of collision more energy goes into plasma jets. At some critical angle the entire energy is converted into hot cumulative jets. It is important to note that the enhanced heating and jet formation are processes occurring *after the shock collision*; even without collision, the regular 'behind-shock' heating occurs in any case of reconnecting magnetic flux tubes.

The processes described above also occur if colliding shocks are produced by neighboring reconnection events. In a 3-D geometry there may occur a violent event of cylindrical focusing, when the initial shock-front is close to a figure of revolution (Guderley, 1942; Landau and Lifshits, 1987). In the case of a cylindrical cumulation, strongly collimated two-directional hot jets of material are formed in the intersection point. There is not only a considerable increase of temperature, but also an increase of the density behind the reflected shock fronts. As shown by Tarbell *et al.* (1999), estimates of the temperature increase, predicted jet velocities, and height of efficient energy release correspond to transition region parameters.

We present observational data that confirm and extend the earlier findings, strongly supporting the efficiency of the scenario: reconnection of flux tubes in the photosphere → shocks produced by the sling-shot effect → gradient acceleration and collision of shocks → energy release in the form of jets and transients. We believe that this process may serve as a natural mechanism of coupling between the dynamics of the photospheric network and energetic events in upper layers of atmosphere.

3. Observations

JOP 72 was designed to study waves and transient phenomena in the chromosphere and transition region with very high time resolution. In the observations described here, SUMER, MDI, and TRACE were pointed at the same region of quiet sun near disk center. Time series were collected on 16 May 1998, by each instrument at the highest possible cadence for the observing mode used. MDI took continuum filtergrams, velocities, and magnetograms in its high resolution mode (0.6 arc sec pixels, 1.2 arc sec resolution), with the usual 60 s cadence. TRACE ran a special program optimized for cadence (15 s) in its 1700, 1600, and 1550 Å bands. The field of view (128 × 320 arc sec) was restricted to include the neighborhood of the SUMER slit but minimize telemetry and cycle time. These three images were combined using the technique of Handy *et al.* (1999) to make C IV images, proportional to the integrated intensity in the 1548 and 1551 Å lines. The TRACE images were co-aligned with the MDI magnetogram at each time step using a cross-correlation between the 1700 Å intensity and the absolute value of the magnetogram; slight stretching was also done to match the image scale and aspect angle differences between SOHO and TRACE. The result was three co-aligned movies of the Sun as seen from SOHO with a pixel scale of 0.5 arc sec.

The SUMER spectra described here were taken with a 1 × 360 arc sec slit moving to track solar rotation and ∼ 12 s cadence. We show here profiles and integrated intensities of the C II 1037.0 Å and O VI 1037.6 Å lines. Co-alignment with TRACE was achieved first by making the two-dimensional space-time image of integrated O VI intensity. Then an equivalent C IV image was synthesized by assuming a slit position and extracting the proper column from the TRACE image at each time step. These two images were compared, first visually and then by a cross-correlation technique, to determine the true position of the SUMER slit in the TRACE (and hence MDI) datacubes. The bright transient events seen in both C IV and O VI made this co-alignment possible, and we estimate its accuracy to be better than 2 arc sec. The SUMER slit position was found to deviate from the stated position by 1 arc sec EW and 6 arc sec NS, with respect to the MDI coordinate system. The MDI high-resolution magnetogram of the studied region is shown in Figure 3 (the second panel) together with the TRACE images in 1700 Å (the first panel) and C IV line (the third panel) and the SUMER spectra (fourth panel) in two

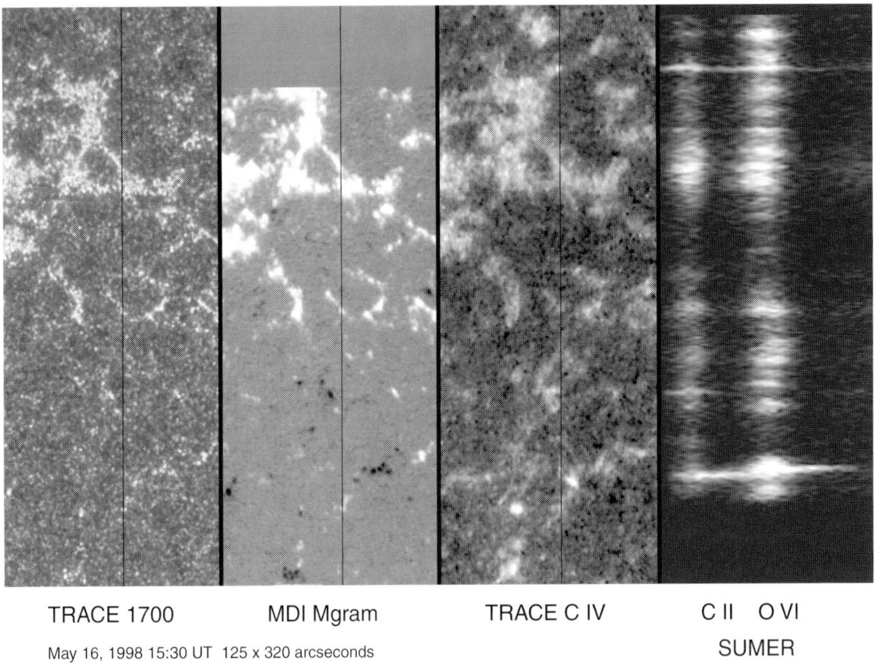

Figure 3. Snapshot of the studied region: first and third panels are the TRACE images in UV continuum and C IV line ($\sim 10^5$ K); the second panel is the MDI magnetogram; the fourth panel shows SUMER spectra in two lines: C II 1037.0 Å corresponding to $\sim 2.5 \times 10^4$ K (*left*) and O VI 1037.6 Å corresponding to $\sim 3 \times 10^5$ K (*right*). The SUMER slit position is indicated on three images.

lines, C II (left) and O VI (right). The imprint of the magnetic field pattern in the photosphere is clearly seen in the 1700 Å and C IV images showing the unambiguous connection between the photospheric magnetic fields and significant emission in the chromosphere and transition region. We will show below that plasma flows in the chromosphere and transition region also often correlate with dynamic events in the photosphere. For quantitative studies we picked several areas along the position of the SUMER slit, shown in Figure 4 under the numbers 1–7. Figure 4 shows a magnetogram rebinned to 1 arc sec resolution, with uneven circles marking the supergranular pattern. Using the high cadence measurements, we have computed the time variation of magnetic flux in chosen areas and the corresponding intensity curves in the C IV (TRACE), C II and O VI (SUMER) lines.

3.1. Typical Set of Explosive Events

Examples of a typical set of events are shown in Figure 5. The upper panel corresponds to region No. 1 (also visible at the bottom of Figure 3). Reduction of magnetic flux is followed by appearance of bright transients in O VI and C IV. Two strong events are seen at about 55 and 100 min. In both events, the main C IV emission occurs over the canceling magnetic features, but it extends far beyond

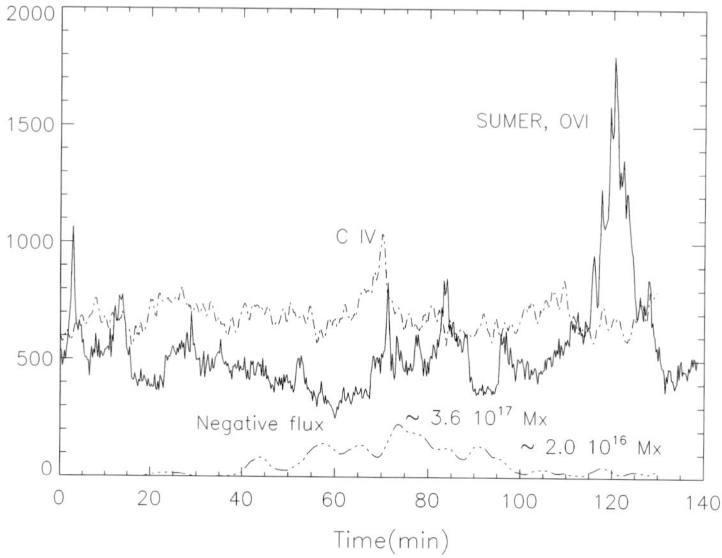

Figure 6. Region No. 7: strong jets at about $t = 125$ min and slightly enhanced background heating may be the case where the entire cumulative energy is deposited into the plasma flows, and slightly enhanced heating is regular 'behind-shock' increase of temperature.

occur in the region No. 7 (Figure 6). Slightly enhanced emission in the C IV line indicates the regular 'behind-shock' heating corresponding to the increase of the background temperature in the chromosphere and transition region. A strong sporadic event in the form of a very symmetrical two directional jet with average velocity of 80 km s^{-1} at about $t = 125$ min may be explained by Gudereley's cylindrical focusing.

3.2. Effects of Converging Supergranular Flows

The supergranular topology and corresponding flows play very important roles in electro-mechanical coupling between the photosphere and chromosphere/transition region. Vertices where three supergranules meet are especially interesting. There are clear differences in detail depending on whether magnetic flux concentrations near the vertex are weak or not detectable (like regions No. 2 and No. 6 in Figure 4) or strong (like regions No. 4 and No. 5).

At a vertex converging flows will attract all possible magnetic elements, including the smallest. Ground-based observations (Berger and Title, 1996) obtained with the resolution as high as 0.2 arc sec (less than 200 km in diameter), show that cancellation, merging and splitting processes among the tiny flux tubes persist to the limit of resolution. The process of heating and jet formation caused by reconnection of the smallest possible elements should have two general features: (1) the shock convergence and formation of jets may occur at lower altitude (because of small magnetic fluxes), i.e., in cooler emission lines; and (2) the generated flows will

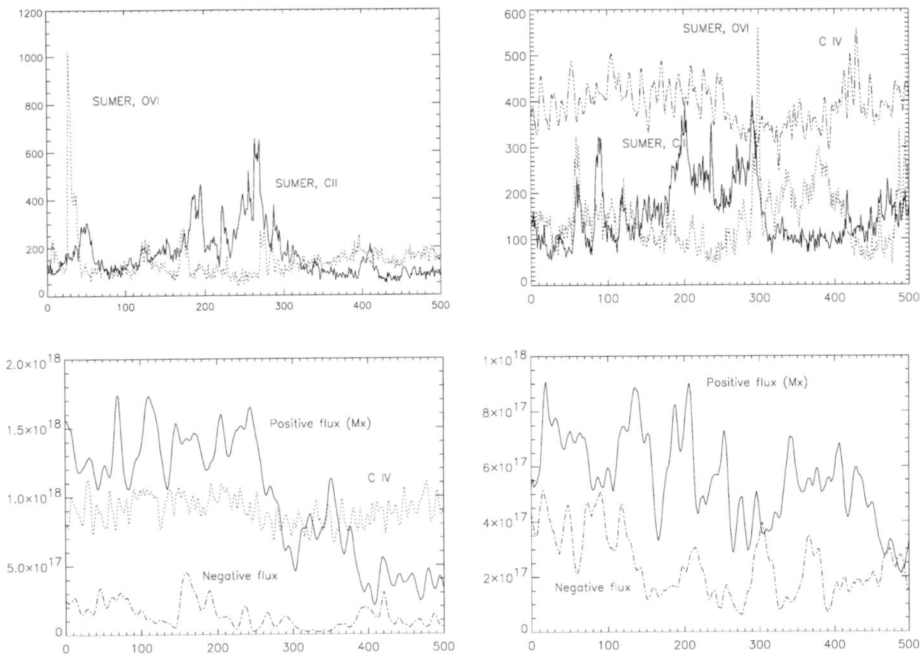

Figure 7. The effect of a vertex with weak magnetic fields: slit passing through the region of converging supergranules with weak magnetic field concentrations reveals the generation of multiple flows in cooler, C II line, supporting the idea that the energy produced by reconnecting weak flux tubes flowing into the vertex is deposited at lower altitudes (*left panels* correspond to region No. 6, *right panels* to region No. 2).

reoccur frequently in the same place, because the long lifetime of supergranules make a long-lived attracting center. On the other hand, if there is a strong magnetic flux concentration near the attracting center, the multiple flows may be observed at higher altitudes and in hotter emission lines.

Examples of a vertex with very weak magnetic field are regions No. 2 and No. 6. Plots of the dynamic events in these regions are shown in Figure 7. Left panels are region No. 6 and right panels region No. 2 (in these plots time is in the movie frames, with $\sim$ 13 s steps). Lower panels for both cases show the variation of magnetic flux (for clarity we put the C IV emission for No. 2 in the upper panel and for No. 6 in lower panel). One can see that the estimates of magnetic flux have an irregular oscillatory character: small flux tubes sliding towards a vertex experience apparent cancelation. Generated flows are seen predominantly in the C II line. The strong explosive event in region No. 2 at $t = 300$ may have an origin associated with the strong downflows observed a few moments earlier in C II line. This type of event is quite common: appearance of sporadic explosive events and complex dynamical structures with multiple velocities in regions of maximum downflows, has been also reported earlier (Kjeldseth-Moe *et al.*, 1988; Dere *et al.*, 1988; Dere,

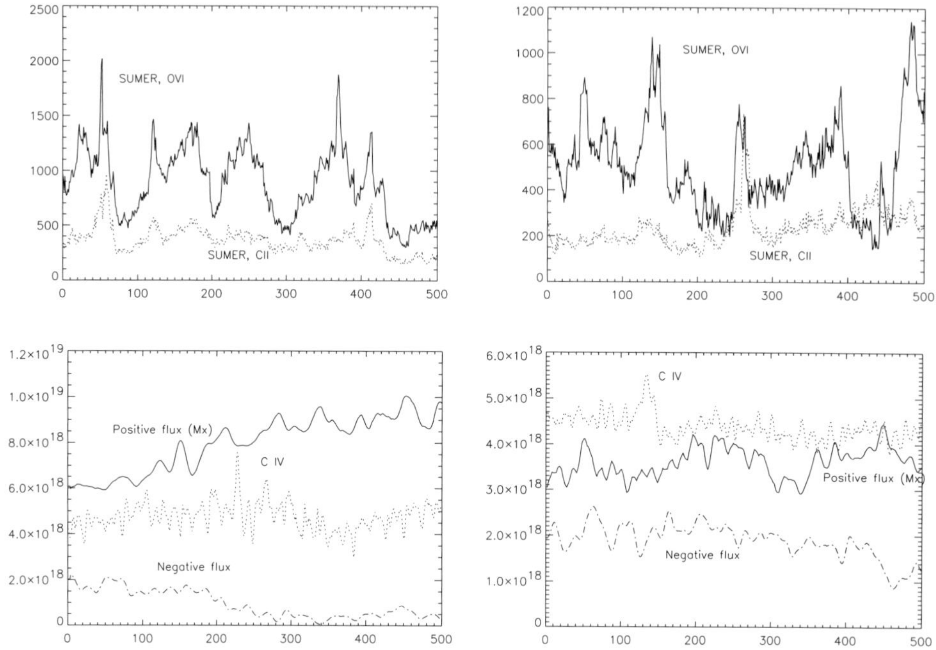

Figure 8. The effect of a vertex with strong magnetic fields: same as in Figure 7 but in the presence of strong magnetic field. The energy avalanche produced by reconnecting strong magnetic flux concentrations and enhanced by converging downflows leads to generation of strong multiple flows in the hotter O VI, line (*left panels* correspond to Region No. 4, *right panels* to region No. 5).

Bartoe, and Brueckner, 1989). In the next section we discuss a mechanism which may explain this type of explosive event.

If in the region of converging supergranules strong magnetic flux concentrations are present, the series of reconnections releases magnetic energy at higher altitudes which manifests itself in hot and dense multiple flows observable in the hotter O VI line. Indeed, these processes are well demonstrated by the evolution of regions No. 4 and No. 5 (see Figure 4). Figure 8(a) shows plots of the observations of these regions. One can see the series of hot plasma flows and some explosive events during the entire time of observation. As predicted by our proposed coupling mechanism, the hydromagnetic activity of flux concentrations is strongly enhanced by the supergranular vertex which provides a continuous supply of small scale magnetic elements to interact with longer-lived strong magnetic field. Note that in both regions the magnetic flux is quite strong and either polarity magnetic flux experience strong fluctuations, reaching changes as large as 2×10^{18} Mx. As shown earlier, cancellation of magnetic flux of 10^{17} Mx is enough to generate strong and dense jets with 2.5×10^5 K temperature. Very interesting, similar events are observed in both regions: at time intervals $t = 50$–80 in region No. 4 (left panels) and $t = 250$–280 in region No. 5 (right panels) quite significant jets in the C II line are seen with some delay after jets appearing in the O VI line. Both of them

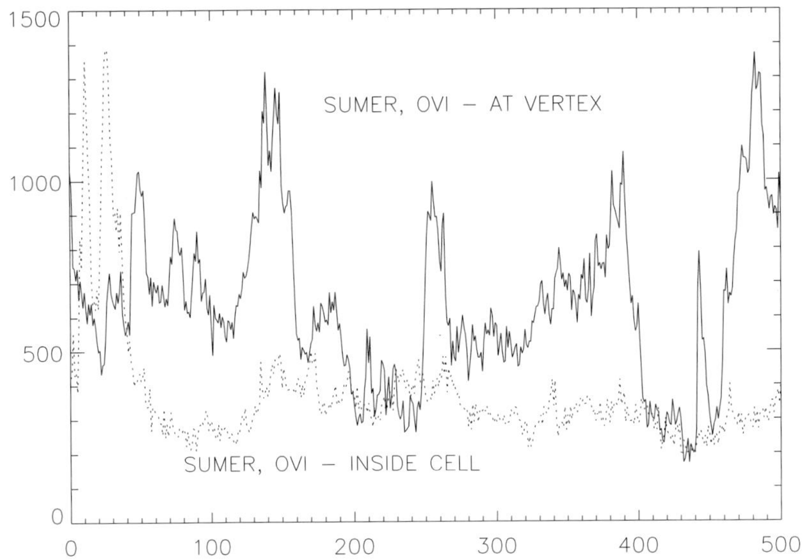

Figure 9. SUMER data in O VI line for two regions: region No. 3 located inside the cell (*dotted line*), and the vertex region No. 4 (*solid line*). The comparison clearly shows the greater activity at the vertex due to the larger supply of flux flowing into it.

have an explosive character. The latter is another example where the O VI jet is observed 1500 – 2000 km from the bright center of C IV emission in the TRACE image and from the canceling photospheric magnetic features. Note that very little energy goes here into pure heating: emission in C IV line shows only slightly enhanced 'behind shock' heating. A different example are the events occurring in time interval $t = 120–180$ in region No. 5 and $t = 220–280$ in region No. 4, where the energy is distributed between the heat (C IV bright transient) and strong jets. To illustrate one more time the effect of a vertex, we superimpose the O VI intensities for the vertex region No. 4 and cell interior region No. 3, which is quite distant from the supergranular lane (Figure 9). One can see that during more than 2 hours nothing spectacular happens in region No. 3. The explosive event seen in the very beginning of our observation time span is probably a typical sporadic event like the ones shown in Figures 5 and 6.

4. Evolution of Generated Flows: Theoretical Aspects

As shown above, strong hydromagnetic activity of the photospheric flux tubes producing a magnetic energy avalanche provides direct electro-mechanical coupling between the dynamic events in the photosphere and transition region. Two major forms of the energy release, impulsive heating (radiative transients – microflares, blinkers) and plasma flows each evolve further in their own way. The radiative transients are symptomatic of direct heating of the atmosphere, which results in sharp temperature jump in the transition region. Possible generation of thermo-magnetic and entropy waves will be discussed elsewhere.

Evolution of the plasma flows may proceed in several different ways depending on their parameters and local properties of the atmosphere. Observational evidence for extreme fine structure in the transition region, much smaller that 1 arc sec in size has been reported earlier (e.g., Dere *et al.*, 1988; Kjeldseth-Moe *et al.*, 1994) and supported by high-resolution images taken by the TRACE (Schrijver *et al.*, 1999; Berger *et al.*, 1999).

Plasma consisting of highly intermittent magnetic fields, density and flows is known to exhibit complex dynamics. A large class of nonlinear events is caused by various kinds of flow instabilities (Ryutova, 1988), which we discuss below. There are also observable macroscopic effects caused by flows whose velocity is below the threshold of any instability: for example, such flows lead to strong enhancement of the Alfvén wave dissipation and appearance of 'patchy' pattern in the energy deposition (Ryutova and Habbal, 1995).

The onset of the instabilities occurs mainly at two thresholds with respect to velocity: (1) at a flow velocity u larger than the upper threshold, u_{c2}, a linear hydrodynamic instability similar to Kelvin–Helmholtz instability (KH) occurs; (2) at a flow velocity less than u_{c2} (when the system is stable with respect to KH instability) but larger than some u_{c1} (lower threshold), an instability of negative energy waves (NEW) develops putting the system in strongly nonlinear stage which includes a regime of dissipative and explosive instabilities (Ryutova, 1988; Ryutova *et al.*, 1998). Note that the Kelvin–Helmholtz instability develops at a linear stage and its threshold is usually supersonic in the absence of magnetic field and becomes super-Alfvénic when the magnetic field is involved. Dissipative instabilities connected with NEW are *below* the KH instability, and require weaker flows to trigger the energy and momentum exchange between the background magnetic structures and generated flows.

Instability of negative energy waves is possible only in energetically open systems, i.e., non-conservative systems with sources and sinks of energy. In such systems any perturbation, including Alfvén waves will have a negative energy if in some range of parameters the derivative of their dispersion function $D(\omega, k)$ ($D(\omega, k) = 0$), with respect to ω becomes negative: $\partial D(\omega, k)/\partial\omega < 0$. In this range of parameters the energy of a system with negative energy waves is lower than without them, and any kind of losses, i.e., any mechanism of subtraction of their energy (dissipative effects, interaction with other waves, etc.), leads to the growing of their amplitudes (Sturrock, 1960; Kadomtsev, Mikhailovski, and Timofeev, 1964; Dikasov, Rudakov, and Ryutov, 1965). This in turn results in the onset of strongly nonlinear processes whose form is determined by the dispersive properties of the system, degree of nonlinearity and dissipative effects and is described by a generalized KdV-Burgers' equation:

$$\frac{\partial\psi}{\partial t} + c\frac{\partial\psi}{\partial x} + \alpha\psi\frac{\partial\psi}{\partial x} + \mu\frac{\partial^3\psi}{\partial x^3} = \epsilon\hat{L}(\psi)\,. \tag{3}$$

Here c is a phase velocity of linear oscillations, ψ is the amplitude of perturbation e.g. it may be a velocity, $\psi = v$, or transverse magnetic field, $\psi = B_y^2 + B_z^2$ like in magnetic kink perturbations, or $\psi = B_x$ in the Alfvén wave perturbations; α is a coefficient of nonlinearity, and μ is a dispersion coefficient. $\hat{L}$ is the operator describing the influence of non-conservative factors, like mass flows as a source, radiative or dissipative damping as sinks, etc.

In the absence of dissipation and sources when $\hat{L}(\psi) = 0$, a stationary solution of Equation (3) is either a shock wave, at large values of α and weak dispersion, or a soliton, if dispersion is strong enough to compete with nonlinear growth. If $\epsilon\hat{L}(\psi)$ is small (as it is in many practical cases), the solution of Equation (3) can be found by assuming that parameters of a stationary solution (a shock or a soliton) vary slowly in time, which allows us to obtain an evolutionary equation for the amplitude of a shock or solitary wave, H:

$$\frac{\mathrm{d}H}{\mathrm{d}t} = q\langle\psi\hat{L}(\psi)\rangle\,. \tag{4}$$

The brackets indicate the time average. In many cases Equation (4) can be represented as follows (see, e.g., Dodd *et al.*, 1982, and also below):

$$\frac{\mathrm{d}H}{\mathrm{d}t} = \sigma(u,\alpha,\mu)H + \kappa(u,\alpha,\mu)H^n + \mathcal{O}(H)\,, \tag{5}$$

where $n > 1$, and

$$\sigma(u,\alpha,\mu,) \sim -\frac{\partial D(\omega,k)}{\partial\omega}\,. \tag{6}$$

For positive energy waves $\partial D(\omega,k)/\partial\omega > 0$ *always*, and σ is a *damping rate*, $\sigma < 0$. Thus in the beginning, when nonlinearity is weak (the second term in (5) is small), and a linear term is leading, positive energy waves experience damping close to exponential; very quickly nonlinear terms become negligibly small, and waves damp out in a regular exponential way.

For negative energy waves $\partial D(\omega,k)/\partial\omega < 0$, and σ becomes a *growth rate*, $\sigma(u,\alpha,\mu,) > 0$. One can see from Equation (5) that in the initial stage (when, again, nonlinear terms are smaller than the first term), the amplitude of the negative energy waves grows exponentially. Very quickly nonlinear terms become much larger than the linear term and Equation (5) acquires a form of the 'explosion equation'(we drop here higher power nonlinear terms):

$$\frac{\mathrm{d}H}{\mathrm{d}t} = \kappa(u,\alpha,\mu)H^n, \qquad n > 1\,. \tag{7}$$

In the absence of higher nonlinear terms this equation has an integral $H \sim 1/(t - t_{\mathrm{expl}})^{1/(n-1)}$. Thus, in a few inverse growth rates negative energy waves experience explosive growth: the amplitudes of all perturbations reach infinitely large values in finite time (Sturrock, 1966; Dikasov, Rudakov, and Ryutov, 1965; Coppi, Rosenbluth, and Sudan, 1969). In reality, of course, higher nonlinear effects turn on and stabilize the explosive growth of perturbations. The detailed process is determined by the specific physical conditions of medium. For the solar photosphere this process has been studied in detail and used to explain the properties of moving magnetic features in the sunspot regions by Ryutova *et al.* (1998). It was shown (see also below) that nonlinear kink excited by sheared flows becomes a stable soliton propagating with the characteristic velocities, sizes and time-scales typical of moving magnetic features (Harvey and Harvey, 1973).

In our case of the fine intermittent magnetic structures in the chromosphere/transition region and sheared flows along the magnetic field, the most readily excited and fastest growing perturbation is also a kink mode with the natural phase velocity $c_{k0} = B/\sqrt{4\pi(\rho_i + \rho_e)}$, where ρ_i and ρ_e are the plasma densities inside and outside magnetic flux respectively. In the presence of flows (Ryutova, 1988)

$$c_k = \frac{1}{1+\eta}\left[u \pm \sqrt{\eta[(1+\eta)v_A^2 - u^2]}\right], \qquad \eta = \rho_i/\rho_e . \tag{8}$$

When the flow velocity exceeds the threshold

$$u > u_c^{(2)} = v_A\sqrt{1+\eta}\,, \tag{9}$$

the system becomes unstable with respect to linear modes; this is a typical Kelvin–Helmholtz instability. It follows from (8) that unstable perturbations correspond to $\mathrm{Re}(\omega/k) = u/(1+\eta) > 0$, and therefore, propagate *upstream*. This means that if generated flows 'blow along' a certain magnetic structure, the excited perturbations will subsequently propagate in the same direction as the flows. A typical frequency of these perturbations is $\omega \sim kv_A\sqrt{1+\eta}$. For an estimate of this frequency we replace η by its approximate equivalent in terms of plasma beta: $\eta \simeq \beta/(1+\beta)$. This estimate is limited, of course, by the assumption of the total pressure balance, and is, therefore, quite rough. We adopt the following parameters: $T = 10^5$ K, $n = 10^{10}$ cm^{-3}, $B = 10$ G; then $v_A = 218$ km s^{-1} and $\beta = 0.03$. For $k =$ 0.5–1.0 Mm^{-1} ($k^{-1} \sim$ 200–100 km), the frequency of excited wave is $\omega =$ 1.1–2.2 s^{-1}. It is important, that these waves are excited *locally*, wherever the generated flows fulfill the condition (9).

Given the wide variety of flows generated in the chromosphere/transition region, high frequency kink and Alfvén waves in a broad frequency interval will be easily excited. The wave generated by the upflows will propagate upward in the corona. The greatest advantage of these waves compared to those excited by convective motions is that they have much higher frequencies and can be excited directly at any place in the chromosphere and transition region.

Below the threshold (9) when the system is stable with respect to the KH instability, the lower branch of the dispersion relation (8) corresponds to negative energy waves. The energy of these waves is proportional to (Ryutova, 1988)

$$W \sim (\eta v_A^2 - u^2)\,, \tag{10}$$

which at $u > v_A\sqrt{\eta}$ becomes negative. Thus, the interval of the velocities leading to excitation of negative energy waves is $v_A\sqrt{\eta} \leq u < v_A\sqrt{1+\eta}$, or in terms of plasma beta:

$$v_A\sqrt{\frac{\beta}{1+\beta}} \leq u \leq v_A\sqrt{\frac{1+2\beta}{1+\beta}}\,. \tag{11}$$

Note that negative energy waves propagate *against* the generated flows, i.e., in regions dominated by *downflows* the waves propagating upward (with $\omega > 0$) and having parameters determined by (11) will have a negative energy.

The equation for these waves has a form of Equation (3); the operator $\hat{L}(\psi)$ and all the coefficients are specified in the paper by Ryutova *et al.* (1998) (see also Ryutova and Sakai, 1993). We discuss here only an evolutionary equation of the type of Equation (7), for nonlinear stage of the perturbation amplitude which has a following form (see Equation (20) in Ryutova *et al.*, 1998):

$$\frac{\mathrm{d}H}{\mathrm{d}t} = a(\alpha, \mu)H^{3/2} - b(\alpha, \mu)H^2 \tag{12}$$

with

$$a = \frac{2.92}{\pi}\gamma\left(\frac{\alpha}{12\mu}\right)^{1/2}, \qquad b - \frac{16}{15}\frac{\alpha}{12\mu}\nu\,, \tag{13}$$

where μ is a dispersion coefficient, γ and ν are dissipative coefficients, and α is a coefficient of nonlinearity.

As mentioned above in the stage when higher nonlinearities are still weak, the first nonlinear term provides explosive growth of perturbations, and from Equation (12) we have

$$H \simeq \frac{H_0}{(1 - t/t_{\mathrm{expl}})^2}\,, \tag{14}$$

where the explosion time is $2/(aH_0^{1/2})$. As shown in Ryutova *et al.* (1998), under photospheric conditions this time is quite large, i.e. the second term in (12) stabilizes the explosive growth of the amplitude before the actual 'explosion' occurs. In this case Equation (12) describes a stable soliton, which explains quite well the observed properties of moving magnetic features.

To estimate the explosive time and range of velocities corresponding to NEWs in the transition region we use the same parameters as above ($T = 10^5$ K, $n = 10^{10}$ cm^{-3}, $v_A = 218$ km s^{-1} and $\beta = 0.03$). With these parameters the interval (11) becomes:

$$40\ \mathrm{km\ s^{-1}} \leq u \leq 220\ \mathrm{km\ s^{-1}}\,. \tag{15}$$

For small plasma beta, $\eta \simeq \beta$, and the phase velocity of the kink mode in absence of flow is $c_{k0} \simeq v_A$. In the presence of flows,

$$c_k \simeq u - \sqrt{\beta(v_A^2 - u^2)}\,. \tag{16}$$

Now we can specify the parameter $a(\alpha, \mu)$ simplifying expressions for μ and γ in the approximation of small β , e.g., $\beta < 0.1$ is already enough (see Equations (8) –(10) in Ryutova *et al.*, 1998):

$$\mu \simeq \frac{R^2}{2\beta v_A^2}[\beta(v_A^2 - u^2)]^{3/2}\,, \quad \gamma \simeq \frac{\pi}{4\beta}\delta[\beta(v_A^2 - u^2)]^{1/2}\,, \quad \alpha \simeq \tfrac{3}{4}\,; \tag{17}$$

here R is a radius of magnetic flux, and δ is measure of its diffusive boundary, which may be taken as $0.01R$. For the interval of velocities (15) these parameters give an estimate for the time interval corresponding to development of explosive instability $t_{\mathrm{expl}} \simeq 30$–$100$ s. From the observational point of view, this means that the explosive instability develops very fast in the region of downflows if the parameters meet the required conditions.

It is worthwhile to mention one more macroscopic and observable effect which may occur at flow velocities far below the threshold of any instability. As shown in Ryutova and Habbal (1995), such flows considerably modify the dissipation of shear Alfvén waves, affecting both the magnitude and the height of maximum dissipation. In the case of downflows the height of Alfvén wave dissipation is *independent* of the Reynolds number: at any height where the flow velocity becomes equal to the local Alfvén velocity, the wave comes to extinction and gives off its energy completely. This process is accompanied by the radial redistribution of the energy input across the magnetic structure. The location of the energy deposition is determined by those radii where the sum, $v_A(r) + u(r)$, has an extremum (Figure 10). The extrema of this function correspond to the energy escape channels, and the regions of steepest gradients are those where the strongest energy absorption occurs, thus creating a mosaic pattern in the emitting regions. This effect can explain the high variability and 'patchy' picture of the emission observed around 10^5 K.

5. Summary

When comparing the signatures of the proposed mechanism of electro-mechanical coupling between the photosphere and the chromosphere/transition region with the observations, we will follow the order of observed properties listed in the Introduction.

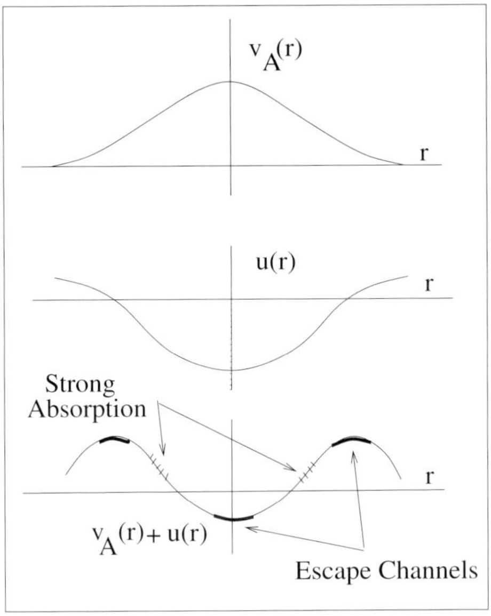

Figure 10. The sketch of slightly different magnetic field and flow profiles. The extrema of the phase velocity correspond to the energy escape channels, and the regions of steepest gradients are those where the strongest energy absorption occurs (After Ryutova and Habbal, 1995).

(1) Enhanced emission in the C IV line is generally co-spatial with the magnetic pattern in the photosphere. This is true all the time for the quasi-steady or 'background' component of the emission, regardless of transients events. During transients, the emission takes more complex shapes including blobs, jets, arcs and even spirals, which are not as tightly correlated with the photospheric pattern. These shapes can change completely on time scales of tens of seconds and extend more than 5000 km beyond the photospheric magnetic features. We believe that a basic process providing this emission is the cascade of shock waves produced by colliding and reconnecting flux tubes in the photosphere. Some of the background emission may also be explained by regular 'behind-shock' heating which occurs whenever flux tubes collide and reconnect. Non-collinearity of photospheric flux tubes both enhances their collision rate and enables reconnection for like polarity elements as well as opposite polarities. The continuous supply of flux tubes in the 'magnetic carpet' (replaced every $\sim$ 40 h, Title and Schrijver, 1997) ensures the ubiquitous nature of this process and its imprint on the upper atmosphere.

(2) The appearance of bright transients often, but not always, correlates with canceling mixed polarity magnetic elements in the photosphere. Sometimes the reduction of magnetic flux can be measured directly, as in Figure 5. In other cases, transients occur in regions of unipolar flux tubes, suggesting reconnection of oblique components. When opposite polarity elements collide, the bisector of their angle of collision is probably oriented nearly along the line of sight. In this case the post-reconnection shocks are strongly enhanced by gradient acceleration leading

to collision of the shock fronts, causing bright transients and jets in addition to the regular 'behind-shock' heating. The fact that the C IV and O VI emission is much more dynamic than the photospheric magnetograms is also consistent with the proposed mechanism of coupling: in each reconnection process only a limited portion of the magnetic flux is involved. The characteristic size of magnetic flux participating in one elemental act of reconnection is $\sim$ 40–100 km. This means that magnetic flux concentrations several times larger than this are sources of multiple reconnection processes. Transients, including enhanced emission and jets, are also seen in regions with no fields detected with the MDI sensitivity ($\sim 7 \times 10^{16}$ Mx). These may be reconnections of tiny features with diameters less than 100 km.

(3) Blinkers and other bright transients are often accompanied by two directional plasma jets. These jets are sometimes observed in the SUMER spectra thousands of kilometers from the center of the C IV emission, which is near the interacting flux tubes identified in magnetograms. There are at least two conditions which can generate two directional jets: cylindrical self-focusing of shock fronts, and collision of shocks produced by neighboring reconnection processes. A collision of two independent shock fronts may create in the vicinity of the collision a figure of revolution, thus providing the conditions similar to Gudereley's effect for generation of strong two directional plasma jets.

(4) Energy released in transient events is distributed between plasma flows and radiative emission. The observations suggest that stronger emissions correspond to lower velocity jets, and *vice versa*. This observed property is a natural consequence of the proposed mechanism (see Figure 2). There are three regimes depending on the parameters of the colliding shocks: (1) strong and hot jets are formed in the range of parameters above the solid curve; (2) intense radiative transients with weak or no mass flows appear in the range of parameters below the dashed curve; (3) in largest range of parameters corresponding to region between the two curves, energy is distributed between the radiative transients and plasma flows.

(5) Plasma flows are always seen whenever the slit crosses strong magnetic flux tubes, either in quiet sun or plage. They are also seen at vertices of converging flows in the supergranular network. This is direct evidence that interacting magnetic flux tubes cause the energetic events in the upper atmosphere. The vertices of supergranules (sinks in 'cork' movies) are especially active because more magnetic elements of either polarity flow into them than any other points.

(6) If the magnetic flux concentration in the vicinity of a vertex is weak or even below the MDI threshold, multiple flows are seen in the cooler C II line without much activity in the transition region. The vertex collects all possible magnetic flux tubes, including the smallest. In this case, the shock convergence and formation of jets occurs at lower altitudes in denser plasma and therefore shows up in cooler emission lines (see Figure 7).

(7) When a supergranular vertex has a strong magnetic flux concentration, multiple flows and explosive events occur repeatedly in the O VI line. The strong fields cause shock waves to converge in higher altitudes, due the larger initial velocity

which is proportional to Alfvén velocity. This leads to formation of multiple flows at higher temperatures. In either case of strong or weak magnetic flux residing in a vertex, the long lifetime of supergranules provides a long-term sink point, creating the conditions for generation of multiple flows. This can be seen in Figures 7 and 8, where multiple flows are seen during the observation interval.

(8) Explosive events can occur only at lower temperatures (seen in C II and perhaps C IV), only at higher temperatures (O VI and perhaps C IV), or over the whole temperature range observed (C II, C IV, and O VI, or 10^4–3×10^5 K). The energy distribution between heating and plasma flows is an intrinsic feature of our mechanism and corresponds to the wide parameter range between the solid and dashed curves in Figure 2. If the parameters of the colliding shocks are closer to the solid curve, more of the cumulative energy goes to the plasma jets than into pure heating; above the solid curve, all the energy is converted into hot and dense plasma jets. Furthermore, an additional mechanism described in Section 4 predicts the development of explosive events in the presence of downflows without any connection to the radiative transients. For this scenario, provided by the explosive instability of negative energy waves, the necessary conditions are the intermittent structure of the magnetic field and the presence of downflows in a large range of velocities, roughly 40–220 km s^{-1}. These velocities have been observed in earlier experiments and are confirmed by our present observations.

Acknowledgements

We would like to thank Phil Judge and Hardie Peter of HAO for providing the SUMER data, and Zoe Frank for assistance. This work is supported by the TRACE and MDI projects: NAS5-38099 at Lockheed Martin and NAG5-3077 at Stanford University, respectively. SOHO is a mission of international cooperation between ESA and NASA.

References

Berger, T. and Title, A.: 1996, *Astrophys. J.* **463**, 365.

Berger, T. E., De Pontieu, B., Schrijver, C. J., and Title, A. M.: 1999, *Astrophys. J.* **519**, L97.

Brekke, P., Hassler, D. M., and Wilhelm, K.: 1997, *Solar Phys.* **175**, 349.

Coppi, B., Rosenbluth, M. N., and Sudan, R.: 1969, *Ann. Phys.* **55**, 201.

Dere, K. P.: 1994, *Space Sci. Rev.* **70**, 21.

Dere, K. P., Bartoe, J.-D. F., and Brueckner, G. E.: 1989, *Solar Phys.* **123**, 41.

Dere, K. P., Bartoe, J.-D. F., Brueckner, G. E., Cook, J. W., and Socker, D. G.: 1988, *Solar Phys.* **114**, 223.

Dikasov, V. M., Rudakov, L. I., and Ryutov, D. D.: 1965, *Soviet Phys. JETP*, **21**, 1965.

Dodd, R. K., Eilbeck, J. C., Gibbon, J. D., and Morris, H. C.: 1982, *Solitons and Nonlinear Wave Equations*, Academic Press, London.

Guderley, G.: 1942, *Luftfahrtforschung* **19**, 302-312.
Handy, B. N., Bruner, M. E., Tarbell, T. D., Title, A. M., Wolfson, C. J., Laforge, M. J., and Oliver, J. J.: 1998, *Solar Phys.* **183**, 29.
Harrison, R. A. *et al.*: 1997, *Solar Phys.* **170**, 123.
Harvey, K. and Harvey, J.: 1973, *Solar Phys.* **28**, 61.
Innes, D. E., Brekke, P., Germerott, D., and Wilhelm, K.: 1997, *Solar Phys.* **175**, 341.
Kadomtsev, B. B., Mikhailovski, A. B., and Timofeev, A. V.: 1964, *Soviet Phys. JETP* **20**, 1517.
Kjeldseth-Moe, O., Brynildsen, N., Brekke, P., Engvold, O., Maltby, P., Bartoe, J.-D. F., Brueckner, G. E., Cook, J. W., Dere, K. P., and Socker, D. G.: 1988, *Astrophys. J.* **334**, 1066.
Kjeldseth-Moe, O., Brynildsen, N., Brekke, P., and Maltby, P.: 1994, *Space Sci. Rev.* **70**, 89.
Landau, L. D. and Lifshits, E. M.: 1987, *Fluid Mechanics*, Pergamon Press, Oxford.
Mariska, J. T.: 1992, *The Solar Transition Region*, Cambridge University Press, Cambridge.
Moses, D. *et al.*: 1994, *Astrophys. J.* **430**, 913.
Ostrovski, L. A., Rybak, S. A., and Tsimring, L. Sh.: 1986, *Soviet Phys. Usp.* **29**, 1040.
Petschek, H. E.: 1964, *NASA Spec. Publ.*, SP-50, p. 425.
Petviashvili, V. and Pokhotelov, O.: 1992, *Solitary Waves in Plasmas and in the Atmosphere*, Gordon and Breach, Philadelphia, PA.
Ryutova, M. P.: 1988, *Soviet Phys. JETP*, **67**, 1594.
Ryutova, M. and Sakai, J.: 1993, *JETP Lett.* **58**, 507.
Ryutova, M. P. and Habbal, S. R.: 1995, *Astrophys. J.* **451**, 381.
Ryutova, M., Shine, R., Title, A., and Sakai, J. I.: 1998, *Astrophys. J.* **492**, 402.
Schrijver, C. J. *et al.*: *Solar Phys.* **187**, 261.
Sturrock, P. A.: 1960, *J. Appl. Phys.* **31**, 2052.
Sturrock, P. A.: 1966, *Phys. Rev. Lett.* **16**, 270.
Tarbell, T., Ryutova, M., Covington, J., and Fludra, A.: 1999, *Astrophys. J.* **514**, L47.
Title, A. M. and Schrijver, C. J.: 1997, in R. Donahue and J. A. Bookbinder (eds.), *Cool Stars, Stellar Systems, and the Sun*, Astron. Soc. of the Pacific Conf. Series.
Wikstol, O. *et al.*: 1997, in A. Wilson (ed.), *Proc. of 5th SOHO Workshop, The Corona and Solar Wind near Minimum Activity*, ESA SP-404, ESA, Noordwijk, p. 733.

THE MAGNETIC CONNECTIVITY OF MOSS REGIONS

X. P. ZHAO, J. T. HOEKSEMA, A. G. KOSOVICHEV, R. BUSH and P. H. SCHERRER
W. W. Hansen Experimental Physics Laboratory, Stanford University, Stanford, CA 94305

(Received 20 September 1999; accepted 24 January 2000)

Abstract. A novel emission feature resembling moss was first identified in high-resolution TRACE Fe IX/X 171 Å images by Berger *et al.* (1999). The moss emission is characterized by dynamic arc-second scale, bright elements surrounding dark inclusions in images of solar active regions. Patches of moss elements, called moss regions, have a scale of 20–30 Mm. Moss regions occur only above some of magnetic plages that underlie soft X-ray coronal loops. Using the potential field extrapolation of the photospheric magnetic field into the corona, we find that the magnetic field lines in moss-associated magnetic plages connect with adjacent plages with opposite polarity; however, all field lines from mossless plages end in surrounding 'quiet regions'. This result is consistent with the idea that the TRACE moss is the emission from the upper transition region due to heating of low-lying plasma by field-aligned thermal conduction from overlying hot plasma (Berger *et al.*, 1999).

1. Introduction

'Moss' is a novel solar emission pattern first identified in high-resolution Fe IX/X 171 Å images of active regions observed by the Transition Region And Coronal Explorer (TRACE) on 30 May 1998. Moss is characterized by arc-second scale, bright elements surrounding dark inclusions; patches of elements appear spongy or moss-like. Moss elements are confined to layers approximately 1–3 Mm thick with base heights about 2–4 Mm above the photosphere (Berger *et al.*, 1999). SOHO/CDS and TRACE observations show that the low-lying moss plasma has a temperature in the range 0.6–1.5 MK and an electron density of (2–5) $\times 10^9$ cm^{-3} at a temperature of 1.3 MK. The pressure of 0.7–1.7 dynes cm^{-2} is a few times higher than in coronal loops observed in the TRACE Fe IX/X 171 Å passband (Fletcher and De Pontieu, 1999).

Moss elements move and vary in brightness on time scales of one minute or less. The motions are random displacements on spatial scales of 1000 km. 'Moss regions' are patches of dynamic moss elements with a scale of 20–30 Mm. The shape and size of moss regions appear, however, relatively static and usually evolve slowly on time scales of 10–20 hours. Moss regions occur only above certain magnetic plages that underlie the 3–5 MK coronal loops seen in SXT images. Based on the one-to-one correspondence of 171 Å moss regions with soft X-ray bright regions, Berger *et al.* (1999) suggest that the moss in the TRACE Fe IX/X 171 Å bandpass is the emission from the upper transition region due to heating of

The physics of the solar corona and transition region. Part II.
Reprinted from Solar Physics **193,** 2000.

low-lying plasma by thermal conduction along coronal magnetic field lines from overlying hot plasma.

Similar low-lying emission with underlying Hα plage has previously been noted in the normal-incidence soft X-ray. By means of hydrostatic models of coronal arches, the emission pattern in X-ray images can be explained by the characteristics of high-pressure large-scale loops having a thin region of high surface brightness at the base (Peres, Reale, and Golub, 1994). On the other hand, the low-lying unresolved bright regions in SOHO/EIT Fe IX/X 171 Å solar images have been explained as small-scale loops (Moses *et al.*, 1997).

Only some magnetic plages are associated with moss regions. What are the unique characteristics of the moss-associated magnetic plages? Comparing the connectivity of moss-associated and mossless magnetic plages with their surrounding magnetic field may provide some clue for understanding the cause of moss regions.

In the next section we display the simultaneous distribution of the photospheric magnetic field underlying and surrounding the 30 May 1998 moss regions. The computed magnetic field lines connected with moss-associated and with mossless magnetic plages are presented in Section 3. The results are summarized in the last section and discussed.

2. Observations

The 30 May 1998 moss regions occurred over NOAA active region 8227. The top panel of Figure 1 shows the full-disk magnetic image observed by SOHO/MDI on 30 May 1998 at 14:24:30 UT. The bright and dark areas denote positive and negative magnetic polarities. AR 8227 was the only active region in the northern hemisphere, though there were several positive and negative magnetic plage networks in the southern hemisphere. AR 8227 consists of a leading sunspot group and an extensive trailing plage network. The dotted lines outline the boundary of the TRACE 171 Å field of view (FOV) on 30 May 1998 at 14:41:01 UT, which covers AR 8227. The dashed box covers the area where the 30 May 1998 moss regions occurred. It is the same area as Figure 1 of Berger *et al.* (1999) and Fletcher and De Pontieu (1999). As shown in the bottom panel, the negative plage (A) and the positive plage (B) in the solid box are moss-associated and mossless, respectively.

The bottom panel shows the TRACE 171 Å FOV image observed at 30 May 1998 14:41:01 UT. It includes moss regions and coronal loops (see the colorful Figure 1 of Berger *et al.* (1999) for details). The superimposed white and black contours are for magnetic strength of positive and negative polarities, respectively. The contour levels are ± 50, ± 200 and ± 500 gauss. Some magnetic plages are associated with moss regions, as shown by symbol A, and others associated with coronal loops, as shown by C. However, the magnetic plage B is nearly devoid of Fe IX/X 171 Å emission, even though the size and strength distributions for plage B are similar to the nearby plage A.

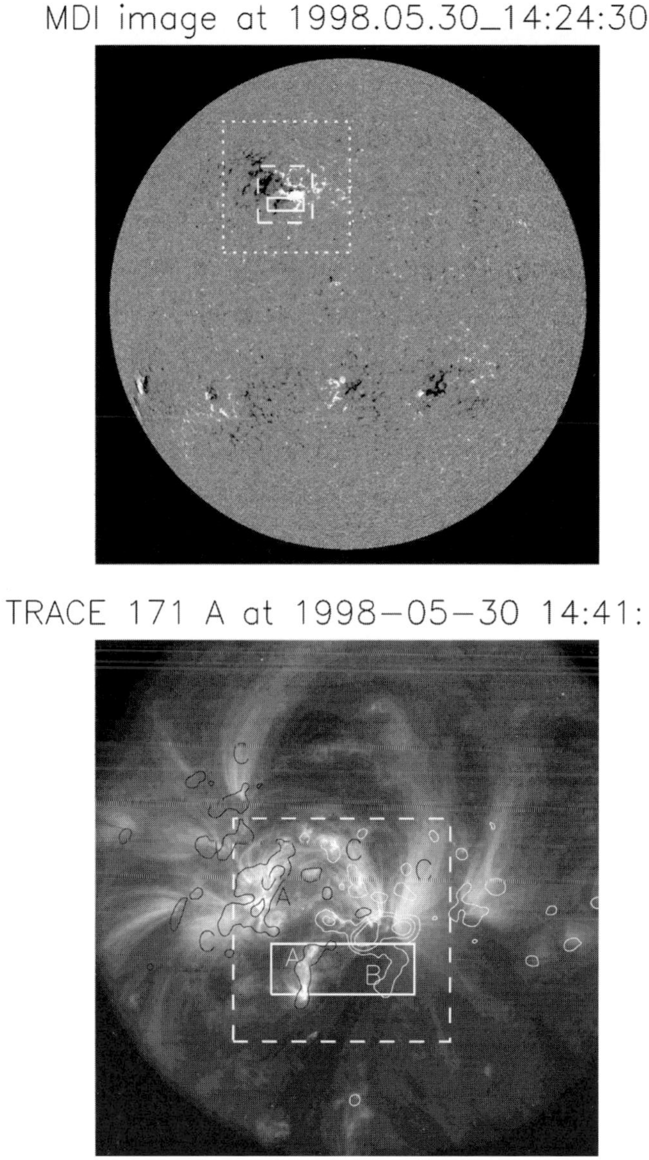

Figure 1. The SOHO/MDI full-disk magnetic image on 30 May 1998 at 14:41:01 UT (*top*) and the TRACE 171 Å FOV image on 30 May 1998 at 14:24:30 UT (*bottom*). The bright and dark areas in the top panel denote the line-of-sight field with positive and negative polarities. The dotted box shows that the NOAA AR 8227 is within the TRACE 171 Å FOV and the dashed box is the area where moss regions occurred. The negative and positive plages in the solid box are the moss-associated and mossless plages. The bottom panel shows the moss regions (A), mossless region (B) and coronal loops (C). The white and black curves are contours of magnetic strength of ± 50, ± 200 and ± 500 G.

Yohkoh/SXT images obtained after 30 May 1998 06:53 UT show that all hot loops in the northern hemisphere occur above and around AR 8227. In the next section, we compute the magnetic connectivity of the moss-associated and mossless magnetic plages with their surrounding magnetic features in the northern hemisphere.

3. The Magnetic Connectivity

Moss regions are spatially associated with certain magnetic plages in the photosphere, though the details of this association are complex on the small scale (Berger *et al.*, 1999). If a loop that passes through a moss region is a flux tube complex, the field configuration in the complex can not be potential. However, there are loops that connect some moss-associated magnetic plages in the TRACE movies, and those loops appear nearly potential. Thus to the first approximation, the connectivity of moss-associated magnetic plages with other magnetic features and the shape of the loops may be approximated using potential field models (Sakurai, 1982; Zhao and Hoeksema, 1994).

Figure 2 displays magnetic field lines (green) computed using the potential field model (Sakurai, 1982). Input to the model is the MDI magnetogram bordered by dotted lines (see top panel of Figure 1). The computation is carried out using spherical geometry based on Sakurai's software package for the potential magnetic field model. The starting footpoint of each field line is selected based on the strength of the observed photospheric field being greater than 50 G. To avoid the confusion of too many field lines in the figure, the total number of field lines is limited to 120. We overlay contours of the line-of-sight magnetic field strength in the photosphere (orange and blue for positive and negative polarities, respectively). The contour levels are ± 10, ± 20, ± 50, ± 100, ± 200, ± 500, ± 1000 G. The thick white lines are contours of TRACE 171 Å intensity that show the areas of moss regions and coronal loops. Comparing Figure 2 with the bottom panel of Figure 1, it appears that coronal loops observed in 171 Å correspond to field lines that connect to distant magnetic plages of the opposite polarity, while moss regions correspond to the field lines that connect adjacent magnetic plages. The mossless plage B is associated with field lines that connect the plage to 'quiet regions' with weaker field of the opposite polarity.

Figure 3 displays computed magnetic field lines (green) within the solid rectangle in Figure 1. Most of the field lines originating within the white outline over magnetic plage A (see also the bottom panel of Figure 1) are connected with the red plage having stronger field strength. The field lines in magnetic plage B all end in the quiet region with weaker field strength and opposite polarity, confirming the above inference.

Figure 4 shows part of the full-disk Hα image observed at BBSO. The Hα plage A corresponds to moss region A in bottom panel of Figure 1, and the thin and short

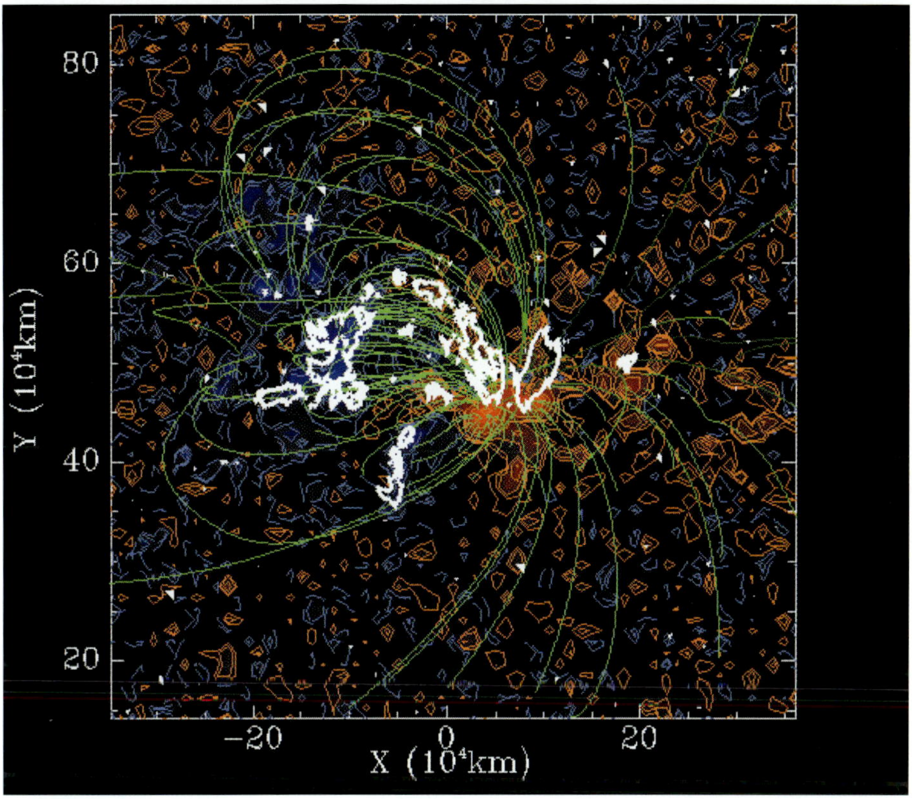

Figure 2. Magnetic field lines (green) computed using the potential field model and the observed photospheric field bordered by the dotted box in top panel of Figure 1. The orange (+) and blue (−) curves are contours of magnetic strength of ±10, ±20, ±50, ±100, ±200, ±500, ±1000 G. The thick white outlines denote the 310 unit intensity contour in TRACE 171 Å, marking moss regions and coronal loops.

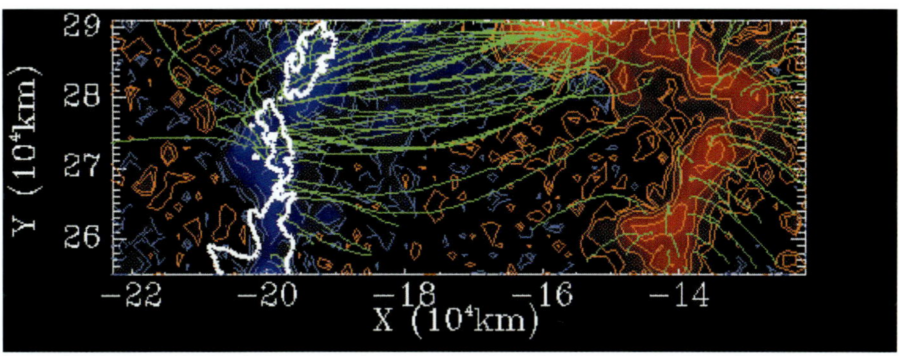

Figure 3. The same as Figure 2 but the field lines are computed using the photospheric field bordered by the solid box in top panel of Figure 1.

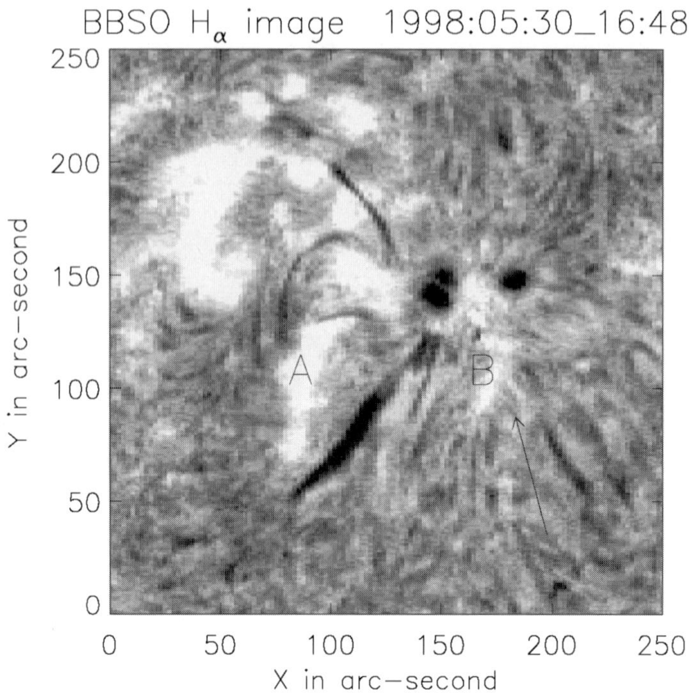

Figure 4. The BBSO Hα image observed at 30 May 1998 16:48 UT. The loops indicated by the dark arrow confirm the model prediction of field lines for the magnetic plage B in Figure 3.

bright lines over magnetic plage B, indicated by the black arrow, correspond to computed field lines that connect with adjacent quiet regions.

4. Summary and Discussion

We have shown that moss-associated magnetic plages connect mainly to adjacent magnetic plages with opposite polarity, while mossless magnetic plages connect only to 'quiet regions' with opposite polarity. As the distance between two magnetic plages increases, the field lines that connect the two plages extend higher and can be called coronal loops. Is there any critical distance that may be used to distinguish the formation of moss from loop emissions?

The small-scale loop model for the low-lying emission assumes that the low coronal loops are nearly isothermal and in the range of 1.2–1.7 MK. The height of those loops is less than 10–20 Mm (Moses *et al.*, 1997), much less than the height of magnetic loops we computed. Thus the adjacent-plage connectivity found here supports the idea that the TRACE moss emission is due to heating of low-lying plasma by field-aligned thermal conduction from overlying hot plasma (Berger *et al.*, 1999). The thermal conduction requires the existence of a gradient in temperature. In addition to the soft X-ray and Hα 6563 Å images, counterparts of the moss region were observed co-temporally on 30 May 1998 by SOHO/CDS in emission lines with temperature higher than 1.5 MK and lower than 0.6 MK (see Figure 1 of Fletcher and De Pontieu (1999)). This supports the existence of the temperature gradient.

Peres, Reale, and Golub (1994) hydrostatic model considers a semicircular coronal loop with constant cross section that is symmetric with respect to its apex, and the maximum plasma temperature is located at the loop apex. However, moss regions occur only at one end of the field lines, even though the field lines from moss-associated plages connect with other adjacent plages with similar strength and opposite polarity. A new physical model is needed to understand the asymmetry.

Acknowledgements

We would thank T. E. Berger for helpful discussions and T. Sakurai for allowing us to use his Software Package for the Computation of Solar Magnetic Field (MAGPACK2). Figure 4 is produced using data from the Big Bear Solar Observatory. This work was supported by NASA Grants NAG5-3077 (SOHO/MDI) and NAS5-38099 (TRACE) subcontract SN30G4920R. SOHO is a mission of international cooperation between ESA and NASA.

References

Berger, T. E., DePontieu, B., Schrijver, C. J., and Title, A. M.: 1999, *Astrophys. J.* **519**, L97.
Fletcher, L. and De Pontieu, B.: 1999, *Astrophys. J.* **520**, L135.
Moses, D. *et al.*: 1997, *Solar Phys.* **175**, 571.
Peres, G., Reale, F., and Golub, L.: 1994, *Astrophys. J.* **422**, 412.
Sakurai, T.: 1982, *Solar Phys.* **76**, 301.
Zhao, X. P. and Hoeksema, J. T.: 1994, *Solar Phys.* **151**, 91.

MAGNETIC ACTIVITY ASSOCIATED WITH RADIO NOISE STORMS

R. D. BENTLEY[1], K.-L. KLEIN[2], L. VAN DRIEL-GESZTELYI[2,*], P. DÉMOULIN[2], G. TROTTET[2], P. TASSETTO[2] and G. MARTY[2]

[1]*Mullard Space Science Laboratory, University College London, Holmbury St. Mary, Dorking, Surrey RH5 6NT, U.K. (rdb@mssl.ucl.ac.uk)*

[2]*Observatoire de Paris, Section de Meudon, DASOP, F-92125 Meudon, Principal Cedex, France*

(Received 9 October 1999; accepted 23 November 1999)

Abstract. As it crossed the solar disk in May and June 1998, AR 8227 was tracked by TRACE, *Yohkoh*, SOHO, and many ground-based observatories. We have studied how the evolution of the magnetic field resulted in changes in activity in the corona. In particular, we examine how the evolving field may have led to the acceleration of electrons which emit noise storms observed by the Nançay Radio Heliograph between 30 May and 1 June 1998, in the absence of any flare. The magnetic changes were related to moving magnetic features (MMFs) in the vicinity of the leading spot and are related to the decay of this spot. Within the limits of the instrumental capabilities, the location in time and space of the radio emissions followed the changes observed in the photospheric magnetograms. We have extrapolated the photospheric magnetic field with a linear force-free approximation and find that the active region magnetic field was very close to being potential. These computations show a complex magnetic topology associated to the MMFs. The observed photospheric evolution is expected to drive magnetic reconnection in such complex magnetic topology. We therefore propose that the MMFs are at the origin of the observed metric noise-storms.

1. Introduction

1.1. Suprathermal Electrons in Quiescent Active Regions

Suprathermal electron populations are ubiquitous in interplanetary space (Lin, 1985, 1997) and in active regions of the Sun, even in the absence of major events such as flares and coronal mass ejections (cf., the review by Klein, 1998). The question of where, and under which conditions these particles are accelerated is largely open. Radio observations at decimetre-to-meter wavelengths are the most sensitive diagnostic of suprathermal electrons in the middle corona. The most frequently observed signature comes from electron populations confined in large-scale coronal loops where they radiate noise storms, consisting of broadband continuum emission (bandwidth comparable to the center frequency) with duration between $\sim$1 hr and several days, and bursts lasting $\leq$1 s with much smaller bandwidth. Noise storms require persisting electron acceleration, presumably to energies rang-

*Also at Konkoly Observatory (Budapest, Pf. 67, H-1525, Hungary), and a visiting professor at University College London.

The physics of the solar corona and transition region. Part II.
Reprinted from Solar Physics **193,** 2000.

ing from a few keV to a few tens of keV, during the whole lifetime of the radio emission.

Noise storms have long been known to be related especially to large spot groups, preferentially with some degree of complexity (cf., review by Elgarøy, 1977, Ch. 1.3), and with the establishment of large loop systems in the middle corona (Stewart, Brueckner, and Dere, 1986; Krucker *et al.*, 1995; references in Klein, 1998). Their onset is often related to distinct changes in the overlying corona (Kerdraon *et al.*, 1983) and to flare-like energy release in the underlying active region (Lantos *et al.*, 1981; Raulin *et al.*, 1991; Raulin and Klein, 1994; Crosby *et al.*, 1996), and their evolution is accompanied by changes of source position indicating the sudden evolution of the coronal structure or of the site of electron acceleration (Malik and Mercier, 1996). This suggests that, like flares, noise storms as a signature of time-extended acceleration of electrons are related with the discontinuous evolution of coronal magnetic fields.

The energy requirements of noise storms are probably below those of coronal heating in active regions (cf., Klein, 1998). The question is then how the energy is converted to create the suprathermal electron populations whose radio emission we observe. Two frequently cited models rely on the interaction of newly emerging flux with the ambient corona, either in the course of magnetic reconnection (Benz and Wentzel, 1981) or through the generation of lower hybrid waves at the interface of rapidly expanding flux and the ambient corona (Spicer, Benz, and Huba, 1981).

1.2. Decay of Sunspots

The best targets to investigate the circumstances under which noise storms occur are non-flaring active regions. Conditions for discontinuous coronal evolution can also be created during the decay of active regions – a quieter and longer-lasting evolutionary phase than the relatively short periods of flux emergence. The decay of sunspots shows two modes: (i) the fast mode (fragmentation), which is mainly seen in small and irregularly shaped spots, and (ii) the gradual mode, which is a slow shrinking, frequently ended by fragmentation when the sunspot has shrunk to a small size. Spots are frequently surrounded by a so-called ‘moat cells’, a kind of modified supergranular cell, which ensures the stability of the spots (Meyer *et al.*, 1974). Moats can develop around both preceding (p) and following (f) spots even when they are still growing (Vrabec, 1974). Spots which have no moat cell die by quick fragmentation, while those with a well-developed moat cell shrink gradually.

Magnetic elements move across moat cells, radially away from the spot, with a velocity of in the range 0.1–2 km s^{-1}. Harvey and Harvey (1973) named these ‘moving magnetic features’ or MMFs. The MMFs have either the same magnetic polarity as the central sunspot, or they are bipoles (Zhang *et al.*, 1992). Maltby (1975) found that their outward flow appears to be concentrated in channels. The outer edge of the moat cells form a boundary that has the same polarity as the sunspot in its center, and the leading polarity of the outward moving bipoles has a polarity opposite to that of the spot (Note: sometimes only the leading polarity is condensed enough to be visible). When the bipolar pairs reach the moat boundary the leading one merges and cancels with a moat feature and the two disappear.

MMF activity is observed to be related to X-ray transient brightenings (Shimizu, 1994), surge and X-ray jet activity (Canfield *et al.*, 1996), and even hard X-ray emitting micro-flares (Nitta, 1997). MMFs are believed to represent flux which is 'peeling off' the main flux tube, and is being swept away by the moat (supergranular) velocity field (Harvey and Harvey, 1973). Their puzzling bipolar nature may be attributed to a twist of the peeling-off flux bundle. Then, some of the twisted parts may emerge and form several bipoles or MMFs. Spruit, Title, and van Ballegooijen (1987) attribute the MMF phenomenon to the emergence of U-loops, which are horizontal loops linking two remote ARs. As flux is peeling off the spot, the so-called 'sea-serpent process' (cf., Spruit, Title, and van Ballegooijen, 1987, Figure 5) breaks down the horizontal rope creating small magnetic loops which are carried away by the moat flow.

1.3. Aim of the present work

In this paper we study one active region with only few reported Hα flares during several days around its central meridian passage. As it crossed the solar disk in May and June 1998, active region 8227 was tracked for more than two weeks as part of an observing campaign by the TRACE, *Yohkoh* and SOHO spacecraft, as well as many ground-based observatories. The campaign was the subject of two Co-ordinated Data Analysis Workshops (CDAWs) where members of the teams that participated in the campaign came together to undertake joint analysis of the data. This study evolved from work that was started at the CDAW held at MEDOC, in Orsay in March 1999.

Magnetography using SOHO/MDI, EUV imaging with TRACE, and multi-frequency imaging with the Nançay Radio Heliograph were used to determine how changes in activity in the corona were related to the evolution of the magnetic field. In particular, we have examined how the evolving field may have led to the suprathermal electrons whose emissions were observed by the Nançay Radio Heliograph between 29 May and 1 June 1998. The images from TRACE, together with other coronal images, allow us to follow candidate loops confining the suprathermal electrons back to elements of magnetic field in the photosphere and are crucial in that they provide an insight as to where in the lower atmosphere it may be important to look for related changes.

2. Observations

This study uses data from the Michelson Doppler Imager (MDI; Scherrer *et al.*, 1995) and Extreme-ultraviolet Imaging Telescope (EIT; Delaboudinière *et al.*,

1995) on the SOHO mission, the Transition Region and Coronal Explorer (TRACE, Handy *et al.*, 1999), the Soft X-ray Telescope (SXT; Tsuneta *et al.*, 1991) on *Yohkoh*, and Nançay Radio Heliograph (NRH; Kerdraon and Delouis, 1997).

MDI made observations in the Ni I line at 6768 Å; longitudinal magnetograms were constructed by measuring the Doppler shift of the line separately with right and left circularly polarized wave plates. The difference between these two is a measure of the Zeeman splitting and is roughly proportional to the magnetic flux density, the line-of-sight component of the magnetic field averaged over the resolution element. Since the SOHO spacecraft was in a halo orbit around the Sun – Earth L1 Lagrangean point, it made uninterrupted observations of the Sun. MDI was therefore able to obtain high-quality synoptic observations of the full-disk line-of-sight magnetic field every 96 min. These were interleaved with the helioseismology observations to ensure minimal impact on MDI's primary observing goal. The 1024×1024 pixel magnetogram images are made up of 2 arc sec pixels.

For most of the campaign, TRACE was taking EUV images at 171 Å with a cadence of between 40 and 60 s. These were interspersed with images in white-light to allow the accurate co-alignment of data from TRACE with observations made by other instruments. The 1024×1024 pixel images cover only a 8.5×8.5 arc min field of view and are made up of 0.5 arc sec pixels. The TRACE observatory was in a high inclination (97°), low-Earth orbit and observes the Sun continuously for 9 months of the year. TRACE was tracking the active region and taking EUV images at all times outside of the South Atlantic Anomaly, although some of the images taken when the spacecraft at high latitudes are heavily affected by high-energy particle tracks. In some cases, full-disk images taken by SOHO/EIT at 171 Å were used to fill data gaps and also to interpret features that extended outside of the TRACE field of view.

Radio images of the solar corona were taken by the Nançay Radio Heliograph (2° E, 47° N) at the frequencies 164, 237, 327, 410 and 432 MHz, from ~08:10 to 15:30 UT each day. In the 2D imaging mode, correlation of the 15×24 antennas of a T-shaped array provided a beam size at half maximum of 1.9×1.3 arc min at noon, and 2.1×1.3 arc min two hours later (at 410 MHz). The cadence was 2 images per second. Pointing was checked by daily observations of the Cyg A radio galaxy. Solar images were interpolated such as to give a pixel size of 15 arc sec. To within this limit the pointing is accurate at frequencies $\geq$237 MHz. There are larger pointing errors at 164 MHz, but this frequency is not used for the following discussions.

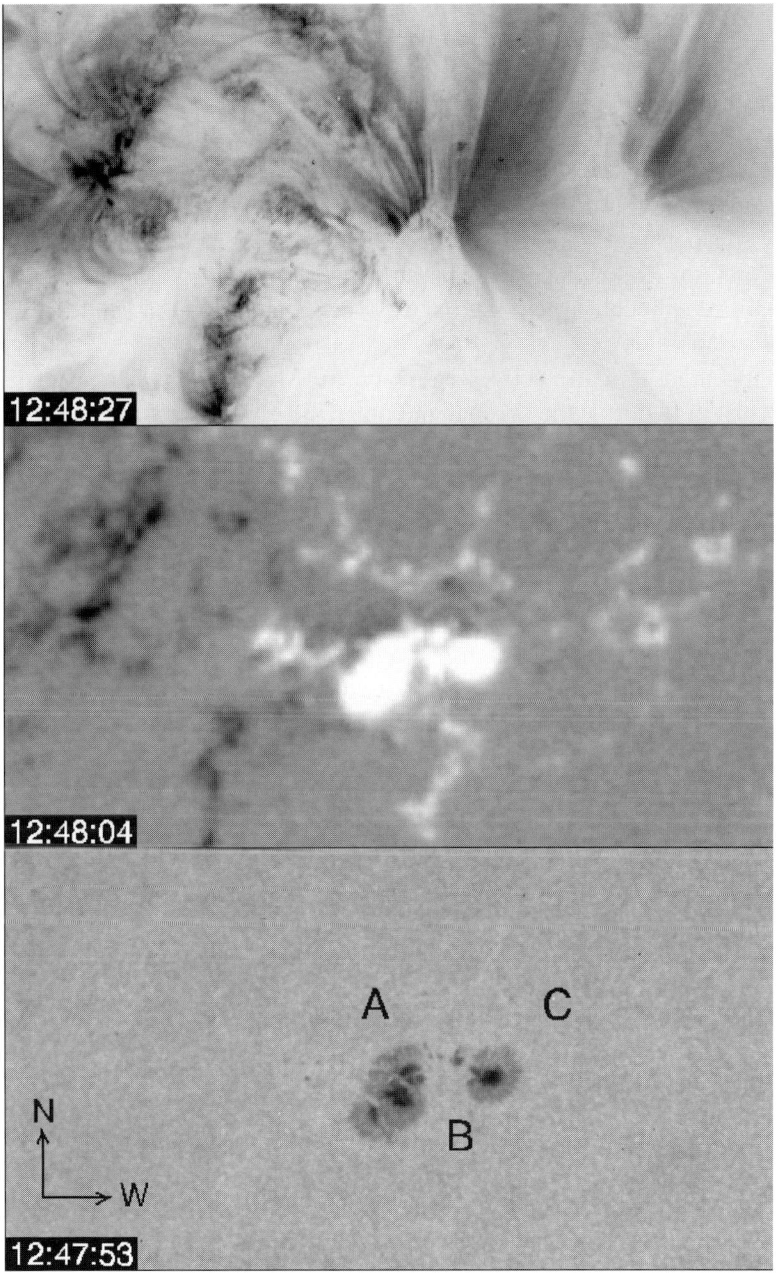

Figure 1. Overview of the structure of AR 8227 on 30 May 1998 showing co-aligned images from TRACE 171 Å (*top*), SOHO/MDI (*middle*) and TRACE white-light (*bottom*). The field of view of each panel is 300 × 160 arc sec; dark shading in the magnetogram shows negative (southern) polarity. The three regions of magnetic evolution (A, B and C) are marked on the bottom panel. The locations of specific features in region C are picked out in Figure 3. The orientation of this and other figures is shown in the bottom panel (north is to the top).

3. Results and Discussion

3.1. Overview

The active region observed during the campaign had first emerged on the previous rotation as AR 8214. By the rotation under study (Carrington rotation 1936), it was a mature region (AR 8227) consisting of two well developed leading (positive polarity) sunspots that were in the process of breaking up, and a dispersed spotless negative polarity following part. The active region was no longer visible on the next rotation, although a new region was seen slightly to the south at the same Carrington longitude.

The appearance of the active region on 30 May 1998 is shown at different wavelengths in Figure 1; the coronal and magnetic evolution of the region between 30 May and 1 June are shown in Figures 2 and 3. Between 30 May and 1 June significant changes in the appearance of the active region can be seen in the MDI data. The size of the 'double-headed' leading spot had diminished and the clump of field between the two spots had elongated into a north-south oriented bar-like feature. Also, where as there were small patches of negative field visible to the north of the spots and almost none to the south on 30 May, these had largely disappeared to the north and there was an intense patch to the south on 1 June. In the TRACE images, bright loop footpoints that were associated with the northeast edge of the eastern spot on 30 May have disappeared by 31 May. On 1 June, TRACE showed low-lying loops connecting the patch of negative field back to the eastern spot. Changes were also seen in the north-western edge of the frames, where a bright loop system seen on 30 May gradually faded over the following days. Figure 3 shows that the small-scale photospheric magnetic features evolve from day to day in this region.

The magnetic activity studied here is principally in three areas around the positive polarity sunspot pair. These are marked on the white-light image of Figure 1: to the north of the sunspot pair and inside the moat boundary (A), to the south of the sunspot pair (B), and to the west and northwest of the western sunspot (C). For the first few days after the active region rotated onto the disk, a number of moving magnetic features (MMFs) were seen in region A; these were flowing northwards away from the sunspot pair. As the active region continued to degenerate (as seen by MDI), the outflow of MMFs was less pronounced in region A and was replaced by a similar outflow to the south in region B.

Noise storm emission was observed in the vicinity of this active region at all NRH frequencies between 29 May and 2 June. The storm probably started on 28 May, when occasional brightenings were visible especially at the higher frequencies. The start and end of the radio emission are, however, uncertain since noise storms from other active regions were prominent until 28 May and after 1 June. One noise storm source was observed to persist above the leading polarity of the active region. It remained essentially stable on 30 and 31 May, and shifted southward on 1 June. The most conspicuous changes in the radio emission were two short-lived storms north-westward of the long-lived component, which appeared for $\sim (1-1.5)$ hours on 30 and 31 May.

The continuous stream of high-resolution magnetograms with a regular cadence that were provided by the SOHO/MDI permit us to track the evolution of the solar photospheric magnetic field in a way that was not possible before. When combined with EUV data from TRACE, for the first time it is possible to see how structures in the corona are related to the underlying magnetic field, and how changes in that field affect the overlying structures. In the following text we attempt to identify features seen by MDI that are related to changes in the source configuration of radio noise storms.

3.2. On and Before 30 May

The first sign of the returning active region was emission from high coronal loops seen by *Yohkoh*/SXT on 22 May. The region crossed onto the disk on 24 May and reached central meridian on 31 May.

From 28 May onwards, the region was far enough onto the disk for the magnetic structure to be clearly visible in MDI magnetograms. When the MDI images are viewed as a movie, the dynamic nature of the region's magnetic evolution becomes evident. To the north of the leading sunspot pair (region A), particularly to the north and east of its eastern spot, a general outflow of material is visible. This is seen as a continuous series of small negative polarity moving magnetic features (MMFs); we have attempted to demonstrate this in Figure 4.

On 30 May, from 04:48 UT a larger and more visible MMF (black feature; negative magnetic polarity) was seen moving northwards, this was last seen as it merged with the positive polarity moat boundary in the magnetogram at 14:24 UT. The motion of this larger MMF can be seen in the frames in the right hand panel of Figure 4, on the right of the frames. The MMF followed a similar evolution to a feature seen in the same area on 29 May (just right of center in the frames in the left panel of Figure 4).

In TRACE 171 Å images, a low-lying extended loop structure associated with this MMF was visible, and an EUV bright point appeared as the feature approached the surrounding boundary of positive field. The bright point can be seen in the top frame of Figure 1 just to the left of the fan of loops associated with the leading spot (the MMF is visible in the middle frame). The magnetic field associated with the structure seen in TRACE has been modeled and is discussed in Section 3.5.

The observations from the Nançay Radio Heliograph show a noise storm located in projection to the north of the positive polarity spot pair. A snapshot of the 410 MHz source is given by the eastern set of iso-intensity contours in Figure 5 (top left). Starting near 12:20 UT at 410 MHz, occasional brightenings were observed north-west of this noise storm. The brightness increased and dominated the pre-existing storm from 12:57 to 14:08 UT – the source's location is shown

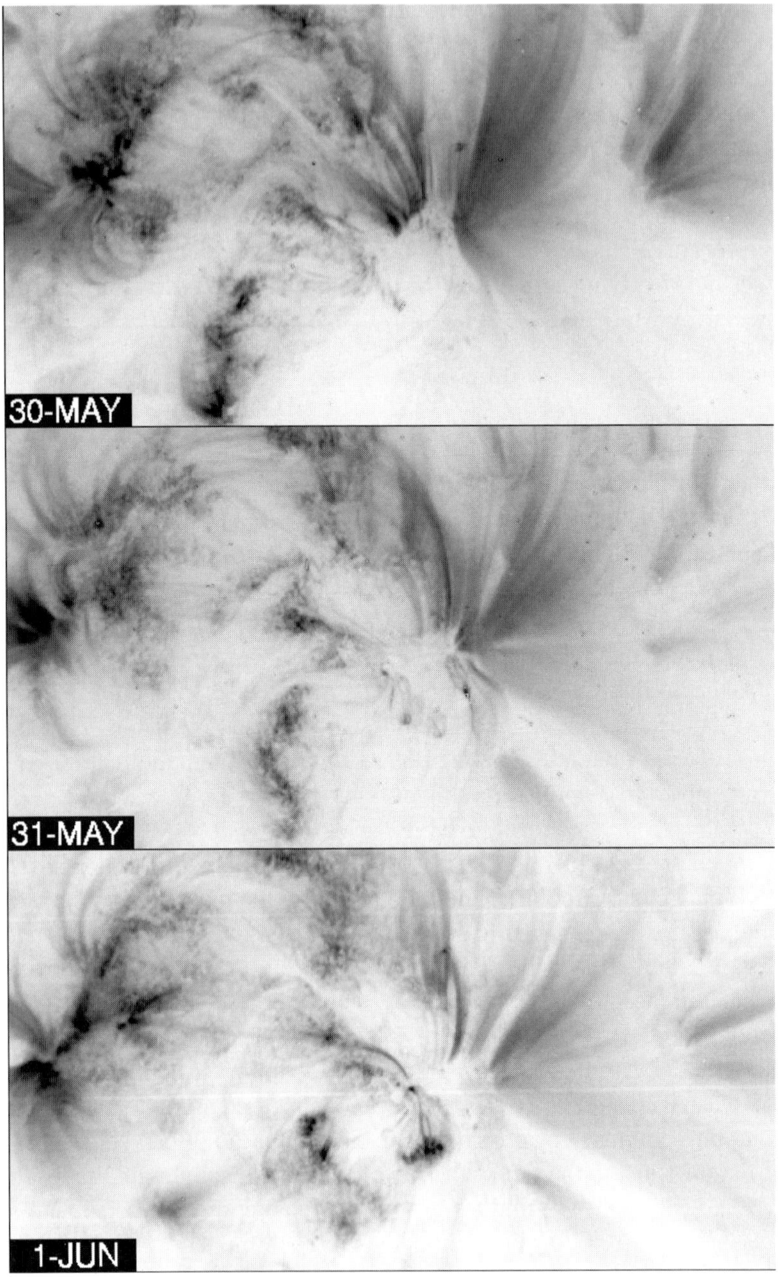

Figure 2. Active region 8227 as observed by TRACE at 171 Å between 30 May and 1 June. The active region has been tracked by the analysis software so that it always appears in the center of the 300 × 160 arc sec field of view. The image times are 30 May 1998 at 12:48:27 UT, 31 May 1998 at 12:51:15 UT, and 1 June 98 at 12:15:10 UT; they match the times of the magnetograms shown in Figure 3. (Note: the TRACE images are plotted with reverse intensity to emphasize less intense structures.)

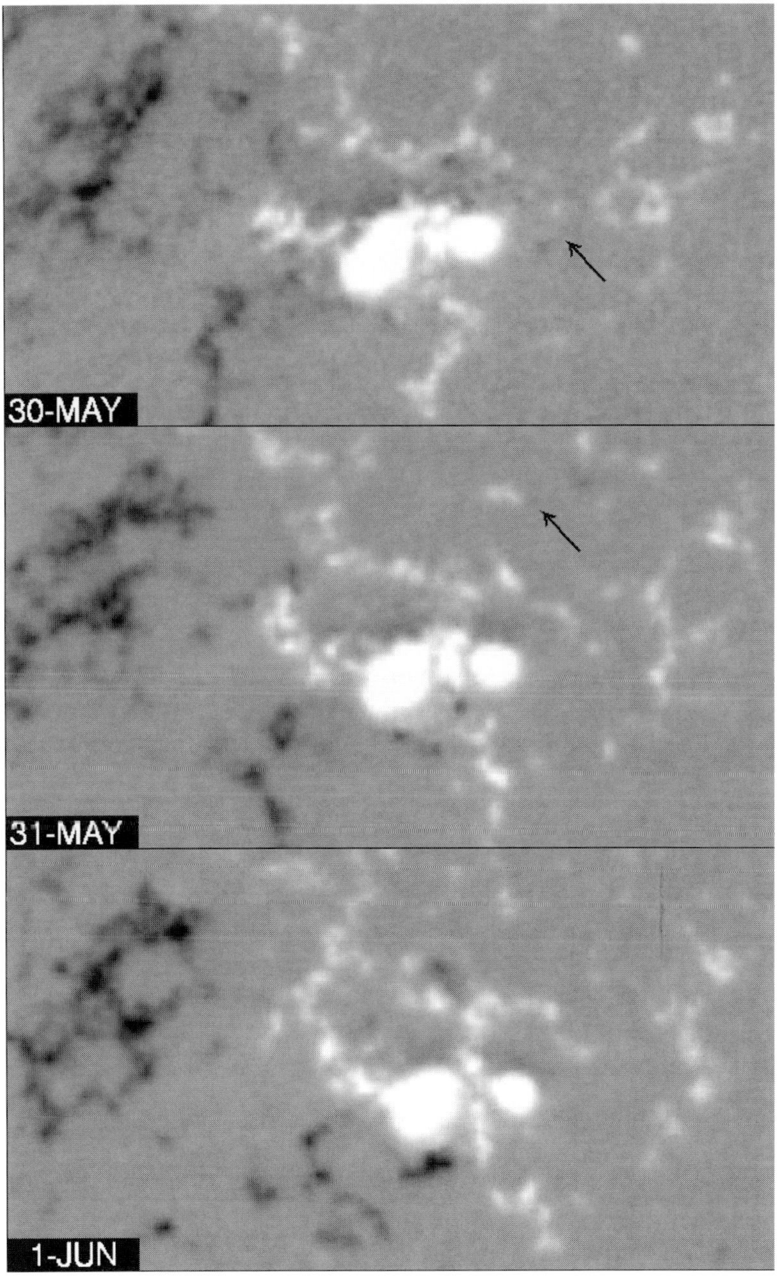

Figure 3. Magnetograms of active region 8227 taken by SOHO/MDI. The center and size of the field of view of the panels of this figure matches those of Figure 2. The image times are 30 May 1998 at 12:48:04 UT, 31 May 1998 at 12:51:04 UT, and 1 June 1998 at 12:51:04 UT. The arrows in the top two panels indicate the general location of features whose evolution is detailed in Figures 6 and 7.

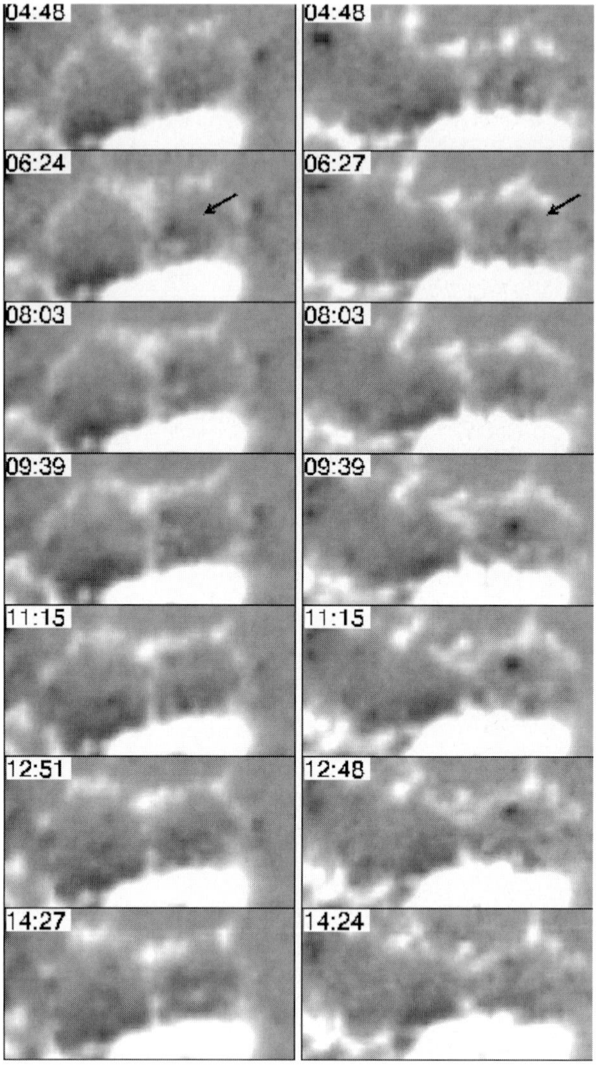

Figure 4. Time sequence of images taken by SOHO/MDI showing the moving magnetic features (MMFs; marked with arrows) in region A (cf., Figure 1) on 29 May (*left column*) and on 30 May (*right column*). The images have a cadence of ~96 minutes and a field of view of 100 × 50 arc sec.

by the western set of contours in Figure 5 (top left). Both radio sources were completely circularly polarized in the same sense (left handed, corresponding to ordinary mode emission with respect to a magnetic field with positive polarity, which was the dominant polarity of the leading spot). Its duration and polarization show that the new radio emission is also a noise storm. After 14:08 UT the old noise storm reappeared at the same flux density level as before 12:20 UT, and the new noise storm was no longer visible at 14:38 UT. Although the old noise storm was not detected between 13 and 14 UT, we believe that it persisted throughout this

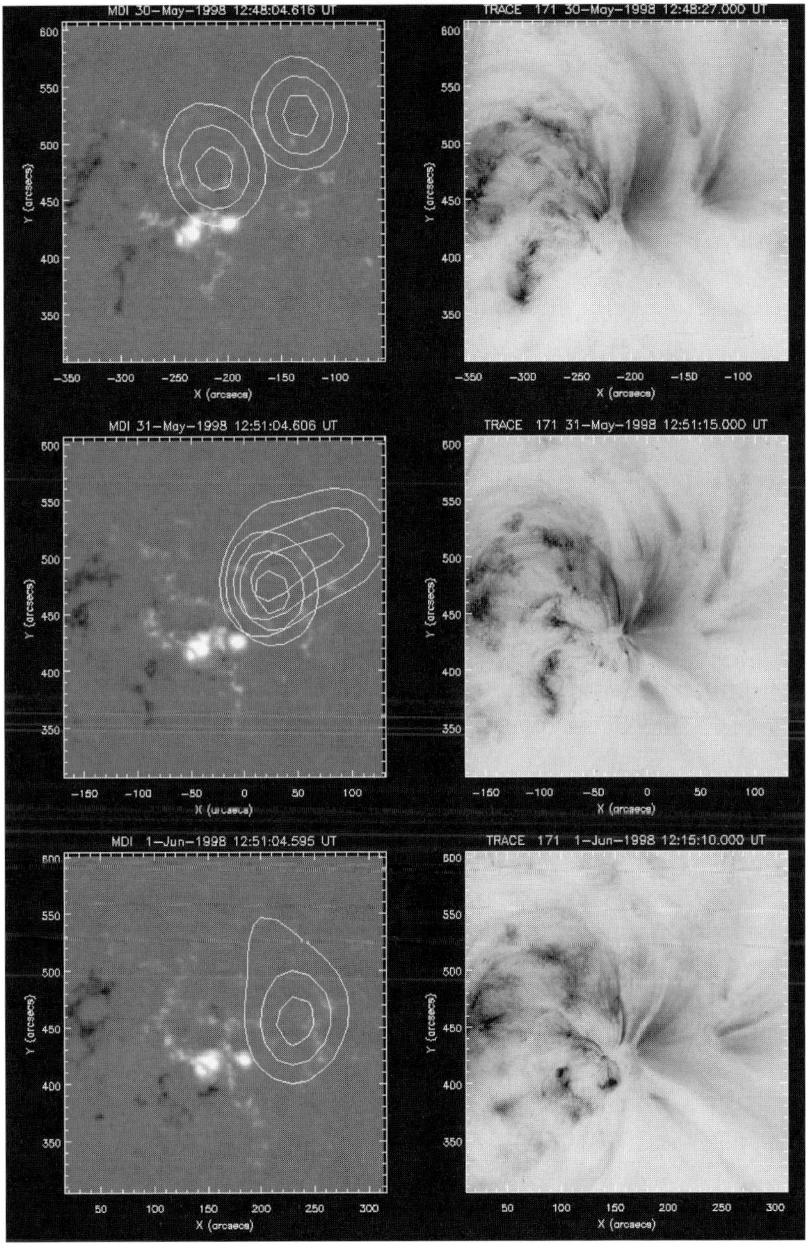

Figure 5. Snapshot maps of the radio noise storms observed by Nançay Radio Heliograph between 30 May and 1 June are plotted as contours on the SOHO/MDI images of about the same time (left-hand panels); the appearance of the region in TRACE 171 Å is shown in the right-hand panels. The 410 MHz Nançay Radio Heliograph images were at 12:08:34 and 13:12:34 UT on 30 May, 12:29:30 and 12:50:50 UT on 31 May, and 12:10:40 UT on 1 June; the contour levels are 50, 70 and 90% of maximum intensity. The radio images were integrated over 128 s such that the burst and continuum components of the noise storms are not distinguished. (The scales on the axes are arc sec from disk center.)

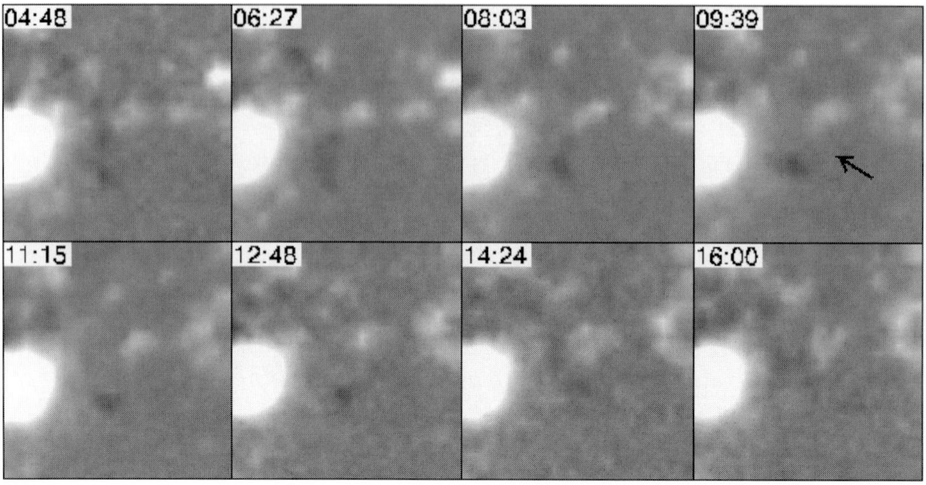

Figure 6. Magnetic activity observed in region C (cf., Figure 1) by SOHO/MDI on 30 May that is believed related to the appearance of a short-lived noise storm seen by the Nançay Radio Heliograph. The two small fragments of opposite polarity (marked with arrow; cf. Figure 3) are closer from 12:48 UT and are in the process of merging at 14:24 UT; the peak field strengths (averaged on the magnetograph pixel) of the fragments at 11:15 UT are −60 and 116 G. The western part of the leading spot is visible on the left of the frames. The images have a cadence of ~96 min and a field of view of 60 × 60 arc sec.

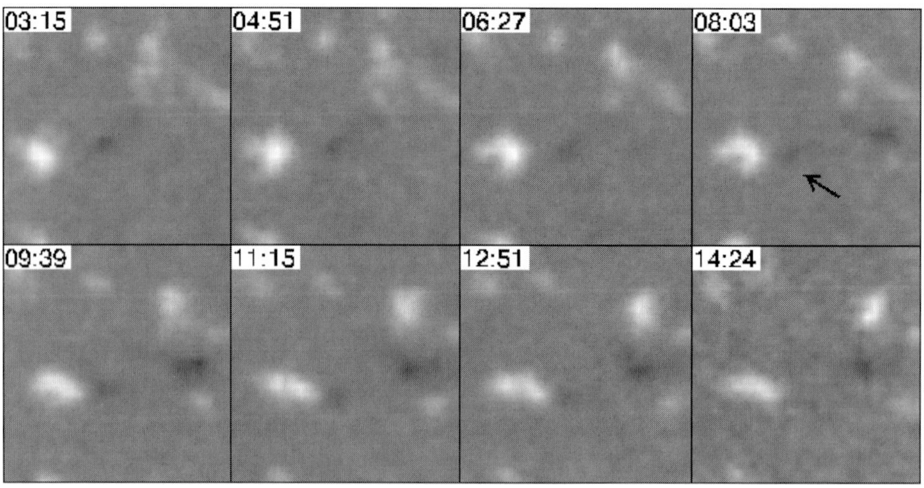

Figure 7. Magnetic activity in the northern part of region C on 31 May believed to be related to the temporary noise storm from 12:09 to 12:44 UT. The feature to the left of center merges between 12:51 and 14:24 UT; the peak field strengths of its components at 09:39 UT are −50 and 195 G. The western leading spot is just below the bottom left of the frames. The images have a cadence of ~96 min and a field of view of 60 × 60 arc sec.

time interval. A similar evolution to that at 410 MHz was observed at all Nançay Radio Heliograph frequencies. The two sources were not clearly separated at 237 and 164 MHz.

Comparison with the TRACE images shows that the noise storm sources were located in two different loop systems anchored in the leading part of the active region. If we assume that the long-lived storm was at the same position from 29 May to the morning of 1 June, in a reference system rotating with the angular speed of the active region, the 410 MHz source was located about 0.07 $R_{\odot}$ above the photosphere. Therefore some projection effect must be taken into account when comparing the noise storm source with the magnetograms. The thermal electron density in the 410 MHz source was $\approx 2 \times 10^9$ cm^{-3}, as inferred from the generally accepted idea that the radiation is emitted at the local plasma frequency. This is comparable to average electron densities in the loops observed by TRACE in this active region on 30 May (Fletcher and de Pontieu, 1999). The lower frequencies are emitted in more dilute plasma, at greater height, which is not seen by TRACE. It is noteworthy that *Yohkoh*/SXT does not see these structures, but only the hot loop tops above the center of the active region (cf., Berger *et al.*, 1999, Figure 1).

Although the similar timing might suggest that the temporary noise storm occurred in conjunction with the EUV bright point and the larger MMF discussed above, the radio source location – west of the leading spot and projecting onto a different loop system than the one that overlay the bright point argues for an association with region C of the magnetograms. In Figure 6, two short-lived magnetic fragments of opposite polarity can indeed be seen to the west of the leading spot in region C. They moved toward each other and merged at the time that the temporary noise storm was observed in the overlying corona. TRACE did not show any enhanced emission from this area at this time.

3.3. 31 May

On 31 May, when AR 8227 crossed central meridian, the noise storm was still visible in the Nançay Radio Heliograph data – see the nearly circular contours in the middle panel of Figure 5. It was slightly further to the west than the previous day, but this is probably due to the change in projection as the Sun rotates. Between 12:09 and 12:44 UT another noise storm emerged to the north-west of the main one. This never became dominant and at its peak was only as strong as the existing storm. At the time of its peak intensity, the source configuration is shown by the elongated contours in Figure 5 (center left). The new noise storm was not observed at 237 and 164 MHz. The relative location of the new noise storm with respect to the pre-existing one was the same as on 30 May.

Again a short lived magnetic feature was observed in region C – see Figure 7. The fragments of the feature to the left of center start to move together at 08:03 UT and merge between 12:51 and 14:24 UT. The motion in this area appears to be reasonably wide spread since at the same time the fragment of negative field on the right of the frame appeared to be rotating around the positive fragment to its north. In TRACE 171 Å images, loop footpoints brightened a little around the time that the fragments were merging.

3.4. 1 June

By the latter part of 31 May, the magnetic outflow to the north of the spot pair (region A) had all but disappeared and had been replaced by a similar outflow to the south (region B). Extended loop structures similar to the one seen in TRACE on May 30 were also seen connecting the larger MMFs back to the sunspots.

A single noise storm source was observed on 1 June. Again there was a small displacement to the west thought to be due to the differing projection as the Sun rotates (Figure 5, bottom left). The most prominent change was a systematic southward drift of the centroid position during the observing period. The displacement was marginally visible at 410 MHz (Figure 5), but attained 0.08 $R_{\odot}$ at 237 MHz, which clearly exceeds the dispersion of north-south positions observed on previous days. The marginal shift at 410 MHz and the larger shift at 237 MHz, i.e., higher in the corona, suggests that the whole structure that confines the noise storm source shifted southward. We propose this is related to the change in location of the outflow seen in the magnetogram data (from region A to region B).

No temporary storm was observed on 1 June. In passing, we note that the fading of loops seen in the north-western part of the TRACE images (Figure 2) suggests a reduced coronal heating in those regions where the temporary storms had been observed on previous days.

3.5. Proposed magnetic configuration

Changes of the magnetic field in the active region under discussion consist mainly of MMFs in different regions around the leading spots. In order to understand how they affect the coronal magnetic field configuration, we analyze the MMF on 30 May, which involved structures that were large enough and contained dense enough plasma to be seen. We believe that the many other smaller MMFs seen had similar magnetic structures associated with them, but did not contain enough plasma to be visible.

We performed a linear force-free field extrapolation of the MDI longitudinal magnetograms. A comparison of the computed field lines with TRACE data (Figure 8) shows that the coronal magnetic field was close to being potential. Thus, the diffuse and concentrated electric currents (the latter in current sheets) were both too weak to significantly deform the magnetic field. The large scale connectivity of AR 8227 was a simple bipole. However, at a smaller scale a complex magnetic topology was present for each MMF with opposite polarity to the two leading spots. This complex topology is shown for the larger MMF of 30 May in Figure 9; a side view is shown in the sketch in Figure 10.

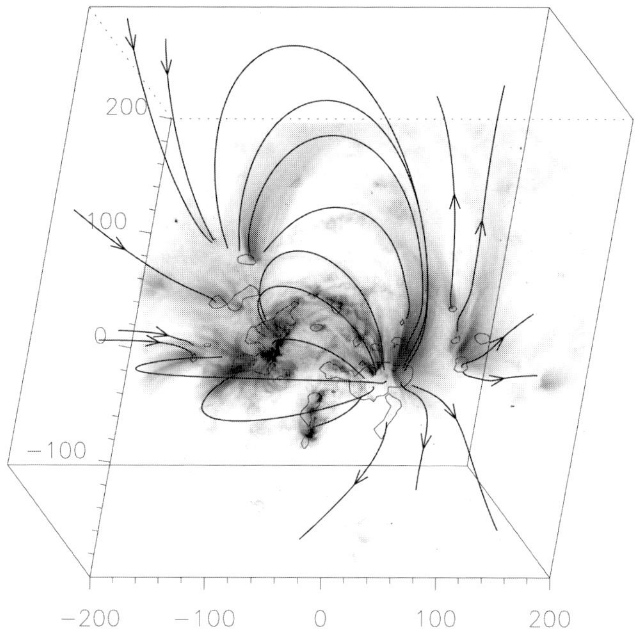

Figure 8. A potential model of the active region overplotted on a TRACE image. The MDI magnetogram of 12:48:04 UT is outlined with the iso-contour 100 G (the solid contour is for both positive and negative values). The box axes are parallel to the local solar frame. The field lines which exit the box are marked by an arrow. The scales on this and the following two figures are in Mm from an arbitrary point in the middle of the active region.

We propose below a physical mechanism which helps to understand that the observed photospheric evolution is the underlying common cause of the radio and EUV observations. The combination of a complex magnetic topology and photospheric footpoint motions created a favorable situation to drive magnetic reconnection as explained below. Initially the MMF was close to the strong spot and mainly connected to it magnetically. Meanwhile, the positive moat boundary (at the top of Figure 9) was connected to the following polarity by long field lines (partially visible in Figure 9, but fully present in Figure 8). As the MMF started to move northward, it induced concentrated electric currents, then magnetic reconnection, between the two sets of structures (see Figure 10). In this way, the MMF progressively became connected to the moat boundary (short field lines in the middle of Figure 9), then finally canceled with part of the moat flux. The magnetic reconnection created two kinds of new connections: a short one consisting of the moat boundary to the MMF, and a much longer one from the spot to the following polarity.

The energy deposited in the short loops increased both their temperature and density (by chromospheric evaporation) making them brighter - this is the plausible interpretation of the brightening observed by TRACE. However, the long reconnected loops had a much larger volume and a similar increase of temperature and

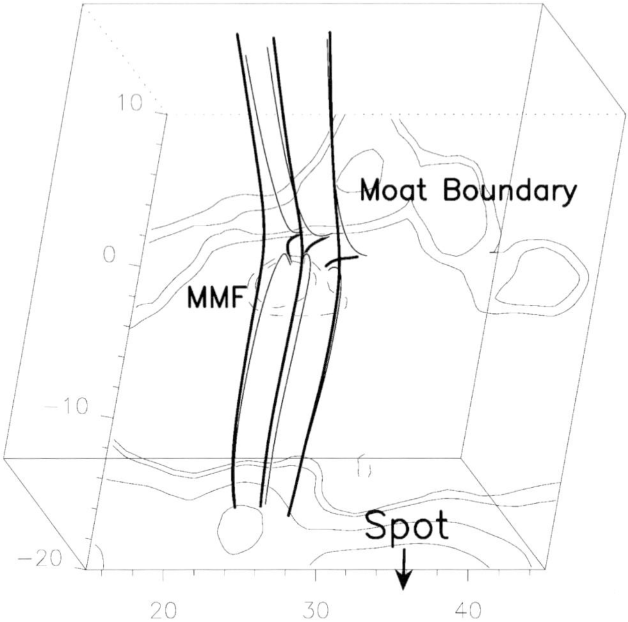

Figure 9. A zoom in the potential model of the field associated with the MMF observed on 30 May. The MDI magnetogram of 12:48 UT is shown as a contour map; the solid contours refer to the positive longitudinal component of the magnetic field, and the dashed contours to the negative field (values are 50, 100, 250, 500 G). The four kinds of connectivities, as discussed in the text, are shown at three locations. Reconnected field lines are drawn thicker. Note the very short field lines connecting the MMF to the moat boundary. The central one corresponds to the EUV bright point seen with TRACE (cf., Figure 1 top panel).

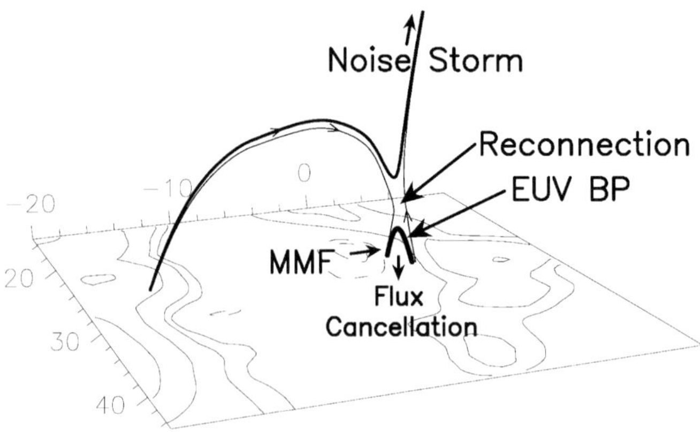

Figure 10. Sketch showing the basic connectivity implied in the magnetic reconnection. It shows the four kinds of connectivities with the central set of field lines shown in Figure 9. The leading polarity is on the left, the MMF is in the center and the moat is on the right. Magnetic reconnection is driven by the approach of the MMF towards the moat boundary. It forms short loops connecting the MMF to the moat boundary, and long loops connecting the leading polarity to the following one.

density would require a proportionally larger input of energy. Because of this strong asymmetry in the reconnection process, the long reconnected loops are expected to be much dimmer than the short loops (as observed in X-ray bright points or flares, e.g., Mandrini *et al.*, 1996; Manoharan *et al.*, 1996), and not even apparent in the very low energy events studied here.

The radio emission, however, occurs only in the long loops, because the density in the short reconnected loops is expected to be too high for the frequency range used by the Nançay Radio Heliograph. Further discussion of the topological and reconnection aspect is presented in Mandrini *et al.* (1996) for a more energetic event (X-ray bright point), but with a similar magnetic topology (in that case the energy is released by the emergence of a new magnetic bipole instead of an MMF).

The above scenario allows us to understand the physical link between magnetic evolution and the radio and EUV observations. We propose that the same mechanism (magnetic reconnection induced by photospheric motions) is valid for all the MMFs described above even though the small MMFs are expected to result in less heating so that even short reconnected loops may not be visible in TRACE.

4. Conclusions

The comparison of photospheric magnetic features, coronal plasma structures, and the sites of emission from suprathermal electrons in the active region studied here suggests an intimate relationship between the evolution of magnetic features in the low atmosphere and the generation and evolution of the suprathermal electrons and their radio emission.

A long-lived (several days) and two short-lived ($\leq$2 hr) radio noise storms observed by the Nançay Radio Heliograph between 30 May and 1 June are proposed to be related to moving magnetic features (MMF) associated with the decay of the active region sunspots. A long-lived outflow of magnetic polarities is observed to the north of the active region. These features provide a plausible indication for persistent energy conversion and electron acceleration, but due to the long lifetime of both the principal noise storm and the magnetic activity no correlated time evolution between the radio emission and the magnetic features can be demonstrated. We emphasize, however, that a conspicuous southward shift of the noise storm position occurred when the outflow of magnetic features stopped north of the active region and started south of it. Moreover the two short-lived noise storms on 30 and 31 May are related in time and connected in space (within the uncertainties of the alignment of the radio maps and the magnetograms) with new canceling magnetic features.

The possible identification of magnetic restructuring on time scales $\leq$1 hr agrees with the former finding that short-lived noise storms may be accompanied by soft X-ray or white-light brightenings (e.g., Lantos *et al.*, 1981; Kerdraon *et al.*, 1983; Raulin and Klein, 1994; Crosby *et al.*, 1996). Although some EUV emission was observed at the time of the canceling magnetic fragments on 31 May, none appeared to be associated with a similar feature on 30 May. However, in such low energy events an EUV counterpart is expected only if enough energy can be deposited in a small enough volume.

We have extrapolated the photospheric magnetic field with a linear force-free approximation and find that the active region magnetic field was very close to being potential. These computations show a complex magnetic topology associated to the MMFs. The observed photospheric evolution is expected to drive magnetic reconnection in such a complex magnetic topology. We therefore propose that the MMFs are at the origin of the observed metric noise storms.

In the data presented, the noise storms are associated to the displacement of photospheric parasitic polarities without significant emergence of new magnetic flux. These observations are therefore closer to a scenario where the electrons are accelerated in the course of magnetic reconnection (cf., Benz and Wentzel, 1981) than to the model of particle acceleration at the interface between rapidly emerging flux and pre-existing coronal structures advocated by Spicer, Benz, and Huba (1981). We note, however, that the basic ingredients for magnetic reconnection to occur are the same in both cases: a complex magnetic topology together with an evolution forced from below. Emerging magnetic flux is not necessary for a noise storm to exist. The emergence of new flux induces the displacement of photospheric parasitic polarities; but it also brings new magnetic flux and free energy into the corona, so it is expected to trigger, on average, more energetic events.

Acknowledgements

This study evolved from work that was started at the CDAW held at MEDOC, in Orsay in March 1999, and we thank the organizers of that meeting. Much of this analysis was only possible because of the SolarSoft suite of analysis software; particular thanks to Dominic Zarro and Sam Freeland. The TRACE data was provided by the *Yohkoh* Data Archive Centre (YDAC).

RDB is supported under the MSSL Rolling Grant by the UK Particle Physics and Astronomy Research Council (PPARC). LvDG acknowledges the Hungarian Government grants OTKA T-026165 and AKP 97-58 2,2.

The Nançay Radio Observatory is funded by the French Ministry of Education, the CNRS and the Région Centre.

References

Benz, A. O. and Wentzel, D. G.: 1981, *Astron. Astrophys.* **94**, 100.

Berger, T. E., de Pontieu, B., Schrijver, C. J., and Title, A. M.: 1999, *Astrophys. J.* **519**, L97.

Canfield, R. C., Reardon, K. P., Leka, K. D., Shibata, K., Yokoyama, T., and Shimojo, M.: 1996, *Astrophys. J.* **464**, 1016.

Crosby, N., Vilmer, N., Lund, N., Klein, K.-L., and Sunyaev, R.: 1996, *Solar Phys.* **167**, 333.
Delaboudinière, J.-P., Artzner, G. E., Brunaud, J., Gabriel, A. H., Hochedez, J. F., Millier, F., Song, X. Y., Au, B., Dere, K. P., Howard, R. A., Kreplin, R., Michels, D. J., Moses, J. D., Defise, J. M., Jamar, C., Rochus, P., Chauvineau, J. P., Marioge, J. P., Catura, R. C., Lemen, J. R., Shing, L., Stern, R. A., Gurman, J. B., Neupert, W. M., Maucherat, A., Clette, F., Cugnon, P., and van Dessel, E. L.: 1995, *Solar Phys.* **162**, 291.
Elgarøy, Ø.: 1977, *Solar Noise Storms*, Pergamon Press, Oxford.
Fletcher, L. and de Pontieu, B.: 1999, *Astrophys. J.* **520**, L135.
Handy, B. N., Acton, L. W., Kankelborg, C. C., Wolfson, C. J. et al.: 1999, *Solar Phys.*, in press.
Harvey, K. L. and Harvey, J. W.: 1973, *Solar Phys.* **28**, 61.
Kerdraon, A. and Delouis, J.-M.: 1997, in G. Trottet (ed.), 'Coronal Physics from Radio and Space Observations', *Lecture Notes in Physics* **483**, Sringer-Verlag, Berlin, p. 192.
Kerdraon, A., Pick, M., Trottet, G., Sawyer, C., Illing, R., Wagner, W. and House, L.: 1983, *Astrophys. J.* **265**, L19.
Klein, K.-L.: 1998, in C. Alissandrakis and B. Schmieder (eds), 'Three-dimensional Structure of Solar Active Regions', *Astron. Soc. Pacific Conf. Series* **155**, 182.
Krucker, S., Benz, A. O., Aschwanden, M. J., and Bastian, T. S.: 1995, *Solar Phys.* **160**, 151.
Lantos, P., Kerdraon, A., Rapley, G. G., and Bentley, R. D.: 1981, *Astron. Astrophys.* **101**, 33.
Lin, R. P.: 1985, *Solar Phys.* **100**, 537.
Lin, R. P.: 1997, in G. Trottet (ed.), 'Coronal Physics from Radio and Space Observations', *Lecture Notes in Physics* **483**, Springer-Verlag, Berlin, p. 93.
Malik, R. K. and Mercier, C.: 1996, *Solar Phys.* **165**, 347.
Maltby, P.: 1975, *Solar Phys.* **43**, 9.
Mandrini, C. H., Démoulin, P., van Driel-Gesztelyi, L., Schmieder, B., Cauzzi G., and Hofmann A.: 1996, *Solar Phys.* **168**, 115.
Manoharan, P. K., van Driel-Gesztelyi, L., Pick, M., and Démoulin, P.: 1996, *Astrophys. J.* **468**, L73.
Meyer, F., Schmidt, H. U., Weiss, N. O., and Wilson, P. R.: 1974, *Monthly Notices Royal Astron. Soc.* **169**, 35.
Nitta, N.: 1997, *Astrophys. J.* **491**, 402.
Raulin, J. P. and Klein, K.-L.: 1994, *Astron. Astrophys.* **281**, 536.
Raulin, J. P., Kerdraon, A., Klein, K.-L., Trottet, G., Willson, R. F., and Lang, K. R.: 1991, *Astron. Astrophys.* **251**, 298.
Scherrer, P. H., Bogart, R. S., Bush, R. I., Hoeksema, J. T., Kosovichev, A. G., Schou, J., Rosenberg, W., Springer, L., Tarbell, T. D., Title, A., Wolfson, C. J., Zayer, I. and the MDI Engineering Team: 1995, *Solar Phys.* **162**, 129.
Shimizu, T. 1994, in T. Sakurai, T. Hirayama and G. Ai (eds.), *Proc. of the 2nd Japan-China Seminar on Solar Phys.*, Nat. Astron. Obs. of Japan, 193.
Spicer, D. S., Benz, A. O., and Huba, J. D.: 1981, *Astron. Astrophys.* **105**, 221.
Spruit, H. C., Title, A. M., and van Ballegooijen, A. A.: 1987, *Solar Phys.* **110**, 115.
Stewart, R. T., Brueckner, G. E., and Dere, K. P.: 1986, *Solar Phys.* **106**, 107.
Tsuneta, S., Acton, L., Bruner, M., Lemen, J., Brown, W., Caravalho, R., Catura, R., Freeland, S., Jurcevich, B., Morrison, M., Ogawara, Y., Hirayama, T., and Owens, J.: 1991, *Solar Phys.* **162**, 37.
Vrabec, D.: 1974, in R. G. Athay (ed.), 'Chromospheric Fine Structure', *IAU Symp.* **56**, 201.
Zhang, H., Ai, G., Wang, H., Zirin, H., and Patterson, A.: 1992, *Solar Phys.* **140**, 307.

A KINETIC MODEL OF CORONAL HEATING AND ACCELERATION BY ION-CYCLOTRON WAVES: PRELIMINARY RESULTS

PHILIP A. ISENBERG, MARTIN A. LEE and JOSEPH V. HOLLWEG
Department of Physics and Institute for the Study of Earth, Oceans and Space, University of New Hampshire, Durham, NH 03824, U.S.A.

(Received 16 September 1999; accepted 19 December 1999)

Abstract. We present a kinetic model of the heating and acceleration of coronal protons by outward-propagating ion-cyclotron waves on open, radial magnetic flux tubes. In contrast to fluid models which typically insist on bi-Maxwellian distributions and which spread the wave energy and momentum over the entire proton population, this model follows the kinetic evolution of the collisionless proton distribution function in response to the combination of the resonant wave-particle interaction and external forces. The approximation is made that pitch-angle scattering by the waves is faster than all other processes, resulting in proton distributions which are uniform over the resonant surfaces in velocity space. We further assume, in this preliminary version, that the waves are dispersionless so these resonant surfaces are portions of spheres centered on the radial sum of the Alfvén speed and the proton bulk speed. We incorporate the fact that only those protons with radial speeds less than the bulk speed will be resonant with outward-propagating waves, so this rapid interaction acts only on the sunward half of the distribution. Despite this limitation, we find that the strong perpendicular heating of the resonant particles, coupled with the mirror force, results in substantial outward acceleration of the entire distribution. The proton distribution evolves towards an incomplete shell in velocity space, and appears vastly different from the distributions assumed in fluid models. Evidence of these distinctive distributions should be observable by instruments on *Solar Probe*.

1. Introduction

Wave-driven models of the fast solar wind date back at least to the work of Hollweg (1973) . The idea is basically to heat and accelerate solar wind protons from the cyclotron-resonant dissipation of outward-propagating ion-cyclotron (or Alfvén) waves. A number of fast solar wind properties in interplanetary space are consistent with the expectations of this interaction. The ions in the fast solar wind are hot, and the ion magnetic moments are found to increase with radial distance from the Sun, implying substantial ion heating. Additionally, the cores of the ion distributions are seen to be anisotropic with $T_\perp > T_\parallel$, in agreement with the perpendicular heating expected from the damping of ion-cyclotron waves. Finally, the heavy ions in the fast wind are seen to flow faster and have higher temperatures than the protons, implying a resonant interaction with the more intense waves at the lower heavy ion cyclotron frequencies. It is particularly noteworthy that recent observations by the UVCS instrument on the SOHO spacecraft have found all of these properties to exist in coronal holes as close to the Sun as 1.5 R_s (Kohl *et al.*, 1998, 1999; Esser

The physics of the solar corona and transition region. Part II.
Reprinted from Solar Physics **193,** 2000.

et al., 1999), emphasizing the likely importance of this interaction in the generation of the fast solar wind flow.

There have been many models of this wave heating in the very low corona which have led to reasonable results for the fast solar wind. Most of these models, while making various assumptions about the source of the waves and the nature of the heating, treat the coronal ion populations as fluids (e.g., Hollweg, 1986; Hollweg and Johnson, 1988; Isenberg, 1990; McKenzie, Banaskiewicz, and Axford, 1995; Axford and McKenzie, 1996; Esser *et al.*, 1997; Tu and Marsch, 1997; Marsch and Tu, 1997; Li *et al.*, 1999; Cranmer, Field, and Kohl, 1999). Specifically, each ion species is assumed to have a Maxwellian or bi-Maxwellian structure in velocity space, and the energy and momentum deposition by the waves is assumed to be shared over the entire distribution. In this manner, these models increment the fluid moments U, $T_{\parallel}$, and $T_{\perp}$, (the bulk flow speed and the temperatures parallel to and perpendicular to the magnetic field, respectively) at increasing radial distance from the Sun without allowing the distribution to deviate from a bi-Maxwellian shape.

However, it is well known that the cyclotron resonance with outward-propagating ion cyclotron waves only takes place for ions in a limited portion of velocity space (Rowlands, Shapiro, and Shevchenko, 1966; Dusenbery and Hollweg, 1981). In particular for protons, these waves are only resonant with particles in the sunward half of the distribution, $v_{\parallel} < U$, and it is these slower protons only that can respond to the perpendicular heating by the waves. Furthermore, coronal ions are essentially collisionless for heliocentric radii $r \gtrsim 1.5\ R_s$. Thus, in reality the resonant wave-particle interaction will strongly distort the ion distributions in velocity space and collisions will not be able to restore a smooth bi-Maxwellian state.

Recently, there has been some effort to extend the bi-Maxwellian treatment to include higher-order moments of the ion distribution (Olsen and Leer, 1999; Li, 1999). Although the distributions in these works are still expansions about bi-Maxwellians, these models are able to display some of the distortion expected in the distribution from the combination of perpendicular heating and the collisionless mirror force in the expanding magnetic field. However, these models do not incorporate any specific characteristics of the resonant cyclotron interaction. Instead, they utilize ad-hoc expressions to designate the amount of perpendicular heating as a function of radial position and model the resulting evolution of the plasma. Li (1999) includes heating or cooling in the parallel direction as well, but the magnitude of the effect is still arbitrarily chosen and not related to properties of the ion-cyclotron waves.

There have also been several previous studies of the kinetic effects of the ion cyclotron resonance in the low corona. Cranmer *et al.* (1997) and Hollweg (1999a, b) used a single-particle description to represent the average response of an ion distribution to the large-scale and resonant forces. Hollweg (1999c) explored the consequences of restricting the resonance to the sunward protons by considering a two-proton representation, where only the slower 'proton' was allowed to interact with the waves. These models have provided considerable insight, but this technique cannot be extended indefinitely.

A full kinetic treatment has also been introduced by Tam and Chang (1999), although this work, too, considers wave heating in the perpendicular direction only and the model results published thus far have not been particularly realistic.

This paper presents a progress report on the construction of a new kinetic model for a full distribution of coronal protons. In this model, the resonance properties of each particle are determined from its parallel speed $v_{\|}$ with respect to a self-consistently determined bulk speed U of the entire distribution. We are motivated in this effort by the following questions: First of all, can we in fact generate a realistic fast solar wind by heating only the sunward half of the proton distribution? If so, how much wave energy is required to accomplish this? Finally, what does the resulting proton distribution look like, particularly in light of possible *in-situ* measurements by instruments on *Solar Probe*? This preliminary report is necessarily incomplete, but we expect the completed model to provide the answers to these questions.

In the following sections, we will give a brief description of the model assumptions, outline the derivation of the model equations, and present some preliminary results obtained to date. There remains much work to be done, and we will close with a list of extensions and improvements we plan to include in future versions of this model.

2. Description of the Model

We are interested in constructing a system of equations to model the proton distribution in the low solar corona, where the sunward half of the distribution ($v_{\|} < U$, where U is the bulk flow speed of the entire proton distribution) responds to the resonant interaction with intense outward-propagating ion-cyclotron waves and the anti-sunward half of the distribution ($v_{\|} > U$) does not.

The essence of the resonant wave-particle interaction with transverse waves as described by quasilinear theory is a diffusion in particle pitch angle which conserves the particle energy in the reference frame moving with the phase speed of the resonant wave (Kennel and Engelmann, 1966; Rowlands, Shapiro, and Shevchenko, 1966). To make the problem tractable, we assume here that the pitch-angle scattering of the resonant coronal protons is much faster than all other processes in the system. Thus, the sunward protons in our model will maintain a uniform velocity-space density on constant energy surfaces centered on the wave phase speed, as depicted in Figure 1.

As an initial simplification in the development of our model, we currently take the resonant waves to be dispersionless, even though the restriction that resonant protons have $v_{\|} < U$ is a consequence of wave dispersion. In this case, the phase

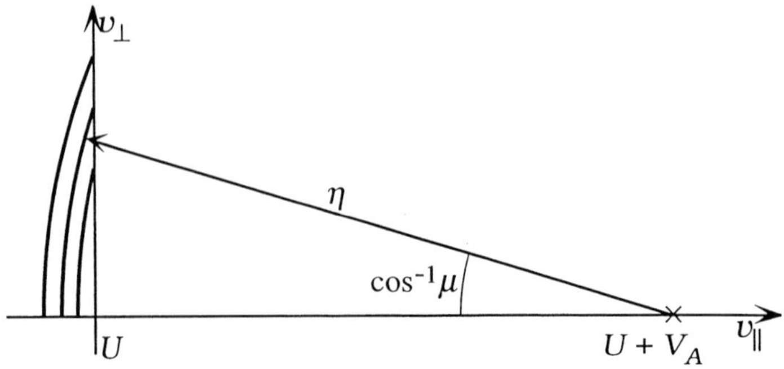

Figure 1. The constant energy resonant surfaces in our model. The resonant surfaces are confined to the sunward hemisphere in velocity space, $v_\parallel < U$. In the dispersionless case used here, the surfaces are sections of spheres centered on $v_\parallel = U + V_A$. The variable η is the radius of each surface and serves to label the surfaces in the model.

speed of the wave is $\omega/k = U(r) + V_A(r)$, independent of wavenumber k, and the constant energy surfaces are spherical sections of radius η, as drawn in Figure 1.

This rapid scattering of protons over the available resonant surface should be a reasonable approximation in the low corona, since the maximum range of the resonant pitch-angle cosine, denoted μ, is small. If we define a characteristic scale η_0, such that the shells $\eta < \eta_0$ contain most of the sunward particles, we see from inspection of Figure 1 that the maximum change in μ is

$$\Delta\mu = 1 - V_A/\eta_0 \approx 1 - V_A/(V_A + v_{\parallel\, th})$$
$$\approx \sqrt{\beta_\parallel} \lesssim 0.03 \quad \text{near} \quad 2\ R_s\ .$$

We note further that this approximation results in the maximum possible energization of the protons through the resonant cyclotron interaction. Once the protons fill the resonant surfaces uniformly, there is no way for them to tap any more energy from the waves at that point.

Another useful aspect of this approximation is that it does not require any spectral information about the waves. The model simply assumes that enough resonant wave energy is present to maintain the uniform proton density on the resonant surfaces. Furthermore, this energy absorbed by the particles can be calculated from the model. Thus, rather than the usual assumption of a specific quantity and spectrum of waves as an input, we find that the amount of wave energy required to drive the wind in this manner will be an output of our model.

The protons everywhere in the model respond to the external forces of gravity, the charge-separation electric field, and the mirror force in the expanding radial magnetic field. It is primarily this mirror force which transforms the perpendicular heating by the waves into outward motion. In contrast to the sunward protons, the protons in the anti-sunward region of velocity space ($v_\parallel > U$) are unaffected by the wave-particle interaction and simply undergo single-particle motion in the external fields.

3. Derivation of the Model Equations

In this section, we obtain the equations which define our model and briefly sketch their derivation. We start with the stationary guiding-center equation for a gyrotropic proton distribution $f(r, v_r, v_\perp)$ in a heliocentric radial magnetic field, where the velocities parallel and perpendicular to the field $(v_r, v_\perp)$ are defined in the rest frame of the distribution moving at the radial bulk speed $U(r)$ (e.g., Skilling, 1971; Kulsrud, 1983; Isenberg, 1997):

$$(U + v_r)\frac{\partial f}{\partial r} - \frac{1}{2}(U + v_r)\frac{\mathrm{d}\ \ln\ A}{\mathrm{d}r}v_\perp\frac{\partial f}{\partial v_\perp} + \left[F(r) - (U + v_r)U' + \right.$$
$$\left. + \frac{v_\perp^2}{2}\frac{\mathrm{d}\ \ln\ A}{\mathrm{d}r}\right]\frac{\partial f}{\partial v_r} = \frac{\partial}{\partial \mu}\left[\left(1 - \mu^2\right) D(\mu)\frac{\partial f}{\partial \mu}\right] S(-v_r)\,. \tag{1}$$

Here, we define the effective gravitational force per unit mass,

$$F(r) = \frac{-GM_s}{r^2} + \frac{e}{m}E(r)\,, \tag{2}$$

which incorporates the charge-separation electric field E, the area of a magnetic flux tube $A(r) \sim B^{-1}$, and the relative bulk acceleration of the distribution $U' = \mathrm{d}U/\mathrm{d}r$. The terms on the left-hand side of (1) are the radial convection, the decrease of $v_\perp$ due to magnetic moment conservation in a decreasing magnetic field, and the radial evolution of the distribution due to the gravity plus electric field, the acceleration of the reference frame, and the mirror force, respectively. The right-hand side of (1) represents the usual quasilinear pitch-angle diffusion centered on the wave speed $U + V_\mathrm{A}$, but affecting only the protons in the sunward hemisphere $v_r < 0$. The step function, S, on the right-hand side of (1) splits the model velocity space into two parts, each described by a separate equation and joined by a boundary condition at $v_r = 0$.

To obtain the equation for the sunward distribution f_-, we first change the velocity-space variables from $(v_r, v_\perp)$ to (η, μ) as defined in Figure 1. The variable η essentially acts as a label for each constant energy surface. The approximation of very rapid pitch-angle scattering then requires that

$$\left.\frac{\partial f_-}{\partial \mu}\right|_{\eta=\mathrm{constant}} = 0. \tag{3}$$

We apply (3) and then eliminate the pitch-angle variable by integrating over the sunward range of μ to yield an equation for $f_-(r, \eta)$ which describes the slow motion of the resonant surfaces under the action of the external forces,

$$\left(U - \frac{\eta - V_A}{2}\right)\frac{\partial f_-}{\partial r} - \left\{\frac{1+x}{2}F - \left[\frac{1+x}{2}(U + V_A) - \frac{\eta}{3}(1 + x + x^2)\right]\times \right.$$
$$\left. \times (U' + V_A') + \frac{\eta}{6}(2 - x - x^2)(U + V_A)\frac{\mathrm{d}\ \ln\ A}{\mathrm{d}\,r}\right\}\frac{\partial f_-}{\partial \eta} + \tag{4}$$
$$+\left[F - UU' + \frac{\eta^2 - V_A^2}{2}\frac{\mathrm{d}\ \ln\ A}{\mathrm{d}\,r}\right]\frac{f_+(v_r = 0) - f_-}{\eta - V_A} = 0\,,$$

where we define $x = V_A(r)/\eta$. We note that the influence of the anti-sunward particles enters through the quantity $f_+(v_r = 0)$ appearing in the last term. Although (4) looks complicated, it remains a first-order, linear equation for f_-.

The protons in the anti-sunward hemisphere undergo single-particle motion in the external gravitational, electric, and magnetic fields of the system. This motion is described by (1) with zero on the right-hand side, which is equivalent to Liouville's equation for f_+,

$$\left.\frac{\partial f_+}{\partial r}\right|_{\Gamma,\rho=\text{constant}} = 0\,, \tag{5}$$

where the total energy per mass of a particle in the reference frame of the Sun is

$$\Gamma = \tfrac{1}{2}(U + v_r)^2 + \tfrac{1}{2}v_\perp^2 - \textstyle\int^r F\ \mathrm{d}r' \tag{6}$$

and the particle's magnetic moment is $\rho = 0.5\ mv_\perp^2/B(r)$.

The boundary condition which relates the solutions of f_- and f_+ across $v_r = 0$ is obtained by considering the particle transport across the boundary. This transport is controlled by the forces in the scatter-free anti-sunward hemisphere, since the rapid pitch-angle scattering in the sunward hemisphere supplies or absorbs whatever particles are called for by these forces. The transport in the anti-sunward hemisphere is given by

$$\left.\frac{\mathrm{d}v_r}{\mathrm{d}r}\right|_{v_r=0+} = \frac{F - UU' + \dfrac{v_\perp^2}{2}\dfrac{\mathrm{d}\ \ln\ A}{\mathrm{d}r}}{U}\,. \tag{7}$$

The first two terms in the numerator of (7) cause particle motion toward the boundary, representing the action of the effective gravity which pulls particles into the boundary and of the bulk acceleration which pushes the boundary into the particles. The third term in the numerator represents the mirror force, which propels particles away from the boundary to larger v_r. It is clear from (7) that there exists a critical value of $v_\perp = v_{\perp c}$ at which these forces balance. For $v_\perp > v_{\perp c}$, the mirror force dominates the decelerating forces and the anti-sunward particle motion is away from the boundary. These particles are supplied by the rapid scattering in the sunward hemisphere so the value of f_+ at the boundary is given by f_- at the equivalent $v_\perp$. Thus, in this region, the quantity $f_+(v_r = 0) - f_- = 0$ and the last term on the left-hand side of (4) vanishes. Conversely when $v_\perp < v_{\perp c}$, the anti-sunward particles flow into the boundary, providing a source of particles to the sunward hemisphere. In this region, $f_+(v_r = 0)$ must be calculated from the velocity-space densities upstream along the scatter-free characteristics of the anti-sunward equation (5).

Equations (4) and (5), with the boundary condition described above, form the mathematical construction of our kinetic model. In the next section, we present results of some preliminary calculations of this model.

4. Preliminary Results

As an initial test of our model, we investigate the solutions in a spherically expanding flux tube, taking $A(r) = r^2$ and the Alfvén speed V_A to fall as r^{-1}. We also set the charge-separation electric field to simply reduce the gravitational force by a factor of two. (The self-consistent modifications of the Alfvén speed and the electric field due to the faster-than-spherical decrease in density in the accelerating flow will be included in later versions of the model.) The calculation is initiated at $r = 2\ R_s$, a point beyond the region of proton collisions. We take $V_A = 7000$ km s^{-1} there (Hollweg, 1999b), and start with a proton distribution having a parallel thermal extent of $\pm$ 50 km s^{-1} and a bulk flow speed of $U = 100$ km s^{-1}.

Figure 2 shows the resulting model proton distributions for selected radial positions. The distribution in each panel is shown in the proton rest frame at the specified position, related to the inertial (solar) reference frame by the self-consistently calculated flow speed U. The color plots depict the value of log f, normalized in each plot to the maximum value in the distribution. The colors range logarithmically from red at the maximum value down to blue at 10^{-4} of the maximum. Values of f less than 10^{-4} of the maximum are not plotted.

Figure 2(a) shows the starting distribution at $r = 2.0\ R_s$. This model cannot treat the evolution of the distribution from collisional conditions, so we must start with the sunward protons uniformly distributed on the constant energy surfaces in the wave frame. We choose to start with a spherical velocity space distribution in the form of an inverted parabola confined to a radius of 50 km s^{-1} (note the different scales for $v_\parallel$ and $v_\perp$) and we then smear out the sunward half of the protons to a distribution consistent with our model assumption as shown.

Figure 2(b) shows the proton distribution a short distance further from the Sun, at $r = 2.06\ R_s$. The most striking development is the unusual 'spike' of particles slanting off to the upper right in the figure. This structure is the result of protons with high $v_\perp$ in the starting distribution which enter the scatter-free region in the anti-sunward hemisphere of velocity space and are pushed to higher values of $v_\parallel$ by the mirror force, while the conservation of their magnetic moment causes a systematic decrease in their $v_\perp$. This spike appears to be merely a transient result

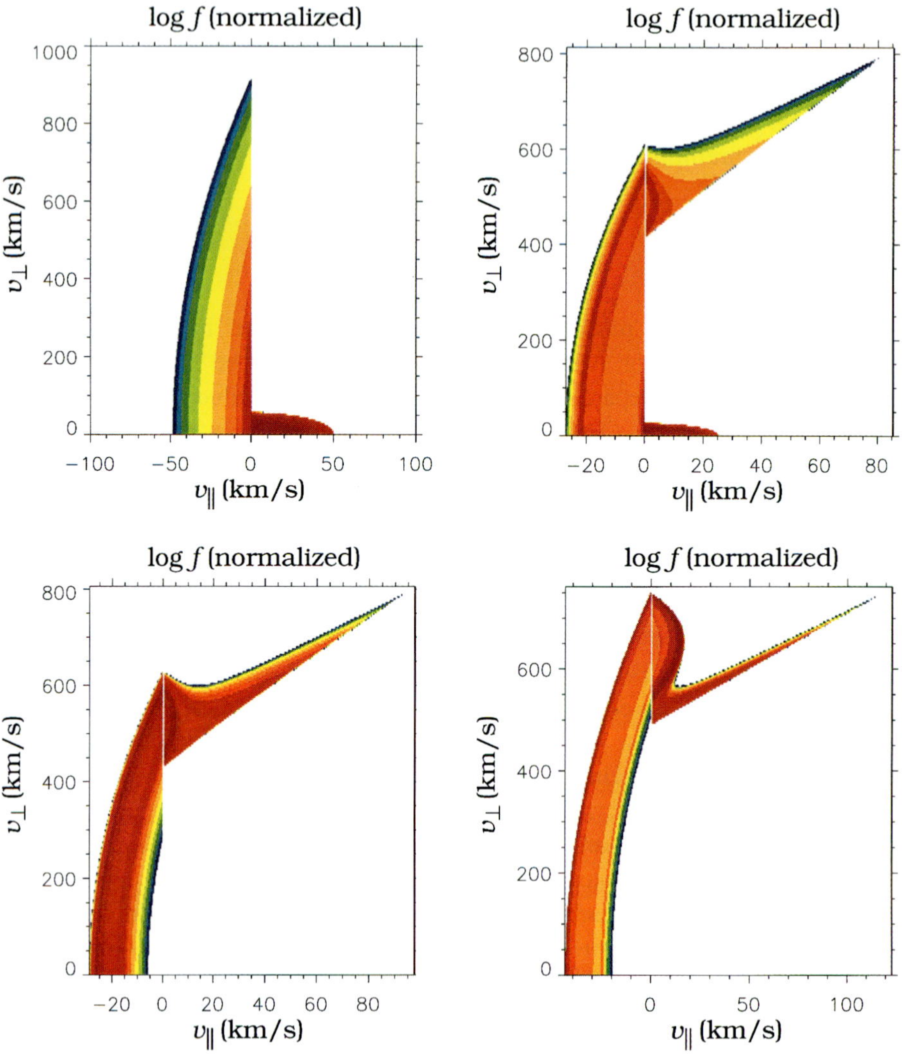

Figure 2. Model proton distribution functions for selected radial positions. The distribution in each panel is shown in its own rest frame, related to the inertial (solar) reference frame by the self-consistently calculated flow speed U. The colors indicate the value of log f, normalized in each plot to the maximum value in the distribution. The colors range logarithmically from red at the maximum value down to blue at 10^{-4} of the maximum. Values of f less than 10^{-4} of the maximum are not plotted. (a) The starting distribution at $r = 2.0\ R_s$, $U = 100$ km s^{-1}. (b) $r = 2.06\ R_s$, $U = 133$ km s^{-1}. (c) $r = 2.08\ R_s$, $U = 147$ km s^{-1}. (d) $r = 2.2\ R_s$, $U = 229$ km s^{-1}.

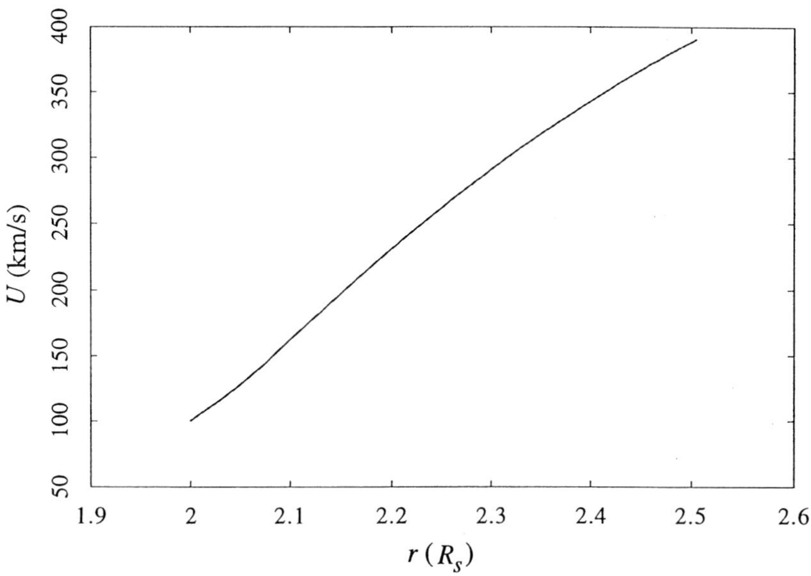

Figure 3. Proton bulk flow speed as a function of heliocentric radius.

of our starting distribution, since we see in the subsequent figures that it is eventually absorbed into the remainder of the distribution. More important to notice in Figure 2(b) is that the anti-sunward thermal 'core' of particles is falling into the sunward hemisphere under the action of gravity and the acceleration of the sunward protons. At this radial position, the bulk flow speed has increased to 133 km s^{-1}.

By the time the plasma has reached $r = 2.08\ R_s$, shown in Figure 2(c), the anti-sunward 'core' has been completely incorporated into the sunward distribution and an evacuated 'bubble' has developed at small velocities. The bulk flow speed at this point is $U = 147$ km s^{-1}.

Finally, Figure 2(d) shows the distribution at $r = 2.2\ R_s$, where the bulk flow speed is $U = 229$ km s^{-1}. Here, we see indications of the form that the proton distribution will ultimately take (as long as the waves are sufficiently intense to maintain our model assumption). The sunward portion of the distribution has formed an elongated shell, empty in the center, and extending to increasing values of $v_\perp$. The anti-sunward portion (apart from the vanishing spike) has developed into a rounded 'cap', approximately centered on the critical value of $v_\perp$ where the action of the scatter-free forces changes sign. The density of this cap must be somewhat higher than that of the shell, since its contribution to the parallel flow speed in this frame must be equal and opposite to that of the more extended shell. It is clear that this distribution is significantly different from a bi-Maxwellian.

It is also clear that this model has no problem accelerating the entire proton population. Figure 3 shows the bulk flow speed for this preliminary calculation, extended to $r = 2.5\ R_s$. The acceleration is quite strong, and we expect that flow speeds equivalent to those of the fast solar wind will be easily attainable within several solar radii.

5. Concluding Remarks

In this paper, we have introduced a new kinetic model of coronal protons interacting with intense outward-propagating ion-cyclotron waves, under the assumption of rapid pitch-angle scattering over the resonant portion of the distribution. We have presented the governing equations of the model and results of a preliminary calculation.

This paper is in the nature of a progress report: we are still in the process of developing this model, and there is quite a lot to do to make it more realistic. Among the improvements to be made in the near future are: including the self-consistent radial dependence of V_A and the charge-separation electric field, using a more careful choice for the starting distribution, incorporating a super-radial expansion of the model flux tube, extending the calculation to larger radial values, and analyzing the model demands for the wave energy supplied to the particles.

Eventually, we plan to include the full effects of wave dispersion into the model. The phase speed of ion-cyclotron waves falls steadily to zero as the wavenumber increases to infinity, and it is these high-wavenumber oscillations which are resonant with the small-$v_{\parallel}$ protons found in the low corona. The effect of dispersion on the resonant surfaces of our model will be to significantly reduce their perpendicular extent (see Isenberg and Lee, 1996, for an example in another context), and this will likely modify our quantitative results substantially. In particular, the enormous (and somewhat unrealistic) perpendicular speeds and radial acceleration obtained in this model should be considerably smaller in a dispersive calculation. It will be interesting to see if there are important qualitative changes as well.

It will also be important to extend this kinetic model to treat heavy ions in the corona, if possible. The resonant cyclotron effects on heavy ions are thought to be much stronger than they are on protons, and a full understanding of the wave-driven fast solar wind will require an equivalent description of the resonant interaction with them.

Finally, we emphasize that this model predicts a proton distribution in the low corona whose shape is distinctly different from the expectations of previous models. This shape will persist until the resonant ion-cyclotron wave power becomes too weak to maintain the rapid pitch-angle scattering. At that point, the distribution will presumably collapse toward a more compact shape, but this process will still take place on a length scale comparable with the expansion scale of the flux tube. If the resonant cyclotron interaction is, in reality, the dominant heating and acceleration process generating the fast solar wind, it seems likely that clear evidence of these shell proton distributions will be observable by the instruments to be carried on the *Solar Probe* mission.

Acknowledgements

The authors are grateful for valuable conversations with S. R. Cranmer, T. G. Forbes, J. L. Kohl, and B. J. Vasquez. This work was supported in part by the NASA Sun–Earth Connections Theory Program under grants NAG5-1479 and NAG5-8228 to the University of New Hampshire. Further support was provided by NASA grant NAG5-7097 and NSF grant ATM-9633366. JVH is also grateful to UVCS/ SOHO for travel support.

References

Axford, W. I. and McKenzie, J. F.: 1996, in D. Winterhalter, J. T. Gosling, S. R. Habbal, W. S. Kurth and M. Neugebauer (eds.), *Solar Wind Eight*, AIP, Woodbury, NY, p. 72.

Cranmer, S. R., Field, G. B., and Kohl, J. L.: 1999, *Astrophys. J.* **518**, 937.

Cranmer, S. R., Field, G. B., Noci, G., and Kohl, J. L.: 1997, in A. Wilson (ed.), *Correlated Phenomena at the Sun, in the Heliosphere, and in Geospace*, ESA, Noordwijk, p. 89.

Dusenbery, P. B. and Hollweg, J. V.: 1981, *J. Geophys. Res.* **86**, 153.

Esser, R., Habbal, S. R., Coles, W. A., and Hollweg, J. V.: 1997, *J. Geophys. Res.* **102**, 7063.

Esser, R., Fineschi, S., Dobrzycka, D., Habbal, S. R., Edgar, R. J., Raymond, J. C., Kohl, J. L., and Guhathakurta, M.: 1999, *Astrophys. J.* **510**, L63.

Hollweg, J. V.: 1973, *Astrophys. J.* **181**, 547.

Hollweg, J. V.: 1986, *J. Geophys. Res.* **91**, 4111.

Hollweg, J. V.: 1999a, *J. Geophys. Res.* **104**, 505.

Hollweg, J. V.: 1999b, *J. Geophys. Res.* **104**, 24,781.

Hollweg, J. V.: 1999c, *J. Geophys. Res.* **104**, 24,793.

Hollweg, J. V. and Johnson, W.: 1988, *J. Geophys. Res.* **93**, 9547.

Isenberg, P. A.: 1990, *J. Geophys. Res.* **95**, 6437.

Isenberg, P. A.: 1997, *J. Geophys. Res.* **102**, 4719.

Isenberg, P. A. and Lee, M. A.: 1996, *J. Geophys. Res.* **101**, 11,055.

Kennel, C. F. and Engelmann, F.: 1966, *Phys. Fluids* **9**, 2377.

Kohl, J. L., Noci, G., Antonucci, E., Tondello, G., Huber, M. C. E., Cranmer, S. R., Strachen, L., Panasyuk, A. V. *et al.*: 1998, *Astrophys. J.* **501**, L127.

Kohl, J. L., Esser, R., Cranmer, S. R., Fineschi, S., Gardner, L. D., Panasyuk, A. V., Strachen, L., Suleiman, R. M. *et al.*: 1999, *Astrophys. J.* **510**, L59.

Kulsrud, R. M.: 1983, in A. A. Galeev and R. N. Sudan (eds.), *Basic Plasma Physics I*, North-Holland, New York, pp. 115–145.

Li, X.: 1999, *J. Geophys. Res.* **104**, 19,773.

Li, X., Habbal, S. R., Hollweg, J. V., and Esser, R.: 1999, *J. Geophys. Res.* **104**, 2521.

Marsch, E. and Tu, C.-Y.: 1997, *Astron. Astrophys.* **319**, L17.

McKenzie, J. F., Banaszkiewicz, M., and Axford, W. I.: 1995, *Astron. Astrophys.* **303**, 45.

Olsen, E. L. and Leer, E.: 1999, *J. Geophys. Res.* **104**, 9963.

Rowlands, J., Shapiro, V. D., and Shevchenko, V. I.: 1966, *Soviet Phys. JETP* **23**, 651.

Skilling, J.: 1971, *Astrophys. J.* **170**, 265.

Tam, S. W. Y. and Chang, T.: 1999, *Geophys. Res. Lett.* **26**, 3189.

Tu, C.-Y. and Marsch, E.: 1997, *Solar Phys.* **171**, 363.

SECIS: THE SOLAR ECLIPSE CORONAL ECLIPSE IMAGING SYSTEM

K. J. H. PHILLIPS[1], P. D. READ[1], P. T. GALLAGHER[2], F. P. KEENAN[2], P. RUDAWY[3], B. ROMPOLT[3], A. BERLICKI[3], A. BUCZYLKO[3], F. DIEGO[4], R. BARNSLEY[5], R. N. SMARTT[6], J. M. PASACHOFF[7] and B. A. BABCOCK[7]

[1] *Space Science Department, Rutherford Appleton Laboratory, Chilton, Didcot, Oxon, OX11 0QX, U.K.*

[2] *Department of Physics, Queen's University Belfast, Belfast BT7 1NN, N. Ireland, U.K.*

[3] *Astronomical Institute, University of Wrocław, Wrocław, Poland*

[4] *Department of Physics and Astronomy, University College London, London WC1E 6BT, U.K.*

[5] *JET Joint Undertaking, Abingdon, Oxon. OX14 3EA, U.K.*

[6] *National Solar Observatory/Sacramento Peak, NM 88349, U.S.A.*

[7] *Williams College, Williamstown, MA 01267, U.S.A.*

(Received 14 October 1999; accepted 5 November 1999)

Abstract. The Solar Eclipse Coronal Imaging System (SECIS) is an instrument designed to search for short-period modulations in the solar corona seen either during a total eclipse or with a coronagraph. The CCD cameras used in SECIS have the capability of imaging the corona at a rate of up to 70 frames a second, with the intensities in each pixel digitised in 12-bit levels. The data are captured and stored on a modified PC. With suitable optics it is thus possible to search for fast changes or short-period wave motions in the corona that will have important implications for the coronal heating mechanism. The equipment has been successfully tested using the Evans Solar Facility coronagraph at National Solar Observatory/Sacramento Peak and during the 11 August 1999 eclipse at a site in north-eastern Bulgaria. The instrument is described and preliminary results are outlined.

1. Introduction

The problem of exactly why the solar coronal temperature is so high ($\sim 10^6$ K) remains unknown despite more than 50 years of theoretical and observational work (Phillips, 1995). It is known that magnetic fields are involved, so that somehow magnetic energy is converted to thermal energy. Currently, the debate centres on whether the energy to heat the corona derives from dissipation of magnetohydrodynamic (MHD) waves (e.g., Hollweg, 1981) or from numerous small-scale magnetic reconnections ('nanoflares', Parker, 1988). There are theoretical arguments and observational evidence for both mechanisms.

Wave heating or nanoflare heating might result in short-term variability of coronal structures, for example white-light or X-ray loops. This has been searched for with coronagraphs observing in the Fe XIV 'green' line (at 530.3 nm, emitted at a temperature of $\sim 2 \times 10^6$K). Wood *et al.* (1998) observed significant variations in compact loop structures with timescales of about 30 minutes in data from the LASCO coronagraph on board the SOHO spacecraft. Also, Stenborg *et al.* (1998)

The physics of the solar corona and transition region. Part II.
Reprinted from Solar Physics **193,** 2000.

have used a new ground-based German-Argentinian MICA coronagraph to look for variability in both the green and 'red' lines (the latter at 637.5 nm being due to the cooler Fe X ion): MICA is capable of imaging the corona every minute.

Theoretical studies of MHD wave heating show that only very high frequency ($\gtrsim$ 0.5 Hz) waves are capable of significant heating of coronal loops (Porter *et al.*, 1994). Such frequencies are higher than what is in practice possible to examine from spacecraft because of telemetry rate limitations, so that instrumentation must at present be ground-based, and so operate either during total solar eclipses or with coronagraphs. High-frequency intensity modulations were first searched for during the total solar eclipse of February 1980, when Pasachoff and Landman (1984) found some evidence of periodicity, with enhanced power in the 0.5-2 Hz frequency range (about 1% of the total) when measuring through fibre optics feeding filters and photomultipliers. Their result was confirmed in more refined observations in the June 1983 eclipse (Pasachoff and Ladd, 1987; see also Pasachoff, 1997). More recently, Pasachoff, Babcock and collaborators have used a CCD camera with 16-bit digitization and a frame rate of 10 Hz through coronal-green-line and continuum filters at the 1998 total eclipse in non-photometric skies and at the 1999 total eclipse in photometric skies. Possible short-period oscillations have also been detected by Singh *et al.* (1997), who observed the continuum at $\sim$ 550 nm, and by Rušin and Minarovjech (1994), who observed the red and green lines.

Searches with the Evans Solar Facility coronagraph at National Solar Observatory/Sacramento Peak (Sac Peak) for green-line modulations in both intensity and velocity have been made by Koutchmy *et al.* (1983). These observations were made using repeated spectral scans across the green line. Evidence was found for periodicity with periods of 43 s, 80 s, and 300 s.

Aside from periodic modulations, white-light observations made with a very large optical telescope, the CFHT instrument, during the 1991 eclipse visible from the Mauna Kea observatory (Vial *et al.*, 1992; Koutchmy *et al.*, 1994) show the presence of rapid material motions in the form of a moving white-light cloud or 'plasmoid' with dimensions of 3 to 4 arcsec. The motion was complex, with the plasmoid apparently being heated up, splitting into two components, and finally merging with the ambient corona after about 4 min. Random transient variations in the localized brightness of complex loop systems have also been observed in the green-line (Smartt, Zhang, and Smutko, 1993) and in soft X-rays (for example, Shimizu *et al.*, 1992).

In summary, there is observational evidence for fast wave motions and fast changes in small structures such as plasmoids in the corona. It is important that these are studied further to see if these are important for coronal heating. Here we describe an instrument we have developed, called the Solar Eclipse Coronal Imaging System (SECIS), which performs rapid imaging of the white-light corona with CCD cameras and the purpose of which is to look for short-period oscillations or fast changes. As well as instrumental details we will outline preliminary results obtained from it.

2. The SECIS Experiment: Outline

The SECIS instrument has been used in three different set-ups over the past two years. A preliminary, first version of SECIS was built for the 26 February 1998 eclipse visible in the Caribbean. In this case a single achromatic lens focused sunlight from a small heliostat on to the chip of an analogue CCD camera controlled by an adapted PC. A power failure immediately before totality prevented images from being recorded. On a second occasion, with newly delivered digital CCD cameras, we used the 40 cm coronagraph of the Evans Solar Facility (ESF), Sac Peak, obtaining green-line images of portions of the corona with a frame rate of between 20 and 70 s^{-1}. The PC grabbing the images had been further modified since our Caribbean expedition to deal with the digital output from the cameras. Comparisons with images from the *Yohkoh* and *Transition Region and Coronal Explorer (TRACE)* spacecraft confirm the reality of the coronal structures we recorded with the SECIS instrument. Finally, for the 11 August 1999 eclipse, we used the two digital cameras, together with a high-precision heliostat, telescope and other optics, to record the eclipsed Sun in white light (i.e., viewing prominences as well as coronal structures) and in the green line (only coronal structures). This experiment was entirely successful, and we recorded some 12728 frames in the two channels at a rate of 44 frames s^{-1} for the 2 min 22 s of eclipse totality.

3. CCD Cameras and Acquisition of Data

The CCD cameras, manufactured by EEV (Chelmsford, U.K.), used in this experiment are high-performance cameras designed specifically for scientific and machine vision applications. The image sensor is a monochrome 512 × 512 pixel frame transfer CCD. This device has square (15 μm ×15 μm) pixels, and can be driven at a non-interlaced frame rate up to 70 frames s^{-1}. This allows these devices to provide higher temporal resolution than standard CCIR TV cameras, with the additional benefit of equal horizontal and vertical spatial resolution. The camera electronics operate the cameras in an 'asynchronous' mode, where a trigger pulse from the control electronics commands one of the cameras to capture an image at a precise moment in time and with a precise exposure period. This feature allows the two cameras to capture accurately synchronised images. Each camera digitises the signal from the CCD to 12 bits (though the least significant two bits of the data are reckoned to be noise), and provides a dynamic range of over 1000:1.

A computer system supplied by Carr Crouch Computer Company (Maidenhead, U.K.) captures the synchronised video streams from the two CCD cameras and reconstitutes the video images, storing them on a computer network for more detailed analysis. It has dual Pentium processors and 128 megabytes of memory. Up to about four minutes of data can be captured from the two cameras for frame rates of about 50 frames s^{-1} – some 8 GB of data – and so the system is capable of operating for the totality period of most solar eclipses. The exposure time and frame frequency are adjustable and values are supplied by the computer operator who also commands the system to begin data acquisition. The video data are packed and temporarily stored in two circular buffer sets in main memory. When full, the data are transferred to one of four 9-GB disk drives. The work load is shared by the two processors. To allow the operator to supervise the task, two small live images are displayed and the buffer-filling status is shown on the computer screen.

The digital video data can be copied to compact digital tapes on the computer for security. The image processing software allows the replay of the video, and the cropping to sub-sequences and regions of interest. These selections can then be exported in a variety of formats including FITS.

The computer system, interfaces, and camera power supplies are contained in a medium-sized PC cabinet suitable for carrying into the field. The cabinet is fitted with extra cooling to allow operation in warmer environments than normal. Power to the entire system, complete with light-weight monitor, is through an uninterruptable power supply unit which provides eight minutes of operation in the event of a fault in the main electricity supply.

4. Instrumentation: ESF Coronagraph

We obtained observations from a four-week run on the ESF coronagraph, from late August 1998. We made observations of a field approximately $0.5^\circ \times 0.5^\circ$ using the two CCD cameras in an optical bench lay-out as indicated in Figure 1. Light from the coronagraph (aperture 40 cm) was directed to the cameras via a field lens producing almost parallel light. This light beam was then split by a pellicle beam splitter, the direct beam incident on camera CCD2 (see Figure 1) passing through a 0.1-nm passband green line filter ('Narrow' filter in Figure 1) whose wavelength was checked with a spectrophotometer. The reflected beam from the beam splitter was incident on the other camera (CCD1) via a 10-nm-wide filter ('Broad' filter). On occasion it was necessary to use neutral density (ND) filters to reduce light levels on the cameras to an acceptable level. One of the cameras malfunctioned during the course of these observations, so in the event we made observations with camera CCD2 and in the green-line filter only.

Despite interruptions by cloudy weather, coronal active regions were successfully observed in early September. The best quality data were obtained for a region on the north-west limb of the sun on 5 September 1998. On this occasion the data were recorded at 68 frame rates in 15-s-long intervals, with transparent ('coronal') skies (sky brightness at 50 millionths of the photospheric surface brightness). The

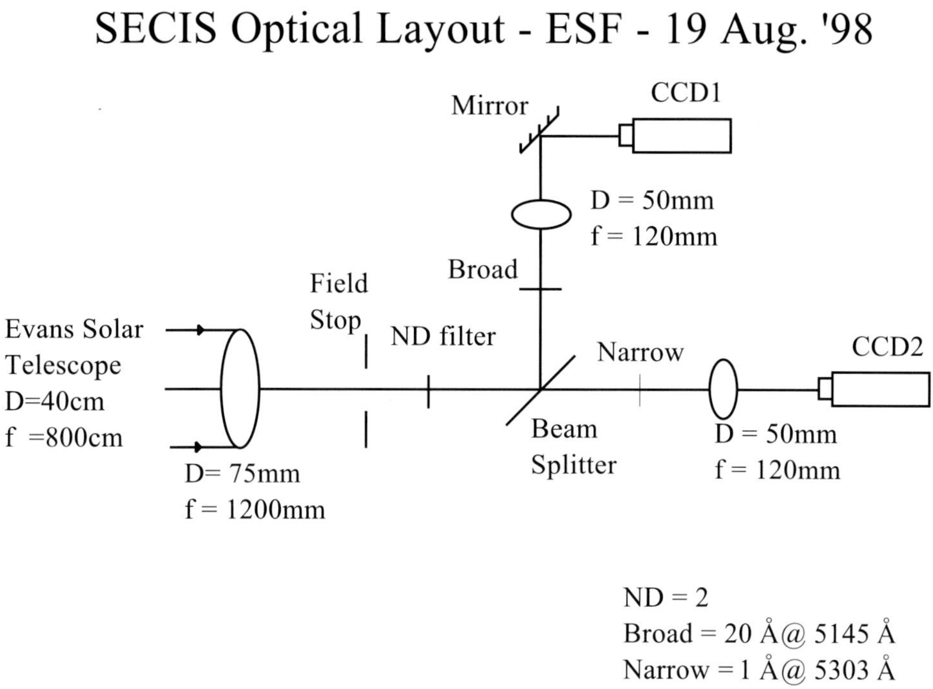

Figure 1. The optical scheme for the SECIS experiment while at the Evans Solar Facility, National Solar Observatory/Sacramento Peak. Light from the coronagraph is incident from the right. The apertures and focal lengths of the various lenses are indicated (D and f). 'Broad' and 'Narrow' indicate the filters used, Narrow being the green-line (0.1 nm bandpass) filter. In the event only camera CCD2 with the narrow-band filter was used. (Drawing: P. Gallagher.)

data were converted to FITS format images and analyzed with Interactive Data Language (IDL) procedures. Figure 2 (left-hand panel) shows the active region recorded with the camera through the green-line filter at about 14:30 U.T. Near-simultaneous images of this active region from the *Yohkoh* Soft X-ray Telescope (SXT) and from the TRACE extreme ultraviolet telescope are also shown. Figure 2 and more clearly in the movie made from the SECIS images one can clearly see several individual loops. Light curves of features near the solar limb show intensity increases, particularly in the brightest loop structure (Region 'A' in the figure) but at present we cannot certainly state whether these features are real or are due to effects of slight mispointing of the coronagraph allowing tiny amounts of photospheric emission to enter the light path. These instrumental effects are under investigation. With this in mind, the preliminary Fourier power spectra shown in Figure 3 are for two regions, 'A' near the solar limb and within a coronal loop structure, the other ('B') some distance away from the solar limb where the emission is entirely sky brightness. These power spectra show hardly any excess power for region B but do show excess power at frequencies less than about 5 Hz for region A, though

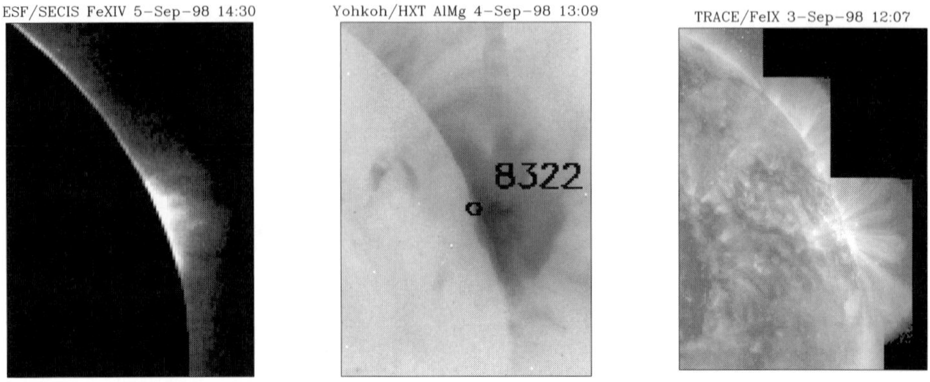

Figure 2. Comparison of a SECIS Fe XIV green line image taken at the Evans Solar Facility, Sacramento Peak, NM with X-ray data from *Yohkoh* and an EUV image from TRACE. (P. T. Gallagher.)

there appears to be no specific feature over and above the expected 1/(frequency) decrease.

5. Instrumentation: 1999 August 11 Eclipse

We now describe the SECIS instrumentation used for the 11 August 1999 eclipse. The equipment basically consisted of a high-precision heliostat which directed sunlight into the optics and cameras of the SECIS instrument, all enclosed in a light-tight box.

5.1. Heliostat

The heliostat was designed and constructed at the Astronomical Institute of the Wrocław University. The instrument design was optimised for providing a stable horizontal beam of sunlight for the SECIS optics. There are three main mechanical modules: a pedestal, a main axis unit with a driving mechanism, and a fork with a mirror container (Figure 4).

The heliostat pedestal has an adjustable bottom frame which can be attached to a stable base, in this case large blocks sunk into the ground. The pedestal may be rotated and tilted, allowing easy adjustment of the pedestal orientation to the local horizon and meridian. The main axis unit consists of three co-axial steel tubes, connected by high-precision ball bearings, and a horizontal axis. The horizontal axis connects the pedestal with the external tube, which acts as a body for a rotating fork of the heliostat. The whole main axis unit can be inclined to the pedestal with a double-nutted bolt at an angle equal to the latitude of the observing site. The middle tube is rotated by a diurnal motion driving system. The driving system consists of a stepper motor (1600 steps per rotation), a three-stage planetary gear (ratio 1 : 256),

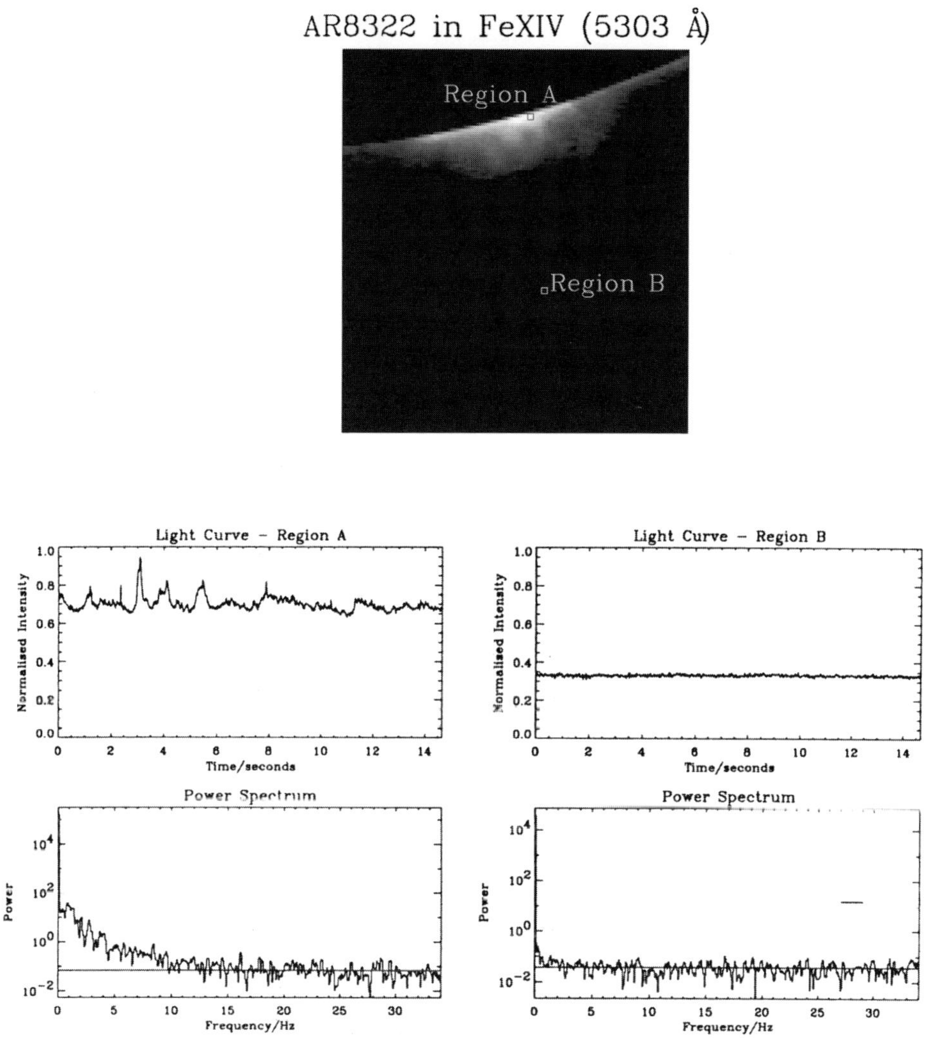

Figure 3. Initial Fourier results from Sacrament Peak data. (P. T. Gallagher.)

an overload coupling and a high-precision Zeiss power screw. The power screw pulls a steel tape around the perimeter of a disk mounted to the middle tube. There are correction units allowing fine adjustments in hour angle and declination. The heliostat was powered by an uninterruptable power supply, capable of supplying the instrument for more than 20 min. The base dimensions are 80 × 50 cm, the height approximately 1 m, and the total weight is nearly 90 kg. For easy handling of the equipment, the three main modules of the heliostat can be (and for the Bulgarian expedition were) transported separately and connected at the observing site.

The heliostat mirror, made by the Turnov Optical Workshop, Czech Republic. is of rigid low-expansion glass (Simax), with external diameter 265 mm (clear

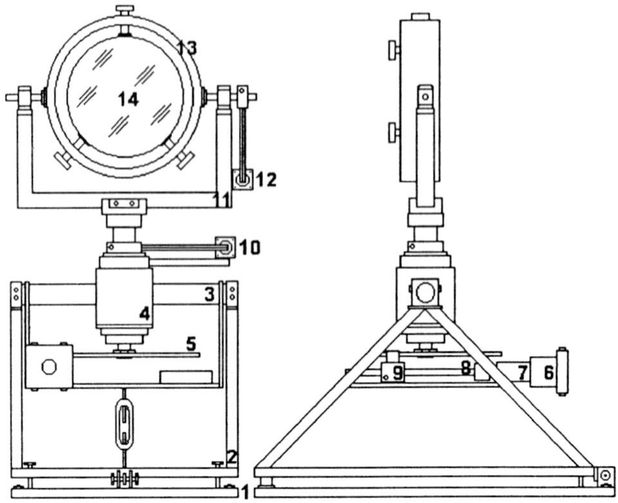

Figure 4. Schematic diagram of the heliostat used in the SECIS experiment. 1 - base of the pedestal, 2 - pedestal, 3 - horizontal axis, 4 - main axis unit, 5 - disk of the driving mechanism, 6 - diurnal motion stepper motor, 7 - planetary gear, 8 - power screw, 9 - power screw nut, 10 - hour angle correction unit, 11 - fork, 12 - declination correction unit, 13 - mirror container, 14 - mirror.

diameter 250 mm) and thickness 41 mm. It is flat to 1/10 wave at 546 nm, and the coating is aluminium and silicon oxide. The reflectance at the wavelengths we used in the experiment is 0.89.

5.2. The SECIS instrument

The two CCD cameras used at Sac Peak were used for the August 11 eclipse, the faulty one having been repaired. A Meade Schmidt-Cassegrain $f/10$ telescope, 200 mm aperture and horizontally mounted on an optical bench (as were all other optical components), was used to receive sunlight from the heliostat. The optical bench was attached with adjustable screws to a trestle-like aluminium table which, like the heliostat base, rested on blocks sunk into the ground for maximum stability and freedom from vibrations. Light from the telescope was then collimated and the collimated beam incident on a pellicle beam splitter at 45° to the optical axis. The transmitted beam was directed to a 0.5 nm-bandpass green-line interference filter mounted on a graduated rotatable holder then onto one of the cameras. The surface of the filter (manufactured by Barr Associates) is treated with ion-assist deposition, with refractory oxide coating, enabling the central wavelength to be held constant over a wide range of operating temperatures: the wavelength shift with temperature is only 0.002 nm/°C. The reflected beam was directed to the second camera without a filter. We considered inserting a broad-band filter in the general wavelength region of the green line but decided not to do this so that prominences could be recorded. This turned out to be a wise decision as there were several very intense

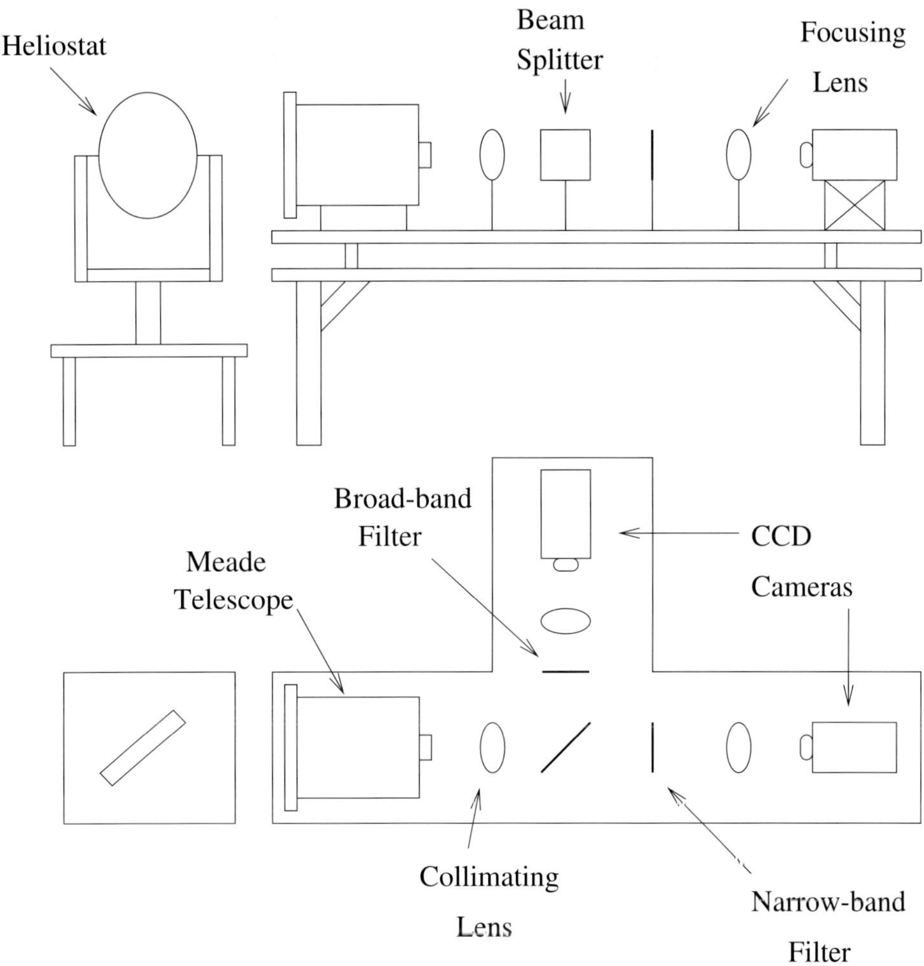

Figure 5. Plan and side-view sketches of SECIS and the Wrocław heliostat. (P. T. Gallagher.)

prominences visible during the eclipse so that fast changes in these structures can be searched for.

With pre-eclipse tests and from calculations of expected coronal intensities, we found that the two cameras could operate at between 40 and 50 frames s^{-1} with a field of view measuring $0.5^{\circ} \times 0.5^{\circ}$. We selected which portion of the solar corona was to be viewed by inspecting SOHO and *Yohkoh* images taken shortly before the eclipse. The spatial resolution was approximately that of the pixel size, viz. $3.5'' \times 3.5''$ arc.

5.3. Eclipse site and pre-eclipse preparations

Our observing site for the eclipse was at a Bulgarian Army Air Defence establishment at Shabla, about 60 km north of Varna on the Black Sea coast. Electrical

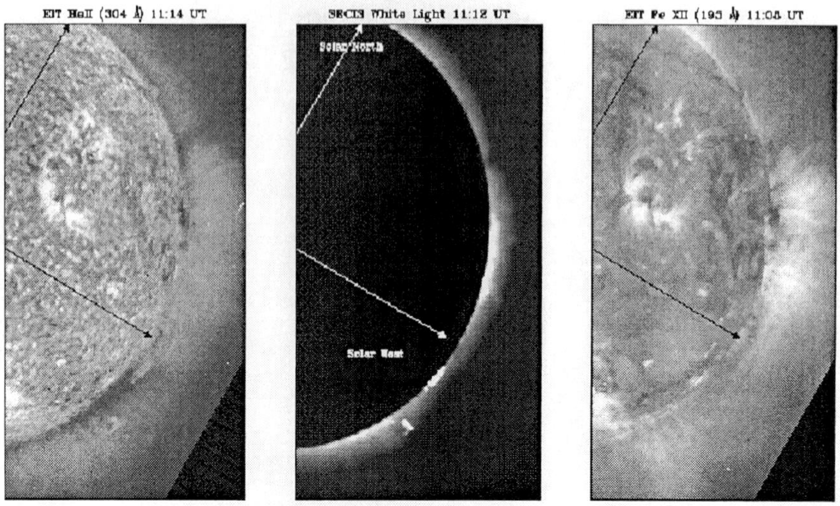

Figure 6. One of the 12 728 images from the SECIS instrument during the 1999 August 11 eclipse (white-light channel), showing coronal structures and prominences, compared with SOHO EIT images in He II (left) and Fe XII (*right*).

power was supplied from nearby buildings of the establishment, with backup generator available in case of power interruption. The site was chosen for us by prior arrangement with Prof. V. N. Dermendjiev and colleagues at the University of Sofia. It is within a few hundred metres from the sea, almost exactly on the centre line of the eclipse path which at this point was about 120 km wide. The weather conditions were expected to be very good on the basis of information for previous years during mid-August. According to Dermendjiev *et al.* (1998), for the month of August, the number of clear days there is 17.6 (out of 31), and the mean diurnal temperature is 22.1 °C. This information agrees with Espenak and Anderson (1997) who give the probability of seeing the eclipse as 63% for this location. The time of mid-eclipse was 11 : 11 UT (local time 14:11), with the sun at a very high altitude.

A number of pre-eclipse preparations were necessary. Alignment of the heliostat was done a few days before the eclipse, just after assembling the instrument (see Pasachoff and Livingston, 1984, for generalized procedure). On 5 August the heliostat pedestal was levelled with a spirit level and roughly set up for the site's local meridian with north point determined from 5000-year-old but still reliable gnomon technique. We found a direction of the local meridian by observing the shadow of a straight, vertical steel rule at the moment of the true solar transit using *Astronomical Ephemeris* data and DCF77 atomic clock radio signals. The precise inclination of the hour axis to the local horizon and its alignment to the local meridian plane were made during night observations on 6 August. We observed transits of some bright stars by the meridian and by the first vertical with a small optical telescope attached to the mirror axis. The azimuth of the reflected beam of sunlight at the moment of the solar eclipse can be determined simply from the

Sun's declination during the event and the latitude of the observing site (so this procedure is subject to a very small variation from day to day). Between 6 and 9 August, using a distant screen as a target, we confirmed that the reflected beam was constant in direction. The telescope and other optical components of the SECIS instrument were aligned with a small pocket laser on their optical bench and next the whole assembly was aligned using sunlight on the day of eclipse. Since the telescope's field of view was much smaller than the diameter of the Sun's image, we fine-adjusted the heliostat mirror to obtain the required part of the sun on the cameras.

In the event, the sky was almost completely cloudless throughout the day of 11 August. A slight breeze before totality died down immediately before totality, so that we are confident that there was no instrument motion due to wind vibrations.

5.4. Preliminary results

With excellent weather and with the heliostat, computer, cameras and optical components performing flawlessly, we were able to record a total of 12 728 images (6364 from each camera) in the 2 min 22 s of eclipse totality. We chose a section of the sun's limb from near the north pole to the south-west. As we found from our images, this included a large coronal active region just north of the west limb, a probable set of post flare loop prominences (just south of the west limb), and a small suspended loop prominence off the south-west limb. These features have been recorded on several photographs and other electronic images of the eclipsed sun. In Figure 6 we show one of the images in the white-light channel (so showing prominences as well as coronal structures). This is compared with SOHO images from the Extreme-ultraviolet Imaging Telescope (EIT) in the He II (30.4 nm) chromospheric emission line and Fe XII (19.5 nm) coronal emission line. The He II image shows the prominences that the SECIS white-light channel recorded while the Fe XII image shows the coronal active regions also seen in the SECIS image.

Preliminary analysis suggests that the instrument performed well throughout the period of eclipse totality and all images from both cameras are of high quality. The images need to be processed for detailed analysis, with corrections for slight image motion and for flat-fielding and dark current (data for the latter were recorded immediately after eclipse totality). The image motion is in the form of a small (few arcseconds) continuous drift, but no perceptible jitter, over the period of the eclipse totality. Once the corrections are applied, we will begin searches for periodic modulations using Fourier analysis techniques and for rapid changes including travelling plasmoids. This is expected to take place in the next few months.

Acknowledgements

We thank EEV (Chelmsford, U.K.) who manufactured the CCD cameras and whose staff have continued to be on call for trouble-shooting. We are particularly indebted to Mr Scott York for help with the camera technicalities in writing this paper. Mr Steve Crouch and colleagues at Carr Crouch Computer Company (4C's) are thanked for their very valuable contribution to this project in providing hardware and software enabling the data from the cameras to be recorded, and for providing information for this paper. Mr E. G. Capucci did a great deal of the SECIS hardware and we acknowledge his valuable contribution. We are grateful for help provided by staff at the Evans Solar Facility, Sac Peak, and to the NSO Telescope Allocation Committee for awarding us time on the Facility. PTG thanks the NSO for the Scholarship which allowed him to work at Sac Peak during the summer of 1998. We are very grateful to the Polish State Committee for Scientific Research (KBN) who generously provided funds (grant number 2 P03D 005 15) to allow PR, BR, and AB to build the heliostat. KJHP and PTG are also grateful to colleagues at the Astronomical Institute, University of Wrocław for hospitality during a visit there for instrument testing in May 1999. JMP and BAB thank the US National Science Foundation, the National Geographic Society, and NASA. Prof. Vladimir N. Dermendjiev and Dr Maria Madjarska (University of Sofia, Bulgaria) and Prof. Dimitar Mishev (Bulgarian Academy of Sciences) are thanked for their tireless help during the eclipse expedition to Shabla and for helping us in the transport of our equipment. The staff of the Bulgarian Army Air Defence Establishment are thanked for their help in setting up our base in Shabla. We are grateful to several colleagues for technical help and for providing encouragement, notably Dr Serge Koutchmy (L'Institut d'Astrophysique, Paris) and Dr Richard Wade (RAL). The hardware and staff time involved in the SECIS experiment was financed through the RAL Facilities, Technology, and Studies (FTS) line and from RAL Instrumentation Department. Assistance with our web sites and illustrations and other activities was generously given by Mr George Hanoun who has given a lot of time over the years at RAL.

Further Information

The following World Wide Web sites are useful for providing further information about SECIS and the 1999 eclipse:

http://ast.star.rl.ac.uk/secis http://www/eclipse99.org.uk/
(General information about the 1999 August 11 eclipse. The SECIS instrument, heliostat, and preliminary results from Sac Peak are described under 'Eclipse Science'/'What the Scientists are doing' in this site.)

http://www.bbc.co.uk/eclipse *(For general information about the 1999 August 11 eclipse and the SECIS instrument.)*

References

Dermendjiev, V. N., Mishev, D. I., Tsanev, V. I., and Popov, G. K.: 1998, in *Advances in Solar Physics Euroconference*, ASP Conference Series 155, 403.

Espenak, F., and Anderson, J.: 1997, *Total Solar Eclipse of 1999 August 11*, NASA Ref. Publ. 1383.

Hollweg, J.: 1981. *Solar Phys.* **70**, 25.

Koutchmy, S., Zugzda, Y. D., and Locans, L.: 1983, *Astron. Astrophys.* **120,** 185.

Koutchmy, S. *et al.*: 1994, *Astron. Astrophys.* **281,** 249.

Parker, E. N.: 1988, *Astrophys. J.* **330,** 474.

Pasachoff, J. M.: 1997, in Mouradian and Stavinschi (eds), *Theoretical and Observational Problems Related to Solar Eclipses*, Kluwer Academic Publishers, Dordrecht, p. 181.

Pasachoff, J. M. and Ladd, E. F.: 1987, *Solar Phys.* **109,** 365.

Pasachoff, J. M. and Landman, D. A.: 1984, *Solar Phys.* **90,** 325.

Pasachoff, J. M. and Livingston, W.: 1984, *Appl. Optics* **23,** 2803.

Phillips, K. J. H.: 1995, *Guide to the Sun*, Ch. 5, Cambridge University Press, Cambridge.

Porter, L. J., Klimchuk, J. A., and Sturrock, P. A.: 1994, *Astrophys. J.* **435**, 482.

Rušin, V., and Minarovjech, M.: 1994, *IAU Symp.* **144**, 487.

Shimizu, T., Acton, L.W., Tsuneta, S., Lemem, J.R., and Uchida, Y.: 1992, *Publ. Astron. Soc. Japan* **44,** L147.

Singh, J. *et al.*: 1997, *Solar Phys.* **170,** 235.

Smartt, R. N., Zhang, Z., and Smutko, M. F.: 1993, *Solar Phys.* **148,** 139.

Smith, J. M., Roberts, B., and Oliver, R.: 1997, *Astron. Astrophys.* **327,** 377.

Stenborg, G., Schwen, R., Srivastava, N., Inhester, B., Podlipnik, B., Rovira, M., and Francile, C.: 1998, private communication.

Vial, J.-C., Koutchmy, S., and the CFHT Team.: 1992, ESA SP-344, p. 87

Wood, B. E., Karovska, M., Cook, J. W., Brueckner, G. E., Howard, R. A., Korendyke, C. M., and Socker, D. G.: 1998, *Astrophys. J.* **505,** 432.

THE SOHO–STELLAR CONNECTION

THOMAS R. AYRES
Center for Astrophysics and Space Astronomy, University of Colorado, Campus Box 389 [CASA], Boulder, CO 80309-0389, U.S.A. (E-mail: ayres@casa.colorado.edu)

(Received 24 September 1999; accepted 17 November 1999)

Abstract. The 'solar–stellar connection' bridges the daytime and nighttime communities; an essential link between the singular, but detailed, views of our Sun, and the broad, but coarse, glimpses of the distant stars. One area in particular – magnetic activity – has profited greatly from the two way traffic in ideas. In that spirit, I present an evolutionary context for coronal activity, focusing on the very different circumstances of low-mass main-sequence stars like the Sun, compared with more massive stars. The former are active mainly very early in their lives, whereas the latter become coronal only near the end of theirs, during the brief incursion into the cool half of the Hertzsprung–Russell diagram as yellow, then red, giants. I describe tools at the disposal of the stellar astronomer; especially spectroscopy in the ultraviolet and X-ray bands where coronae leave their most obvious imprints. I compare *HST* STIS spectra of solar-type dwarfs – ζ Dor (F7 V), an active coronal source, and α Cen A (G2 V), near twin of the Sun – to the *SOHO* SUMER UV solar atlas. I also compare the STIS line profiles of the active coronal dwarf to the corresponding features in the mixed activity 'hybrid chromosphere' bright giant α TrA (K2 II) and the archetype 'non-coronal' red giant Arcturus (α Boo; K2 III). The latter shows dramatic evidence for a 'cool absorber' in its outer atmosphere that is extinguishing the 'hot lines' (like Si IV λ1393 and N V λ1238) below about 1500 Å; the corona of the red giant seems to lie *beneath* its extended chromosphere, rather than *outside* as in the Sun. I present an early taste of the moderate resolution spectra we can expect from the recently launched *Chandra* X-ray Observatory (*CXO*), and contemporaneous STIS high resolution UV measurements of the *CXO* calibration star Capella (α Aur; G8 III + G1 III). Last, I describe preliminary results from a May 1999 observing campaign involving *SOHO* SUMER, *TRACE*, and the Kitt Peak Infrared Imaging Spectrometer (IRIS). The purpose was to explore the dynamics of the quiet solar atmosphere through the key 'magnetic transition zone' that separates the kinetically dominated deep photosphere from the magnetically dominated coronal regime. Linking spatially and temporally resolved solar phenomena to properties of the average line shapes (widths, asymmetries, intensity ratios, and Doppler shifts) is a crucial step in carrying physical insights from the solar setting to the realm of the distant stars.

1. Introduction

My contribution – to the Monterey workshop on the solar corona – concerns the $10^{12}-1$ other stars in the Galaxy which are not the Sun. The vast majority of these are low-mass ($\mathcal{M} < 1.5\ \mathcal{M}_{\odot}$) dwarfs, which experience some level of chromospheric and coronal activity throughout most of their lives. The remaining handful of the Galaxy's trillion are moderate-mass ($\mathcal{M} \simeq 1.5–5\ \mathcal{M}_{\odot}$) and high-mass ($\mathcal{M} \simeq 5–100\ \mathcal{M}_{\odot}$) stars, which lack surface convection zones and thus are non-

The physics of the solar corona and transition region. Part II.
Reprinted from Solar Physics **193,** 2000.

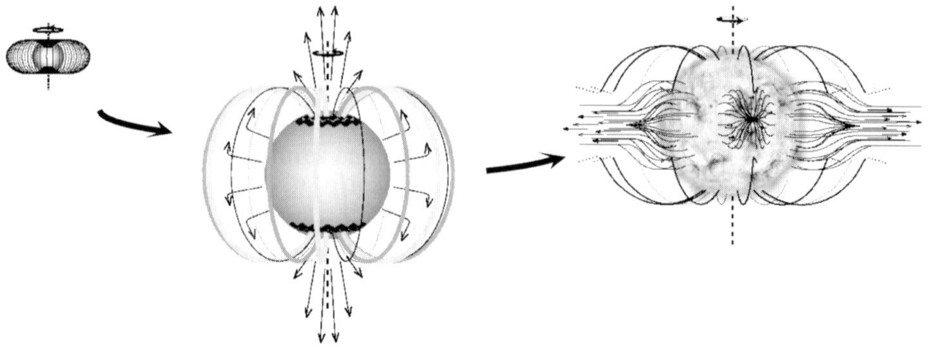

Figure 2. Speculative evolutionary scenario for Hertzsprung-gap stars (Ayres *et al.*, 1998). *Left:* MS phase – rapidly rotating magnetized blue dwarf. *Middle:* X-ray deficient stage – fast rotating, thin convection zone giant with coronal 'magnetosphere' inhibiting angular momentum loss. *Right:* rapid braking zone – deepening convection zone, strengthening dynamo, disruption of magnetosphere, flare-assisted mass flux and spindown.

spectral types to the base of the red giant branch (RGB) at $\sim$ K0; briefly rises to the red giant tip and helium ignition; then returns again to the base of the RGB as a helium burning 'clump' giant, where it resides for the remaining $\sim$20% of its total lifetime (Schaller *et al.*, 1992).

Unlike low-mass dwarfs, moderate-mass stars retain substantial angular momentum at the conclusion of their MS phase, and possibly strong 'fossil' global magnetic fields as well. When such stars first become convective, but not deeply enough to excite dynamo action, the relic dipole could become sufficiently stressed by turbulent flows in the polar regions to light up the extended field lines with hot coronal gas. Nevertheless, the magnetospheric 'dead zone' would inhibit equatorial mass loss, postponing magnetic braking. Soon, however, the persistently deepening convection should not only kindle a solar-like dynamo, but also would shred the roots of the relic bipole. Collapse of the magnetosphere would bare the giant to wind braking; perhaps explaining the sharp drop in stellar rotational velocities seen just beyond G0 III (Gray, 1981). Coronal emissions might be boosted temporarily in the initial dynamo stage, but they inevitably would subside as the star spun down. Figure 2 illustrates a cartoon view of the (speculative) scenario outlined above, from the upper MS through the rapid braking phase (Ayres *et al.*, 1998).

Note the presence in the middle portion of the coronal H–R diagram (Figure 1) of strong X-ray emitters only in the rapid braking zone. Because of the strong correlation between spectral type and age in the giant branch, activity there is a more systematic function of color than on the lower MS. The red giant region, across the 'coronal dividing line' near $T_{\rm eff} \sim 4500$ K (Linsky and Haisch, 1979), contains only weak X-ray sources or upper limits; as does the supergiant region aside from a handful of modestly active objects like β Draconis (G2 Ib; Brown *et al.*, 1984). Indeed, the red giant portion of the X-ray H–R diagram has been dubbed the 'coronal graveyard' (Ayres, Fleming, and Schmitt, 1991). The disap-

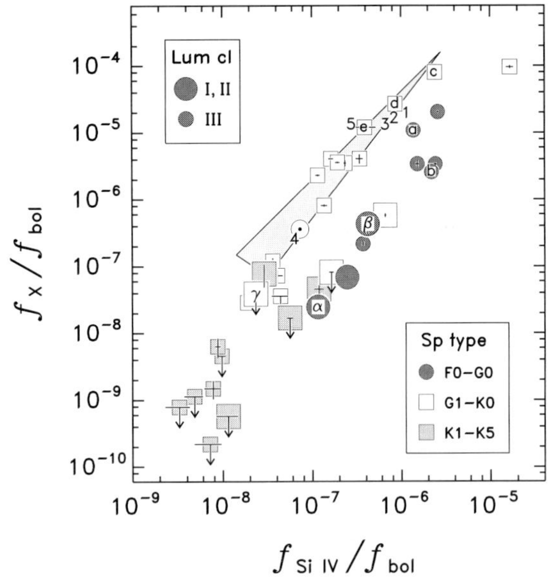

Figure 3. Comparison of 0.2–2 keV X-rays with subcoronal Si IV λ1400. Shaded wedge depicts range occupied by mid-F–early-K dwarfs; average Sun is marked by ⊙. Targets from the STIS cycle 8 cool star survey are designated with Arabic numerals (class V), Roman letters (class III), or Greek letters (classes I–II). Note that the F0–G0 class III ('gap') stars, and most of the supergiants, are 'X-ray deficient' with respect to MS dwarfs and G1–K0 giants. Cluster of objects in lower left corner are 'non-coronal' red giants.

pearance of coronae among the red giants is accompanied by the appearance of circumstellar absorption features in low excitation resonance lines, signaling the development of strong, cool, slow winds (Reimers, 1977). It has been tempting to draw a connection between the two: perhaps the energy that otherwise would power a hot corona instead is depleted by the acceleration of the cool wind (Linsky and Haisch, 1979). Such outflows play a key role in injecting nuclear-processed material into the interstellar medium. Like coronae, the physical origin of the mass flux of cool giants remains one of the central unsolved questions of stellar astrophysics.

On the other hand, there is a curious group of supergiants – the so-called 'hybrid chromosphere' stars – in which the hot lines (i.e., C IV) and cool wind absorptions (in low-excitation species) spectrally coexist (Hartmann, Dupree, and Raymond, 1980). Some of these stars have been detected as weak X-ray sources (Reimers *et al.*, 1996), and even have been caught flaring (e.g., α TrA [K2 II]; Kashyap *et al.*, 1994), a blatant signature of magnetic activity.

2.2. Flux-flux correlations

Figure 3 depicts the correlation between the global X-ray fluxes of a wide variety of F–K stars (as measured in the relatively 'soft' 0.2–2.0 keV band of the Röntgensatellit [*ROSAT*; Trümper *et al.*, 1991]) versus integrated fluxes in the UV emission doublet of Si IV at 1400 Å (as recorded either by the *International Ultraviolet Explorer* [Ayres *et al.*, 1995], or – for the low activity red giants – by the *Hubble Space Telescope* Goddard High Resolution Spectrograph [Ayres *et al.*, 1997])[2].

The flux–flux diagram shows that low mass dwarf stars follow a regular behavior between their coronal and subcoronal emissions, although the slope of the power law is steeper than a simple 1:1 proportionality. Significantly, the Sun traces out a similar power law over its sunspot cycle (Ayres *et al.*, 1996). The G1–K0 giants follow the same trend; but the warmer giants (F0–G0), and most of the supergiants regardless of temperature, appear to follow a parallel track that is 'X-ray deficient' with respect to the F5–K5 MS stars (Simon and Drake, 1989). Exactly where the 'non-coronal' red giants fit into the picture is not clear, because most of them are represented only by X-ray upper limits, and even their Si IV emission has been called into question recently (as discussed in Section 4.3).

2.3. Close binaries

What I have described so far applies only to single stars, yet most of the Galaxy's denizens are found in pairs, some rather close; and triples and quadruples are not especially rare. In a binary system, the inexorable braking of a coronal star can be postponed by infusion of orbital angular momentum into the individual stellar spins through tidal coupling (as described by Bopp and Espenak [1977] in the context of the BY Draconis variables). The large reservoir of angular momentum in the orbit ensures that a system can sustain its coronal emissions over evolutionary timescales, and the short orbital periods of the closest binaries ($\sim$ days, or less) guarantees a robust level of activity. Thus, the most extreme coronal phenomena typically are found in the short-period binary systems, particularly those of the RS Canum Venaticorum variety (Hall, 1978). Similarly, red giants in binaries, like σ Geminorum (K1 III + dG; Engvold *et al.*, 1988), can be hyperactive, even though technically they fall on the non-coronal side of the Linsky-Haisch boundary.

3. Coronal Observing Tools

Unfortunately, few legitimate diagnostics of coronae (or winds) are accessible from the ground. One exception is centimetric radio emission, collected for example with the Very Large Array (VLA) or the Australia Telescope (AT). It is a crucial tracer of transient *nonthermal* bursts from stellar flares (e.g., Neff *et al.*, 1996), as well as the steady thermal emission from the extended wind volume of non-coronal red giants (e.g., α Ori [M2 Iab]; Harper, Brown, and Lim, 2000). 'White-light'

[2] Si IV was selected over the often brighter C IV λ1549 doublet, because the carbon abundances of red giants can be affected unpredictably by dredge-up of CNO burning products during deep convective mixing episodes on the first ascent of the RGB (e.g., Tomkin, Luck, and Lambert, 1976).

TABLE I

Spectrometers to record stellar coronal properties

Epoch	Instrument/mode	Range (Å)	Resolution ($\lambda/\Delta\lambda \times 10^{-3}$)	Grasp (Å)	Comments
1990–97	*HST* GHRS (LOW)	1150–3200	2	290	high $A_{\rm eff}$
	(MED)		25	25–50	
	(ECH)		80	8–16	low $A_{\rm eff}$
1997–	*HST* STIS (LOW)	1140–1740	2.5	610	long slit, high $A_{\rm eff}$
	(MED)	1140–3150	10	40	
	(M-ECH)		30–45	620	
	(H-ECH)		114	270	low $A_{\rm eff}$
1999–	*FUSE*	905–1188	20–35	full range	medium $A_{\rm eff}$
1992–	*EUVE*	70–760	0.3	full range	ISM abs, low $A_{\rm eff}$
1999–	*Chandra* LETG	3–160	1–2 (60–160 Å)	full range	best res, < 1 keV
	HETG	1.2–14 (HEG)	1 (@1 keV)	full range	low $A_{\rm eff}$
		2.5–31 (MEG)	0.5 (@1 keV)		
(1999)	*XMM* RGS	5–35	0.2 (@1 keV)	full range	high $A_{\rm eff}$
(2000)	*Astro-E* XRS	1–30	0.5 @6 keV	full range	best res, > 2 keV

$A_{\rm eff}$ is the effective area of the instrument; high $A_{\rm eff}$ means better sensitivity.

counterparts of stellar flares – occasionally spectacular – are observed in dMe stars using narrow band photometry and spectroscopy (Hawley and Petersen, 1991). Doppler imaging (Vogt and Penrod, 1983) in optical and near-IR absorption lines is essential to infer the sizes and surface distributions of starspots (Strassmeier and Linsky, 1996). Analogous time resolved studies of Hα in fast spinning stars has identified highly extended prominence-like structures (Collier Cameron, 1999). Ca II monitoring (Duncan *et al.*, 1991) likewise is vital for establishing starspot cycles (Baliunas and Vaughan, 1985), stellar rotation periods (Gray and Baliunas, 1997), differential rotation patterns (Donahue and Baliunas, 1992), and active region evolution (Donahue, Dobson, and Baliunas, 1997).

However, to view coronae in their most favorable light, one must go to the ultraviolet and X-ray bands. There, the emissions are directly sensitive to the dominant thermal component of the hot coronal (and subcoronal) gas. Of course, such measurements must be made above the Earth's atmosphere. Tremendous progress has been made over the past two decades in coping with the special instrumental challenges.

Table I lists a number of recent space observatories that operate at ultraviolet and shorter wavelengths, focusing on those with significant spectroscopic capabilities. Spectroscopy is a relatively recent innovation in stellar X-ray astronomy, but is an essential diagnostic tool to dissect thermal structure, abundance patterns, and gas dynamics. At the same time, high-dispersion stellar UV spectroscopy has a long heritage and is considered mature.

Note that the spectral resolving powers of the stellar instruments typically are much higher than routinely applied in the solar setting. The great advantage (!) of stellar astronomy is that one need not worry about either temporal or spatial resolution: stars generally are too faint to allow decent photon statistics over the short time scales thought to govern dynamical phenomena on the Sun, and they are too distant to show discernible surface structure in today's modest aperture orbiting telescopes. Thus, one can boost the remaining instrumental dimension, namely spectral resolution. Indeed, high-dispersion spectroscopy of rapidly spinning stars can serve as a surrogate for high spatial resolution through the technique of Doppler imaging mentioned previously. In some stars, super-rotational broadening (line widths much greater than anticipated from the photospheric $v \sin i$) of far-UV features points to highly extended coronal structures (Ayres *et al.*, 1998).

Note in Table I the obvious hole in spectral coverage immediately below the Lyman edge (912 Å). That is because, of course, the nearby universe is completely opaque between about 700–912 Å (under the best of circumstances, and more typically between 100–912 Å), owing to absorption by atomic hydrogen in the local interstellar medium. Unfortunately, that region of the coronal spectrum bristles with important diagnostics, exploited on the Sun by the early Orbiting Solar Observatory series (Huber *et al.*, 1973), the *Skylab* Apollo Telescope Mount (Reeves, Timothy, and Huber, 1977), and more recently the Solar and Heliospheric Observatory (*SOHO*; Domingo, Fleck, and Poland, 1995), with its Solar Ultraviolet Measurements of Emitted Radiation (SUMER; Wilhelm *et al.*, 1997) and Coronal Diagnostics Spectrometer (CDS; Harrison *et al.*, 1995). Also note the many astronomy missions devoted to spectroscopy in the hard (1–10 Å) and soft (10–100 Å) X-ray bands; while that capability (outside of flares) has been relatively unexploited in solar physics[3].

I now will discuss examples from two of NASA's 'Great Observatories:' the current high performance UV spectrometer on *HST* – Space Telescope Imaging Spectrograph (STIS; Woodgate and Kimble, 1999); and early transmission grating tests of the recently launched *Chandra* X-ray Observatory (*CXO;* Elvis, 1998).

4. Stellar Ultraviolet Spectroscopy

STIS was installed in *Hubble* in early 1997, during the second *HST* servicing mission. STIS has a number of full format (1 K × 1 K) echelle modes that are highly multiplexed spectrally compared with the first generation instrument, Goddard High-Resolution Spectrograph (GHRS; Brandt *et al.*, 1994), which utilized small (500-pixel) linear detector arrays. In just a few *HST* orbits (the pricey commodity

[3]To the consternation of the stellar community, which relies heavily upon solar astronomers to develop and validate plasma diagnostics, particularly in 'new' wavelength regions.

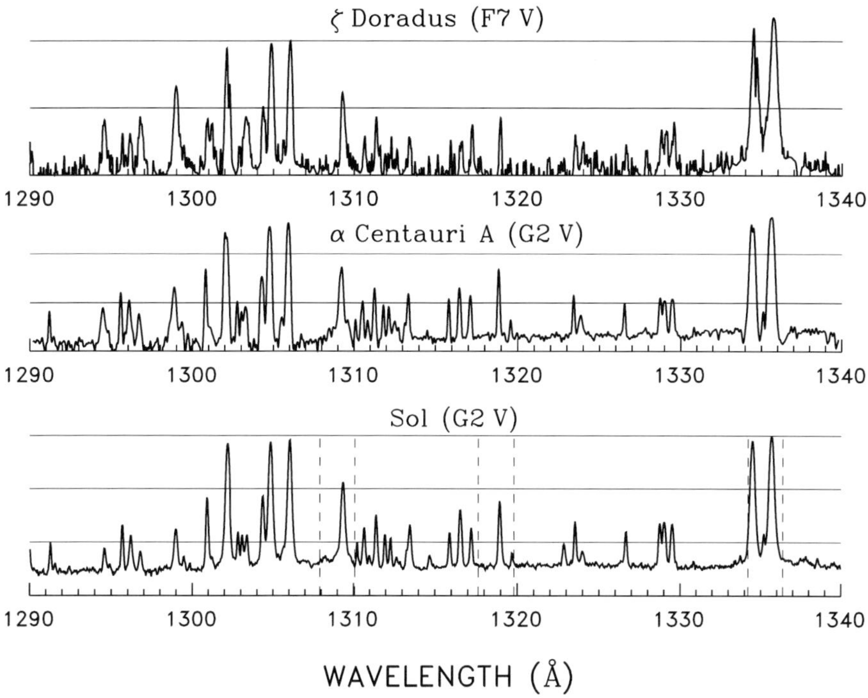

Figure 4. Comparison of far-UV spectra of three solar-like dwarfs in a narrow interval. The flux scales are logarithmic; the thin horizontal lines indicate successive powers of ten. The stellar spectra are from *HST* STIS echelle exposures. The solar spectrum was taken from the *SOHO* SUMER quiet Sun atlas (Brekke, 1996). The vertical dashed lines in the SUMER panel refer to the three 50-pixel spectral 'windows' used in the May 1999 coordinated observing program described in Section 6.

of stellar UV commerce), STIS can obtain the equivalent of a high S/N *atlas* of a bright star between 1150–3000 Å. In current *HST* observing cycle 8 (July 1999 –June 2000), I and my colleagues are conducting an extensive STIS program to acquire high quality UV spectra of thirteen representative late-type stars (righthand panel, Figure 1).

4.1. Solar-type stars

Figure 4 compares recent STIS E140M spectra of the active solar-type dwarf ζ Doradus (F7 V; from the cycle 8 cool star survey), the near solar twin α Centauri A (G2 V), and a quiet Sun segment from the SUMER UV atlas (Brekke, 1996). α Cen was recorded with the E140H high-resolution echelle mode ($R \simeq 1.1 \times 10^5$), as part of an extensive spectral mapping effort (Linsky, Pagano, and Valenti, 2000), but here was smoothed with a FWHM $= 21$ km s^{-1} Gaussian to simulate the SUMER resolution. Aside from a few second order lines in the SUMER trace, the solar and α Cen spectra are nearly identical.

There are two important *instrumental* differences, however. First, as mentioned earlier, STIS can record about 600 Å of the UV spectrum in a single shot, capturing all of the key stellar features simultaneously; whereas SUMER is limited to about 5% of that free spectral range (SUMER, of course, traded spectral grasp for spatial coverage with its long slit). Secondly, STIS carries a Pt/Ne hollow cathode lamp; activated every time the grating carousel is moved to a new setting, and periodically during a long exposure. The 'wavecal' spectra ensure that the radial velocity scale is precise and reproducible.

As a point of interest, I measured narrow low-excitation lines, from multiplets of C I, O I, and Si I having reliable laboratory wavelengths, in the spectra of ζ Dor and α Cen A. My motivation was to check whether the chromospheric gas might show a systematic global velocity shift at a low level. Historically, in 'calibration free' solar spectroscopy (as with SUMER), the low-excitation emissions have been used to establish the velocity zero point (which typically drifts owing to thermal flexing of the instrument). I can report the comforting news that the narrow chromospheric lines in both of the solar-like stars (of widely differing activity levels) appear to fall at the photospheric velocities, to within the measurement errors (fractions of a km s^{-1}). The stellar spectra, particularly the high-resolution α Cen A material, also can be used to check for blends (which might not be obvious at the $R \simeq 1.4 \times 10^4$ resolution of SUMER) and to determine empirical wavelengths for the many other narrow – presumably chromospheric – features lacking reliable laboratory values.

Figure 5 compares selected lines from recent STIS E140M spectra of solar-like ζ Dor, the classical hybrid bright giant α TrA mentioned earlier, and the archetype non-coronal red giant Arcturus (α Boo; K1 III). The velocity scales were obtained from special processing of the STIS echellograms, using software written by the STIS instrument team; the stellar radial velocities have been subtracted. The spectral lines are arranged from bottom to top in order of increasing excitation energy.

Notice that the optically thick chromospheric O I multiplet of ζ Dor falls very close to the photospheric radial velocity (like the narrow chromospheric lines mentioned previously); but the higher excitation emissions of Si IV, C IV, and N V are *red-shifted* by $\sim$6 km s^{-1}. Systematic red shifts of subcoronal lines in stellar UV spectra have been known for a long time (Ayres *et al.*, 1983; Ayres, Jensen, and Engvold, 1988). However, until the advent of *HST* it was extremely difficult to measure such shifts with any confidence in dwarf stars (Wood, Linsky, and Ayres, 1997)[4]. The ζ Dor velocities are not very different from what might be inferred for the disk-average Sun (see P. Brekke's contribution, this workshop). I always have found it remarkable that the TZ red shifts survive in the global UV spectrum of a

[4]Paradoxically, the global red shifts detected in giant and supergiant stars are much larger (tens of km s^{-1}) than in dwarf stars, even though the surface gravities of the former are orders of magnitude less. Naively, one might expect the downflows to scale as free-fall velocities, but the true situation obviously is more complex.

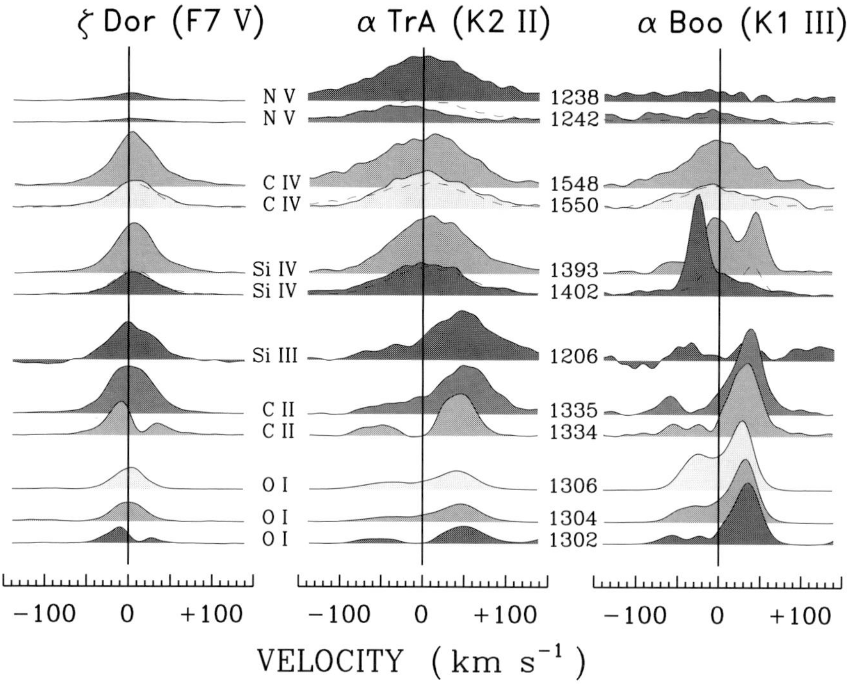

Figure 5. Selected line profiles from STIS E140M exposures of three stars: solar-like ζ Dor, the hybrid chromosphere bright giant α TrA, and the non-coronal red giant α Boo. The latter two are archetypes of their respective classes. In each panel, the vertical line refers to the stellar photospheric velocity. The red-shifted absorptions in the ζ Dor profiles of O I λ1302 and C II λ1334 are interstellar. Similar features are present in α TrA and Arcturus, but are less obvious owing to the prominent blue-shifted wind absorptions. The (linear) flux scales of the 'hot lines' have been adjusted so that C IV λ1548 has about the same peak flux for each star. The sharp structure in the Si IV features of Arcturus are H_2 lines, fluoresced by H I Lyα. The thin dashed line curve over the weaker components of the Si IV, C IV, and N V doublets represents the stronger member scaled down by the optically thin ratio of 2:1.

star. They clearly are an important dynamical signature in both the solar and stellar contexts.

Also notice in the previous figure (Figure 4) that the outer wings of the C II λ1335 doublet are distinctly broader than their counterparts in α Cen A or the Sun. The enhanced widths represent the so-called 'broad components,' commonly seen in the high-excitation far-UV line profiles of coronal stars (Wood, Linsky, and Ayres, 1997). Linsky and Wood (1994) proposed that such features are produced by stellar analogs of the TZ explosive events seen on the Sun, as recorded by the High-Resolution Telescope and Spectrograph sounding rocket series (HRTS; Dere, Bartoe, and Brueckner, 1989) and more recently SUMER (Innes *et al.*, 1997). Careful examination of the Si IV and C IV profiles in Figure 5 also shows the low amplitude extended emission wings. Again we see in the globally average line shapes of a star evidence for what must be a pervasive, highly dynamic surface phenomenon; whose existence undoubtedly bears directly on the issue of mechanical heating in the large scale magnetic network.

4.2. Red giants and supergiants

Figure 5 illustrates a number of key features of stellar activity among the evolved stars.

First, the chromospheric and transition zone (TZ) line widths increase with increasing stellar luminosity (V→III→II). This is analogous to the Wilson–Bappu effect (in Ca II H and K; Wilson and Bappu, 1957), and undoubtedly devolves from the general thickening (and thus increased 'opacity broadening') of the stellar outer atmosphere with decreasing surface gravity (Ayres, 1979).

Second, there are dramatic blueward absorptions in the low-excitation lines of the hybrid star and the red giant. The cutouts in the resonance lines are signatures of low-speed winds, with terminal velocities ~100 km s^{-1}, in contrast to the ~600 km s^{-1} for the fast solar wind, and ~10 km s^{-1} for the massive outflows of M supergiants like α Orionis. One also sees significant *redward* absorptions in O I λ1302, C II λ1334, and possibly also Si III λ1206 of ζ Dor. These features are interstellar. They provide an additional check on the velocity zero point; and are of interest in their own right for studies of the structure of the local interstellar medium and cosmologically significant light element abundances (Linsky, 1998). ISM features are present in the ground state lines of α TrA and Arcturus as well; compare, e.g., C II λ1334 and λ1335.

Third, many of the TZ lines of the hybrid star show evidence of red shifts, without any obvious sign of sharp wind absorption features on the blueward sides of the line profiles. The weakening of the wind absorption in Si III λ1206, compared with the lower excitation species, suggests that the wind temperature is not much more than 2×10^4 K, consistent with the lack of higher temperature absorptions (Harper, 2000). It follows that the 'TZ' lines might form completely independently of the acceleration sites of the cool wind; perhaps not unlike the solar case, where strong coronal emissions are confined to areas of closed magnetic field, whereas the wind emerges from areas that topologically are open to the heliosphere (and have low X-ray emissions; Withbroe, Feldman, and Ahluwalia, 1991).

This view conflicts with that of Dupree and Brickhouse (1998). The authors analyzed low-resolution 900–1200 Å spectra of three hybrid stars from the Shuttle-borne ORFEUS instrument, finding that the O VI lines near 1030 Å were systematically *blue-shifted* with respect to Si III λ1206. They interpreted the relative shifts to indicate formation in an expanding warm ($\sim 3 \times 10^5$ K) outflow (see also Dupree, 1975). Unfortunately, the ORFEUS spectrometer lacked a calibration lamp and thus has an uncertain wavelength scale. With the much higher resolution, wavelength precision, and S/N of the STIS spectra, we see that the Si III feature of α TrA itself is *red-shifted* significantly (owing to the blueward wind absorption), and thus is poorly suited for a velocity reference.

Furthermore, the STIS profile of N V λ1238 is not significantly shifted, and its formation range strongly overlaps that of O VI. The weaker component of the N V doublet, λ1242, does appear to be blue-shifted; but at the same time its strength is much less than one would anticipate scaling from λ1238 according to the effectively thin doublet ratio. Thus, the λ1242 'shift' very likely is due to selective absorption on the redward side of the line. The weaker component of the Si IV doublet, λ1402, also seems to be affected by red-wing absorption. Because the absorption is not replicated at twice the strength in the stronger component of each doublet, we can be confident that the lines are not *self*-absorbed. The extinction probably is from low-excitation species in cooler gas overlying the hot structures, perhaps in the wind itself.

4.3. The Coronal Graveyard

Finally, consider the non-coronal red giant Arcturus. The chromospheric features (O I, C II) look very similar to their counterparts in the hybrid spectrum, aside from the somewhat narrower widths and lower terminal velocity of the wind (50 km s^{-1}, about half that of α TrA). The higher excitation features are strikingly different, however, with the notable exception of C IV (and the intersystem lines Si III] λ1892 and C III] λ1909 at longer wavelengths). Linsky, Brown, and Carpenter (1991) originally reported weak TZ emissions in the non-coronal giant γ Draconis (K5 III) using high-sensitivity measurements with *HST* GHRS. Subsequently, other red giants were detected in the far-UV hot lines, and in X-rays in a few cases as well (see Ayres *et al.*, 1997). Thus, the presence of C IV in Arcturus is not surprising. What is remarkable, however, is the absence of both Si III λ1206 and the N V doublet, relative to what one might expect scaling from α TrA.

N V should be strong in such stars, because first dredge-up tends to significantly enrich the atmosphere in nitrogen at the expense of carbon, mirroring the composition of the CNO burning products advected into the surface layers (Tomkin, Luck, and Lambert, 1976). The N V emission of active clump giants, like β Cet (K0 III; Ayres *et al.*, 1998), tends to dominate the TZ spectrum, having *twice* the flux of C IV. (By that measure, the N V features of α TrA are weaker than they should be, and thus *both* components might be affected by overlying absorption). Furthermore, the Si IV doublet of Arcturus looks rather peculiar: obviously affected by significant distortions, perhaps accidental blends.

Based on an earlier observation of just the λ1393 Si IV component with GHRS, we interpreted the spectral distortions to be *absorptions* by lines from the A5–X0 band of the carbon monoxide 4th-positive system (Ayres *et al.*, 1997). Now, with the higher-resolution STIS echellograms, covering the entire spectrum longward of 1150 Å (especially Si IV λ1402), it is clear that the spectral structure is produced by *emissions* from the B0–X5 band of molecular hydrogen, pumped by H I Lyα. With the full spectral coverage afforded by STIS, we were able to track down the other B0–Xv'' H_2 lines fluorescing from the Lyα pump, estimate the strengths of the B0–X5 group based on the known branching ratios, and correct the Si IV profiles for the molecular contamination. The result was surprising: hardly any flux remained in either component of Si IV, much like the situation seen at N V. Our conclusion is that the high-excitation spectrum shortward of C IV is being strongly attenuated, probably by a broadband continuum. The λ1525 ground state *b–f* edge of atomic silicon is a strong possibility, probably in collusion with the excited state photoionization edge of carbon at λ1240.

Optically thick Si I continuum absorption implies a large column of neutral hydrogen over the hot structures, enough to extinguish any soft coronal X-rays. Perhaps, then, coronae are not dead in the coronal graveyard at all, but merely buried alive. It would be a rather unsolar-like situation, with the chromosphere on top of the corona; and parts of that chromosphere would have to be rather cool, in order to support the fluorescence of H_2 and also CO (which has strong emission bands throughout the far-UV, pumped by the atomic oxygen multiplet at 1305 Å: Ayres, Moos, and Linsky, 1981). The situation should not be so surprising, perhaps, given the extended outer atmospheres of giant stars compared with high gravity dwarfs. Even relatively ‘long’ magnetic loops, generated in the red giant convection zone and erupting into its photosphere, might fail to poke through the extended chromospheric envelope. In somewhat more active stars, like the hybrids, the coronal loops might attain sufficient lengths to surmount the chromosphere and radiate freely into the X-ray-thin wind (e.g., Haisch, Schmitt, and Fabian, 1992).

5. Stellar High-Energy Spectroscopy

Recording the X-ray emissions of late-type stars is a direct way to explore properties of their hot coronae. *Chandra* is the most recent in a long series of astronomical X-ray observatories, and has the highest potential resolving power ($\sim 2 \times 10^3$) in the important 60–160 Å region. Figure 6 shows an early *Chandra* medium-energy grating spectrum, a 74 ks exposure taken 24–25 September 1999 (courtesy *CXO* public archive). The target was the nearby bright binary Capella (α Aur: G8 III + G1 III). The trace displays numerous iron ‘L shell’ transitions, although hydrogenic and helium-like lines Si, Mg, Ne, and O are prominent as well.

The upper panel illustrates a small interval from a 3-orbit *HST* STIS observation of Capella, conducted 12 September 1999 in a coordinated program as part of the *Chandra* ‘Emission Line Project’ led by J. Drake and N. Brickhouse of CfA. The relatively broad feature at 1354.5 Å, nestled between the stronger Cl I λ1351 and O I λ1355 lines, is a blend of a coronal forbidden line of Fe XXI and a chromospheric C I transition of comparable strength. The 8.1 ks exposure was taken at one of the orbital quadratures, when the primary was moving toward the observer at close to its maximum orbital velocity and the secondary star was receding at close to its maximum.

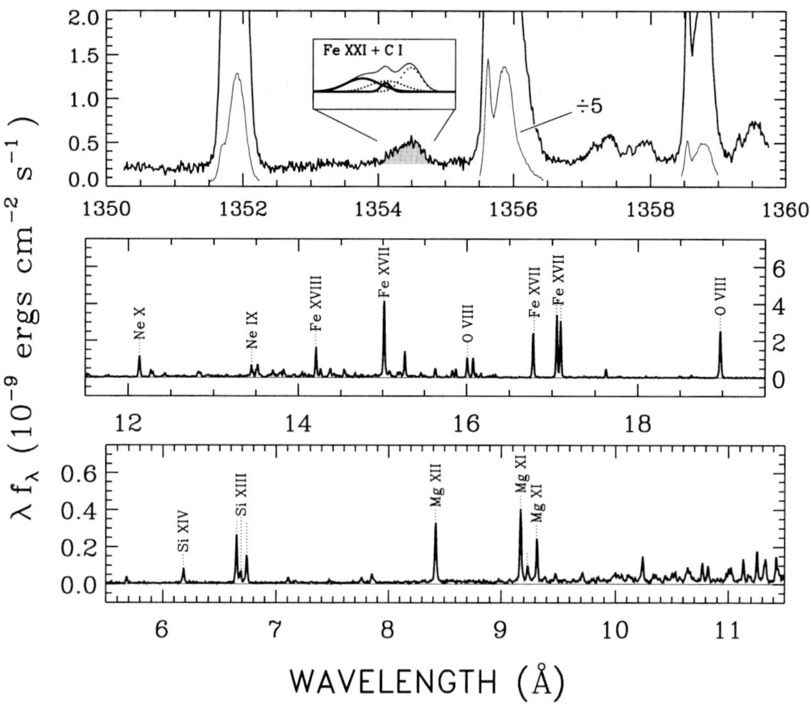

Figure 6. Chandra and STIS observations of Capella, taken during the early calibration phase of the X-ray observatory. The middle and lower panels depict the medium energy channel from a 74 ks exposure with the High Energy Transmission Grating Spectrometer of *Chandra*. The resolution is ~500, about 10× better than recent X-ray spectroscopy missions like *ASCA*. The upper spectrum illustrates a small interval from an 8 ks STIS E140M exposure, showing the faint coronal forbidden line of Fe XXI. The inset is the decomposition of the Fe XXI + C I blend proposed by Linsky *et al.* (1998), based on an earlier GHRS measurement at nearly the same orbital phase.

Note the blueward displaced sharp emission features, and redward broader ones, in the λ1355 and λ1358 profiles, attributed respectively to the slowly rotating G8 III star and the fast spinning G1 III star. Note also the weakness of the blue-shifted primary component in Cl I λ1351, compared with the O I lines. The Cl I feature is radiatively fluoresced by C II λ1335 (Shine, 1983), and the latter is weak in the primary owing to dredge-up dilution of carbon and because the primary has much fainter TZ lines than the secondary (Ayres and Linsky, 1980). The inset shows a multiple Gaussian fit to the Fe XXI + C I blend from a previous deep *HST* GHRS spectrum taken at nearly the same orbital phase (Linsky *et al.*, 1998). The solid curves represent the Fe XXI (left) and C I (right) lines of the primary; the dotted profilees are the corresponding features of the secondary. The C I lineshapes were scaled empirically according to other unblended members of the multiplet. It is clear that the two stars are more equal in their coronal emissions then in their TZ lines (Ayres, Schiffer, and Linsky, 1983): the secondary, in fact, is one of the 'X-ray deficient' Hertzsprung gap giants.

The importance of the STIS measurement is that it reveals unique information concerning the coronal dynamics at 4.6×10^4 resolution, $30\times$ better than current X-ray capabilities. In particular, both of the Fe XXI lines fall at the respective stellar radial velocities at that orbital phase, implying that the coronal gas is bound to the individual stars; rather than, say, located in a colliding wind zone *between* the stars, or freely streaming away from one or the other (Linsky *et al.*, 1998).

6. Making the Solar–Stellar Connection

I have given a few examples of current research in stellar coronae (blatantly slanted toward my own interests, for which I apologize) to impress upon my solar colleagues the diversity of issues facing the stellar community. At the same time, we on the stellar side cannot help but be impressed by the stunning views of the solar 'magnetic junkyard' (attributed to H. E. Mason) from SOHO and the Transition Region and Coronal Explorer (*TRACE*; Handy *et al.*, 1999).

Granted that we find clear evidence of solar coronal phenomena – red shifts, explosive events, spots, flares, etc. – on other stars, how do we go about making the connection back to the Sun; and thus to the underlying physical phenomena which are so much more accessible there?

I will focus on one such effort I and my collaborators carried out this past May (1999), as part of the *SOHO* SUMER 'MEDOC' campaign that month. Our observing program coordinated UV measurements from the *SOHO* and *TRACE* satellites with IR surface mapping (in the 5 μm CO lines) from the Infrared Imaging Spectrometer (IRIS; Ayres and Rabin, 1996) on the McMath-Pierce solar telescope at Kitt Peak.

Our motivation was to obtain spatially resolved measurements of temperature and velocity fields at different elevations in the solar atmosphere, to track the propagation of energy and momentum up through the structures that comprise what one might call the 'magnetic transition zone' (MTZ); the crucial region of the outer atmosphere where kinetic control of the magnetic field in the deep photosphere gives way to magnetic control of the plasma in the canopy and corona. Figure 7 is a cartoon of the MTZ; highly simplified compared with the real Sun, but more realistic than the 'layered' paradigm of classical solar reference models (e.g., Vernazza, Avrett, and Loeser, 1981).

We conducted our program on four consecutive days, 11–14 May 1999, over a four-hour period each day. We were fortunate to have good weather at Kitt Peak, with very exceptional conditions on the mornings of the 12th and 14th. On each day, the $300''$ slit of SUMER was positioned at heliocentric disk center, oriented north-south. Solar rotation carried the surface east–west across the $1''$ wide aperture at a rate of one slit width per 6.5 min. Three ~ 2.4 Å spectral windows (containing Si II λ1309, N I λ1319, both components of C II λ1335, and associated continuum windows; see Figure 4) were read out with a cadence of one set per 45 s, for a period

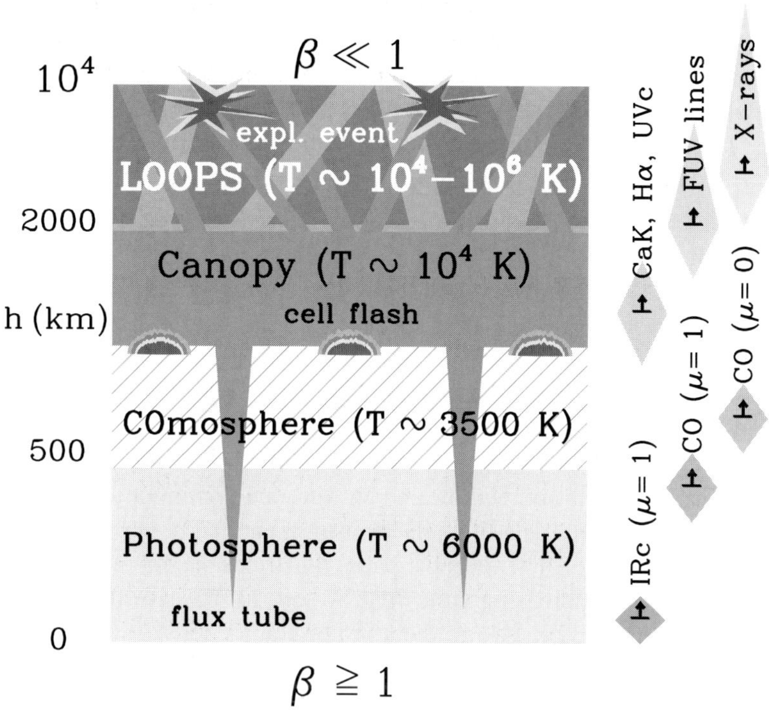

Figure 7. Cartoon of 'magnetic transition zone.' The classical chromosphere between 500 km and 2000 km has given way to a cool zone (the *COmosphere*) on the bottom of a hot *canopy* where the thin *flux tubes,* rising like funnels from the photospheric supergranules, spread out and merge together. From that base spring the large-scale magnetic *loops* that outline the corona. On the underside of the canopy are intermittent *flashes,* primarily over supergranular cell interiors, thought to result from acoustic shocks. *Explosive events* dot the base of the coronal loops where opposite polarity force lines cancel catastrophically. Diamonds at right depict schematic heights of formation of important atmospheric diagnostics ($\mu \equiv \cos\theta = 1$ is disk center; $\mu = 0$ is the extreme limb; 'IRc' is the 5 μm continuum; 'UVc' is the 1300 Å continuum). $\beta \equiv P_{\rm gas}/P_{\rm mag}$ is the ratio of gas to magnetic pressures.

of about 3.6 hr. One thus builds up a short-term record of time variations of velocity and intensity arrayed along the slit length (in the several diagnostics), while solar rotation gradually replaces the scene. Such 'drift scanning' is well matched to the spatial sizes and lifetimes of photospheric *p*-mode oscillations and shorter period chromospheric modes. Prior to the drift scan, a context spatial map in the He I λ584 line – plus first order λ1170 C I continuum – was taken by rastering the slit over an ∼80″ swath, in 1″ steps, starting eastward of, and concluding at, the final slit position. A context spectrum of the full 1024-pixel wavelength interval was taken as well.

Concurrently, *TRACE* snapped off series of 128″ × 320″ images centered on the SUMER slit position, of white light (deep photosphere), C IV λ1550 (hot gas proxy), and the 1700 Å continuum (upper photosphere); a history – with 45 s cadence at 1″ resolution – of intensity fluctuations at different levels of the atmosphere coincident with the SUMER and IRIS measurements. At this writing, the *TRACE* frames still are being analyzed.

The IRIS side consisted of repeated 8 step rasters (1.9″ per step) with the campaign cadence of 45 s. A fixed 2.3 cm^{-1} interval of the CO spectrum near 2143 cm^{-1} (4.7 μm) – containing four strong CO lines, several weaker features, and continuum windows – was recorded. The IRIS raster covered a 13″ × 97″ area centered on the SUMER slit position. Context images taken periodically with a coaligned Ca II K system allowed post facto registration of the IRIS field with the SUMER slit position.

Figure 8 is an example of one of the SUMER drift scans, taken on 14 May. The spectral features in the 290 individual long slit frames (each of which consisted of the three 50 × 360 pixel^2 spectral windows) of the time sequence were measured with a Gaussian fitting procedure, yielding a 1-D run of integrated line intensities, Doppler shifts and widths, and background continuum brightness along the slit length. Concatenating the 1-D measurements of a particular parameter yields a space/time 'map.' The leftmost panel illustrates such a map for the λ1320 continuum (with time increasing from right to left, to preserve the proper spatial orientation of the image; i.e., east to the left, north at the top).

Notice the sprinkling of short vertical streaks throughout the image. These are transient UV continuum brightenings, probably related to the Ca II 'K_{2v} grains' or 'cell flashes' (Rutten and Uitenbroek, 1991). The larger-scale bright areas, elongated in the time dimension, represent persistent surface features. The narrow strip immediately to the right of the continuum drift scan is a time averaged spatial map, in which transient structures have been filtered out, leaving mostly the persistent bright fragments of the supergranulation network. The next field, to the right, depicts relative velocities measured in the narrow N I λ1319 feature. It shows predominantly red shifts over the magnetic areas, and alternating blue shifts and red shifts in the cell interiors. If we believe the 'stellar' calibration, discussed earlier, N I should have a net velocity of zero averaged over the disk, so the *relative* shifts over this large (space/time) field should be equivalent to *absolute* shifts. The FWHM of the corresponding velocity distribution function is 3.5 km s^{-1}.

The next panel to the right depicts the drift scan in C II λ1335, which forms in the low TZ, at least in a layered model. The persistent structures appear to be more diffuse than in the UV continuum, and the small scale 'flashes' are not obvious. The filtered spatial map of C I (to the right of the drift scan) again shows a blurrier scene than the continuum, likely because of the vertical spreading of the network flux tubes into the canopy zone. The C II velocities show larger amplitudes than N I, and a stronger correspondence between red shifts and the bright – presumably magnetic – regions. The FWHM of the C II velocity distribution is 5.5 km s^{-1}.

The upper frames of the diagram illustrate correlations between different pairs of velocity and intensity fluctuations (the latter expressed in terms of equivalent brightness temperatures). The leftmost set depicts the correlations at the same space/time point, while the other frames show the correlation with progressive lags in the time dimension; e.g., $\Delta T_{\text{C II}}$ at $(t = t_0, y = y_0)$ compared with ΔT_{cont} at $(t = t_0 - \Delta t, y = y_0)$. In constructing these diagrams, I applied masks to flag regions considered 'network' according to their brightness and spatial persistence characteristics; only lagged pairs from the cell interiors were included, and thus the correlations should apply exclusively to the 'internetwork' disturbances.

These are very preliminary results, to illustrate that one easily can derive velocity and intensity information at high temporal and spatial resolution; to identify the sources of particular features in spatially and temporally averaged solar line shapes, analogous to stellar observations. However, several peculiarities of these diagrams are worth noting. First, the C II and continuum temperature fluctuations initially show a positive correlation; but when a lag of ~ 200 s or more is imposed, the diagram appears to split into two distinct components. In the velocity-velocity frame, one sees a positive correlation at $\Delta t = 0$, but immediately upon imposition of a one step lag (45 s) the distribution becomes uncorrelated, and remains that way for all larger lags. If the disturbances traveling through the continuum and C II altitudes were pervasive standing waves, one would expect a strong positive correlation at zero lag (atmosphere moving as a unit), a gradual shift to a negative correlation with a lag corresponding to half an oscillation cycle, then back to a positive correlation after a full cycle lag.

The fine-scale disturbances that pepper SUMER continuum images have been identified as chromospheric 3 minute modes (Carlsson, Judge, and Wilhelm, 1997), but a distinct signature of that repetition time in the correlations is elusive. Apparently, the oscillatory modes occupy too little of the surface area to impose a distinct signature on the correlation maps; they are overwhelmed by the other components of the velocity field which apparently are not correlated on few minute time scales.

Also relevant to the solar–stellar discussion is that it is possible to make analogous maps of excess C II line broadening, a signature of explosive events. One can use such maps to estimate the spatial/temporal filling factor of the phenomenon, in order to more quantitatively interpret broad components measured in the globally averaged line shapes of other stars.

Figure 9 depicts corresponding space/time diagrams obtained from the IRIS long slit rasters. The velocity and temperature maps refer to an average over the four strong CO lines in the spectral bandpass. The image on the left side shows a concatenated sequence of the CO rasters (8 steps in each 45 s interval) over a 30 minute period during the morning of 14 May when the seeing was particularly good. The image to the right displays the first 10 rasters of the sequence with the proper spatial scaling and orientation (east is to left; north is up). The upper panels are CO velocity, the middle panels are CO intensity (expressed in equivalent brightness temperature), and the bottom panels are the 2143 cm^{-1} continuum intensity (also in ΔT_{B}).

The 'compressed image' at the left allows one to more clearly visualize the large-scale p-mode 'wavepackets' (which are the dominant velocity and tempera-

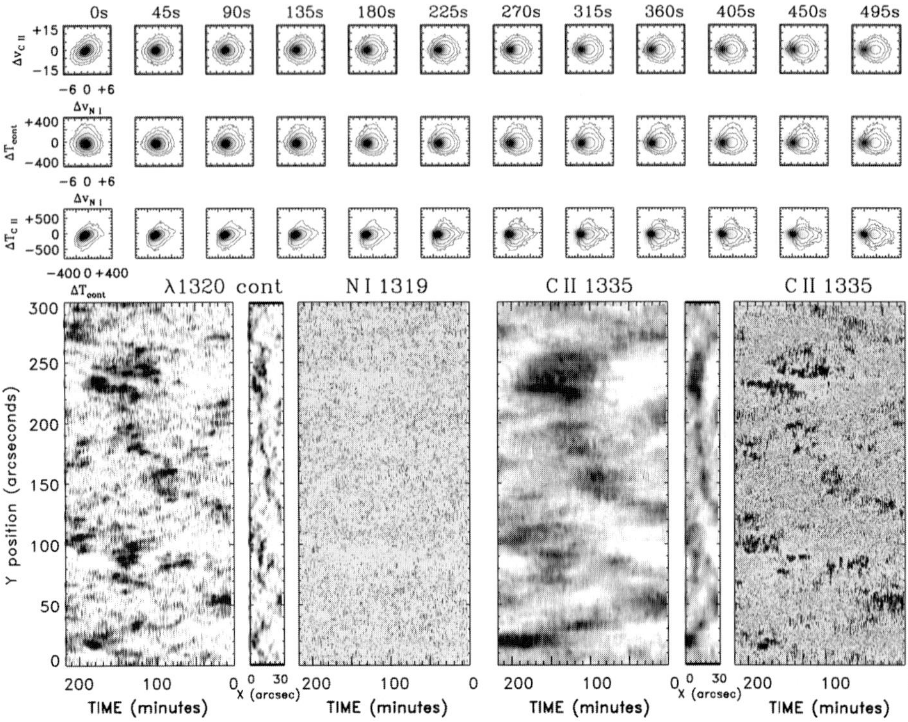

Figure 8. SOHO SUMER drift scans. Leftmost panel depicts the λ1320 continuum brightness (dark represents higher intensity); *y*-axis is along the slit, with north at top. The time axis is reversed to maintain an orientation with east to the left. The thinner frame immediately to the right of the continuum drift scan shows the result of filtering out transient structure, and displaying with the correct east–west spatial proportion. The next frame to the right represents the velocity of N I λ1319: black indicates red shifts ($> +4$ km s^{-1}), medium gray indicates blue shifts (< -4 km s^{-1}), and lighter gray indicates velocities between those limits. The third large panel represents the intensity of C II λ1335, followed by the 'spatial map' and velocity drift scan for that feature (with the same velocity limits as for N I). Top portion of figure depicts correlations between various combinations of the intensities and velocities (velocities are in km s^{-1}, temperatures in K). Leftmost graph shows correlation over full map; subsequent panels depict correlation with successive 1 step (45 s) time lags imposed on the *x*-axis quantity. Central 'image' part of each graph is on linear scale; outer contours are logarithmic.

ture undulations along the 'time' dimension in the diagram). Even with an average along the full 100″ IRIS slit, a clear 5-min signal is seen (which explains why Noyes and Hall [1972] were able to record significant 5-min oscillations in CO intensity with their long slit, single element detector system). However, sitting on top of the 5-min modes is a sprinkling of short period disturbances like those seen in the λ1320 continuum images. Incidentally, no comparable 5-min signal is apparent in the SUMER drift scans.

The *p*-mode wavepackets, where the complex spatial interference pattern creates local maxima (or minima), are of order ∼10″ in diameter, but the shorter

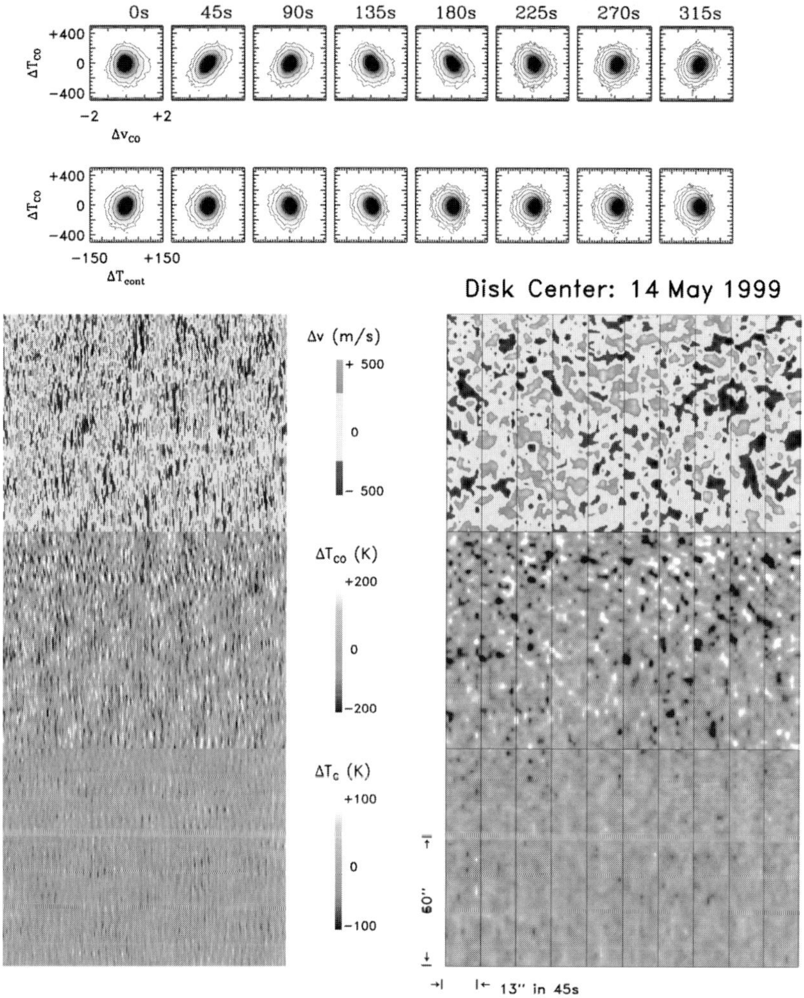

Figure 9. Space–time 'fluctuation' maps and correlation diagrams for CO velocity, CO temperature, and 5 μm continuum temperature from the IRIS part of 1999 coordinated campaign. Data are from morning of 14 May during period of excellent seeing. $\Delta T_{\rm CO}$ and $\Delta v_{\rm CO}$ are averages over four strong lines. Time increases to the right; y-axis is N/S (north is at top); x-axis is E/W (east is to left). Left side is full 30 minute, 40 raster segment (one of seven taken that morning), compressed in x. Right side illustrates first ten 45 s rasters of the segment. In the 'compressed' version, alternating large-scale bands of $\Delta v_{\rm CO}$ and $\Delta T_{\rm CO}$ are due to the 5 min p-modes. In the full resolution version, smaller scale disturbances stand out (they are the short streaks on top of the p-modes in the compressed image). These are the higher frequency waves which have appeared in the past primarily in power spectra of 11 μm OH core intensities obtained with spatial resolution higher than the typical size of the p-mode packets (Deming *et al.*, 1988). Top portion of figure depicts apparent correlation of $\Delta T_{\rm CO}$ versus $\Delta v_{\rm CO}$ (upper series; velocities are in km s^{-1}, temperatures in K) and $\Delta T_{\rm CO}$ versus $\Delta T_{\rm continuum}$ (lower series). Leftmost graph shows correlation over full map; subsequent panels depict correlation with indicated time lags imposed on the x-axis quantity. Central 'image' part of each graph is on linear scale; outer contours are logarithmic.

period disturbances are much smaller in scale, only a few arcseconds. That probably explains why the small aperture ($\sim 2''$) 11 μm OH heterodyne oscillation experiment of Deming *et al.* (1988) saw primarily the 'chromospheric' (i.e., high-frequency) mode, rather than the predominantly 5-min oscillations seen in rapid cadence Fourier transform spectrometer observations of CO using larger fields of view (Ayres and Brault, 1990). With further analysis, I and my collaborators hope to track the small-scale – presumably acoustic – events from the upper photosphere (where the 5 μm CO lines form for a $\mu = 1$ line of sight; see Figure 7), through the magnetic transition zone. These waves are thought to be responsible for discrete heating events (the 'cell flashes') in a shock formation layer close to the canopy height (Carlsson and Stein, 1995). The FWHM of the CO velocity distribution is only ~ 0.7 km s^{-1}, reflecting the lower formation altitude.

Incidentally, the dramatic *off-limb emissions* of CO (Solanki, Livingston, and Ayres, 1994) require a significant amount of cool gas at 'chromospheric' altitudes (Ayres and Rabin, 1996), contrary to the assertions of Kalkofen, Ulmschneider, and Avrett (1999) in a recent paper criticizing the Carlsson–Stein model. The cool high altitude gas is relatively transparent in the strong CO features at normal incidence, and is visible only at the extreme limb – and off-limb – where the long tangential sight-line becomes optically thick.

In the $\Delta T_{\rm CO}$ versus $\Delta v_{\rm CO}$ diagrams, there is a lack of correlation between temperature and velocity at zero lag, but sequentially positive and negative correlations develop with increasing lag over an interval of ~ 300 s, corresponding to the expected behavior of nearly adiabatic evanescent waves high in the photosphere (e.g., Ayres and Brault, 1990).

When we have carefully measured the CO velocity component associated with the short-period disturbances that appear in the UV lines, we then can deduce the relative heights of the CO 'absorbing' and UV emitting layers. In particular, one expects the amplitudes v of the sound waves, if they are roughly adiabatic, to grow as $\rho^{-1/2}$. The increase in velocity amplitudes implies a specific decrease in density, and thus a physical separation in scale heights. Such information will help constrain the thickness of the so-called COmosphere, and serve as a test of wave propagation models.

In the new IRIS data sets, taken under conditions of excellent seeing, we also have searched for a connection between the COmospheric layers and atmospheric flows. Uitenbroek (1998) suggested that the CO molecules might exist preferentially in small patches associated with overshooting convective granules, expansively cooling to low temperatures as they plowed through the upper photosphere. The scenario anticipates an association between cooler temperatures and blue shifts in the strong CO lines at disk center; corresponding to the outward flow necessary to populate even higher layers with cold gas (seen in extreme limb observations). No such correlation is obvious, however, in the disk center CO maps (see Figure 9; '0 s' lag).

We clearly have scratched only the surface of the multiple data sets; we are relying on crucial registration information from the *TRACE* images to tie together the IRIS and SUMER velocity maps on a point-to-point basis (in space and in time). I hope it is clear, nevertheless, that detailed multi-wavelength dissections of the kinematic and thermal structure of large swaths of the solar atmosphere provide a vital link in interpreting spatially and temporally averaged ultraviolet spectra of solar-like stars. The stellar side repays that detailed view by providing access to a rich collection of phenomena across the full range of stellar temperature, gravity, age, and metallicity. The stellar phenomena share the same radiation-hydromagnetic foundations as their often more pallid solar counterparts. It is, after all, the dissection of those fundamental processes that is our ultimate goal.

Acknowledgements

I thank the following, who are collaborators on the solar and stellar sides: W. Curdt, P. Lemaire, J.-C. Vial, and K. Wilhelm (*SOHO*); T. Tarbell and R. McMullen (*TRACE*); D. Rabin and C. Plymate (IRIS); A. Brown, J. Linsky, G. Harper, and the rest of the cycle 8 large project team (STIS). This work was supported by grants from the Space Telescope Science Institute (GO-08280.01-99A); the National Aeronautics and Space Administration (NAG5-3226, NAG5-7234); and the National Science Foundation (AST-9618505). Stellar parameters were collected from the SIMBAD database, operated at CDS, Strasbourg, France; and from VizieR, a joint effort of CDS and ESA-ESRIN. X-ray data were obtained through the High Energy Astrophysics Science Archive Research Center Online Service, sponsored by NASA's Goddard Space Flight Center, or from web pages provided by the *Chandra* X-ray Observatory public archive. Stellar ultraviolet spectra are from active *HST* observing programs or were taken from its public archive, both sponsored by the STScI; operated by the Association of Universities for Research in Astronomy, Inc., under NASA contract NAS5-26555. The solar infrared spectra are from the Kitt Peak facility of the National Solar Observatory, one of the National Optical Astronomy Observatories (NOAO), operated by AURA, under a cooperative agreement with the National Science Foundation.

References

Ayres, T. R.: 1979, *Astrophys. J.* **228**, 509.
Ayres, T. R.: 1997, *J. Geophys. Res.* **102**, 1641.
Ayres, T. R. and Brault, J. W.: 1990, *Astrophys. J.* **363**, 705.
Ayres, T. R. and Linsky, J. L.: 1980, *Astrophys. J.* **241**, 279.
Ayres, T. R. and Rabin, D.: 1996, *Astrophys. J.* **460**, 1042.
Ayres, T. R., Fleming, T. A., and Schmitt, J. H. M. M.: 1991, *Astrophys. J.* **376**, 45.
Ayres, T. R., Jensen, E., and Engvold, O.: 1988, *Astrophys. J. Suppl. Series* **66**, 51.

Ayres, T. R., Moos, H. W., and Linsky, J. L.: 1981, *Astrophys. J.* **248**, 137.
Ayres, T. R., Schiffer, F. H., III., and Linsky, J. L.: 1983, *Astrophys. J.* **272**, 223.
Ayres, T. R., Linsky, J. L., Vaiana, G. S., Golub, L., and Rosner, R.: 1981, *Astrophys. J.* **250**, 293.
Ayres, T. R. *et al.*: 1983, *Astrophys. J.* **274**, 801.
Ayres, T. R. *et al.*: 1995, *Astrophys. J. Suppl. Series* **96**, 223.
Ayres, T. R., Simon, T., Stauffer, J. R., Stern, R. A., Pye, J. P., and Brown, A.: 1996, *Astrophys. J.* **473**, 279.
Ayres, T. R. *et al.*: 1997, *Astrophys. J.* **491**, 876.
Ayres, T. R., Simon, T., Stern, R. A., Drake, S. A., Wood, B. E., and Brown, A.: 1998, *Astrophys. J.* **496**, 428.
Baliunas, S. L. and Vaughan, A. H.: 1985, *Ann. Rev. Astron. Astrophys.* **23**, 379.
Bopp, B. W. and Espenak, F.: 1977, *Astron. J.* **82**, 916.
Brandt, J. C. *et al.*: 1994, *Publ. Astron. Soc. Pacific* **106**, 890.
Brekke, P.: 1996, http://www.uio.no/~paalb/sumer_atlas.html.
Brown, A., Jordan, C., Stencel, R. E., Linsky, J. L., and Ayres, T. R.: 1984, *Astrophys. J.* **283**, 731.
Carlsson, M. and Stein, R. F.: 1995, *Astrophys. J.* **440**, 29.
Carlsson, M., Judge, P. G., and Wilhelm, K.: 1997, *Astrophys. J.* **486**, L63.
Ciaravella, A., Peres, G., Maggio, A., and Serio, S.: 1996, *Astron. Astrophys.* **306**, 553.
Collier Cameron, A.: 1999, in C. J. Butler and J. G. Doyle (eds), *Solar and Stellar Activity: Similarities and Differences,* ASP Conf. Ser. 158, , ASP, San Francisco, p. 146.
Collier Cameron, A., Campbell, C. G., and Quaintrell, H.: 1995, *Astron. Astrophys.* **298**, 133.
Deming, D., Glenar, D. A., Käufl, H. U., and Espenak, R.: 1988, in J. Christensen-Dalsgaard and S. Frandsen (eds), 'Advances in Helio- and Asteroseismology', *IAU Symp.* **123**, 425.
Dere, K. P., Bartoe, J.-D. F., and Brueckner, G. E.: 1989, *Solar Phys.* **123**, 41.
Domingo, V., Fleck, B., and Poland A. I.: 1995, *Space Sci. Rev.* **72**, 81.
Donahue, R. A. and Baliunas, S. L.: 1992, *Astrophys. J.* **393**, L63.
Donahue, R. A., Dobson, A. K., and Baliunas, S. L.: 1997, *Solar Phys.* **171**, 191.
Duncan, D. K. *et al.*: 1991, *Astrophys. J. Suppl. Series* **76**, 383.
Dupree, A. K.: 1975, *Astrophys. J.* **200**, L27.
Dupree , A. K. and Brickhouse, N. S.: 1998, *Astrophys. J.* **500**, L33.
Durney, B. R., Mihalas, D., and Robinson, R. D.: 1981, *Publ. Astron. Soc. Pacific* **93**, 537.
Elvis, M.: 1998, in K. Koyama, S. Kitamoto, and M. Itoh (eds), 'The Hot Universe', *IAU Symp.* **188**, 79.
Engvold, O. *et al.*: 1988, *Astron. Astrophys.* **192**, 234.
Gray, D. F.: 1981, *Astrophys. J.* **251**, 155.
Gray, D. F. and Baliunas, S. L.: 1997, *Astrophys. J.* **475**, 303.
Haisch, B., Schmitt, J. H. M. M., and Fabian, A. C.: 1992, *Nature* **360**, 239.
Hall, D. S.: 1978, *Astron. J.* **83**, 1469.
Handy, B. N. *et al.*: 1999, *Solar Phys.* **187**, 229.
Harper, G. M.: 2000, in R. Rebolo and R. García López (eds), *Cool Stars, Stellar Systems, and the Sun: the 11th Cambridge Workshop*, ASP Conf. Ser. (in press).
Harper, G. M., Brown, A., and Lim, J.: 2000, *Astrophys. J.* (submitted).
Harrison, R. A. *et al.*: 1995, *Solar Phys.* **162**, 233.
Hartmann, L., Dupree, A. K., and Raymond, J. C.: 1980, *Astrophys. J.* **236**, L143.
Hawley, S. L. and Petersen, B. R.: 1991, *Astrophys. J.* **378**, 725.
Huber, M. C. *et al.*: 1973, *Astrophys. J.* **183**, 291.
Hünsch, M. and Schröder, K.-P.: 1996, *Astron. Astrophys.* **309**, L51.
Innes, D. E., Inhester, B., Axford, W. I., and Wilhelm, K.: 1997, *Nature* **386**, 811.
Kalkofen, W., Ulmschneider, P., and Avrett, E. H.: 1999, *Astrophys. J.* **521**, L141.
Kashyap, V., Rosner, R., Harnden, F. R., Jr., Maggio, A., Micela, G., and Sciortino, S.: 1994, *Astrophys. J.* **431**, 402.

Linsky, J. L. (ed.): 1993, *Physics of Solar and Stellar Coronae: G.S. Vaiana Memorial Symposium,* Dordrecht Academic Publishers, Dordrecht, the Netherlands.
Linsky, J. L.: 1998, *Space Sci. Rev.* **84**, 285.
Linsky, J. L. and Haisch, B. M.: 1979, *Astrophys. J.* **229**, L27.
Linsky, J. L. and Wood, B. E.: 1994, *Astrophys. J.* **430**, 342.
Linsky, J. L., Pagano, I., and Valenti, J.: 2000, (in preparation).
Linsky, J. L., Brown, A., and Carpenter, K. G.: 1991, in A. L. Kinney and J. C. Blades (eds), *The First Year of HST Observations,* STScI, Baltimore, 70.
Linsky, J. L., Wood, B. E., Brown, A., and Osten, R. A.: 1998, *Astrophys. J.* **492**, 767.
Maunder, W. E.: 1894, *Publ. Astron. Soc. Pacific* **6**, 125.
Nakariakov, V. M., Ofman, L., DeLuca, E. E., Roberts, B., and Davila, J. M.: 1999, *Science* **285**, 862.
Neff, J. E. *et al.*: 1996, *Astron. Astrophys.* **310**, 173.
Noyes, R. W. and Hall, D. N. B.: 1972, *Astrophys. J.* **176**, L89.
Parker, E. N.: 1970, *Ann. Rev. Astron. Astrophys.* **8**, 1.
Radick, R. R., Lockwood, G. W., Skiff, B. A., and Baliunas, S. L.: 1998, *Astrophys. J. Suppl. Series* **118**, 239.
Reeves, E. H., Timothy, J. G., and Huber, M. C. E.: 1977, *Appl. Optics* **16**, 837.
Reimers, D.: 1977, *Astron. Astrophys.* **57**, 395.
Reimers, D., Hünsch, M., Schmitt, J. H. M. M., and Toussaint, F.: 1996, *Astron. Astrophys.* **310**, 813.
Rutten, R. J. and Uitenbroek, H.: 1991, *Solar Phys.* **134**, 15.
Schaller, G., Schaerer, D., Meynet, G., and Maeder, A.: 1992, *Astron. Astrophys. Suppl. Series* **96**, 269.
Shindell, D., Rind, D., Balachandran, N., Lean, J., and Lonergan, P.: 1999, *Science* **284**, 305.
Shine, R. A.: 1983, *Astrophys. J.* **266**, 882.
Simon, T. and Drake, S. A.: 1989, *Astrophys. J.* **346**, 303.
Simon, T., Drake, S. A., and Kim, P. D.: 1995, *Publ. Astron. Soc. Pacific* **107**, 1034.
Skumanich, A.: 1972, *Astrophys. J.* **171**, 565.
Solanki, S. K., Livingston, W., and Ayres, T.: 1994, *Science* **263**, 64.
Strassmeier, K. G. and Linsky, J. L. (eds): 1996, 'Stellar Surface Structure', *IAU Symp.* **176**.
Tomkin, J., Luck, R. E., and Lambert, D. L.: 1976, *Astrophys. J.* **210**, 694.
Trümper, J., Hasinger, G., Aschenbach, B., Braeuninger, H., and Briel, U. G.: 1991, *Nature* **349**, 579.
Uitenbroek, H.: 1998, *Bull. Am. Astron. Soc.* **30**, 840.
Vaiana, G. S. and Rosner, R.: 1978, ARA&A **16**, 393.
Vernazza, J. E., Avrett, E. H., and Loeser, R.: 1981, *Astrophys. J. Suppl. Series* **45**, 635.
Vogt, S. S. and Penrod, G. D.: 1983, *Publ. Astron. Soc. Pacific* **95**, 565.
Wilhelm, K. *et al.*: 1997, *Solar Phys.* **170**, 75.
Wilson, O. C. and Bappu, M. K. V.: 1957, *Astrophys. J.* **125**, 661.
Withbroe, G. L., Feldman, W. C., and Ahluwalia, H. S.: 1991, in A. N. Cox, W. C. Livingston, and M. S. Matthews (eds), *Solar Interior and Atmosphere,* University of Arizona Press, Tucson, p. 1087.
Wood, B. E., Linsky, J. L., and Ayres, T. R.: 1997, *Astrophys. J.* **474**, 39.
Woodgate, B. E. and Kimble, R. A.: 1999, in J. M. Shull, A. L. Kinney, and J. A. Morse (eds), *Ultraviolet-Optical Space Astronomy Beyond HST,* ASP Conf. Ser. 164, BookCrafters, San Francisco, p. 166.

NEW BOUNDARY INTEGRAL EQUATION REPRESENTATION FOR FINITE ENERGY FORCE-FREE MAGNETIC FIELDS IN OPEN SPACE ABOVE THE SUN

YIHUA YAN[1] and TAKASHI SAKURAI[2]
[1]*Beijing Astronomical Observatory, Chinese Academy of Sciences, Beijing 100012, China*
[2]*National Astronomical Observatory, Mitaka, Tokyo 181, Japan*

(Received 10 September 1999; accepted 28 February 2000)

Abstract. A boundary integral equation to describe a force-free magnetic field with finite energy content in the open space above the solar surface is found. This is a new representation for a 3-D nonlinear force-free field in terms of the boundary field and its normal gradient at the boundary. Therefore the magnetic field observed on the solar surface can be incorporated into the formulation directly and a standard numerical technique, the boundary element method, can be applied to solve the field. A numerical test case demonstrates the power of the method by recovering the analytical solution to the desired accuracy and its application to practical solar magnetic field problems is straightforward and promising.

1. Introduction

There have been a number of studies of solar magnetic field structures which start from a reasonable model and employ the observed data as boundary conditions. The magnetic field in the atmosphere is then recovered so as to understand the physical processes in the solar atmosphere. Both theory and observation have suggested the existence of nonlinear (non-constant-α) force-free fields above the solar surface (Schmahl *et al.*, 1982; Krall *et al.*, 1982; Aly, 1989; Sakurai, 1989; Wu *et al.*, 1990; Metcalf *et al.*, 1995). Due to the well-known difficulties: e.g., non-linearity of the equations, 180° ambiguity of observed transverse fields, and the topologic complexity of the coronal field, etc., more realistic modeling of the field is still an active research area. Presently, no standard method exists, although several proposed methods have been invoked (Sakurai, 1989; McClymont, Jiao, and Mikić, 1997).

The first practical reconstruction method employing observed boundary data was a finite element method based on the variational principle of the non-linear (non-constant-α) force-free field (Sakurai, 1981). The method works well for near-potential fields but has the limitation of slow convergence in highly non-potential situations, though why this was so is not clear (see, e.g., discussions in McClymont, Jiao, and Mikić, 1997). A direct discretization of the force-free equations by the finite difference techniques (Wu *et al.*, 1990) yields a well-known ill-posed feature despite its attractive simplicity in formulation: extrapolation from the photospheric

The physics of the solar corona and transition region. Part II.
Reprinted from Solar Physics **195,** 2000.

surface upwards contains both exponentially decaying and growing solutions, and the latter cannot be eliminated in the formulation (Sakurai, 1989; McClymont, Jiao, and Mikić, 1997). Others choose to quasi-physically evolve the MHD equations to a force-free state approximately by temporal evolution instead of a non-linear iteration procedure (Klimchuk and Sturrock, 1992; Mikić and McClymont, 1994; Roumeliotis, 1996). In this case, a large code and supercomputer resources are generally involved.

All the above-mentioned methods are based on the techniques of solving differential equations. On the other hand, if any physical problem can be expressed by an integral equation as well, the integral technique can be used to solve the problem instead. Applications indicate that integral techniques are advantageous in obtaining the approximate solution because the integral method is actually a summation technique and the errors in some elements in the summation (integral) may be compensated by errors in other elements. Furthermore, not all elements in the summation have equal influence on the correctness of the integral result and the correctness of the elements with major contributions in the integral is more important than that of the elements with minor contributions (see, e.g., discussions in Harrington, 1961, Section 3.11). In addition, integral techniques such as the boundary element method are particularly suitable for dealing with open field or far field problems (Brebbia, Telles, and Wrobel 1984; Yan, Yu, and Shi, 1993).

We have recently found a boundary integral equation representation, for the first time, for the non-constant-α force-free field in the exterior of the Sun (McClymont, Jiao, and Mikić, 1997; Yan and Sakurai, 1997). In this paper, we present a comprehensive study of this general representation for force-free field problems with finite energy content relevant above the solar surface. We first describe the model for the exterior boundary value problem and discuss some useful properties of the field (Section 2). Then in Section 3, we present a new boundary integral equation representation of the solution. The numerical implementation is described in Section 4 and an analytical solution (Low and Lou, 1990) has been used to show how the proposed method works. The formulation can be applied to a variety of solar magnetic field problems and the conclusions are summarized in Section 5.

2. Proposed Exterior Boundary Value Problem and Its Properties

2.1. Solar Force-Free Magnetic Field Model

Let Ω be the open space above the Sun and Γ the photosphere surface, i.e., $\Omega = \{(x, y, z)|\rho = |x\mathbf{i}+y\mathbf{j}+z\mathbf{k}| > R_o\}$, and $\Gamma = \{(x, y, z)|\ \rho = |x\mathbf{i}+y\mathbf{j}+z\mathbf{k}| = R_o\}$, where R_o is the solar radius. However, in the following derivations, there is no need for any *a priori* assumption on the shape of the boundary, thus the conclusions are generally valid. We also set $\bar{\Omega} = \Omega \cup \Gamma = \{(x, y, z)|\rho = |x\mathbf{i} + y\mathbf{j} + z\mathbf{k}| \geq R_o\}$. Throughout this paper we have tacitly assumed that the field, $\mathbf{B}$, satisfies sufficient continuity and/or differential conditions so as to apply Green's theorems (Courant, and Hilbert, 1962; Arfken, 1966).

The equations describing the magnetic field in the region $\bar{\Omega}$ can be obtained as follows.

Above the solar surface Maxwell's equation

$$\nabla \cdot \mathbf{B} = 0 \quad \text{in} \quad \Omega \tag{1}$$

is subject to

$$\nabla \times \mathbf{B} = \alpha \mathbf{B} \quad \text{in} \quad \Omega \tag{2}$$

for a force-free magnetic field, where α is a function of spatial location. By taking the divergence of (2) and substituting in (1), one has

$$\nabla \alpha \cdot \mathbf{B} = 0\,. \tag{3}$$

Thus α is constant along each field line and may be determined from the boundary condition by assuming that the α distributions there are known. We further assume that both α and $\nabla \alpha$ are bounded.

The boundary condition is

$$\mathbf{B} = \mathbf{B}_o \quad \text{on} \quad \Gamma\,. \tag{4}$$

$\mathbf{B}_o$ here denotes known boundary values which can be supplied from vector magnetograph measurements. Since there are three independent variables involved in the general force-free field equations (1) and (2), we use the vector field as boundary conditions and assume that the boundary values are prescribed consistently with the force-free field in space.

Physically the magnetic field should tend to zero identically at infinity. Therefore we present an asymptotic condition which ensures a finite energy content in open space above the Sun,

$$\mathbf{B} = O(\rho^{-2}), \quad \text{when} \quad \rho \to \infty\,. \tag{5}$$

Thus the model proposed here for the non-linear (non-constant-α) force-free magnetic field is an exterior boundary value problem including the field equations (1) and (2), the Dirichlet condition (4), and the asymptotic condition (5), as shown in Figure 1. Obviously this is a mathematic model describing the ideal case of a magneto-hydrostatic equilibrium in the solar corona (see, e.g., discussions in Sakurai (1989) and Low and Lou (1990), etc.)

2.2. Some Properties of the Solution

Up to now there is, unfortunately, no strict mathematical proof of the existence of a general solution of the finite energy non-linear force-free field in open space above the Sun. But observations have uncovered evidence for the existence of non-linear

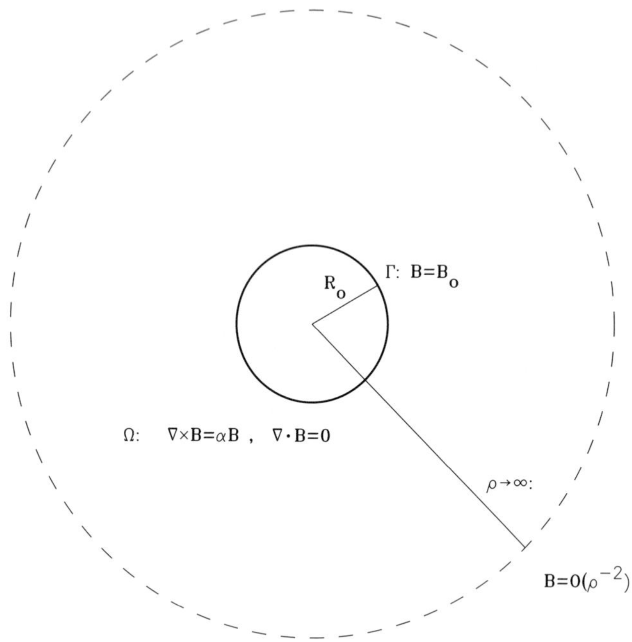

Figure 1. The exterior boundary value problem of the general force-free magnetic field problem.

force-free fields (Schmahl *et al.*, 1982; Krall *et al.*, 1982). Furthermore, several attempts have been made to extrapolate the 3-D nonlinear force-free field using boundary data (Sakurai, 1981; Wu *et al.*, 1990; Klimchuk and Sturrock, 1992; Mikić and McClymont, 1994; Roumeliotis, 1996). Aly (1989) has also investigated the theoretical aspects of the construction of the nonlinear force-free magnetic field from boundary data. In all these cases the physically-existing nonlinear force-free field must have a finite energy content in outer space above the Sun.

According to the knowledge of vector analysis, any vector field can be uniquely specified if its divergence and curl are known and its normal component is given over the boundary (Arfken, 1966). For the force-free magnetic field, **B**, as considered throughout this paper, its divergence is zero as given by Equation (1), but Equation (2) only specifies the components of the curl that are perpendicular to the direction of **B**. The other component of the curl that is in parallel with **B** can be expressed as $\nabla \times \mathbf{B} \cdot \mathbf{B} = \alpha B^2$ from Equation (2). Therefore, the force-free field, **B**, can be uniquely determined only if we know the distribution of α in space (Hu, 1987). Here we have a non-linear problem and the α function is dependent on the field lines; α may be different according to the different field distributions. For example, if all the field lines in the space are anchored at the boundary, the α distribution may be consistently determined from the boundary condition. If, however, there are some closed field lines in the space, the α distribution along those closed lines could not be determined from the boundary condition and the field there could not be solved from the governing equation plus the boundary condition.

This latter case causes a serious problem on the uniqueness of the solution (Aly, 1987), and is, fortunately, excluded in the modeling of solar coronal magnetic field problems. Presently, the reconstruction of the solar coronal magnetic fields with the observed photospheric magnetograms as boundary conditions implies that the magnetic field lines are rooted on the boundary (Sakurai, 1981; Wu *et al.*, 1990; Klimchuk and Sturrock, 1992; Mikić and McClymont, 1994; Roumeliotis, 1996). Therefore in this paper we consider that α is a function consistently determined from the boundary condition only and it should be unaltered wherever the same boundary condition is applied. In an ideal case, the magnetic field thus established occupies the outer space above the boundary area and decays to zero by the asymptotic condition (5). Either in the outer space or at the boundary, the vector magnetic field determines the force-free factor α consistently because there are three independent variables involved in the general force-free field Equations (1) and (2).

Before we derive the boundary integral equation representation for the above field, or the proposed exterior boundary value problem, (1), (2), (4), (5), we would like to note that the field has a finite energy content in the open space above the solar surface, and this property can be easily obtained. Therefore, if the asymptotic condition (5) is imposed, we have

$$W = \int_{\Omega} \frac{1}{2\mu} B^2 \, d\Omega < \infty \,, \tag{6}$$

where μ is the magnetic permeability.

The computation of the magnetic energy from boundary data under different coordinates can be found in many papers (Low, 1982; Sakurai, 1989, etc).

In the next section, we derive a boundary integral equation representation of the solution of the exterior boundary value problem, (1), (2), (4), (5).

3. Boundary Integral Equation Representation of the Field

3.1. Boundary Integral Equation of the Solution

Let us choose a sufficiently large radius, R, forming a sphere,

$$S_R = \{(x, y, z) | \, |(x - x_i)\mathbf{i} + (y - y_i)\mathbf{j} + (z - z_i)\mathbf{k}| = R\}$$

to bound a finite volume, Ω'. The volume, Ω', is then bounded internally by Γ and externally by the sphere, S_R. For the sake of clarity, in the following derivations we use B to denote each component of $\mathbf{B}$.

In order to apply Green's theorem, e.g., Equation (9) below, two functions must be involved (Courant and Hilbert, 1962; Arfken, 1966). Therefore we consider also a reference function, Y, chosen as

$$Y = \frac{\cos(\lambda r)}{4\pi r}\,, \tag{7}$$

where $r = |\rho - \rho_i|$ with $\rho = x\mathbf{i} + y\mathbf{j} + z\mathbf{k}$, $\rho_i = x_i\mathbf{i} + y_i\mathbf{j} + z_i\mathbf{k}$ is the distance between a variable point and a fixed point i, both in Ω, and λ is a parameter dependent on the location of point i only, and to be found by the arguments given later. The reference function satisfies the equation

$$\nabla Y + \lambda^2 Y = \delta_i\,, \tag{8}$$

where δ_i is the Dirac function defined at the point, i, in Ω. Note that the point, i, can be at any location in Ω and on Γ as will be explicit later when the numerical procedure is implemented in Section 4.1. In general, the solution of Equation (8) is not unique. Since the force-free field is not a wave problem and a complex representation is inadequate, we are to seek the real results that have finite energy content in open space above the Sun. Therefore we have chosen Y in its present form as defined by Equation (7) in our analysis because it can ensure that the magnetic field decays to zero by the asymptotic condition (5). In the case that $\lambda^2 < 0$ occurs under certain situations, we will choose $Y = \exp(-|\lambda|r)/4\pi r$ (Arfken, 1966), which still satisfies (8) and guarantees that the above property and the following derivations are valid.

As discussed in Section 2.2, for the reconstruction of a solar coronal force-free field one in general assumes that all field lines are rooted with either one foot or both feet on the solar surface and the α distribution can thus be determined accordingly.

The function, Y, however, fails to satisfy the necessary condition of continuity at the point i, where $r = 0$. To exclude the singularity, a small sphere, $S_\epsilon = \{(x, y, z)||(x - x_i)\mathbf{i} + (y - y_i)\mathbf{j} + (z - z_i)\mathbf{k}| = \epsilon\}$, of radius ϵ, is circumscribed about the point i as a center. The volume is then bounded internally by Γ and externally by S_R excluding the small volume, V_ϵ, bounded by S_ϵ, as shown in Figure 2. Within $\Omega'' = \Omega' - V_\epsilon$ both B and Y now satisfy the requirements of Green's second identity and furthermore $\nabla^2 Y + \lambda^2 Y = 0$. Thus we have

$$\int_{\Omega''} (Y\nabla B - B\nabla^2 Y)\,\mathrm{d}\Omega = \oint_{\Gamma + S_R + S_\epsilon} \left(Y\frac{\partial B}{\partial n} - B\frac{\partial Y}{\partial n} \right) \mathrm{d}\Gamma\,. \tag{9}$$

A similar treatment of the integrals in (9) to that in classical exterior boundary value problems (Stratton, 1941; Courant and Hilbert, 1962) gives the following results.

Over the small sphere, S_ϵ, we have

$$\frac{\partial B}{\partial n} = -\frac{\partial B}{\partial r}, \qquad \frac{\partial Y}{\partial n} = -\frac{\mathrm{d}Y}{\mathrm{d}r} = \frac{\lambda\epsilon \sin(\lambda\epsilon) + \cos(\lambda\epsilon)}{4\pi c^2}\,. \tag{10}$$

Since ϵ is constant, the surface integral over the sphere, S_ϵ, is

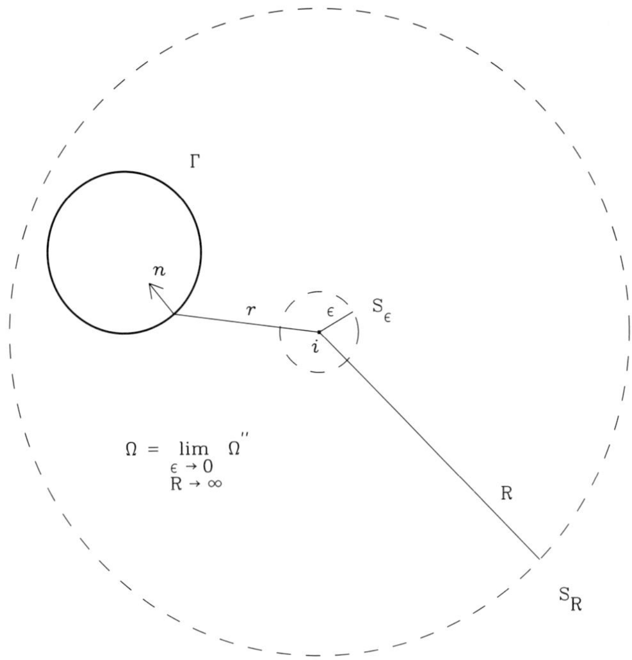

Figure 2. The geometry for application of Green's second identity in the exterior of the Sun.

$$\oint_{S_\epsilon} \left(Y \frac{\partial B}{\partial n} - B \frac{\partial Y}{\partial n} \right) \mathrm{d}\Gamma =$$

$$\begin{aligned} &= -\frac{\cos(\lambda\epsilon)}{4\pi\epsilon} \oint_{S_\epsilon} \frac{\partial B}{\partial r} \mathrm{d}\Gamma - \frac{\lambda\epsilon \sin(\lambda\epsilon) + \cos(\lambda\epsilon)}{4\pi\epsilon^2} \oint_{S_\epsilon} B \, \mathrm{d}\Gamma \\ &= -\frac{\cos(\lambda\epsilon)}{4\pi\epsilon} 4\pi\epsilon^2 \overline{\frac{\partial B}{\partial r}} - \frac{\lambda\epsilon \sin(\lambda\epsilon) + \cos(\lambda\epsilon)}{4\pi\epsilon^2} 4\pi\epsilon^2 \overline{B} \,, \end{aligned} \tag{11}$$

where $\overline{B}$ and $\overline{\partial B/\partial r}$ are mean values of B and $\partial B/\partial r$ on S_ϵ, respectively. Therefore we finally obtain the contribution of the small sphere, in the limit $\epsilon \to 0$, as follows:

$$\lim_{\epsilon \to 0} \oint_{S_\epsilon} \left(Y \frac{\partial B}{\partial n} - B \frac{\partial Y}{\partial n} \right) \mathrm{d}\Gamma = -B_i \,, \tag{12}$$

where B_i is the value of B at point i.

Whereas over the sufficiently large sphere, S_R, after some calculation, we obtain

$$\frac{\partial Y}{\partial n} = \frac{\mathrm{d}Y}{\mathrm{d}r} = -\frac{\lambda R \sin(\lambda R) + \cos(\lambda R)}{4\pi R^2} \,. \tag{13}$$

Since R is constant, the contribution of the sphere, denoted by U, is

$$
\begin{aligned}
U &= \oint_{S_R} \left(Y \frac{\partial B}{\partial n} - B \frac{\partial Y}{\partial n} \right) \, \mathrm{d}\Gamma = \\
&= \oint_{S_R} \left[\frac{\cos(\lambda R)}{4\pi R} \frac{\partial B}{\partial n} + \frac{\lambda R \sin(\lambda R) + \cos(\lambda R)}{4\pi R^2} B \right] \mathrm{d}\Gamma \\
&= \frac{\cos(\lambda R)}{4\pi R^2} \oint_{S_R} \left(R \frac{\partial B}{\partial n} + B \right) \mathrm{d}\Gamma + \frac{\lambda \sin(\lambda R)}{4\pi R^2} \oint_{S_R} RB \, \mathrm{d}\Gamma \, .
\end{aligned} \tag{14}
$$

For U to vanish, it is sufficient that the above two integrals vanish. Therefore we have

$$
R \frac{\partial B}{\partial n} + B = O(R^{-1}) \quad \text{and} \quad RB = O(R^{-1}) \, , \tag{15}
$$

which in the limit $R \to \infty$ is identical with the asymptotic condition (5) since the first condition in (15) can be deduced from the second one.

The volume integral of (9) is reduced through substituting (1), (2), and (5) (where we only consider the case for $B = B_x$ but it also applies for other components of $\mathbf{B}$),

$$
\begin{aligned}
&\int_{\Omega''} (Y \nabla^2 B_x - B_x \nabla^2 Y) \, \mathrm{d}\Omega = \\
&= \int_{\Omega''} \{ Y[(\nabla(\nabla \cdot \mathbf{B}) - \nabla \times (\alpha \mathbf{B}))_x + \lambda^2 B_x] - B_x (\nabla^2 Y + \lambda^2 Y) \} \, \mathrm{d}\Omega \\
&= \int_{\Omega''} Y[\lambda^2 B_x - \alpha^2 B_x - (\nabla \alpha \times \mathbf{B})_x] \, \mathrm{d}\Omega = \Psi_i(\lambda) < \infty ,
\end{aligned} \tag{16}
$$

where we have denoted the volume integral as $\Psi_i(\lambda)$, which is finite due to (5), and bounded α, $\nabla \alpha$ functions when $R \to \infty$, and $(\cdots)_x$ means the x-component of the vector. It should be noted that if the field has a complex topology, α may have discontinuities across the separatrix surfaces and $\nabla \alpha$ has a delta-function on them. In that case the above integral may not be bounded. If, however, the volume containing the separatrix surfaces does not extend to infinity, we could still establish a corresponding integral equation by excluding this volume but adding a volume integral term, which is finite due to the delta-function property, in the equation. In that case we could still apply Green's identity to the corresponding regions and boundary surfaces but the expressions of the governing equation and procedures will be different from the present research. Therefore we do not consider the case in the present study. Nevertheless the present method can be applied to analyze practical problems containing highly non-potential magnetic topology, like quasi-separatrix layers etc., as demonstrated later and in other applications.

Upon introducing the above results, (12), (14), (15), (16), and $\Omega'' \to \Omega$ when $\epsilon \to 0$ and $R \to \infty$ into Equation (9) the field value at any point i is

$$B_i + \Psi_i(\lambda) = \int_\Gamma \left(Y \frac{\partial B}{\partial n} - B \frac{\partial Y}{\partial n} \right) \mathrm{d}\Gamma \,. \tag{17}$$

In order that the field vanishes at infinity, we must have $\lim_{|\rho_i| \to \infty} \Psi_i(\lambda) = 0$. Though it does not imply that $\Psi_i(\lambda) = 0$ in the near field, we may still require that it holds at any point in Ω as follows:

$$\Psi_i(\lambda) = \int_\Omega Y[\lambda^2 B - \alpha^2 B - (\nabla\alpha \times \mathbf{B})_p] \,\mathrm{d}\Omega = 0, \quad p = x, y, \text{ or, } z \,. \tag{18}$$

Thus, λ can be determined in principle and it is a function of the location of the point i. For each component of $\mathbf{B}$ we can get a corresponding reference function Y with the parameters λ_x, λ_y and λ_z defined by (18). If we denote a diagonal matrix $\mathbf{Y}$ as

$$\mathbf{Y} = \mathrm{diag}\left\{Y_x, Y_y, Y_z\right\} = \mathrm{diag}\left\{\frac{\cos(\lambda_x r)}{4\pi r}, \frac{\cos(\lambda_y r)}{4\pi r}, \frac{\cos(\lambda_z r)}{4\pi r}\right\}, \tag{19}$$

we can finally obtain a boundary integral equation representation for the solution of the non-linear force-free field problem ((1), (2), (4), (5)), with finite energy content in open space as follows:

$$\mathbf{B}_i = \int_\Gamma \left(\mathbf{Y} \frac{\partial \mathbf{B}}{\partial n} - \frac{\partial \mathbf{Y}}{\partial n} \mathbf{B} \right) \mathrm{d}\Gamma \,. \tag{20}$$

This equation indicates that if the values of $\mathbf{B}$ and $\partial\mathbf{B}/\partial n$ over Γ are known, the field value $\mathbf{B}$ at any point in open space, Ω, is determined by the integration of products of the proposed reference function, $\mathbf{Y}$, and values of $\mathbf{B}$ and $\partial\mathbf{B}/\partial n$ over Γ. Therefore, together with the criterion (18) from which the anisotropic parameter λ in $\mathbf{Y}$ should be found, it gives an equivalent boundary integral equation representation of the solution of the general force-free field ((1), (2)) in open space, Ω, satisfying the specified conditions over the boundary, Γ, and at infinity (Stratton, 1941; Courant and Hilbert, 1962). Both $\mathbf{B}$ and $\partial\mathbf{B}/\partial n$ given on the boundary, Γ, will over-determine the exterior boundary value problem and we only need $\mathbf{B}$ known on Γ. Next we will determine the unknown boundary parameters of (20).

3.2. Singularity Treatment of Boundary Integral Equation

Let us see what consistent relationship on the boundary the field and its normal differential should satisfy. We proceed as above but now take the point i as fixed at the smooth surface Γ. Thus only a semi-sphere is needed to exclude the singularity of (7) and let Γ' denote the boundary excluding a small neighborhood of the singularity. As in Section 3.1, similar arguments can be carried out. However, in this case, the integral of (12) in the limit $\epsilon \to 0$ reduces to $-\mathbf{B}_i/2$.

Upon introducing this value, and $\Gamma' \to \Gamma$, when $\epsilon \to 0$, a boundary integral equation is obtained:

$$\frac{1}{2}\mathbf{B}_i = \int_\Gamma \left(\mathbf{Y}\frac{\partial \mathbf{B}}{\partial n} - \frac{\partial \mathbf{Y}}{\partial n}\mathbf{B} \right) \mathrm{d}\Gamma \,. \tag{21}$$

The integrals should be taken over the surface Γ excluding an ϵ neighborhood of the singularity, i.e., in a sense of Cauchy principal value integration (Courant and Hilbert, 1962). Note here we are referring to the converged singular integration problem, not a Cauchy boundary value problem which is ill-posed (Sakurai, 1989). In general, the boundary integral equation is expressed as follows:

$$c_i\mathbf{B}_i = \int_\Gamma \left(\mathbf{Y}\frac{\partial \mathbf{B}}{\partial n} - \frac{\partial \mathbf{Y}}{\partial n}\mathbf{B} \right) \mathrm{d}\Gamma \,, \tag{22}$$

where c_i is a constant depending upon the location of the point i. For example, $c_i = 1$ if point i is in volume Ω, and $c_i = \frac{1}{2}$ if point i is on smooth surface Γ.

Thus, a consistent relation of the boundary values and their normal differentials is obtained. This Equation (22) can be solved numerically with the help of a well-established standard procedure: the boundary element method (Brebbia, Telles, and Wrobel, 1984; Yan, Yu, and Shi, 1993), which is described in the next section.

4. Numerical Implementation

4.1. Boundary Element Approximation

Although the integral representation (22) is established in a spherical geometry, there is not any *a priori* assumption on the shape of the boundary in the above derivations. Therefore it can also be applied to a half-space, e.g., $z > 0$, with a part of $z = 0$ plane being the boundary. In this case, only a portion of the big sphere S_R and $z = 0$ plane are needed to bound the volume and a similar derivation can be carried out. Finally, we could obtain the same expression (22) with the boundary integration taken over a part of the $z = 0$ plane when the asymptotic condition (5) is applied.

In Cartesian coordinates let $\Omega = \{(x, y, z)|z > 0\}$ be the open space and $z = 0$ correspond to the solar surface. A magnetogram provides the boundary condition of $\mathbf{B}$ in a rectangle region $\Gamma = \{(x, y, z)|0 \le x \le L_x, 0 \le y \le L_y, z = 0\}$, and we assume $\mathbf{B} \equiv 0$ when $(x, y, 0) \notin \Gamma$.

For the boundary element method to solve Equation (22) numerically, we first subdivide Γ into a number of elements and the approximation functions are assumed over each element. In our case the element dimensions should be similar to the observational resolution. Probably the most common way is to use iso-parametric elements (usually bi-quadratic shape functions) to represent both the geometry and functions on each element (Brebbia, Telles, and Wrobel, 1984). A local intrinsic coordinate system (ξ, η) is defined for each element, as shown in

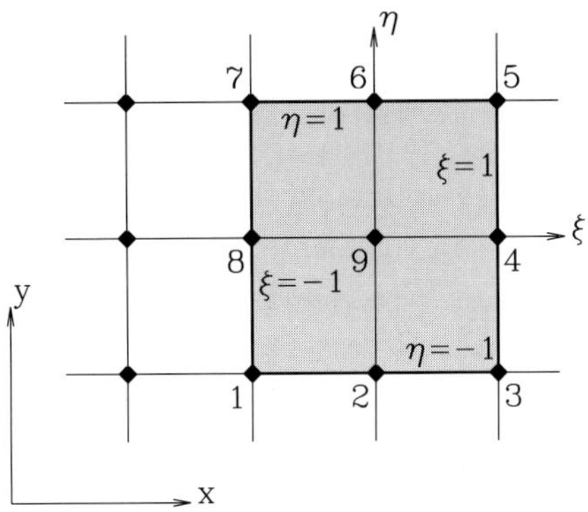

Figure 3. Description of a 9-point bi-quadratic element on the boundary with tangential directions (ξ, η) forming local intrinsic coordinates in the unit area $\{(\xi, \eta)| - 1 \leq \xi \leq 1, -1 \leq \eta \leq 1\}$. The local nodes are numbered as $k = 1, 2, \ldots, 9$ as indicated by '•'.

Figure 3, where the normal direction of the boundary outward from the volume concerned is defined by the right-hand convention. Both the boundary location, and **B** and $\partial\mathbf{B}/\partial n$ functions over each element in the intrinsic system (ξ, η) are represented by the sums of products of nodal values and shape functions. Thus

$$\begin{aligned} x(\xi, \eta) &= \sum_{k=1}^{9} N_k(\xi, \eta) x_k^e, \\ y(\xi, \eta) &= \sum_{k=1}^{9} N_k(\xi, \eta) y_k^e, \\ z(\xi, \eta) &= \sum_{k=1}^{9} N_k(\xi, \eta) z_k^e, \end{aligned} \tag{23}$$

and

$$\mathbf{B}(\xi, \eta) = \sum_{k=1}^{9} N_k(\xi, \eta) \mathbf{B}_k^e, \qquad \frac{\partial \mathbf{B}}{\partial n} = \sum_{k=1}^{9} N_k(\xi, \eta) \left(\frac{\partial \mathbf{B}}{\partial n}\right)_k^e, \tag{24}$$

where x_k^e, y_k^e, and z_k^e, are nodal global coordinates, $\mathbf{B}_k^e$ and $(\partial\mathbf{B}/\partial n)_k^e$ are nodal values of **B** and $\partial\mathbf{B}/\partial n$ respectively, superscript e denotes the element, and $N_k(\xi, \eta)$ are the bi-quadratic shape functions defined as follows:

$$
\begin{aligned}
N_k(\xi,\eta) &= \frac{1}{4}\xi\eta(\xi+\xi_k)(\eta+\eta_k), \quad k=1,3,5,7,\\
N_k(\xi,\eta) &= \frac{1}{2}\eta(1-\xi^2)(\eta+\eta_k), \quad k=2,6,\\
N_k(\xi,\eta) &= \frac{1}{2}\xi(\xi+\xi_k)(1-\eta^2), \quad k=4,8,\\
N_k(\xi,\eta) &= (1-\xi^2)(1-\eta^2), \qquad k=9,
\end{aligned}
\tag{25}
$$

in which (ξ_k, η_k) are nodal intrinsic coordinates as shown in Figure 3. Introducing these representations into Equation (22), and transforming from the surface to the intrinsic coordinates (ξ, η), we obtain the following equation for a boundary node, i:

$$
\begin{aligned}
&c_i \begin{Bmatrix} B_{xi} \\ B_{yi} \\ B_{zi} \end{Bmatrix} + \sum_{e=1}^{M}\sum_{k=1}^{9}\left[\int_{-1}^{+1}\int_{-1}^{+1}\begin{bmatrix} \frac{\partial Y_x}{\partial n} & 0 & 0 \\ 0 & \frac{\partial Y_y}{\partial n} & 0 \\ 0 & 0 & \frac{\partial Y_z}{\partial n} \end{bmatrix} N_k J \,\mathrm{d}\xi\,\mathrm{d}\eta\right]\begin{Bmatrix} B_{xk}^{e} \\ B_{yk}^{e} \\ B_{zk}^{e} \end{Bmatrix} = \\
&= \sum_{e=1}^{M}\sum_{k=1}^{9}\left[\int_{-1}^{+1}\int_{-1}^{+1}\begin{bmatrix} Y_x & 0 & 0 \\ 0 & Y_y & 0 \\ 0 & 0 & Y_z \end{bmatrix} N_k J \,\mathrm{d}\xi\,\mathrm{d}\eta\right]\begin{Bmatrix} \left(\frac{\partial B_x}{\partial n}\right)_k^e \\ \left(\frac{\partial B_y}{\partial n}\right)_k^e \\ \left(\frac{\partial B_z}{\partial n}\right)_k^e \end{Bmatrix},
\end{aligned}
\tag{26}
$$

where M is the number of total boundary elements, $N_k = N_k(\xi, \eta)$, and $J = J(\xi, \eta)$ denotes the Jacobian. In the above Equation (26), the reference functions Y_x, Y_y and Y_z contain the parameters λ_x, λ_y and λ_z corresponding to the present point, i, which need to be determined. We have derived in Section 3 that they satisfy Equation (18) which needs information in the whole space. Since it is a complicated nonlinear equation it is not obvious that (18) has a solution λ_i at each point i in Ω. But in the following we propose an iterative procedure so that λ_i can always be defined during the iteration and the goal is to find an approximation solution of the field.

Now, we at first assume an initial value of the parameters λ_x, λ_y and λ_z so that the integrand in Equation (26) can be calculated for the first approximation of the field, **B**.

Constructing Equation (26) for each boundary node by changing the location of the point i over the boundary with the corresponding parameters λ_x, λ_y and

λ_z initially assumed, we generate a set of simultaneous equations. This set of equations can be written in a matrix form, with the result of each integration in (26) forming a coefficient of the following matrices:

$$\begin{bmatrix} \mathbf{H_x} & \mathbf{0} & \mathbf{0} \\ \mathbf{0} & \mathbf{H_y} & \mathbf{0} \\ \mathbf{0} & \mathbf{0} & \mathbf{H_z} \end{bmatrix} \begin{Bmatrix} \hat{\mathbf{B}}_\mathbf{x} \\ \hat{\mathbf{B}}_\mathbf{y} \\ \hat{\mathbf{B}}_\mathbf{z} \end{Bmatrix} = \begin{bmatrix} \mathbf{G_x} & \mathbf{0} & \mathbf{0} \\ \mathbf{0} & \mathbf{G_y} & \mathbf{0} \\ \mathbf{0} & \mathbf{0} & \mathbf{G_z} \end{bmatrix} \begin{Bmatrix} \hat{\mathbf{Q}}_\mathbf{x} \\ \hat{\mathbf{Q}}_\mathbf{y} \\ \hat{\mathbf{Q}}_\mathbf{z} \end{Bmatrix}, \tag{27}$$

where we have used Q to denote $\partial B/\partial n$, and $\{\hat{\mathbf{B}}\} = \{\hat{\mathbf{B}}_\mathbf{x}, \hat{\mathbf{B}}_\mathbf{y}, \hat{\mathbf{B}}_\mathbf{z}\}^T$ and $\{\hat{\mathbf{Q}}\} = \{\hat{\mathbf{Q}}_\mathbf{x}, \hat{\mathbf{Q}}_\mathbf{y}, \hat{\mathbf{Q}}_\mathbf{z}\}^T$ are vectors of nodal values. In the present case as the boundary is plane, e.g., $z = 0$ surface, Equation (27) actually reduces to

$$\frac{1}{2}\begin{bmatrix} \mathbf{I} & \mathbf{0} & \mathbf{0} \\ \mathbf{0} & \mathbf{I} & \mathbf{0} \\ \mathbf{0} & \mathbf{0} & \mathbf{I} \end{bmatrix} \begin{Bmatrix} \hat{\mathbf{B}}_\mathbf{x} \\ \hat{\mathbf{B}}_\mathbf{y} \\ \hat{\mathbf{B}}_\mathbf{z} \end{Bmatrix} = \begin{bmatrix} \mathbf{G_x} & \mathbf{0} & \mathbf{0} \\ \mathbf{0} & \mathbf{G_y} & \mathbf{0} \\ \mathbf{0} & \mathbf{0} & \mathbf{G_z} \end{bmatrix} \begin{Bmatrix} \hat{\mathbf{Q}}_\mathbf{x} \\ \hat{\mathbf{Q}}_\mathbf{y} \\ \hat{\mathbf{Q}}_\mathbf{z} \end{Bmatrix}, \tag{28}$$

where $\mathbf{I}$ is the identity matrix, and the Jacobian $J(\xi, \eta)$ in (26) reduces to merely a constant: one quarter of the element area (Yan, Yu, and Shi, 1993). At each node $\mathbf{B}$ is known (as given by the boundary condition). Solution of Equation (28) permits unknown values to be determined properly.

Therefore we have obtained a first approximation for the boundary solution, and the field in space can be calculated from (22), or its discrete form (26), with $c_i = 1$ as a first approximation as well. Then we may modify the parameters λ_x, λ_y and λ_z in terms of the three components of the field $\mathbf{B}$ and the initial value of λ's as follows:

$$\begin{aligned} \lambda_x^2 &= \int_\Omega Y\left(\alpha^2 B_x + \frac{B_x B_y}{B_z}\frac{\partial\alpha}{\partial x} + \frac{B_y^2 + B_z^2}{B_z}\frac{\partial\alpha}{\partial y}\right) \mathrm{d}\Omega \Big/ \int_\Omega Y B_x \,\mathrm{d}\Omega\,, \\ \lambda_y^2 &= \int_\Omega Y\left(\alpha^2 B_y - \frac{B_x^2 + B_z^2}{B_z}\frac{\partial\alpha}{\partial x} - \frac{B_x B_y}{B_z}\frac{\partial\alpha}{\partial y}\right) \mathrm{d}\Omega \Big/ \int_\Omega Y B_y \,\mathrm{d}\Omega\,, \\ \lambda_z^2 &= \int_\Omega Y\left(\alpha^2 B_z + B_y\frac{\partial\alpha}{\partial x} - B_x\frac{\partial\alpha}{\partial y}\right) \mathrm{d}\Omega \Big/ \int_\Omega Y B_z \,\mathrm{d}\Omega, \end{aligned} \tag{29}$$

where α is obtained from Equations (2) and (4),

$$\alpha = \left(\frac{\partial B_y}{\partial x} - \frac{\partial B_x}{\partial y}\right) \Big/ B_z\,, \tag{30}$$

and we have substituted (3) into (29) to eliminate $\partial\alpha/\partial z$ so that the divergence free condition (1) holds.

With the updated parameters λ_x, λ_y and λ_z from (29), the coefficient matrices in (28) can be re-generated and the solution of (28) gives a new result of nodal unknowns' vector $\{\hat{\mathbf{Q}}\}$. Therefore an iterative procedure is readily available and the approximate solution $\{\hat{\mathbf{Q}}\}$ of boundary unknowns is found if the difference between two consecutive solutions is small.

Note that the volume integrals in (29) do not increase the scale (or dimension) of the matrix Equation (28) but contribute additional computations in the iterative procedure. These volume integrals can be evaluated by the finite element technique (Sakurai, 1981). In this case, we may choose a sufficiently large box with a 3-d grid above the problem region. Only the quantities at the grid points are calculated and the approximation functions can be assumed over each sub-volume, or finite element. Therefore, the volume integrals can be obtained by summing up all contributions from each sub-volume.

4.2. Computational Procedure

The procedure of the boundary element method (BEM) for boundary solutions can be summarized as follows.

(1) Set $k = 0$. Subdivide the problem region (on the photosphere, $z = 0$) into elements, and prescribe function values at nodes from the boundary condition at the photosphere. For each node, the λ_x, λ_y and λ_z values are assigned with an initially-assumed value, e.g., the averaged value of α distributions over boundary, or, $\lambda^{(k)} = \bar{\alpha}$.

(2) Fix each node in turn as a field point to calculate the contributions of integrals in Equation (26), element by element, by Gaussian quadrature rule. Then assemble the global matrices $\mathbf{G_x}$, $\mathbf{G_y}$, $\mathbf{G_z}$ from the contributions of all elements to establish the system Equation (28) of the BEM.

(3) Substitute the boundary condition into Equation (28) and solve the system by Gaussian elimination or Crout decomposition method to find nodal unknowns on the photosphere, $\{\hat{\mathbf{Q}}^{(k)}\}$.

(4) For a sufficiently large box with a sparsely-distributed 3-d grid (not necessarily compatible with the observational resolutions on the boundary) above the problem region, calculate the quantities in the volume integral of (29) according to (22) while the nodal values of both the field and the approximation of its normal gradient over the boundary, $\{\hat{\mathbf{Q}}^{(k)}\}$, are now known. Therefore the updated value, $\lambda^{(k+1)}$ for each boundary node is obtained from (29).

(5) If $k = 0$, set $k = k + 1$ and go to step (2). If $k > 0$, evaluate whether $||\hat{\mathbf{Q}}^{(k)} - \hat{\mathbf{Q}}^{(k-1)}||$ is small. If it is the case, choose $\hat{\mathbf{Q}}^{(k)}$ as the numerical solution of the nodal values over the boundary; otherwise set $k = k + 1$ and go to step (2).

Based on the above numerical approximation of the boundary integral Equation (22), another iterative procedure for calculating the nonlinear force-free field at any point in open space above the Sun can be written as follows when the nodal values of both the field and its normal gradient over the boundary are known:

Step 1. *Set $k = 0$, and choose initial values of λ_x, λ_y and λ_z for calculating the field, $\mathbf{B}^{(k)}$, at the present location by Equation (22) or its discrete form (26) with $c_i = 1$.*

Step 2. *Modify the λ_x, λ_y and λ_z values from Equation (29) and re-calculate the field by (22) or its discrete form (26) with $c_i = 1$ to obtain a new field value $\mathbf{B}^{(k+1)}$.*

Step 3. *If $||\mathbf{B}^{(k+1)} - \mathbf{B}^{(k)}|| \leq \epsilon$, stop the iterative procedure for the current point, where ϵ is a small positive number. Otherwise set $k = k + 1$ and repeat* **Step 2.**

4.3. A NUMERICAL TEST CASE

A class of nonlinear force-free magnetic fields in closed form has been proposed (Low and Lou, 1990) and these fields satisfy the proposed boundary value problem ((1), (2), (4), (5)). Therefore we have employed one of the solutions in our numerical test. The magnetic field then takes the form

$$\mathbf{B} = \frac{1}{r^3}\left[-\frac{dP}{d\mu}\hat{\mathbf{r}} + \frac{P}{(1-\mu^2)^{1/2}}\hat{} + \frac{aP^2}{(1-\mu^2)^{1/2}}\hat{\phi}\right] \tag{31}$$

in spherical coordinates, where $\mu = \cos(\theta)$, and $P = P_{1,1}(\mu)$ is the eigenfunction corresponding to the eigenvalue $a^2 = a_{1,1}^2 = 0.425$ (Low and Lou, 1990). We use the same convention as in Low and Lou (1990), i.e., the point source in spherical coordinates is located at $(0, 0, -l)$ in Cartesian coordinates (x, y, z), and the angle with the z-axis in z-x plane is Φ. Those parameters are fixed as $l = 0.3$ and $\Phi = \pi/4$. Therefore the $z = 0$ plane corresponds to the boundary, or solar surface, and the half-space $z > 0$ is the problem domain representing the solar atmosphere.

Following the convention in Low and Lou (1990), we also only consider the situation in $\{(x, y)| -1 \leq x \leq 1, \ -1 \leq y \leq 1\}$, which is the problem region corresponding to an active region on the Sun. In the numerical computation, we have used a mesh of 31×31 nodes covering a computing boundary region which is 4 times the problem region, or only a 15×15 mesh covering the problem region. However, we have used 500 discrete points for $\mu \in [-1, 1]$ in the interval for the computation of the eigenfunction $P_{1,1}$ to assure that the obtained values of the nodal magnetic field are accurate enough. Figure 4 shows the eigenfunction obtained by a 4th-order Runge–Kutta method.

We have employed two approaches in the numerical test. The first one is to follow the procedure as described in Section 4.2. Obviously this will take more computer time but gives higher accuracy. In the nonlinear iteration for any field point, if the result of the right-hand side of Equation (29) is negative, we choose the form of the reference function as $Y = \exp(-|\lambda|r)/4\pi r$ (Arfken, 1966), which still guarantees that all the properties and the derivations concerned are valid. The 3-D grid adopted for calculating (29) was $15 \times 15 \times 5$, covering a volume of size $\{(x, y, z)| -2 \leq x, y \leq 2, 0 < z \leq 1.2\}$. The second one is to arbitrarily set the

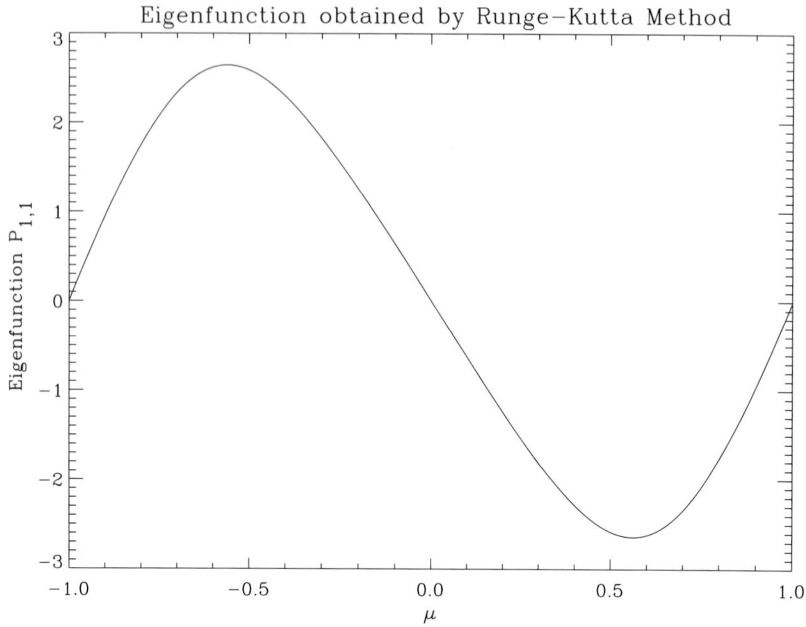

Figure 4. The eigenfunction $P_{1,1}$ with $a_{1,1}^2 = 0.425$ obtained by a 4th-order Runge–Kutta method with 500 discrete points in $[-1, 1]$ interval.

value of λ, for example, to the average value of the boundary α distribution, and still we assume that the constraint (18) holds. In this case (18) does not hold necessarily but merely contributes an error term in (17) to the field **B**. The discrepancy in this assumption is checked later when the solution is evaluated. Our experience indicates that with properly-chosen λ values this error can be negligibly small, as demonstrated later and in Wang, Yan, and Sakurai (2000).

To simulate the real observations by both approaches, we have added normally distributed noise to the magnetic field components on the $z = 0$ boundary. We have assumed that $\frac{1}{20}$ of peak value is 5 times the noise level for the longitudinal component and 3 times noise level for the transverse components. Therefore for the first approach, we have considered two cases: with and without noise added to the boundary data whereas for the second approach, we have considered three cases: $\lambda = \bar{\alpha}$ and $\lambda = 2\bar{\alpha}$ without noise, and $\lambda = \bar{\alpha}$ with noise added to the boundary data.

Figure 5 shows the distribution of the longitudinal field B_z along the z-direction above a fixed point $(0.1, -0.1)$ in (x, y) plane. The solid line indicates the analytical solution; the '$\Diamond$' and '+' symbols indicate the numerical results by the first approach with and without noise added to the boundary values, respectively; and the dashed line indicates the numerical results by the second approach, in which the differences due to the three cases being analyzed arc invisible here. This shows the capability of the new representation to approximate the exact field. When the

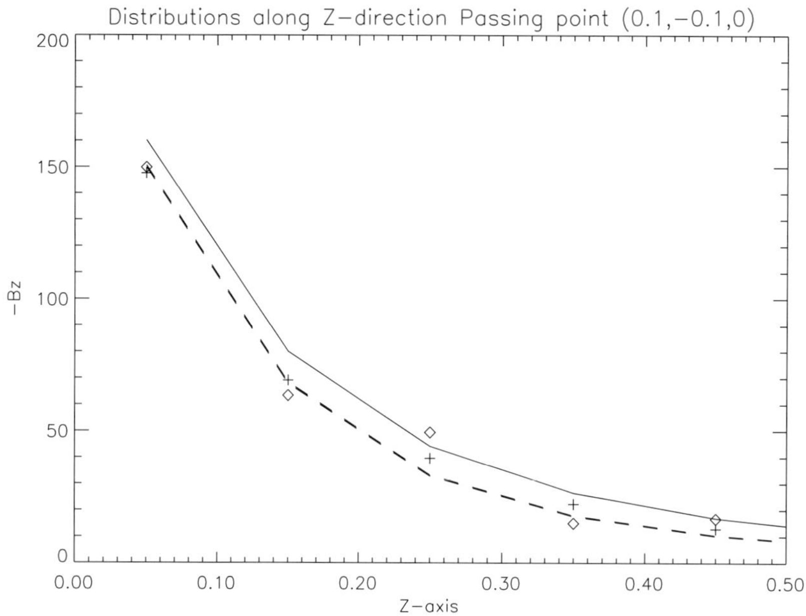

Figure 5. The vertical distribution of the longitudinal field B_z along the z-direction above a fixed point $(0.1, -0.1)$ in the (x, y) plane. The solid line indicates the analytical solution; the '$\Diamond$' and '+' symbols indicate the numerical results by the first approach with and without noise added to the boundary values, respectively; and the dashed line indicates the numerical results by the second approach for $\lambda = \bar{\alpha}$ with and without noise, and $\lambda = 2\bar{\alpha}$ noise-free cases whose differences are invisible in this figure.

boundary noise was added, it can be seen that the calculated results appear as variations about the analytical solution when the volume iteration was executed; but for the other case the boundary noise has little influence on the reconstructed fields above the boundary surface, and the results are not sensitive to the choice of λ. While closer agreements with the analytical solution may be achieved by the first approach, it seems that there exists a systematic bias error between the numerical and analytical solutions using the second approach.

Figure 6 shows the magnetic field lines obtained by the second approach and the analytical solution. These nodes are chosen as footpoints if their longitudinal field values are greater than a certain threshold, e.g., about $\frac{1}{20}$ of the peak longitudinal field. Therefore if the peak value corresponds to 2000 G, this threshold is about 100 G. In the computations, however, the threshold has been chosen even lower, or about one thirtieth of the peak value. The calculated magnetic field lines forming local loops are shown in Figure 6(a). The analytical results starting from the same footpoints are shown in Figure 6(b). When the parameter λ is increased by a factor of two, the calculated field lines are almost identical, as shown in Figure 6(c). Figure 6(d) shows the extrapolated results when the normal distributed noise is added to the boundary data over the $z = 0$ plane. In these figures the thick solid (dotted)

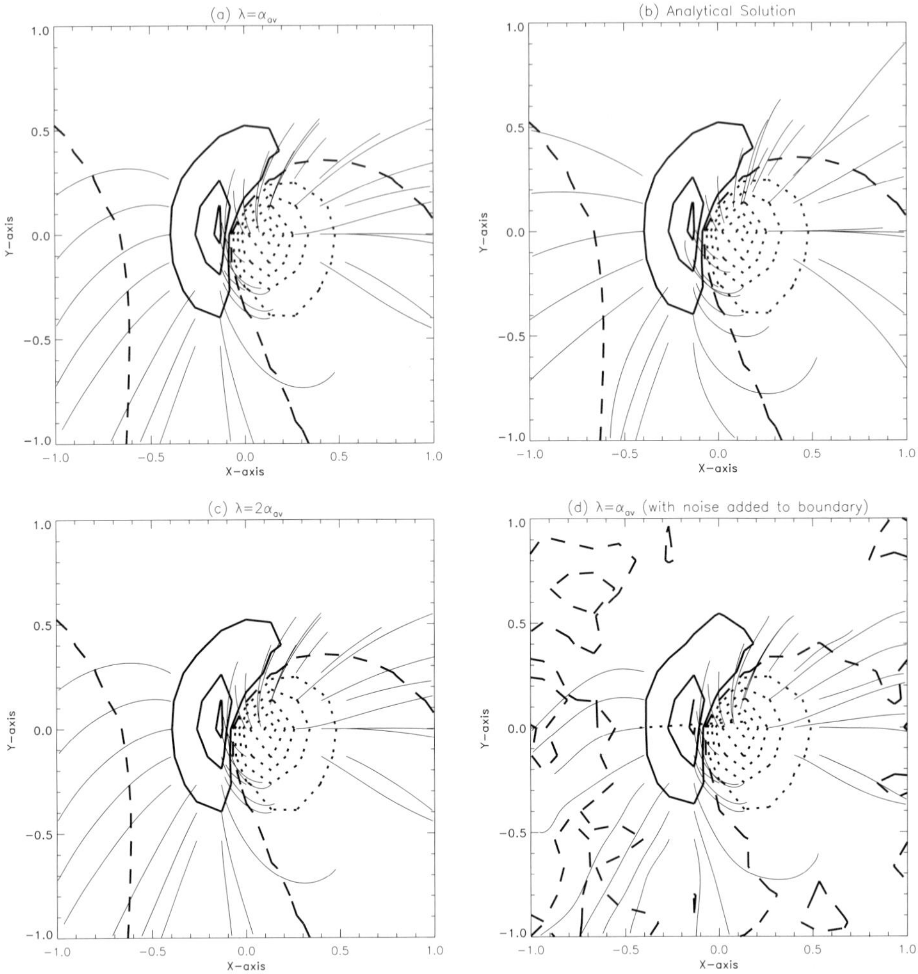

Figure 6. The magnetic field lines projected on $z = 0$ plane with the boundary distribution of the longitudinal field component B_z. (a) The extrapolated field lines with $\lambda = \bar{\alpha}$. (b) The exact field lines of the analytical solution. (c) The extrapolated field lines with $\lambda = 2\bar{\alpha}$. (d) The extrapolated field lines with $\lambda = \bar{\alpha}$ but with a noise of the level $\sigma_\parallel = \max |B_z|_{z=0}/100$ added to the longitudinal component and a noise of $\sigma_\perp = \max |B_z|_{z=0}/60$ to the transverse components.

line indicates positive (negative) polarity, the dashed line indicates the neutral line, or $B_z = 0$. The contours are $\pm 0.1, 0.3, 0.5, 0.7, 0.9$ of the peak longitudinal field, or $\max |B_z|_{z=0}$, and the thin solid lines are projected field lines.

From these figures one can see that in the main spot areas the calculated and analytical field lines agree quite well morphologically, though there are some discrepancies which are rather obvious mainly in the vicinity of the problem region, or weak field areas. Because of the fact that most solar activity such as flares happen in the former areas near the neutral lines, the above results are interesting for solar physics research. In all the above computations we have checked, $|\nabla \cdot \mathbf{B}|/|\mathbf{B}|$

either averaged over different layers or point-wise at the positions given above. We found that these values have a maximum value of about 5% in the length unit compatible with practical resolutions. That is, if the whole boundary region is 200 Mm$\times$200 Mm, we have the length unit being 1 Mm. The values decrease to less than 1% as distance increases.

The above results indicate that the representation of the proposed boundary integral equation for the non-linear force-free field problem is valid and useful for solar physics research (Yan and Sakurai, 1997). For a real magnetogram, with a properly chosen λ-parameter and data reductions, a quantitative evaluation of the divergence-free condition, force-free condition and the criterion (18) has been carried out by the second approach and the results show that these conditions are satisfied with high precision (Wang, Yan, and Sakurai, 2000).

In the above numerical test the total CPU time elapsed was about one hour by the second approach for each computation on a SUN workstation 3/50 M, but it took much longer time to find boundary solutions when the first approach was applied. The time needed for computing the field in space was, however, relatively small.

5. Conclusions

A new representation of the boundary integral equation for the solution of the finite energy non-constant-α force-free magnetic field in open space above the solar surface is provided. Thus we have established the equivalence between the exterior boundary value problem ((1), (2), (4), (5)), and the boundary integral equation representation (22) with the criterion (18) for λ. Therefore, the non-constant-α force-free magnetic field can, for the first time, be expressed in terms of its values and its normal gradients over the boundary.

The present model can be numerically solved by a well-established method: the boundary element method, and the computational procedure in Cartesian coordinates is provided. With appropriate assumptions, an iteration is eventually implemented for the nonlinear boundary solution. If λ functions are not iterated, the solutions are approximate solutions, but can be a practical model field. A test case has been demonstrated to show how the proposed boundary integral equation can be used numerically to recover an analytical solution in the physical domain. A certain amount of noise has been superposed on the prescribed boundary distribution of the vector magnetic field, and the extrapolated field can recover the noise-free field in the main spot area quite well.

Therefore, it can be used as a practical and convenient tool in studying solar physics events. The application of the proposed method to the observed magnetograph data is now under way and are reported elsewhere in following papers (e.g., Wang, Yan, and Sakurai, 2000).

Acknowledgements

Y. Y. thanks Dr R. Rosner for helpful discussions on his previous work, which motivated the present result. He acknowledges Drs A. N. McClymont and B. C. Low for critical and helpful comments on this work. He also thanks the Ministry of Education, Science and Culture of Japan for support to NAOJ as a Foreign Research Fellow. Some of the computations were carried out while Y. Y. was an Alexander von Humboldt Fellow at Astronomy Institute, Universität Würzburg. The work of Y. Y. is also supported by the Natural Science Foundation of China and the Education Bureau of the Chinese Academy of Sciences. The authors thank Drs A. Hujeirat, M. Smith and many others for reading and improving the English of this paper, and the anonymous referee for useful suggestions in revising this paper.

References

Aly, J. J.: 1987, in R. Beck and R. Gräve (eds.), *Interstellar Magnetic Fields*, Springer-Verlag, Berlin, p. 240.
Aly, J. J.: 1989, *Solar Phys.* **120**, 19.
Arfken, G.: 1966, *Mathematical Methods for Physicists*, Academic Press, New York.
Brebbia, C. A., Telles, J. C. F., and Wrobel, L. C.: 1984, *Boundary Element Techniques*, Springer-Verlag, Berlin.
Chandrasekhar, S.: 1961, *Hydrodynamic and Hydromagnetic Stability*, Oxford University Press, London.
Courant, R. and Hilbert, D.: 1962, *Methods of Mathematical Physics*, Vol. II, Interscience Publishers, New York.
Harrington, R. F.: 1961, *Time-Harmonic Electromagnetic Fields*, McGraw-Hill, New York.
Hu, W.: 1987, *Cosmical Magnetohydrodynamics*, Science Press, Beijing (in Chinese).
Klimchuk, J. A. and Sturrock, P. A.: 1992, *Astrophys. J.* **385**, 344.
Krall, K. R., Smith, J. B., Jr., Hagyard, M. J., West, E. A., and Cumings, N. O.: 1982, *Solar Phys.* **79**, 59.
Low, B. C.: 1982, *Solar Phys.* **77**, 43.
Low, B. C. and Lou, Y. Q.: 1990, *Astrophys. J.* **352**, 343.
McClymont, A. N., Jiao, L., and Mikić, Z.: 1997, *Solar Phys.* **174**, 191.
Metcalf, T. R., Jiao, L., McClymont, A. N., Canfield, R. C., and Uitenbroek, H.: 1995, *Astrophys. J.* **439**, 474.
Mikić, Z. and McClymont, A. N.: 1994, in K. S. Balasubramaniam and G. Simon (eds.), *Solar Active Region Evolution: Comparing Models with Observations*, A.S.P. Conf. Proc., San Francisco, p. 225.
Roumeliotis, G.: 1996, *Astrophys. J.* **473**, 1095.
Sakurai, T.: 1981, *Solar Phys.* **69**, 343.
Sakurai, T.: 1989, *Space Sci. Rev.* **51**, 11.
Schmahl, E. J., Kundu, M. R., Strong, K. T., Bentley, R. D., Smith, J. B., Jr., and Krall, K. R.: 1982, *Solar Phys.* **80**, 233.
Stratton, J. A.: 1941, Electromagnetic Theory, McGraw-Hill, New York.

Wang, H. N., Yan, Y., and Sakurai T.: 2000, *Solar Phys.* submitted.

Wu, S. T., Sun, M. T., Chang, H. M., Hagyard, M. J., and Gary, G. A.: 1990, *Astrophys. J.* **362**, 698.

Yan, Y. and Sakurai T.: 1997, *Solar Phys.* **174**, 65.

Yan, Y., Yu, Q., and Shi, H.: 1993, in J. H. Kane, G. Maier, N. Tosaka and S. N. Atluri (eds.), *Advances in Boundary Element Techniques*, Springer-Verlag, Berlin, p. 447.

SYNOPTIC MAPS OF SOLAR WIND: COMPARISON OF IPS DATA AND THE NOAA/SEC VERSION OF THE WANG AND SHEELEY MODEL

BALA BALACHANDRAN
Instituto de Geofisica, Universiad Nacional Autónoma de México, Coyoacán, D.F. 04510 México

(Received 20 September 1999; accepted 23 March 2000)

Abstract. Since the 1970s, the Solar-Terrestrial Environment Laboratory, Japan, has been publishing synoptic maps of solar wind velocity prepared using the technique of interplanetary scintillation. These maps, known as V-maps, are useful to study the global distribution of solar wind in the heliosphere. As the Earth-orbiting satellites are unable to probe regions outside the ecliptic, it is important to exploit the scope of interplanetary scintillation to study the solar wind properties at these regions and their relation with coronal features. It has been shown by Wang and Sheeley that there exists an inverse correlation between rate of magnetic flux expansion and the solar wind velocity. The NOAA/Space Environment Center daily updated version of the Wang and Sheeley model has been used to produce synoptic maps of solar wind velocity and magnetic field polarity for individual Carrington rotations. The predictions of the model at 1 AU have been found to be in good agreement with the observed values of the same. The present work is a comparison of the synoptic maps on the source surface using the interplanetary scintillation measurements from Japan and the NOAA/SEC version of the Wang and Sheeley model. The two results agree near the equatorial regions and the slow solar wind locations are consistent most of the times. However, at higher latitudes within $\pm 60°$, the wind velocities differ considerably. In the Wang and Sheeley model the highest speed obtained is ~ 600 km s^{-1} whereas in the IPS results velocities as high as 800 km s^{-1} have been detected. The paper discusses the possible causes for this discrepancy and suggestion to improve the agreement between the two results.

1. Introduction

Ever since the first spacecraft detected solar plasma in the interplanetary medium, the solar wind and the heliospheric magnetic field have been studied extensively leading to several important results and discoveries. However, Earth-orbiting spacecraft are limited to $\pm 7.25°$ near the solar equator. There were several outer planet missions like *Pioneer* and *Voyager* attaining latitudes as high as 30° but they were all at heliocentric distances of 5 AU or beyond where the solar wind is highly evolved. *Ulysses*, on the other hand, due to its high orbital inclination, returned inward to about 1.3 AU at perihelion and circled the Sun passing both the poles at $\pm 80°$. This polar pass of *Ulysses* is known as the 'fast latitude scan' and made the in situ measurements of solar wind plasma in these regions for the first time. However, launching satellites like *Ulysses* is very rare and so, the properties of out-of-ecliptic solar wind remain unexplored. One way to probe these regions is to employ remote sensing techniques such as interplanetary scintillation (IPS) or

The physics of the solar corona and transition region. Part II.
Reprinted from Solar Physics **195,** 2000.

comet observation. The latter is rather infrequent, whereas using the former one can probe out-of-ecliptic solar wind all the time within 1 AU (Hewish, Scott, and Willis, 1964; Kakinuma, Washimi and Kojima, 1973; Coles *et al.*, 1978). With an appropriate choice of the frequency of observation, IPS can give information about the solar wind very close to the Sun as well. Therefore, IPS, though indirect, provides a rather unique way of obtaining information about the solar wind at all latitudes within 1 AU and at all times.

Interplanetary scintillation is analogous to optical twinkling of stars, scaled to radio frequencies. Radio signals from compact radio sources such as pulsars are modulated by the irregularities in the solar wind. The scattered radio waves interfere with each other as they propagate to the Earth forming a diffraction pattern that drifts across the observer's plane with solar wind speed. A set of suitably spaced radio receivers on Earth detect these diffraction patterns with a time lag. A cross-correlation analysis of signals between pairs of antennae gives the velocity of the diffraction pattern. This velocity is assumed to be the solar wind velocity at the point of closest approach to the Sun (P-point) of the line-of-sight to the scintillating radio source. The P-point approximation has been shown to be nearly valid for most observations beyond 0.5 AU, especially for the slow solar wind (Watanabe and Kakinuma, 1972; Coles *et al.*, 1978; Coles and Kaufman, 1978).

The solar wind velocity V_R estimated by IPS at various distances in the heliosphere is projected back to the source surface (situated at a distance of 2.5 $R_\odot$ from the center of the Sun, within which all the magnetic field is assumed to be derived from a potential and there is no current; at the source surface, all the fields are radial) along the Archimedean spiral using the relations

$$\Phi_0 = \Phi_R + \frac{R\Omega}{V_R}, \qquad \Theta_0 = \Theta_R \,, \tag{1}$$

where, Φ_0 and Φ_R are the longitudes, and Θ_0 and Θ_R the latitudes, on the source surface and at a distance R from the Sun, respectively. Ω is the angular speed of solar rotation. These velocities are then used to make a synoptic map (V-map) in Carrington longitude and heliographic latitude. However, the source surface is situated within the acceleration region of the solar wind. Therefore, the application of Equations (1) all the way to the source surface could be misleading.

Wang and Sheeley (1990) obtained a statistically significant inverse correlation between the solar wind velocity and the rate of expansion of magnetic flux. That is, fast solar wind originates from regions with small magnetic divergence whereas slow solar wind emanates from regions where the magnetic divergence is high. The former corresponds to central regions of coronal holes, while the latter corresponds to the boundaries of open field regions overlying closed field regions. For details of the results and the physical basis for the relationship refer to Wang and Sheeley (1991) and Wang (1993). At the NOAA Space Environment Centre (NOAA/SEC), Boulder, a modified version of Wang and Sheeley model has been used to make synoptic maps similar to the V-maps from IPS observations and to predict the velocity and magnetic field polarity at 1 AU. The present paper is a comparative study of the results obtained by the two methods.

2. Data and the Preparation of V-maps

Using their multi-station IPS facility, the Solar–Terrestrial Environment Laboratory (STELab), Nagoya University, Japan, has been preparing and publishing V-maps since the 1970s. These maps are useful in studying the global structure, distribution and evolution of solar wind in the heliosphere. Recently, using the Wang and Sheeley model, Arge and Pizzo (1998) have obtained similar maps, making use of the relation between magnetic flux expansion factor and velocity of solar wind. The data used in the present study is as follows: Carrington rotations CR 1894–1896 cover a period from 23 March to 12 June, 1995; CRs 1898–1900, from 10 July to 29 September, 1995 and CR 1925–1927 include the period from 15 July–5 October, 1997. The data for these V-maps were superposed over three Carrington rotations indicated in each set. The V-maps computed using IPS measurements as well as Wang and Sheeley model for these periods are shown in Figures 1, 2, and 4. Since the period of study was near solar minimum, very little evolution of solar wind was expected from one rotation to the next and most of the long-lived, large-scale structures were assumed to be retained. Due to many reasons, the velocities beyond $\pm 60^\circ$ may not be reliable and therefore, excluded from the present study. The location of the coronal holes on the source surface and the footpoints mapped on the photosphere using the potential field model of Hoeksema (for the Carrington rotation CR 1898) is depicted in Figure 3. The computation includes all the multi-poles up to order 9.

In the following subsections, the two methods are described in some detail and the latitudinal profile obtained from these V-maps is introduced. A detailed discussion of the results is presented in Section 3.

2.1. Predicted V-maps of NOAA/SEC using Modified Wang and Sheeley Model

The rate at which the flux tube expands in solid angle between the photosphere and the source surface can be represented mathematically as

$$f_s = \left(\frac{R_\odot}{R_{ss}}\right)^2 \frac{B_r(\text{phot})}{B_r(ss)}, \tag{2}$$

where $B_r(\text{phot})$ and $B_r(ss)$ are the radial components of magnetic fields at the photosphere and the source surface, and $R_\odot$ and $R_{ss} = 2.5\ R_\odot$ are the photospheric and source-surface radii, respectively. Wang and Sheeley (1990, 1994) found that the computed values of f_s were associated with different ranges of solar wind speed at 1 AU. However, Arge and Pizzo (1999) obtained the following relation between the flux expansion factor and the solar wind velocity by iteration:

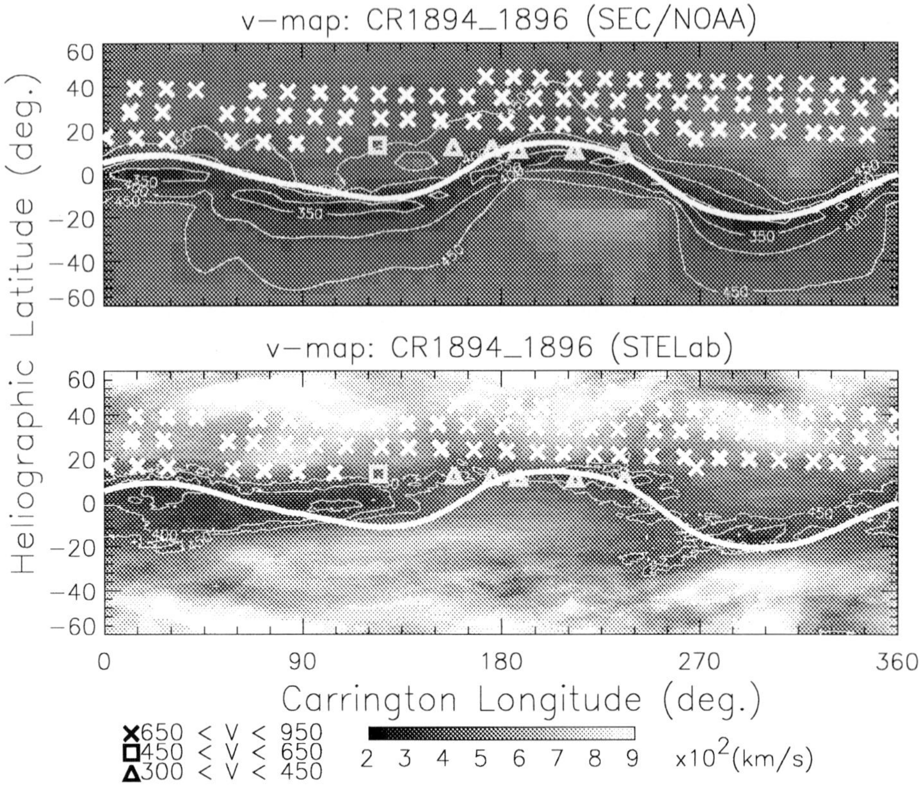

Figure 1. The V-map for superposed CRs 1894–1896 (1995). The upper panel is the predicted V-map of NOAA/SEC using the Wang and Sheeley model while the lower panel is the V-map obtained at STELab, Japan, using IPS data. The thick white solid line is the magnetic neutral line for CR1894. The thin white contours represent the slow solar wind ($V \leq 450$ km s^{-1}) plotted at an interval of 50 km s^{-1}. The *Ulysses* data projected back to the source surface during the same period is over-plotted with various symbols.

$$V_{sw} = 267.5 + 410.0/f_s^{(1.0/2.5)} \tag{3}$$

where, V_{sw} is the speed of the solar wind and f_s is the flux expansion factor obtained using Equation (2). At NOAA/SEC, the Wang and Sheeley model has been updated using the daily magnetograms from the Wilcox Solar Observatory and used to obtain V-maps and magnetic field polarity (http://sec.noaa.gov/~narge/). The velocity, calculated according to Equation (3), and the magnetic field polarity predicted by the model at 1 AU were found to be in good agreement with the *in situ* measurements (Arge and Pizzo, 1998).

The V-maps prepared using the NOAA/SEC version of the Wang and Sheeley model and Equation (3) for Carrington rotations 1894–1896 and 1898–1900 (1995) and 1925–1927 (1997) are depicted in the upper panels in Figures 1, 2, and 4. Superposed is the magnetic neutral line obtained from the potential field model of Hakamada, obtained at STELab, (thick white line), for CR 1894 in Fig-

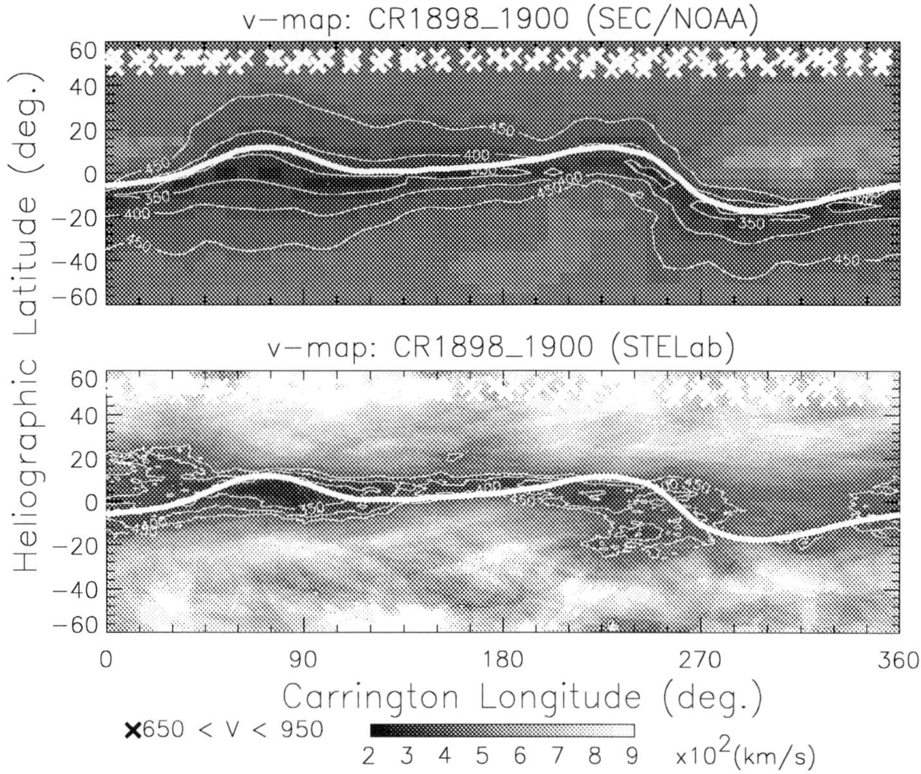

Figure 2. The same as Figure 1 but for CRs 1898–1900 (1995). The neutral line is for CR 1898.

ure 1, CR 1898 in Figure 2, and CR 1925 in Figure 4, respectively. The thin white lines are contours of velocities $\leq$ 450 km s^{-1} and are plotted at intervals of 50 km s^{-1}. The symbols represent the velocities observed by *Ulysses* during the same period beyond the Earth's orbit (between about 1.35 and 1.75 AU), projected back to the source surface, using Equation (1). Different ranges of velocities are depicted by different symbols: triangles are $300 < V < 450$ km s^{-1}, squares are $450 < V < 950$ km s^{-1} and crosses are $650 < V < 950$ km s^{-1}.

2.2. Synoptic maps using IPS data

The IPS facility at STELab consists of four stations, operating at a frequency of 327 MHz (Asai *et al.*, 1995). It is capable of observing about 30 radio sources daily and employs the technique of cross-correlation between pairs of antennae to obtain the solar wind speed. In the technique of IPS, the solar wind velocity is measured as an integration of all the speeds along the line-of-sight to the radio source, weighted by the distribution of density fluctuation. Therefore, the velocity is biased and is different from the real velocity at P-point. Applying the technique of Computer Assisted Tomography (CAT), the bias caused by the line-of-sight integration can be minimized as has already been shown by its application to the IPS data from STELab (Jackson *et al.*, 1988; Kojima *et al.*, 1998). For the application of tomography, both the solar wind velocity and the density fluctuation are necessary. In the present study, the latter is assumed to have a power-law dependence on the former, of the form

$$\delta N_e(z) = V^{-\gamma}(z) \,. \tag{4}$$

Following the studies by Kakinuma, Washimi, and Kojima (1982) on the speed dependence of the density fluctuations, a value of 0.5 has been chosen for γ. For the successful application of tomography, a large number of data points ($\sim$500) per Carrington rotation is necessary. During the period chosen for the present study, the data was insufficient to make V-maps for individual Carrington rotations and hence the data over three rotations were superposed. Since the periods of study (1995, 1997) are near solar minimum, very little evolution of solar wind was expected from one rotation to the next and most of the long-lived, large-scale structures were assumed to be retained. The V-maps obtained for Carrington rotations CRs 1894 –1896, CRs 1898–1900 and CRs 1925–1927 are presented in the lower panels in Figures 1, 2, and 4. Also, the source surface magnetic neutral line (thick white solid line) and velocity contours (thin white lines), as in the upper panel, are overplotted. The crosses in Figure 2 represent the velocities detected by Ulysses during the same period, in the range $650 \leq V \leq 950$ km s^{-1}. Due to certain technical problems, the *Ulysses* data for CRs 1925–1927 could not be plotted.

2.3. Latitudinal Profile

In a study using IPS data for 15 years, Rickett and Coles (1991) showed that the yearly averaged solar wind velocity increases sharply with heliographic latitude during low and declining phases of solar activity whereas the velocity remains uniformly low at solar maximum. A careful analysis of the V-maps published by the IPS group at STELab, Japan, reveals the same tendency. That is, the sharp latitudinal gradient of velocity observed during quiet periods is no longer there during disturbed periods when the slow and fast winds are randomly distributed in the heliosphere (Kojima and Kakinuma, 1990).

Figures 5–7 depict the latitudinal profiles of solar wind velocity obtained from the IPS measurements (solid line) and the modified Wang and Sheeley model (dashed line), for the same rotations shown in Figures 1, 2, and 4. The data for these figures were averaged over the entire longitude range. The curve from the IPS data is consistent with the earlier works of Rickett and Coles but that from Wang and Sheeley model is far from agreement. This (dashed line) resembles an active period of the Sun. The difference between the two curves are very significant and suggests the limitation of the Wang and Sheeley model in handling the high latitudes.

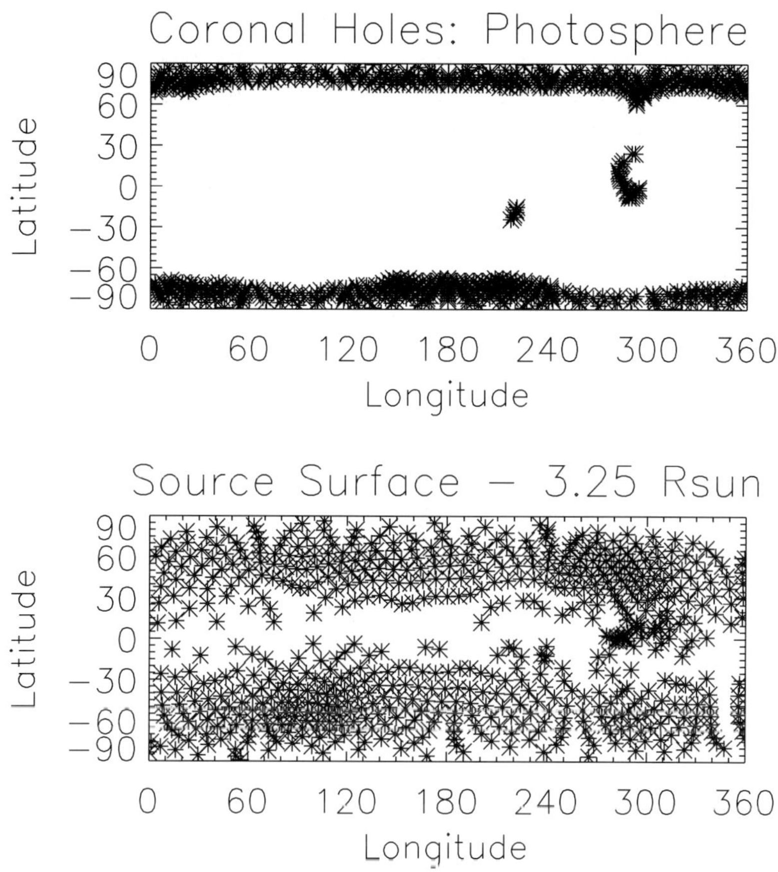

Figure 3. The location of coronal holes mapped on the photosphere (*upper panel*) and on the source surface (*lower panel*) using the potential field model of Hoeksema, for the Carrington rotation CR 1898. This rotation corresponds to that in Figure 2. The symbols represent the location of the open field lines on the photosphere and the source surface in the upper and lower panels, respectively.

3. Discussion of the Results

The solar wind in the heliosphere has two distinct components, the slow and the fast. The former has typical velocities less than or equal to 450 km s^{-1} while the velocities of the latter is typically 500 km s^{-1} and higher. The high-speed solar wind originates from open magnetic field regions in the corona, known as coronal holes and is rather steady. On the other hand, the slow solar wind, originating from regions above the closed magnetic fields, like helmet streamers, is highly structured and evolves significantly from rotation to rotation, making it an interesting and important feature to pursue. The presence of the slow and fast components of the solar wind is very striking in the IPS results (lower panel), whereas the high-speed component in the range 650 to 850 km s^{-1} is missing in the Wang and Sheeley model of NOAA/SEC (upper panel). Also, the latitudinal width of the slow solar

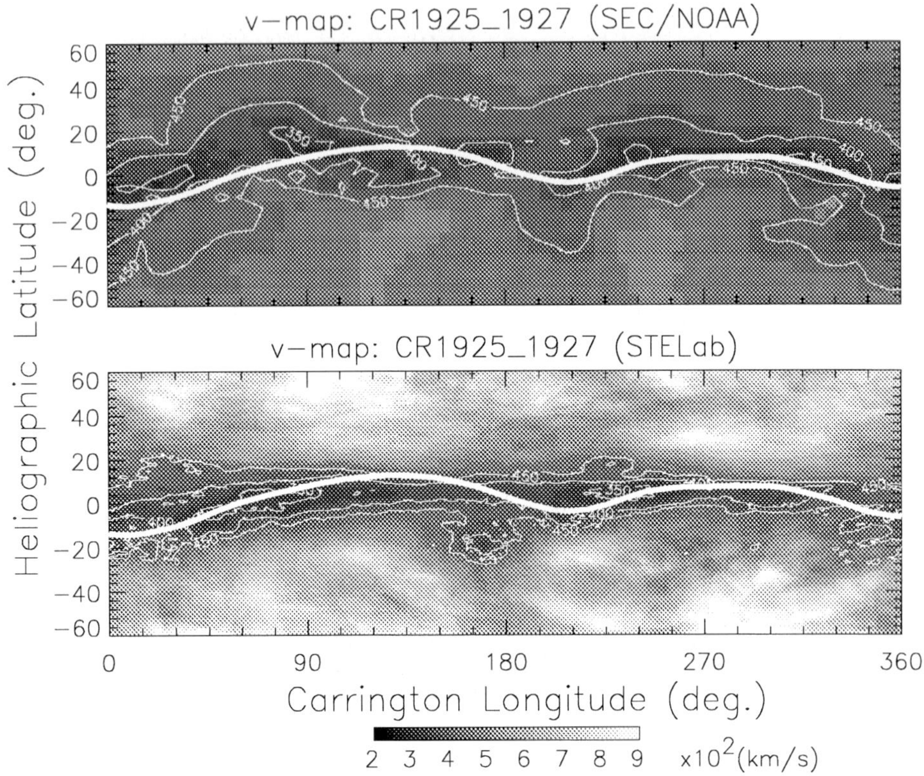

Figure 4. The same as Figure 1 but for CRs 1925–1927 (1997). The neutral line is for CR 1925.

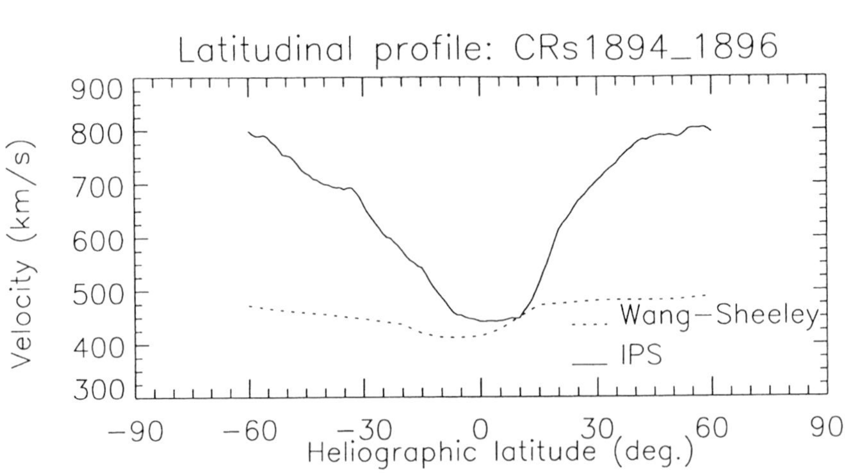

Figure 5. The latitudinal profiles of the solar wind velocity on the source surface averaged over the entire longitude. The solid line represents the IPS measurements and the dashed line, the NOAA/SEC version of Wang and Sheeley model.

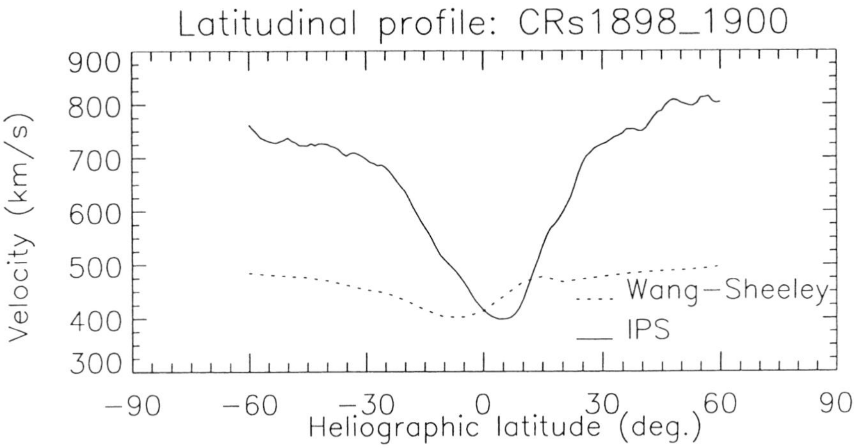

Figure 6. The same as Figure 5 but for CRs 1898–1900.

wind belt is somewhat larger in the Wang and Sheeley model of NOAA/SEC than in the IPS results. However, the longitudinal locations of the slow solar wind in the two V-maps coincide almost everywhere, except in the 90–130° longitude in Figure 1; at this region, the slow solar wind is located below the neutral line in the Wang and Sheeley model while in the IPS results they are above the neutral line. This general agreement in the longitudinal locations of the slow solar wind in both the V-maps somewhat contradicts the conclusions in Wang *et al.* (1997) where they have pointed out that their model predicts too much fast wind. However, they have also pointed out that this could be corrected for by allowing streams to interact in the model. The results shown in this paper are from the NOAA/SEC version of Wang and Sheeley model, where they have devised a simple method to include the stream–stream interaction (Arge and Pizzo, 1999).

On an average, the slow solar wind is confined within ±20° in all three Figures 1, 2 and 4, which is a little larger than the slow solar wind belt observed by *Ulysses* at about 1.34 AU: ±18° (Phillips *et al.*, 1995), This seems to be consistent with other observations where the width of a streamer belt on the solar surface appeared to have a latitudinal width of about 60°, narrowed down to about 45° at 1.74 $R_{\odot}$ when observed by *Yohkoh* as well as Mauna Loa coronameter (Gosling *et al.*, 1995). That is, the width of the slow speed has possibly reduced from the source surface to 1.34 AU at the time of *Ulysses* observation.

In the Wang and Sheeley model of NOAA/SEC (upper panels in Figures 1 and 2) the slow solar wind is a thin, wavy band near the solar equator and is one, single, continuous belt. During CRs 1894–96 (Figure 1), the slow-wind belt appears to be sinusoidal while during CRs 1898–1900 (Figure 2), it is parallel to the solar equator except in the longitudinal range 240–360°, where it appears to be pushed to the higher southern latitudes by fast wind from the north. Also, in Figure 2, in the longitude band 30–90°, the slow solar wind is broader in latitude than in

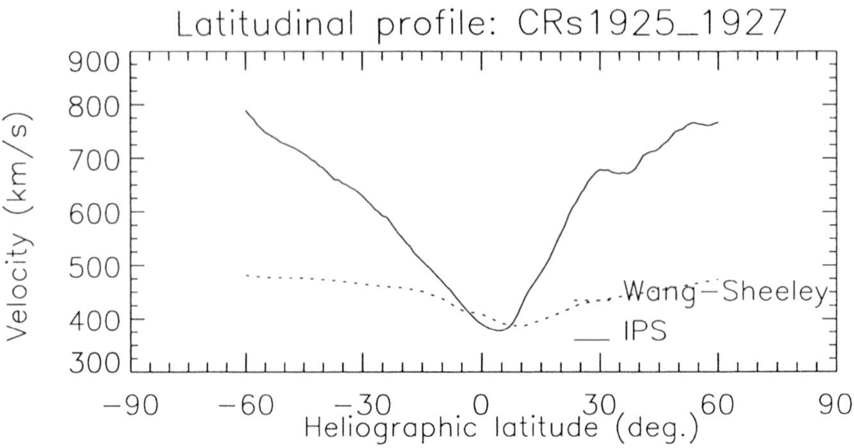

Figure 7. The same as Figure 5 but for CRs 1925–1927.

Figure 1. All these features are contrary to what is seen in the IPS results – lower panels in Figures 1 and 2. Here, the slow solar wind appears as disjoint regions near the solar equator, the region in between being occupied by high-speed wind ($\sim$ 550 km s^{-1}). In the lower panel (IPS) in Figure 2, a high-speed stream is seen to be crossing the equator near the longitude $\sim$ 280–320°. Note that at the same longitude in the upper panel (Wang and Sheeley model of NOAA/SEC), the slow-wind belt is being pushed to higher southern latitudes. From Figure 3, it is clear that there exists a coronal hole at nearly the same longitude near the equator which could be the source of the fast stream that crossed the equator in Figure 2. Figure 3 is obtained using the potential field model of Hoeksema and it is rather surprising to see the absence of a high stream crossing the equator at this location in the Wang and Sheeley model of NOAA/SEC.

In Figure 4, the slow solar wind belt is rather disjoint at a couple of places near the equator in the Wang and Sheeley model of NOAA/SEC, different from the previous two data sets, indicating an evolution of the slow solar wind with solar activity. Surprisingly, this is absent in the IPS results and the slow-wind belt is rather parallel to the equator and continuous. Also, the slow solar wind has a slightly larger latitudinal extent in the upper panel, as indicated by the velocity contours. As in the previous figures, the high latitude fast streams are absent in the Wang and Sheeley model.

Figures 5–7, the latitudinal profile of solar wind velocity, show the striking difference between the two V-maps. Note that there is no fast wind at high latitudes with velocities greater than 500 km s^{-1} in the NOAA/SEC version of Wang and Sheeley model (dotted line). Here, the curve is rather flat with a small dip near the equator. Also, there is a marked north-south asymmetry in all the figures. At the same time, the IPS results show sharp velocity gradients, consistent with previous results (Rickett and Coles, 1991). The period of study, *viz* the years 1995 and 1997, is one with low activity, and so, the latitudinal profile is expected to be as sharp as the IPS results. The trend seen in the Wang and Sheeley model is in contrast to the *Ulysses* observations as well. During its fast latitude scan, *Ulysses* observed velocities as high as 850 km s^{-1} (excluding CMEs) at around 60° latitude in both the hemispheres, but at a heliocentric distance of $\sim$ 1.5 AU (Phillips *et al.*, 1995), which are represented by crosses in Figures 1 and 2. It also observed a belt of slow solar wind within a range of ±18°. And, in general, the wind speed exhibited a positive latitudinal gradient away from the solar equator in either hemisphere, consistent with the earlier studies (Rickett and Coles, 1991).

In short, the predicted V-map of NOAA/SEC does not agree very well with either the IPS measurements or the Ulysses observations at latitudes higher than ±30 or 35°. The possible causes of this discrepancy are discussed in the following section.

4. Concluding Remarks

The immediate conclusion from Figures 1–7 is that the two V-maps are very much in agreement with each other near the solar equator. This is consistent with an earlier study of Arge and Pizzo (1998), where they have shown that the solar wind speed and the magnetic field at 1 AU predicted by the NOAA/SEC version of the Wang and Sheeley model was in agreement, in general, with the spacecraft observations. However, there is a large discrepancy at the higher latitudes, indicating that the model has difficulties in handling the regions where the magnetic field is predominantly open. However, this need not strictly be the implication as there are other possibilities that could be the cause of the observed discrepancy. At various stages of data analysis as well as modeling, a few important physical aspects were overlooked or ignored as a first approximation, which could introduce some errors in the final results. Below is a list of them with a brief elaboration on each.

1. The absence of heliospheric current sheet (HCS) in the Wang and Sheeley model.

As pointed out by Wang and Sheeley (1994), the potential field source surface model does not include the effect of heliospheric current sheet and therefore, the expansion of the magnetic flux tube in the region $R \geq R_{ss}$ is not correctly represented by the model. This, the omission of the current sheet, introduces some uncertainties in the latitudinal and longitudinal positions of the flux tubes for $R \geq R_{ss}$ and in turn, in the locations of the predicted wind streams. How the absence/presence of a current sheet is going to affect the high velocity distribution and even the very values of them at high latitudes needs to be tested.

2. The bias in the estimation of solar wind velocity from IPS.

In the technique of IPS, the solar wind velocity is obtained as an integration of all the velocities along the line of sight to the radio source, perpendicular to it. This includes contribution from faster and slower streams than the one crossing the P-point and causes an underestimation of the solar wind speed at low latitudes and an overestimation at higher latitudes (Kojima *et al.*, 1998). However, the *Ulysses* observation of velocities as high as 850 km s^{-1} at high latitudes substantiates the IPS values. Moreover, the bias caused by the line-of-sight integration is reduced to a large extent by the application of tomography.

3. The inadequacy of the technique of inverse mapping of the velocity to the source surface.

To prepare the V-maps of solar wind on the source surface, the velocities estimated by IPS between 0.1 and 0.9 AU are mapped back using Equation (1). There are three important aspects that are neglected here: first of all, it assumes little latitudinal variation during the outward propagation of the solar wind. That is, the plasma flow is taken to be radial and all non-radial components are neglected. However, this is not strictly true as is evident in the recent LASCO coronagraphic observations of helmet streamers, which are non-radial (see, for example, January 1997, Data: High Altitude Observatory/NCAR). In such cases, a radial inverse mapping would definitely lead to a wrong location as the origin of these plasma streams. Second, the interaction between fast and slow streams is totally neglected. The velocities estimated by IPS are the resultant of, perhaps, the interaction of fast and slow streams, though such interactions within 1 AU are infrequent. In such cases, it is not easy to de-convolve the true velocity and to trace back to the real source on the corona. Finally, the very location of the source surface: at 2.5 $R_\odot$. This region is well within the sonic point (where the solar wind attains the supersonic value) within which the heliocentric distance dependence of the acceleration of both fast and slow wind are still not well understood. The relation is certainly different from that governed by Equation (1) and the application of it will introduce considerable uncertainty.

The non-radial nature of plasma emission and stream-stream interaction are more of a local nature, as they are infrequent in their occurrence, still not negligible. On the other hand, the inverse mapping in the region within 30 $R_\odot$ deserves special attention since it is a global problem being important to the spacecraft data as well. The *Ulysses* data (obtained beyond 1 AU) shown in Figures 1 and 2 were projected onto the source surface using the same relation Equation (1). That is why, perhaps, IPS data is in better agreement with the *Ulysses* data than the Wang and Sheeley model, which, of course, reassures the credibility of the IPS data. As mentioned previously, IPS is a powerful technique in inferring the solar wind properties at all latitudes in the heliosphere and at all times. Therefore, it is important to exploit the possibilities of IPS more effectively.

In the wake of the above discussion, it appears that the major source of error could be the inverse mapping. Perhaps, it is not possible to have velocities as high as 800 km s^{-1} on the source surface, as predicted by the NOAA/SEC version of Wang and Sheeley model. The high velocities present in the IPS data could be an artifact of the 'wrong' inverse mapping from say, a region of 30 $R_\odot$, to the source surface. The IPS V-map in the present form could be valid only outside 0.3 AU or so, and not at a region so close to the Sun as the source surface. In other words, instead of taking all the way back to the source surface, the IPS V-maps should have been prepared at a distance of about 60 or 100 $R_\odot$ so as to provide a more realistic picture of the global structure of solar wind. Proceeding further backwards using Equation (1) may take the solar wind not to the source surface but, perhaps, somewhere deeper within. Though Wang and Sheeley model of NOAA/SEC has certain limitations of its own, the results are more acceptable than the IPS results as the velocities are unlikely to be the same within the sonic point as outside. Therefore, at this point, the present discussion is concluded by emphasizing the importance of trying to improve the inverse mapping technique, especially that inside a region of say, 30 $R_\odot$, so as to be consistent with the acceleration of solar wind in this region.

Acknowledgements

The author is deeply indebted to Drs V. J. Pizzo and C. N. Arge, NOAA/SEC, Boulder, USA, for kindly providing her with the results of their Wang and Sheeley model and the many discussions. She also wishes to thank Prof. M. Kojima for the IPS and magnetic neutral line data. Drs J. Burkepile and T. E. Holzer, are acknowledged for many discussions at HAO, Boulder, Colorado. Thanks are also due to Dr Silvia Bravo, UNAM, México.

Wilcox Solar Observatory data used in this study was obtained via the web site http://quake.stanford.edu/wso/ by courtesy of J. T. Hoeksema.

References

Arge, C. N. and Pizzo, V. J.: 1998, in S. R. Habbal, R. Esser, J. V. Hollweg and P. A. Isenberg (eds.), *Solar Wind 9, Proceedings of the 9th International Solar Wind Conference*, Nantucket Island, Massachusetts, U.S.A., 5–9 October 1998, p. 569.

Arge, C. N. and Pizzo, V. J.: 1999, *J. Geophys Res.*, in press.

Asai, K., Ishida, Y., Kojima, M., Maruyama, K., Misawa, H., and Yoshimi, N.: 1995, *J. Geophys. Res.* **47**, 1107.

Coles, W. A. and Kaufman, J. J.: 1978, *Radio Sci.*. **13**, 591.

Coles, W. A., Harmon, J. K., Lazarus, A. J., and Sullivan, J. D.: 1978, *J. Geophys. Res.* **83**, 3337.

Dennison, P. A. and Hewish, A. A.: 1967, *Nature* **213**, 343.
Gosling, J. T., Bame, S. J., Feldman, W. C., McComas, D. J., Phillips, J. L., Goldstein, B., Neugebauer, M., Burkepile, J., Hundhausen, A. J., and Acton, L.: 1995, *Geophys. Res. Lett.* **22** 3329.
Hewish, A., Scott, P. F., and Willis, D.: 1964, *Nature* **203**, 1214.
Jackson, B. V., Hick, P. L., Kojima, M., and Yokobe, A.: 1998, *J. Geophys. Res.* **103**, 12049.
Kakinuma, T., Washimi, H. and Kojima, M.: 1973, *Publ. Astron. Soc. Japan* **25**, 271.
Kakinuma, T., Washimi, H., and Kojima, M.: 1982, in M. A. Shea *et al.*, *Proc. of the ISTP Symposium on Solar Radio Astronomy, Interplanetary Scintillation and Coordination with Spacecraft*, Spec. Rep. 233, p. 153, Air Force Geophys. Lab. Hanscom, Air Force Base Mass.
Kojima, M. and Kakinuma, T.: 1990, *J. Geophys. Res.* **92**, 7269.
Kojima, M., Tokumaru, M., Watanabe, H., Yokobe, A., Asai, K., Jackson, B. V., and Hick, P. L.: 1998, *J. Geophys. Res.* **103**, 1981.
Phillips, J. L., Bame, S. J. Barnes, A., Barraclough, B. L., Feldman, W. C., Goldstein, B. E., Gosling, J. T., and Hoogeveen, G. W.: 1995, *Geophys. Res. Lett.* **22**, 3301.
Rickett, B. J. and Coles, W. A.: 1991, *J. Geophys. Res.* **96**, 1717.
Wang, Y.-M.: 1993, *Astrophys. J.* **410**, L123.
Wang, Y.-M. and Sheeley, N. R., Jr.: 1990, *Astrophys. J.* **355**, 726.
Wang, Y.-M. and Sheeley, N. R., Jr.: 1991, *Astrophys. J.* **372**, L45.
Wang, Y.-M. and Sheeley, N. R., Jr.: 1994, *J. Geophys. Res.* **99**, 6597.
Wang, Y. -M., Sheeley, N. R., Jr., Phillips, J. L., and Goldstein, B. E.: 1997, *Astrophys. J.* **488**, L51.
Watanabe, T. and Kakinuma, T.: 1972, *Publ. Astron. Soc. Japan* **24**, 459.

TOPOLOGY OF MAGNETIC FIELD AND CORONAL HEATING IN SOLAR ACTIVE REGIONS

II. *The Role of Quasi-Separatrix Layers*

HUANING WANG[1,2,3], YIHUA YAN[1,2,3,4], TAKASHI SAKURAI[2] and MEI ZHANG[1,3]

[1]*Beijing Astronomical Observatory, Datun Rd. No. 20A, Chaoyang District, Beijing 100012, China*
[2]*National Astronomical Observatory, 2-21-1 Osawa, Mitaka, Tokyo 181-8588, Japan*
[3]*National Astronomical Observatories, Datun Rd. No. 20A, Chaoyang District, Beijing 100012, China*
[4]*Chinese Academy of Sciences, Beijing University Joint Beijing Astrophysics Center, Beijing 100871, China*

(Received 20 September 1999; accepted 20 April 2000)

Abstract. The photospheric vector magnetic fields, Hα and soft X-ray images of AR 7321 were simultaneously observed with the Solar Flare Telescope at Mitaka and the Soft X-ray Telescope of *Yohkoh* on October 26, 1992, when there was no important activity in this region. Taking the observed photospheric vector magnetic fields as the boundary condition, 3D magnetic fields above the photosphere were computed with a new numerical technique. Then quasi-separatrix layers (QSLs), i.e., regions where 3D magnetic reconnection takes place, were determined in the computed 3D magnetic fields. Since *Yohkoh* data and Mitaka data were obtained in well-arranged time sequences during the day, the evolution of 3D fields, Hα features and soft X-ray features in this region can be studied in detail. Through a comparison among the 3D magnetic fields, Hα features and soft X-ray features, the following results have been obtained: (a) Hα plages are associated with the portions of QSLs in the chromosphere; (b) diffuse coronal features (DCFs) and bright coronal features (BCFs) are morphologically confined by the coronal linkage of the field lines related to the QSLs; (c) BCFs are associated with a part of the magnetic field lines related to the QSLs. These results suggest that as the likely places where energy release may occur by 3D magnetic reconnection, QSLs play an important role in the chromospheric and coronal heating in this active region.

1. Introduction

Solar observations covering a wide spectral range have revealed that there are many loop structures in the corona. It is believed that these loop structures delineate magnetic field lines because of high electrical conductivity and anisotropic thermal conduction. In that sense, the heating processes are mainly confined in magnetic flux tubes, which might correspond to the observed loop structures. According to soft X-ray observations, there is a correlation between the surface magnetic field strength and the coronal X-ray brightenings in the spatial domain (Golub *et al.*, 1980) and in the long-term temporal domain (Hara, 1997). One could expect that the heating rate is some increasing function of magnetic strength. However, X-ray emission is almost missing in sunspot umbra (Pallavicini *et al.*, 1979). This fact

The physics of the solar corona and transition region. Part II.
Reprinted from Solar Physics **197,** 2000.

indicates that the magnitude of heating does not only depend on magnetic field strength.

Several heating mechanisms have been proposed during the last decades. All of these mechanisms can be divided into two classes: (a) the dissipation of MHD waves by phase mixing or resonant absorption; (b) the formation and dissipation of thin current sheets.

Theoretical and observational studies have demonstrated that magnetic energy deposit and release are intrinsically related to topological features of solar magnetic fields. Null points of the X-type in the solar magnetic fields were regarded as the most important topological features for magnetic reconnection (Giovanelli, 1947, 1948; Parker, 1957; Sweet, 1969; Priest and Forbes, 1986; Somov, 1986; Lau and Finn, 1990; Priest and Titov, 1996; Galsgaard and Nordlund, 1997). Many observational facts support this viewpoint (Mandrini *et al.*, 1991, 1993; Démoulin *et al.*, 1993, 1994; Wang and Wang, 1996; Wang, 1997; Wang, Wang, and Qiu, 1999; Filippov, 1999). Recently, Priest and Démoulin (1995) suggested a theory on 3D magnetic reconnection without null points: In a large class of 3D magnetic fields without null points there are locations, called 'quasi-separatrix layers (QSLs)', where the field-line linkage changes drastically and then concentrated electric currents are formed by smooth boundary motions. These QSLs are the relevant generalization of normal magnetic separatrices to configurations without null points. There are several observational studies on the QSLs related to solar flares (Mandrini *et al.*, 1996; Démoulin *et al.*, 1997a, b; and Gaizauskas *et al.*, 1998), which demonstrate the importance of the magnetic field topology for flares.

Démoulin and Priest (1997) made use of this theory to explain heating processes in the corona. In the case of coronal heating, since photospheric magnetic fields are fragmented into a large number of thin flux tubes, there exists a large number of very thin QSLs in the corona. When a QSL is thin enough, a turbulent resistivity is triggered, and the QSL then rapidly evolves into a dynamic current sheet that releases energy by fast 3D magnetic reconnection. Thus, the fragmentation of the photospheric magnetic field stimulates the dissipation of magnetic energy in the corona. This kind of heating mechanism falls in class (b).

The photospheric vector magnetic fields, Hα and soft X-ray images of active region NOAA 7321 were simultaneously observed with the Solar Flare Telescope at Mitaka (Sakurai *et al.*, 1995) and the Soft X-ray Telescope of *Yohkoh* (Tsuneta *et al.*, 1991) on 26 October 1992, when there was no important activity in this region.

Figure 1 shows two soft X-ray images in which bright coronal features (BCFs) are surrounded by diffuse coronal features (DCFs). Based on *Yohkoh* soft X-ray observations, Priest *et al.* (1998) suggest that the heating mechanism of DCFs is attributed to class (b). However, the heating mechanism of BCFs remains unknown. Falconer *et al.* (1997) believe that a kind of hybrid heating process including classes (a) and (b) is a viable possibility for some of these coronal features. They found that the heating of BCFs in solar active regions, namely enhanced coronal heat-

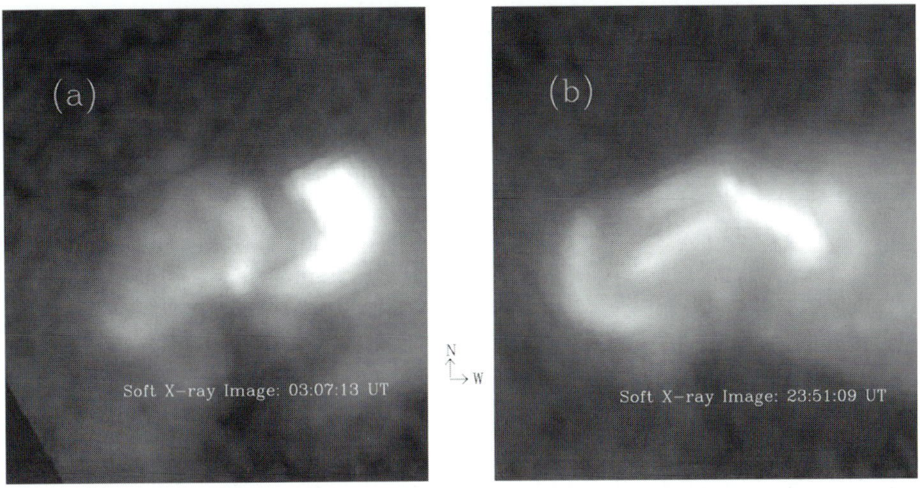

Figure 1. Soft X-ray images of AR 7321 observed with the filter Al.1 of *Yohkoh* SXT on 26 October 1992. Both frames have been normalized to the same brightness scale. The threshold of (a) and (b) is 1000 DN s^{-1}. The brightness in the whitest areas exceeds 4500 DN s^{-1}. Field of view: 200×200 arc sec^2. Location: (a) S26 W04; (b) S26 W15.

ing, is closely related to the presence of strong magnetic shear along neutral lines associated with the main polarity fields in these regions.

Figure 2 shows two photospheric vector magnetograms corresponding to the soft X-ray images in Figure 1. Since coronal magnetic fields can be reconstructed with the numerical technique proposed by Yan and Sakurai (2000) and Wang, Yan and Sakurai (2000), QSLs in the coronal fields will be figured out according to their mathematical definition and their role in the coronal loop heating will be investigated.

Brief reports on this research work were presented by Sakurai and Wang (2000) and Wang, Sakurai, and Yan (2000). In this paper we focus our comprehensive study on the role of QSLs. In Section 2 computations of QSLs from reconstructed 3D magnetic fields will be introduced; in Sections 3 and 4 comparisons between QSLs and features in Hα and soft X-ray images will be shown, and finally summary and discussion will be presented in Section 5.

2. QSLs in the 3D Magnetic Fields

Taking the observed photospheric vector magnetic fields as the boundary condition, we computed 3D force-free magnetic fields above the photosphere with a new numerical technique proposed by Yan and Sakurai (1997, 2000). On 26 October 1992, 150 frames of vector magnetograms were obtained with the Solar Flare Telescope at Mitaka. We select two of them as the boundary conditions for the reconstruction of 3D magnetic fields above the photosphere. The reconstructed 3D fields have

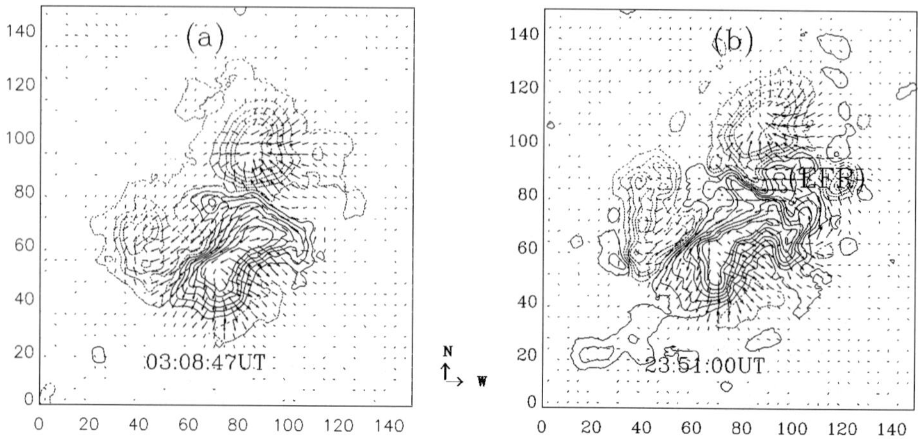

Figure 2. Mitaka vector magnetograms of AR 7321 on 26 October 1992. The strength of the line-of-sight field component is mapped by the contours (solid lines, positive polarity; dotted lines, negative polarity; 80, 320, 640, 960, 1280, 1600 and 1960 G levels). The strength and orientation of the transverse field component is shown by the arrows, the shortest arrows are for 150 G, the longest for 1600 G or more. EFR = emerging flux region. Field of view: 200 × 200 arc sec^2. Location: (a) S26 W04; (b) S26 W15.

been identified to be in agreement with divergence-free and force-free conditions (Wang, Yan, and Sakurai, 2000). In order to determine locations of QSLs in the 3D fields, we employ a method suggested by Priest and Démoulin (1995).

The location of QSLs in a plane at $z = z_o$ is determined by values of $N(x, y)$,

$$N(x, y) = \left[\left(\frac{\partial p_x}{\partial x}\right)^2 + \left(\frac{\partial p_x}{\partial y}\right)^2 + \left(\frac{\partial p_y}{\partial x}\right)^2 + \left(\frac{\partial p_y}{\partial y}\right)^2\right]^{1/2}, \tag{1}$$

where $p_x = x2 - x1$, $p_y = y2 - y1$, which are two components of a vector determined by two foot points of a field line (Démoulin *et al.*, 1996). In fact, N is the norm of the displacement gradient tensor

$$M = \begin{pmatrix} \frac{\partial p_x}{\partial x} & \frac{\partial p_x}{\partial y} \\ \frac{\partial p_y}{\partial x} & \frac{\partial p_y}{\partial y} \end{pmatrix}, \tag{2}$$

which has significant physical meaning in 3D magnetic reconnection (Priest and Démoulin, 1995).

Figures 3 and 4 show the intersections of QSLs with a plane in the chromosphere ($z = 1.5$ arc sec) and the field lines connecting the areas with $N > 10$. These field lines indicate the coronal linkage of the field lines related to the QSLs. When a null point appears in the plane at $z = 1.5$ arc sec, the value of N becomes very large. In fact, there was a null point in the reconstructed 3D fields, which was

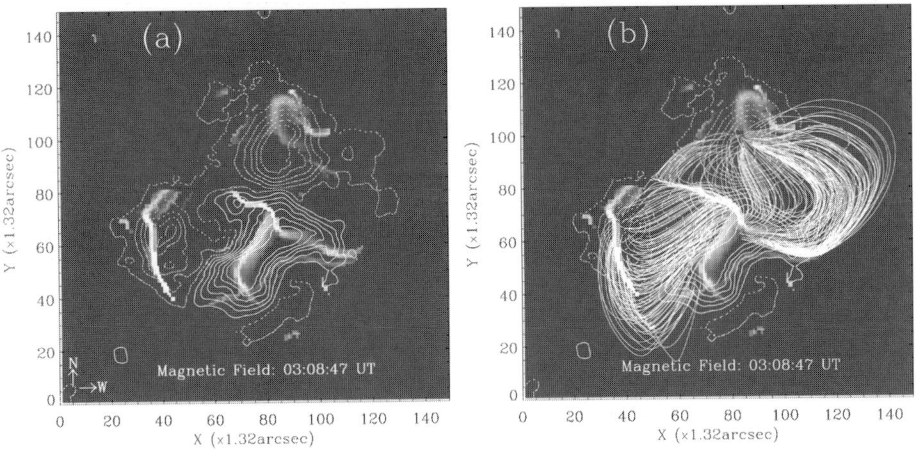

Figure 3. The coronal linkage of the field lines related to the QSLs obtained from 3D magnetic fields in AR 7321 at 03:08:47 UT of 26 October 1992. The contour maps are the same as those in Figure 2. (a) Gray map of $N(N_{\max} = 102)$ at the chromospheric level. This map indicates the intersection of QSLs with the chromospheric surface. (b) Field lines connecting the areas with $N > 10$. These lines form 3D QSLs above the chromosphere.

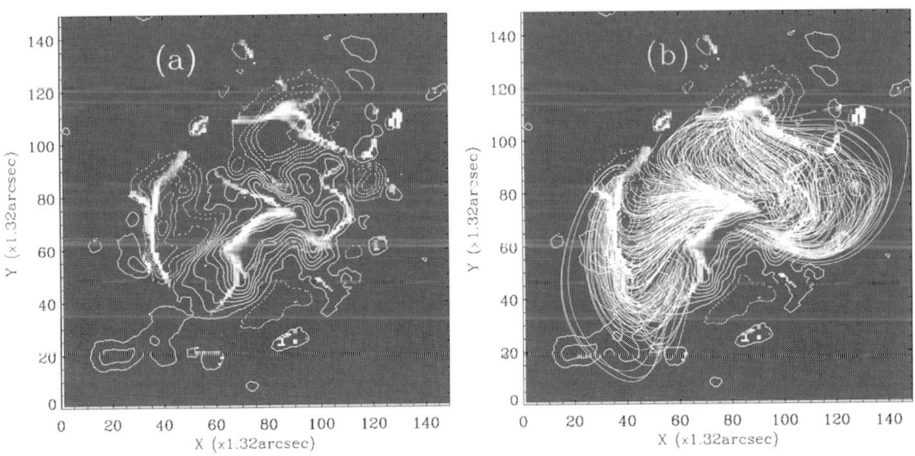

Figure 4. The coronal linkage of the field lines related to the QSLs obtained from 3D magnetic fields in AR 7321 at 23:51:00 UT of 26 October 1992. The contour maps are the same as those in Figure 2. (a) Gray map of $N(N_{\max} = 91)$ at the chromospheric level. This map indicates the intersection of QSLs with the chromospheric surface. (b) Field lines connecting the areas with $N > 10$. These lines form 3D QSLs above the chromosphere.

located between two main negative polarities and close to the chromosphere. This is the reason why the maximum value of N is about 100. It should be pointed out that since the spatial resolution of the magnetograms is larger than 1.5 arc sec, the observed boundary field was smoothed to be a continuous field. The values of N, therefore, are smoothly distributed on the plane in the chromosphere.

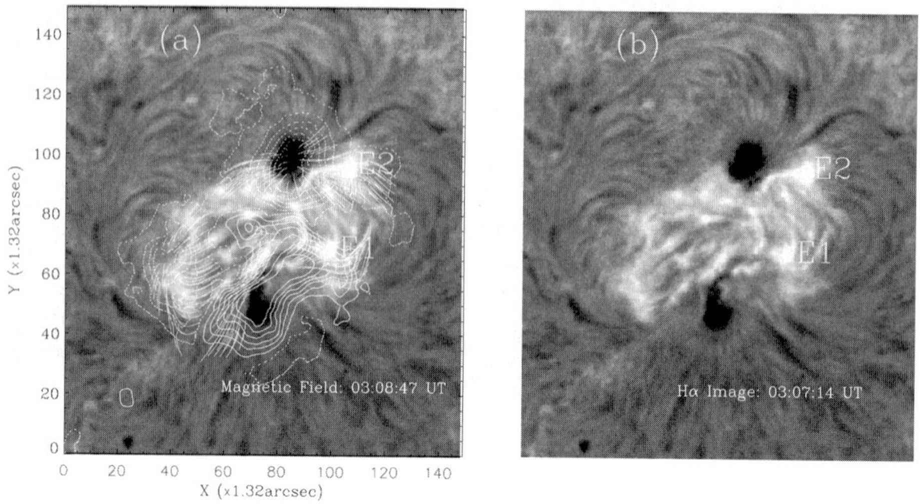

Figure 5. Comparison between QSLs and Hα image of AR 7321 at 03:08:47 UT on 26 October 1992. (a) Field lines showing the shape of QSLs below a height of $z = 4$ arc sec. The contour maps are the same as those in Figure 2. The footpoints of the field lines forming the QSLs are uniformly distributed within the area with $N > 10$ in the chromosphere, where Hα plages appear. (b) Hα image. The bright points, E1 and E2, are the footpoints of the bright feature in Figure 1(a).

3. QSLs and Hα Features

A QSL forms a 3D shell extending from the photospheric surface, which should have its observational effect on different solar atmospheric layers. First of all, we make a comparison between QSLs and Hα features.

Figures 5 and 6 show short portions of the field lines forming the QSLs in the chromosphere. In order to investigate the relationship between the QSLs and the chromosphere, all field lines forming the QSLs are cut off at the height of $z = 4$ arc sec. It can be seen that most of these short field lines are located in the Hα plage areas. In Figure 6, three dark Hα loops are over the height of $z = 4$ arc sec and their footpoints are anchored in the intersections of the QSLs with the plane at $z = 1.5$ arc sec. These facts suggest that the QSLs play an important role in the chromospheric heating: Hα plages are caused by the heating processes associated with QSLs.

However, the Hα bright points, E1 and E2, which have been identified with the footpoints of the BCF in Figure 7(b), are located within the intersection of the QSLs with the chromosphere. With their compact intensity distribution and fast evolution, E1 and E2 are evidently different from the plages. This implies that the enhanced coronal heating might be different from the heating processes related to Hα plages.

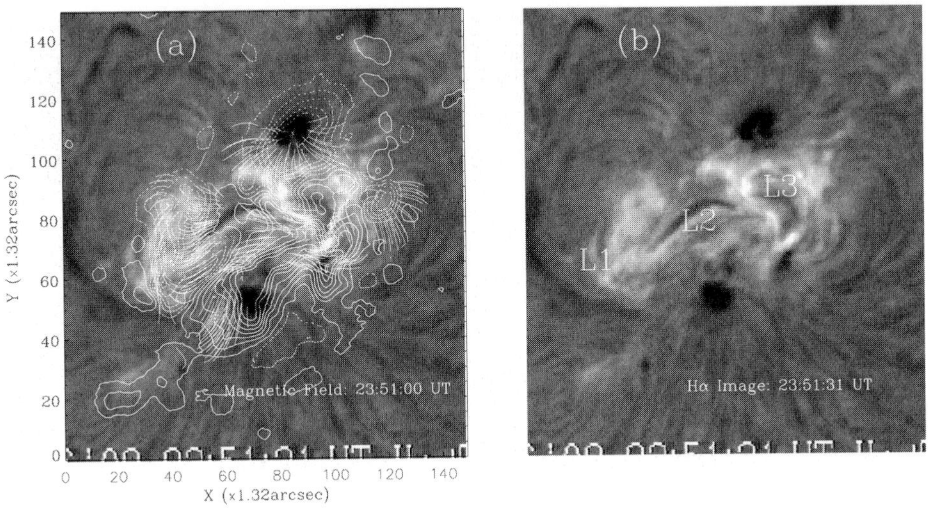

Figure 6. Comparison between QSLs and Hα image of AR 7321 at 23:51:00 UT on 26 October 1992. (a) Field lines showing the shape of QSLs below a height of z = 4 arc sec. The contour maps are the same as those in Figure 2. The footpoints of the field lines forming the QSLs are uniformly distributed in the area with $N > 10$ in the chromosphere, where Hα plages appear. (b) Hα image. The Hα loops, L1, L2 and L3, are in correspondence with three soft X-ray loops in Figure 1(b).

4. QSLs and Soft X-Ray Features

Next, we compare the QSLs with the soft X-ray features observed by *Yohkoh*. Figure 7 shows that the BCF with footpoints E1 and E2 is embraced by the field lines forming the QSLs. Figure 8 shows that three soft X-ray loops are also embraced by this kind of field lines. The intensity of soft X-rays from these loops, however, are different. L3 has the strongest X-ray emission, while L1 the weakest. Since we have only 2D soft X-ray images, the heating rate within the volume confined by the coronal linkage of the field lines related to the QSLs cannot be evaluated. Furthermore, there were no multiple wavelength observations due to the 'quiet' state of this active region. Thus we cannot determine temperature distributions in these QSLs by making use of filter ratios.

From Figures 7 and 8, it can be seen that all of BCFs are surrounded by the DCFs, and both BCFs and DCFs are mainly confined by the coronal linkage of the field lines related to the QSLs. This means that the QSLs are closely associated with the regions with soft X-ray emission.

The time interval between Figures 7 and 8 is about 21 hours, during which there was no important activity. We suppose that magnetic fields in this active region evolved slowly through a series of force-free states. There are more than 150 magnetograms and 300 soft X-ray images obtained during this time interval. We selected several magnetograms obtained at different time to reconstruct their 3D coronal fields. By comparing QSLs in these reconstructed fields with soft X-

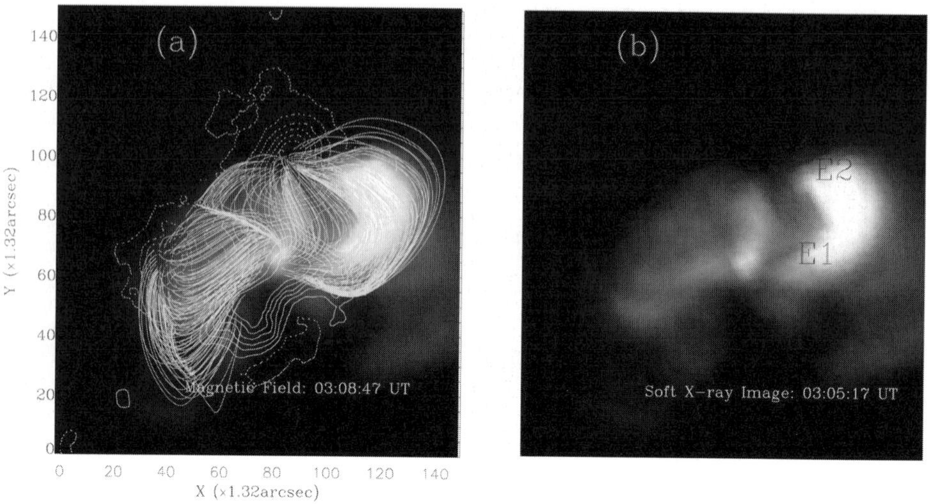

Figure 7. Comparison between QSLs and soft X-ray image of AR 7321 on 26 October 1992. (a) QSLs and soft X-ray image. The contour maps are the same as those in Figure 2. The footpoints of field lines are uniformly distributed in the area with $N > 10$ in the chromosphere. All field lines are not over the height of $z = 40$ arc sec. (b) Soft X-ray image of strongly heated coronal features in AR 7321 on 26 October 1992. E1 and E2 are the locations of the Hα bright points in Figure 5(b).

ray emission at close time, it can be seen that features of soft X-ray emission are morphologically similar to those of QSLs. Since these QSLs were computed from the 3D magnetic fields reconstructed with boundary fields on the photosphere, the morphological evolution of the regions with soft X-ray emission was determined by the variations of the photospheric field.

The facts mentioned above imply that the QSLs play an important role in the coronal loop heating.

However, the BCFs in Figures 7(b) and 8(b) are associated only with a part of the field lines related to the QSLs, in other words, the enhanced coronal heating is not uniformly distributed within the volume bounded by these field lines. This indicates again that the enhanced coronal heating cannot be attributed only to the QSLs. Comparing the magnetogram in Figure 2(a) with that in Figure 2(b), the following differences have been found: (a) The magnetic shear along the neutral-line increased; (b) A large emerging flux region appeared in the western side of this active region. Since there was no important activity during the period from 03:05 UT to 23:52 UT, the configuration of the magnetic fields in this active region evolves slowly during this period. In that sense, the enhanced coronal loop heating might be due not only to the magnetic shear but also to the emerging magnetic flux.

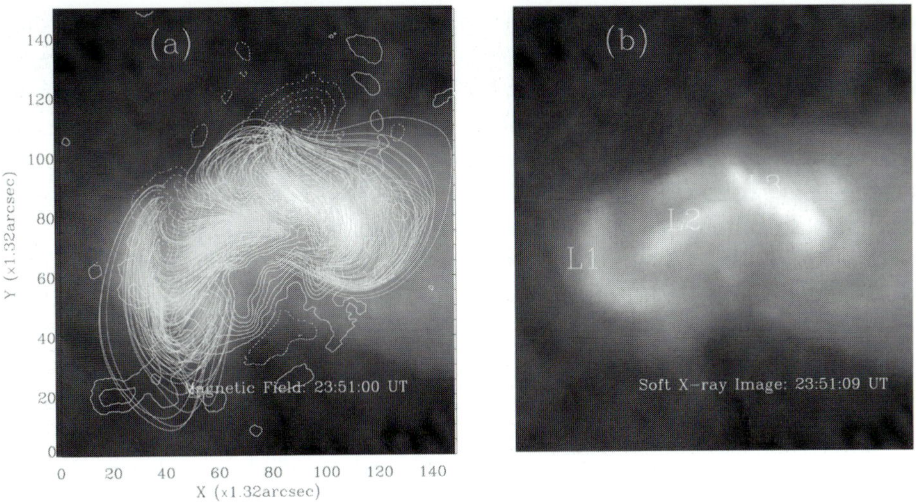

Figure 8. Comparison between QSLs and soft X-ray image of AR 7321 on 26 October 1992. (a) QSLs and soft X-ray image. The contour maps are the same as those in Figure 2. The footpoints of field lines are uniformly distributed in the area with N>10 in the chromosphere. All field lines are not over the height of $z = 40$ arc sec. (b) Soft X-ray image. The soft X-ray loops, L1, L2 and L3, are in correspondence with three Hα loops in Figure 6(b).

5. Summary and Discussion

Taking the photospheric vector magnetic fields in Figure 2 as the boundary conditions, the 3D magnetic fields above the photosphere were computed with the numerical technique suggested by Yan and Sakurai (2000), and then the QSLs were determined in these fields. According to the analysis in Sections 2, 3 and 4, we obtained the following results: (a) the portions of QSLs in the chromosphere are associated with Hα plages; (b) DCFs and BCFs are morphologically similar to the features of QSLs; (c) BCFs are associated with a part of the magnetic field lines related to the QSLs. According to these results we conclude that QSLs play an important role in the chromospheric and coronal heating in the active region NOAA 7321.

It is interesting that the QSLs determined from the 3D magnetic fields inferred from the observed boundary fields define well the domain with soft X-ray emission. The heating mechanism suggested by Démoulin and Priest (1997) emphasizes that the dissipation of magnetic energy in the corona is stimulated by the fragmentation of the photospheric magnetic field. Since we have no vector field observations with high spatial resolution, this heating mechanism needs to be examined further.

QSLs are the regions where 3D magnetic reconnection takes place. If energy is released in the coronal portion of QSLs, then it is transported along the reconnected field lines towards the footpoints of these field lines. In the chromosphere and corona, the linkages of the field lines related to the QSLs define the morphology of Hα and soft X-ray features. This implies that magnetic fields in the volume bounded by these field lines make significant contributions to the chromospheric and coronal heating. As to the BCFs, the energy release may occur along a certain portion of the QSLs due to concentrated currents mainly located in this portion and the photospheric evolution, *e.g.* the magnetic shear and emerging magnetic flux, favorable for energy release there. This implies that enhanced coronal emission might not be uniformly distributed within the volume of QSLs.

The role of QSLs for coronal heating provides an explanation why X-ray emission is almost missing in sunspot umbras (Pallavicini *et al.*, 1979). Since a small part of the magnetic flux in a sunspot is involved in QSLs, the X-ray emission above the sunspot is usually weak.

It is noted that the magnetic shear and emerging magnetic flux are related to the enhanced coronal heating, but the exact heating mechanism in BCFs remains unknown. A comprehensive study of the enhanced coronal heating is necessary in the future.

Acknowledgments

The numerical computations were carried out at Astronomical Data Analysis Center of NAOJ. H. N. Wang is grateful to NAOJ for hospitality during his stay as a Foreign Research Fellow. H. N. Wang and Y. Yan are supported by NSFC grants and CAS grants. The authors are grateful to the referee's valuable suggestions and comments.

References

Démoulin, P., Hénoux, J. C., and Mandrini, C. H.: 1994, *Astron. Astrophys.* **285**, 1023.

Démoulin, P., Van Driel-Gesztelyi, L., Schmieder, B., Hénoux, J. C., Csepura, G., and Hagyard, M. J.: 1993, *Astron. Astrophys.* **271**, 292.

Démoulin, P., Mandrini, C. H., Hénoux, J. C., and Machado, M. E.: 1994, *Solar Phys.* **150**, 221.

Démoulin, P., Hénoux, J. C., Priest, E. R., and Mandrini, C. H.: 1996, *Astron. Astrophys.* **308**, 643.

Démoulin, P., Henoux, J. C., Mandrini, C. H., and Priest, E. R.: 1997a, *Solar Phys.* **174**, 73.

Démoulin, P., Bagalá, L. G., Mandrini, C. H., Hénoux, J. C., and Rovira, M. G.: 1997b, *Astron. Astrophys.* **325**, 305.

Falconer, D. A., Moore, R. L., Porter, J. G., Gary, G. A., and Shimizi, T.: 1997, *Astrophys. J. Sup.* **482**, 519.

Filippov, B.: 1999, *Solar Phys.* **185**, 297.

Gaizauskas, V., Mandrini, C. H., Démoulin, P., Luoni, M. L., and Rovira, M. G.: 1998, *Astron. Astrophys.*, **332**, 353.

Giovanelli, R. G.: 1947, *Monthly Notices Royal Astron. Soc.* **107**, 338.

Giovanelli, R. G.: 1948, *Monthly Notices Royal Astron. Soc.* **108**, 163.

Galsgaard, K. and Nordlund, A.: 1997, *J. Geophys. Res.* **102**, 231.

Golub, L., Mason, C., Rosner, R., Serio, S., and Vaiana, G. S.: 1980, *Astrophys. J. Sup.* **238**, 343.

Hara, H.: 1997, in T. Kosugi *et al.* (eds.), *Observational Plasma Astrophysics: Five years of Yohkoh and Beyond*, Kluwer Academic Publishers, Dordrecht, Holland.
Lau, Y. T. and Finn, J. M.: 1990, *Astrophys. J. Sup.* **350**, 672.
Mandrini, C. H., Démoulin, P., Hénoux, J. C., and Machado, M. E.: 1991, *Astron. Astrophys.* **250**, 541.
Mandrini, C. H., Rovira, M. G., Démoulin, P., Hénoux, J. C., Machado, M. E., and Wilkinson, L. K.: 1993, *Astron. Astrophys.* **272**, 609.
Mandrini, C. H., Démoulin, P., Van Driel-Gesztelyi, L., Schmieder, B., Cauzzi G., and Hofmann, A.: 1996, *Solar Phys.* **168**, 115.
Pallavicini, R., Vainana, G. S., Totani, G., and Felli, M.: 1979, *Astrophys. J. Sup.* **229**, 375.
Parker, E. N.: 1957, *J. Geophys. Res.* **62**, 509.
Priest, E. R. and Démoulin, P.: 1995, *J. Geophys. Res.* **100**, A12, 23443.
Priest, E. R. and Forbes, T. G.: 1986, *J. Geophys. Res.* **91**, 5579.
Priest, E. R. and Titov, V. S.: 1996, *Phil. Trans. Roy. Soc. London* **A354**, 2951.
Priest, E. R., Foley, C. R., Heyvaerts, J., Arber, T. D., Culhane, J. L., and Acton, L. W.: 1998, *Nature* **393**, 545.
Sakurai, T. and Wang, H. N.: 1999, *Solar Physics with Radio Observations*, Proceedings of Nobeyama Symposium 1998, NRO Report, in press.
Sakurai, T., Ichimoto, K., Nishino, Y., Shinoda, K., Noguchi, M., Hiei, E., Li, T., He, F., Mao, W., Lu, H., Ai, G., Zhao, Z., Kawakami S., and Chae, J.: 1995, *Publ. Astron. Soc. Japan* **47**, 81.
Somov, B. V.: 1986, *Astron. Astrophys.* **163**, 210.
Sweet, P. A.: 1969, *Ann. Rev. Astron. Astrophys.* **7**, 149.
Tsuneta, S., Acton, L., Bruner, M., Lemen, J., Brown, W., Caravalho, R., Catura, R., Freeland, S., Jurcevich, B., Morrison, M., Ogawara, Y., Hirayama, T., and Owens, J.: 1991, *Solar Phys.* **136**, 37.
Wang, H. N.: 1997, *Solar Phys.* **174**, 265.
Wang, H. N. and Wang, J. X.: 1996, *Astron. Astrophys.* **313**, 285.
Wang, H. N., Sakurai, T., and Yan, Y.: 2000, *Adv. Space Res.* **25**, 1769.
Wang, H. N., Yan, Y., and Sakurai, T.: 2000, *Solar Phys.*, submitted.
Wang, T. J., Wang, H. N., and Qiu, J.: 1999, *Astron. Astrophys.* **342**, 854
Yan, Y. and Sakurai, T.: 1997, *Solar Phys.* **174**, 65.
Yan, Y. and Sakurai, T.: 2000, *Solar Phys.* **195**, 89.

RECENT THEORETICAL RESULTS ON CORONAL HEATING

DANIEL O. GOMEZ[1,*], PABLO A. DMITRUK[1,†] and LEONARDO J. MILANO[2,†]
[1]*Department of Physics, University of Buenos Aires, Ciudad Universitaria, 1428 Buenos Aires, Argentina*
[2]*National Solar Observatory, P.O.Box 62, Sunspot, NM 88349, U.S.A.*

(Received 20 September 1999; accepted 1 March 2000)

Abstract. The scenario of magnetohydrodynamic turbulence in connection with coronal active regions has been actively investigated in recent years. According to this viewpoint, a turbulent regime is driven by footpoint motions and the incoming energy is efficiently transferred to small scales due to a direct energy cascade. The development of fine scales to enhance the dissipation of either waves or DC currents is therefore a natural outcome of turbulent models. Numerical integrations of the reduced magnetohydrodynamic equations are performed to simulate the dynamics of coronal loops driven at their bases by footpoint motions. These simulations show that a stationary turbulent regime is reached after a few photospheric times, displaying a broadband power spectrum and a dissipation rate consistent with the energy loss rates of the plasma confined in these loops. Also, the functional dependence of the stationary heating rate with the physical parameters of the problem is obtained, which might be useful for an observational test of this theoretical framework.

1. Introduction

Recent X-ray and EUV observations of the solar corona reveal a highly structured brightness distribution, which is the consequence of magnetic fields confining the X-ray emitting plasma and governing its dynamics. Observations made with high spatial resolution (Schrijver *et al.*, 1999; Golub *et al.*, 1990; Golub and Pasachoff, 1997) show coronal magnetic loops with a highly filamentary internal structure.

Coronal heating theories have traditionally been classified into two broad categories: (a) *AC or wave models*, for which energy is provided by waves generated at the photosphere and, (b) *DC or stress models*, which assume that energy dissipates in magnetic stresses driven by slow footpoint motions. Although these two concepts seem mutually exclusive, the following assumptions are shared by both classes of models: (i) the ultimate source for coronal heating is the kinetic energy of the photospheric velocity field, (ii) the dissipation mechanism is Joule heating, although viscosity might also play an important role, (iii) the existence of fine structure in the coronal magnetic field is invoked to speed up Joule (or viscous) dissipation. Recent review articles on coronal heating (Narain and Ulmschneider,

*Also at: Instituto de Astronomía y Física del Espacio, C.C.67, Suc.28, 1428 Buenos Aires, Argentina.
†Presently at: Bartol Research Institute, University of Delaware, Newark, DE 19716, U.S.A.

The physics of the solar corona and transition region. Part II.
Reprinted from Solar Physics **195,** 2000.

1996, 1990; Gómez, 1990; Zirker, 1993), explore the various theoretical models in further detail.

As mentioned, the natural candidate for the energy dissipation is Joule heating, but the typical time scale to dissipate coronal magnetic stresses on the length scale of the driving photospheric motions is exceedingly long. This time-scale can be estimated as $l^2/\eta \sim 10^6$ years ($l = 10^3$ km: length scale of photospheric motions, $\eta = 10^3$ cm^2 s^{-1}: resistivity). Therefore, most of the current theories of coronal heating deal with different mechanisms to speed up dissipation (Parker, 1972, 1983; Heyvaerts and Priest, 1983; van Ballegooijen, 1986). Numerical integrations of the MHD equations have also been performed (Mikic, Schnack, and van Hoven, 1989; Longcope and Sudan, 1994), to study the dynamics of coronal loops driven at their bases by convective motions. These simulations show the development of small spatial structures (current sheets), which contribute to enhance dissipation.

One of the promising scenarios is the assumption that the magnetic and velocity fields of the coronal plasma are in a turbulent state (Gómez and Ferro Fontan, 1988, 1992; Heyvaerts and Priest, 1992; Einaudi *et al.*, 1996; Hendrix and van Hoven, 1996; Dmitruk and Gómez, 1997). Since the dynamics of coronal loops is dominated by their magnetic field, turbulent fluctuations of both the velocity and the magnetic fields are generated predominantly in the directions perpendicular to the main magnetic field (for details on the MHD turbulent scenario, see, for instance Biskamp, 1993). In a turbulent regime, energy is transferred from photospheric motions to the magnetic field. This energy cascades toward small scales due to nonlinear interactions, until highly structured electric currents are formed. The development of fine scales to enhance the dissipation of either waves or DC currents, is therefore a natural outcome of turbulent models. On the other hand, the dynamics of the solar photosphere is dominated by its velocity field, and therefore an essentially hydrodynamic turbulence is observed in this region (Nesis *et al.*, 1999).

We present numerical integrations of the reduced magnetohydrodynamic equations, to simulate the dynamics of coronal loops driven at their bases by footpoint motions. These simulations show that after a few photospheric turnover times a stationary turbulent regime is reached, displaying a broadband power spectrum and a dissipation rate consistent with the energy loss rates of the plasma confined in these loops. In Section 2 we briefly describe recent observational results which are relevant to the problem of coronal heating. The reduced MHD approximation is described in Section 3, and various results from recent numerical simulations of the RMHD equations are summarized in Section 4. An interesting scaling law, which relates the heating rate with the physical parameters of the loop is given in Section 5. A statistics of dissipation events arising from time-extended 2D MHD simulations, which displays a slope quite comparable to those reported for flare (Hudson, 1991) and microflare (Shimizu, 1995) events, is shown in Section 6. Finally, in Section 7 we summarize our conclusions.

2. Recent Observational Results

The Transition Region and Coronal Explorer (TRACE) is showing us the spatial structure and dynamics of the solar corona with unprecedented quality. Active regions observed by TRACE are composed by bundles of long and thin loops displaying a highly dynamic behavior, which involve not only kinematic motions of the magnetic structure, but also rapid changes in the temperature and density of individual loops.

The extremely dynamic nature of coronal loops is suggestive of a rather intermittent heating process, both in space and time. Typical time-scales for this variability are a few minutes or less, while spatial structures of all sizes are observed, down to the resolution limit of 700 km (Schrijver *et al.*, 1999). A spatially intermittent heating would qualitatively explain the rich transverse structure, since only those field lines connected to locations where heating pulses occur will light up. On the other hand, a temporally intermittent heating would force these loops to adjust their internal structures, i.e., their densities, temperatures and longitudinal velocities. In some active regions, transverse oscillations of loops have also been observed, and it has been suggested that the kinetic energy associated to these oscillations might play a role in the heating of the region (Nakariakov *et al.*, 1999).

3. Reduced MHD Approximation

In view of the observed features mentioned above, in what follows we concentrate on the theoretical description of a relatively homogeneous bundle of field lines. The loop footpoints are deeply rooted into the photosphere, which moves individual field lines around, thus generating magnetic stresses in the coronal portion of the loop. Two important time-scales arise in this process: the Alfvén time, t_A, which is the time it takes for an Alfvén wave to travel along the loop; and the photospheric time t_p, which is the turnover time of photospheric granules. If $t_A \gg t_p$, i.e., if the driving time-scale is much shorter than the response time of the loop, AC heating processes might be important (Ionson, 1982; Heyvaerts and Priest, 1983; Hollweg, 1985; Ofman, Davila, and Steinolfson, 1995). On the other hand, whenever $t_A \ll t_p$, DC heating is dominant. Under the assumption that a coronal loop reaches a stationary turbulent state, its response to a broad range of photospheric driving frequencies can be modeled by a closure model. At least for photospheric power spectra decreasing as power laws of both the wavenumber and the time frequency, the heating of the loop is dominated by DC heating (Milano, Dmitruk, and Mandrini, 1997) (see also Heyvaerts and Priest (1992) and Gómez and Ferro Fontán (1992) for closure models applied to a magnetic arcade and to a loop driven by stationary footpoint motions, respectively).

In the present paper, the aim is to study the heating of a topologically simple loop, which can be regarded as a sub-structure or building block of the much more

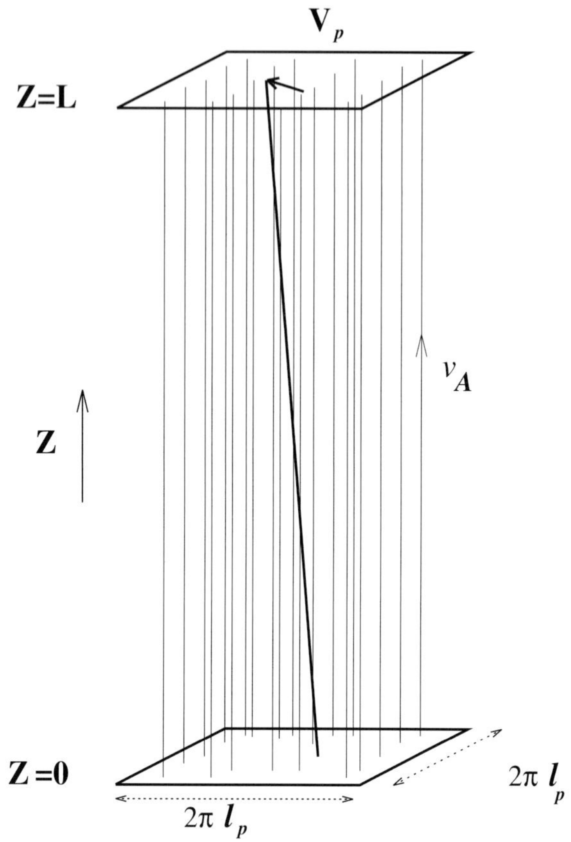

Figure 1. Perturbed field line in a coronal loop. The planes $z = 0$ and $z = L$ represent the photosphere.

complex active regions. The idea behind this approach is that if we manage to explain the heating of simple loops, then we should be able to explain the heating of complex superpositions of these structures, although it seems clear that complex structures might provide extra heating. Furthermore, we take advantage of the (typically) large aspect ratio of observed loops and *straighten them up*, thus neglecting toroidal effects, i.e., effects arising from the curvature of loops. Also, oscillatory motions of the loop as a whole are neglected, as well as its interactions with other magnetic structures.

We therefore consider a simplified model of a coronal magnetic loop with length L and cross section $2\pi l_p \times 2\pi l_p$, where l_p is the length scale of typical photospheric motions. For elongated loops, i.e., such that $2\pi l_p \ll L$, it is reasonable to neglect toroidal effects. The main magnetic field $\mathbf{B}_0$ is assumed to be uniform and parallel to the axis of the loop (the z axis). The planes at $z = 0$ and $z = L$ correspond to the loop footpoints at the photosphere, as shown in Figure 1.

Under these assumptions, we are able to use the reduced MHD approximation (Strauss, 1976), in which the plasma moves incompressibly in planes perpendicular to the axial field $\mathbf{B}_0$, and the transverse component of the magnetic field is small compared to $\mathbf{B}_0$. The very high electric conductivity (frozen field) allows photospheric motions to easily drive magnetic stresses in the corona (Parker, 1972), the field lines twist and bend due to these motions and this generates transverse components of velocity $\mathbf{u}$ and magnetic field $\mathbf{b}$. Therefore:

$$\mathbf{B} = B_0\mathbf{z} + \mathbf{b}(x, y, z, t), \quad \mathbf{b} \cdot \mathbf{z} = 0\,, \tag{1}$$

$$\mathbf{u} = \mathbf{u}(x, y, z, t), \quad \mathbf{u} \cdot \mathbf{z} = 0\,. \tag{2}$$

Since both $\mathbf{b}$ and $\mathbf{u}$ are two-dimensional and divergence-free fields, they can be represented by scalar potentials:

$$\mathbf{b} = \nabla \times (a\mathbf{z}) = \nabla a(x, y, z, t) \times \mathbf{z}\,, \tag{3}$$

$$\mathbf{u} = \nabla \times (\psi\mathbf{z}) = \nabla \psi(x, y, z, t) \times \mathbf{z}\,, \tag{4}$$

where ∇ indicates derivatives in the x, y plane.

The reduced MHD equations for the potentials ψ and a are (Strauss, 1976):

$$\partial_t a = v_A\, \partial_z \psi + [\psi, a] + \eta\nabla^2 a\,, \tag{5}$$

$$\partial_t w = v_A\, \partial_z j + [\psi, w] - [a, j] + \nu\nabla^2 w\,, \tag{6}$$

where vorticity w and current density j relate to the potentials as

$$w = -\nabla^2\psi, \quad j = -\nabla^2 a\,. \tag{7}$$

The brackets $[A, B] = \partial_x A \partial_y B - \partial_y A \partial_x B$ are the standard Poisson brackets and $v_A = B_0/\sqrt{4\pi\rho}$ is the Alfvén velocity (ρ is the plasma density).

Equation (5) describes the advection of the potential a and Equation (6) corresponds to the evolution of vorticity w. The terms $v_A\partial_z$ represent the coupling between neighboring z = constant planes. The role of these terms is to transfer energy from the footpoints into the coronal part of the loop. The ∇^2 terms represent dissipative effects, the constants η and ν being the resistivity and viscosity coefficients. The nonlinear terms are those represented by the Poisson brackets. Their role is to couple normal modes in such a way that energy, and other ideal invariants, can be transferred between them. The fields $\mathbf{u}$ and $\mathbf{b}$ depend on the axial z-coordinate, although their nonlinear interaction proceeds independently on different constant z planes across the loop.

4. Turbulent Heating

Since the Reynolds numbers in coronal active regions are extremely large ($R \sim S \sim 10^{10-12}$) we can expect coronal loops to be in a strongly turbulent regime. Although the theoretical framework for the study of turbulence was first developed for non-magnetic fluids, thanks to the pioneering work of Kolmogorov, the following general description applies to either hydrodynamic or magnetohydrodynamic turbulent regimes. The effect of nonlinear terms is to redistribute excitations from one wavenumber to another in a rather stochastic fashion. Only those excitations at sufficiently large wavenumbers decay as a consequence of dissipative effects. Therefore, a net flow of excitations in Fourier space is established, toward those regions which are deficient with respect to the values calculated from an ideal (without dissipation) model (Montgomery, 1983). As a result, this nonlinear redistribution continuously replenishes the excitations being drained at the large wavenumber region. An increase in Reynolds number only raises the value of the typical wavenumber at which dissipation begins to dominate, but does not inhibit the excitation flow in Fourier space. According to this scenario, three regions in Fourier space (wavenumbers space) can be identified, each of them displaying a different turbulent behavior, as sketched in Figure 2:

(1) *The energy-containing region*: Comprises those modes that are being excited directly by the external driver, which is usually located at low wavenumbers. However, it does not necessarily include the very lowest wavenumbers, which are determined by the size of the system.

(2) *Dissipation region*: Corresponds to those modes where fluctuations are being efficiently quenched by dissipative (viscous or ohmic) effects. This region is normally located at the largest wavenumbers, where the linear dissipative terms become comparable to the nonlinear terms.

(3) *Energy inertial region*: In this region, external forces and dissipation are both negligible. Only nonlinearities play a role, transferring fluctuations from one mode to another, while keeping the total energy constant. This region normally bridges the gap between the low wavenumber energy containing region and the large wavenumber dissipative zone.

Kolmogorov (1941), following heuristic arguments, has shown that when a three dimensional incompressible fluid is submitted to external forcing with a narrow spectrum, a direct energy cascade is generated and a stationary energy spectrum is achieved, displaying the well known $k^{-5/3}$ distribution in the energy inertial region (see Figure 2). Kolmogorov's ideas, mainly based on scaling properties of the ideal equations and on the existence of a net energy flow through the corresponding inertial range, are usually known as *cascade theory* and have been applied to a number of turbulent systems including two and three-dimensional MHD turbulence (Montgomery, 1983). The power spectra predicted by cascade theory for the energy inertial range, have in many cases been confirmed by experiments (Grant, Stewart, and Molliet, 1962; Matthaeus and Goldstein, 1982; Sommeria, 1986) and

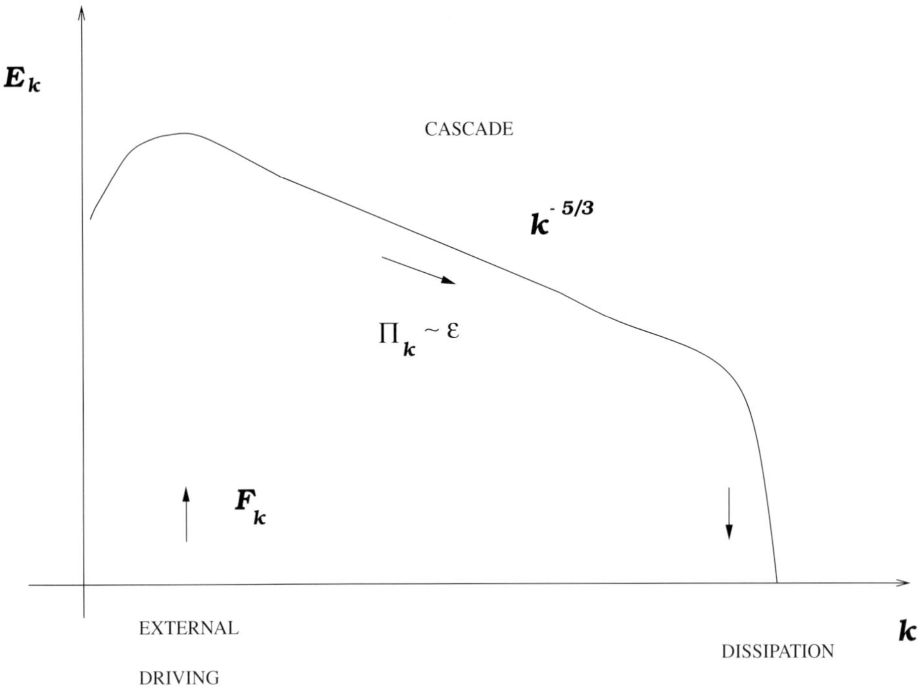

Figure 2. Sketch of the Kolmogorov spectrum for hydrodynamic turbulence.

numerical simulations (Lilly, 1969; Herring and Kraichnan, 1972; Fyfe, Montgomery, and Joyce, 1977; Meneguzzi, Frisch, and Pouquet, 1981; Matthaeus and Lamkin, 1986; Biskamp and Welter, 1989; Politano, Pouquet, and Sulem, 1989).

MHD turbulence displays two rather different regimes, depending on the value of the plasma parameter β, which is the ratio between gas pressure and magnetic pressure. For values of β much larger than unity, the dynamics is dominated by the velocity field. Therefore, the turbulence is essentially hydrodynamic, and corresponds to the Kolmogorov scenario outlined above. This regime probably provides a reasonable description of photospheric motions away from sunspots (see Nesis *et al.*, 1999), where the relatively tenuous magnetic field is simply being advected by the velocity field. On the other hand, the very low β plasma confined in coronal loops, follows a magnetically dominated regime. Turbulent fluctuations are perpendicular to the main (macroscopic) magnetic field $\mathbf{B}_0$, and consequently small scales develop along planes perpendicular to $\mathbf{B}_0$. In this regime, the total energy (i.e., kinetic plus magnetic) follows a direct cascade, and a Kraichnan spectrum is developed in the energy inertial region (for a recent review on MHD turbulence, see for instance Biskamp, 1993). The Kraichnan spectrum, $E_k \simeq k^{-3/2}$, is derived following scaling arguments similar to Kolmogorov's, but including the effect of Alfvén waves traveling along macroscopic magnetic field structures (Kraichnan, 1965). It is important to stress that in this low-β regime, magnetic field lines are not simply being advected by footpoint motions, but also generate their own non-linear response to the forcing, developing small scale structures, as shown below.

We numerically explore the feasibility of a turbulent scenario for coronal loops, describing the internal dynamics of loops through the RMHD approximation. We assume periodicity for the lateral boundary conditions, and specify the velocity fields at the $z = 0$ and $z = L$ photospheric boundaries. In particular, we assume

$$\psi(z = 0) = 0, \quad \psi(z = L) = \Psi(x, y) , \tag{8}$$

where $\Psi(x, y)$ is the stream function which describes stationary and incompressible footpoint motions on the photospheric plane. The strength of this external velocity field is therefore proportional to a typical photospheric velocity V_p ($V_p \approx 1$ km s^{-1}).

To transform Equations (5)–(6) into their dimensionless form, we choose l_p and L as the units for transverse and longitudinal distances ($l_p \approx 10^3$km and $L \approx 10^4$ -10^5 km) and $t_A \equiv L/v_A$ as the time unit ($t_A \approx$ 10–100 s). The dimensionless RMHD equations are:

$$\partial_t a = \partial_z \psi + [\psi, a] + \frac{1}{S}\nabla_\perp^2 a , \tag{9}$$

$$\partial_t w = \partial_z j + [\psi, w] - [a, j] + \frac{1}{R}\nabla_\perp^2 w , \tag{10}$$

where $S = l_p^2/\eta t_A$ and $R = l_p^2/\nu t_A$ are respectively the magnetic and kinetic Reynolds numbers. For typical coronal conditions, these numbers lie within the range $R \approx S \approx 10^{10}$–$10^{12}$. For simplicity, hereafter we will consider the case $S = R$.

For the numerical simulation of these equations, the magnetic vector potential and the stream function are expanded in Fourier modes in each plane (x, y), i.e.,

$$a(x, y, z, t) = \sum_{\mathbf{k}} a_{\mathbf{k}}(z, t)\, e^{i\mathbf{k}\cdot\mathbf{x}} , \tag{11}$$

$$\psi(x, y, z, t) = \sum_{\mathbf{k}} \psi_{\mathbf{k}}(z, t)\, e^{i\mathbf{k}\cdot\mathbf{x}} . \tag{12}$$

The equations for the coefficients $\psi_{\mathbf{k}}(z, t)$ and $a_{\mathbf{k}}(z, t)$ are time-evolved using a semi-implicit scheme (for details, see Dmitruk and Gómez, 1999). Linear terms are treated in a fully implicit fashion, while nonlinear terms are evolved using a predictor-corrector scheme. Also, nonlinear terms are evaluated following a $\frac{2}{3}$ fully de-aliased (Canuto *et al.*, 1988) pseudo-spectral technique. To compute z-derivatives we use a standard method of finite differences in a staggered regular grid (Strauss, 1976).

We model the photospheric boundary motion in Equation (8) as $\Psi_{\mathbf{k}} = \Psi_0 =$ constant inside the ring $3 < kl_p < 4$, and $\Psi_{\mathbf{k}} = 0$ elsewhere, to simulate a

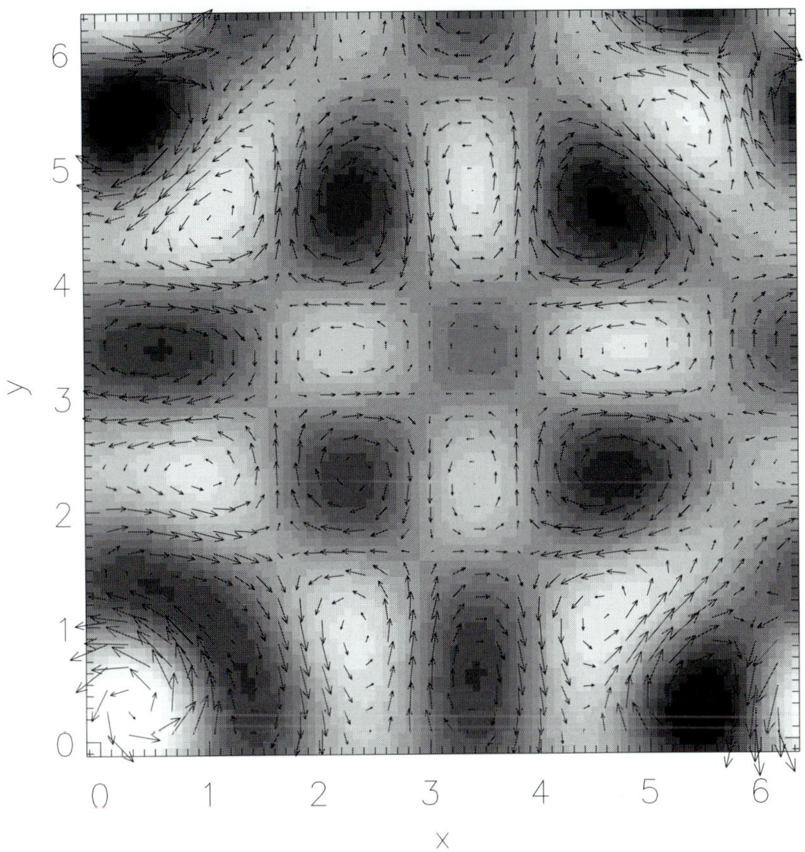

Figure 3. Velocity field applied to the upper boundary ($z = L$). The half-tone shows the stream function and the arrows correspond to the velocity field.

stationary and isotropic pattern of photospheric granular motions of diameters between $2\pi l_p/4$ and $2\pi l_p/3$ (see Figure 3). The typical time-scale associated to these driving motions is the eddy turnover time, which is defined as $t_p = l_p/V_p \approx 10^3$ s. We chose a narrow-band and non-random forcing to make sure that the broadband energy spectra and the signatures of intermittency that we obtained (see below) are exclusively determined by the nonlinear nature of the MHD equations. Resonant interactions between Fourier modes are found to trigger nonlinear instabilities, thus contributing to generate fine spatial structure inside externally driven loops (Gómez, De Luca, and McClymont, 1995). It is important to note that recent observations of solar granulation (see, for instance, Nesis *et al.*, 1999) and references therein), show a very rich dynamical structure, with clear indications of the presence of intermittent hydrodynamic turbulence. However, it seems apparent that if this narrow-band and stationary forcing is able to develop a turbulent regime, more realistic photospheric motions will develop turbulence much easier.

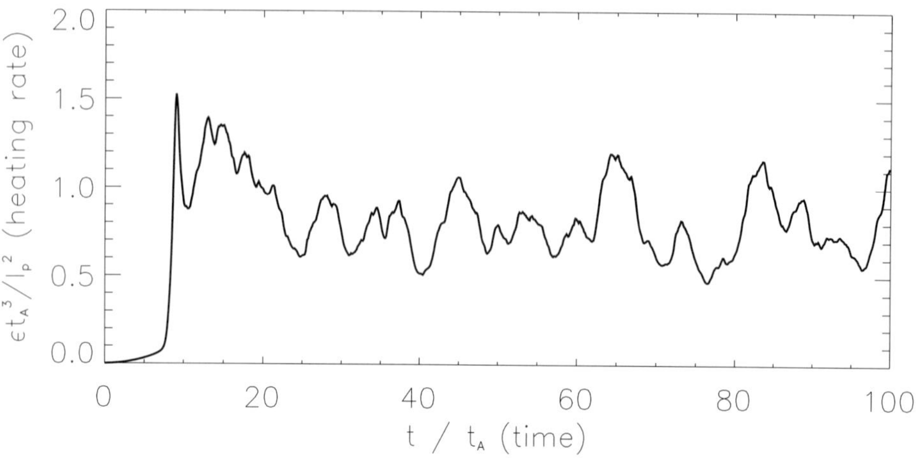

Figure 4. Energy dissipation rate vs time for $t_A/t_p = 0.064$.

We performed a sequence of numerical simulations of Equations (9) and (10) with $192 \times 192 \times 32$ grid points, $S = 1500$ and several values of the ratio t_A/t_p spanning the range [0.01, 0.15] (Dmitruk and Gómez, 1999). The typical behavior of the heating rate ϵ as a function of time, is shown in Figure 4 for the particular case of $t_A/t_p = 0.064$ in dimensionless units. After an initial transient, the heating rate is seen to approach a stationary level.

This level is fully consistent with the heating requirements for coronal active regions of 10^7 erg cm^{-2} s^{-1} (Withbroe and Noyes, 1977). The intermittent behavior of this time series, which is typical of turbulent systems, is also ubiquitous in all our simulations. Once the stationary regime is reached in each of the simulations, we determine the mean value and rms of the departure of the series from the mean.

For both two- and three-dimensional MHD turbulence, the development of a Kraichnan power spectrum is expected, i.e., $E_k \simeq k^{-3/2}$ (Kraichnan, 1965). To compute the energy power spectrum, we performed a single simulation with better spatial resolution ($384 \times 384 \times 32$) and $S = 5000$. The spectrum shown in Figure 5 was taken at $t = 5t_p$, i.e., well in the stationary regime. The obtained slope is consistent with a Kraichnan spectrum. Also, note that the kinetic and magnetic power spectra reach equipartition at large wavenumbers, even though the total kinetic energy is much smaller than the total magnetic energy.

Another important consequence of the energy power spectrum displayed in Figure 5 is the fact that smaller scales dissipate more energy than the larger scales, since $\epsilon_k \sim k^2 E_k \simeq k^{1/2}$. The morphology of small-scale dissipative structures in MHD turbulence has been readily investigated, both in three-dimensional (Mikic, Schnack, and van Hoven, 1989; Longcope and Strauss, 1994; Longcope and Sudan, 1994; Galsgaard and Nordlund, 1996; Milano *et al.*, 1999) and two-dimensional simulations (Matthaeus and Lamkin, 1986; Biskamp and Welter, 1989; Dmitruk, Gómez, and DeLuca, 1998), and have been characterized as elongated current

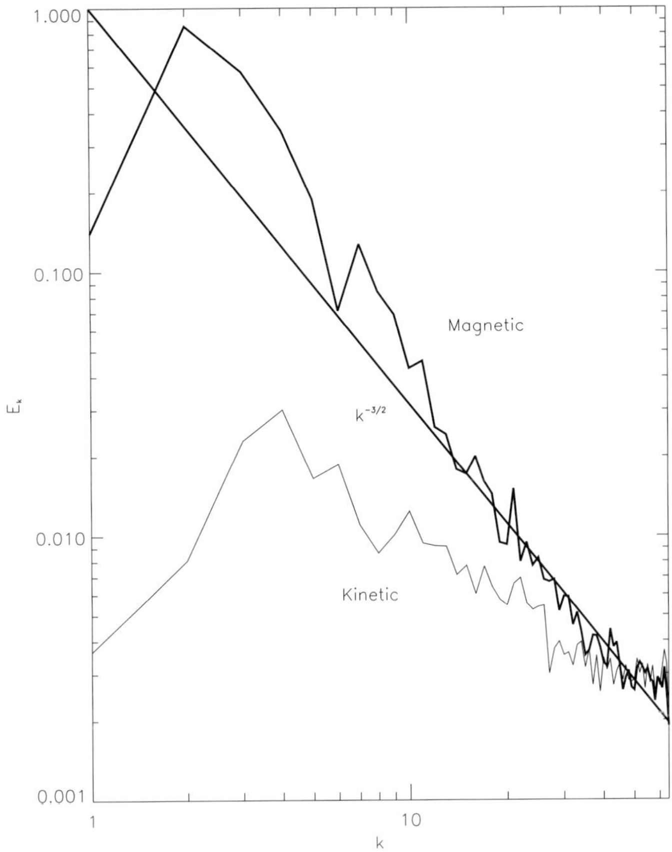

Figure 5. Energy power spectrum for the $384 \times 384 \times 32$ simulation at $t = 5t_p$.

sheets. Figure 6 shows the spatial distribution of the electric current density $j(x, y, z, t)$ for our $384 \times 384 \times 32$ simulation ($t_A/t_p = 0.064$ and $S = 5000$). The presence of elongated current sheets as the most common dissipative structures is also apparent in these simulations.

Figure 7 shows the spatial distribution of current density in a transverse slice at half length of the loop (Figure 7(a)). The arrows correspond to the magnetic field components on that plane. In spite of the artificially small Reynolds number of the simulation, a rich spatial structure can be observed, with current sheets and O-points.

Figure 7(b) also shows the spatial distribution of vorticity for the same slice, and the arrows correspond to the velocity field. The typical quadrupole structures for the vorticity and the alfvenic outflows associated to the current sheets can be observed. Furthermore, these outflows are seen to generate turbulent transients, characterized by Kraichnan energy power spectra (Hendrix and van Hoven, 1996; Milano *et al.*, 1999).

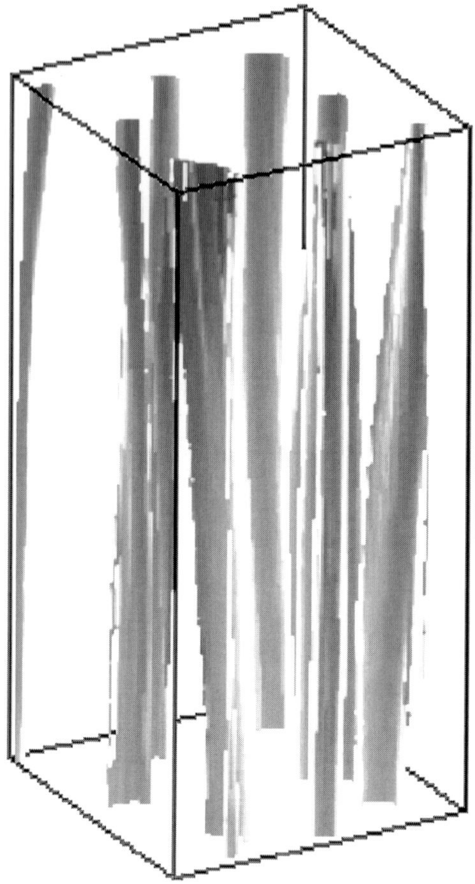

Figure 6. Spatial distribution of currents. The iso-current surface corresponds to $j = 0.3\,j_{\max}$.

From these simulations, we can also derive a turbulent velocity for coronal loops, which is simply the square root of the total kinetic energy (Dmitruk, Gómez, and DeLuca, 1998). These turbulent velocities range between 10 and 20 km s^{-1}. These values are quite comparable to those derived from line broadening of EUV spectral lines, both with SUMER (Seely *et al.*, 1997; Chae, Schuhle, and Lemaire, 1998) and CDS (Harra-Murnion *et al.*, 1999), which in turn confirm earlier measurements (Cheng, Doschek, and Feldman, 1979). These turbulent velocities, however, are noticeably smaller than those derived from SMM data (Saba and Strong, 1991), which presumably correspond to much hotter and larger active regions.

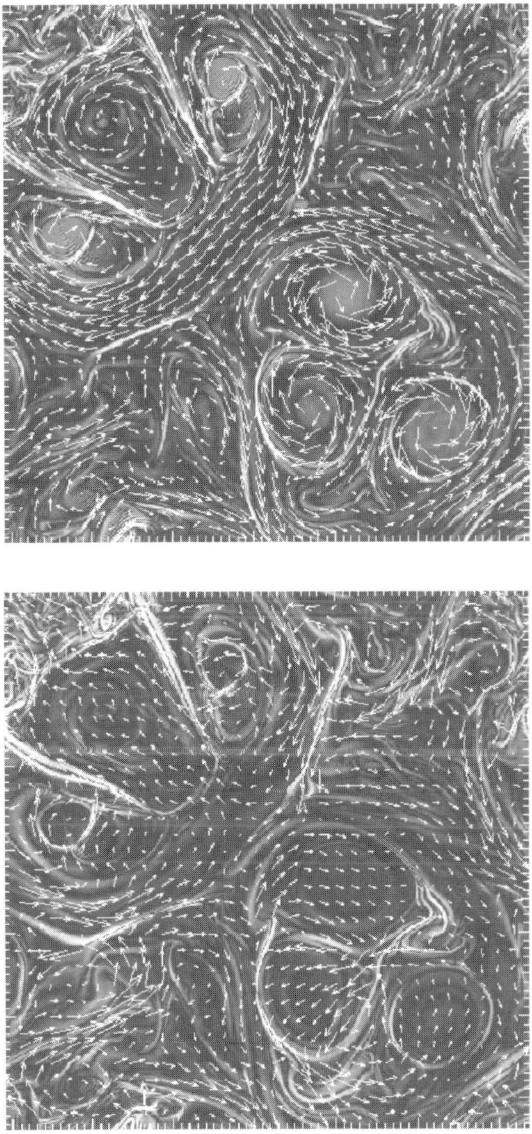

(a)

(b)

Figure 7. (a) Transverse slice at half length of a loop. The half-tone indicates electric current density flowing across this plane. The arrows correspond to the magnetic field components on this plane. (b) Same transverse slice. The half-tone indicates the vorticity component perpendicular to this plane. The arrows correspond to the velocity field.

5. Scaling Law

As mentioned in the previous section, the Reynolds number ($S = R$) is the only dimensionless parameter explicitly present in Equations (9) and (10). The boundary condition (Equation (8)) brings a second dimensionless parameter into play, namely the ratio between the photospheric velocity V_p and our velocity unit $l_p/t_{\rm A}$. Since $V_p = l_p/t_p$, this velocity ratio is equivalent to the ratio between the Alfvén time $t_{\rm A}$ and the photospheric turnover time t_p.

It is essential to note at this point that to integrate Equations (9) and (10) with the boundary condition Equation (8), only two dimensionless parameters are needed: S and $t_{\rm A}/t_p$. Therefore, from purely dimensional considerations, we can derive the following important result: *for any physical quantity, its dimensionless version Q should be an arbitrary function of the only two dimensionless parameters of the problem*, i.e.,

$$Q = \mathcal{F}\left(\frac{t_{\rm A}}{t_p}, S\right) . \tag{13}$$

For the important case of the heating rate per unit mass, i.e., ϵ/ρ, its dimensionless version is

$$Q = \frac{\epsilon}{\rho}\frac{t_{\rm A}^3}{l_p^2} = \mathcal{F}\left(\frac{t_{\rm A}}{t_p}, S\right) . \tag{14}$$

One of Kolmogorov's hypotheses in his celebrated theory for stationary turbulent regimes at very large Reynolds numbers (Kolmogorov, 1941) states that the dissipation rate is independent of the Reynolds number (see also Frisch, 1996). We have already shown that externally driven coronal loops eventually reach a stationary turbulent regime. If we therefore assume that the dissipation rate ϵ is largely independent of the Reynolds number S, we obtain

$$\epsilon = \frac{\rho l_p^2}{t_{\rm A}^3}\mathcal{F}\left(\frac{t_{\rm A}}{t_p}\right) . \tag{15}$$

From a sequence of numerical simulations for different values of the ratio $t_{\rm A}/t_p$ (Dmitruk and Gómez, 1999), we find that the function $\mathcal{F}$ is a power law with a slope $s = 1.51 \pm 0.04$, i.e.,

$$\epsilon \propto \frac{\rho l_p^2}{t_{\rm A}^3}\left(\frac{t_{\rm A}}{t_p}\right)^{3/2} . \tag{16}$$

This result is fully consistent with the prediction arising from a two-dimensional MHD model (Dmitruk and Gómez, 1997), $s_{2D} = \frac{3}{2}$, which is assumed to simulate the dynamics of a generic transverse slice of a loop. This coincidence between both scalings suggests that two-dimensional simulations provide an adequate description of the dynamics of coronal loops (Hendrix and van Hoven, 1996; Matthaeus *et al.*, 1998; Oughton, Gosh, and Matthaeus, 1998), provided that the external forcing in Equations (5) and (6) is approximated as (Einaudi *et al.*, 1996; Dmitruk and Gómez, 1997)

$$f = v_A \partial_z \psi \approx \frac{l_p V_p}{t_A}, \qquad g = v_A \partial_z j \approx 0\,, \tag{17}$$

which stems from the assumption that field lines remain relatively straight while their footpoints are being moved about. In this two-dimensional version of the problem, parameters such as V_p, v_A and L arise only indirectly, through the forcing term f (see Equation (5)). As in the previous section, we also assume: (i) a typical granule length scale l_p for the forcing and, (ii) a dissipation rate ϵ independent of the Reynolds number S. Therefore, from dimensional arguments we find that the heating rate per unit mass (ϵ/ρ) can only depend on the relevant parameters l_p and f in the following fashion:

$$\frac{\epsilon}{\rho} \sim \frac{f^3}{l_p} = \frac{l_p^2}{t_A^3}\left(\frac{t_A}{t_p}\right)^{3/2}, \tag{18}$$

which exactly matches the scaling derived empirically from the sequence of RMHD simulations (see Equation (16)). Alternative laws have been derived by other authors (Hendrix *et al.*, 1996) considering two extra parameters, namely, a mild dependence on the Reynolds number S and a photospheric correlation time (apart from the granules turnover time).

Scaling laws like the one displayed in Equation (16), corresponding to a wide variety of coronal heating theories, have been reviewed in an excellent paper by Mandrini, Démoulin, and Klimchuk (2000). In their study, the authors report a statistical correlation between the lengths of loops and their average magnetic field, based on magnetogram extrapolations for a large number of active regions. This correlation has the form $\langle B \rangle \sim L^{-0.9\pm0.3}$, confirming an earlier result (Porter and Klimchuk, 1995). By combining this law with the approximate dependence of the heating rate like L^{-2} (arising from a static balance between a uniform heating source, thermal conduction and radiative losses), they can test different models of coronal heating. They find that DC models are in generally better agreement with these observational constraints than AC heating models.

6. Statistics of Events

To perform extended time simulations we focus on the dynamics in a given transverse plane, that is, we study the evolution of a generic two-dimensional slice of a loop. To this end, we model the $v_A \partial_z$ terms as indicated in the previous section.

In spite of the narrow forcing and even though velocity and magnetic fields are initially zero, nonlinear terms quickly populate all the modes across the spectrum and a turbulent state develops, with features quite comparable to the ones displayed

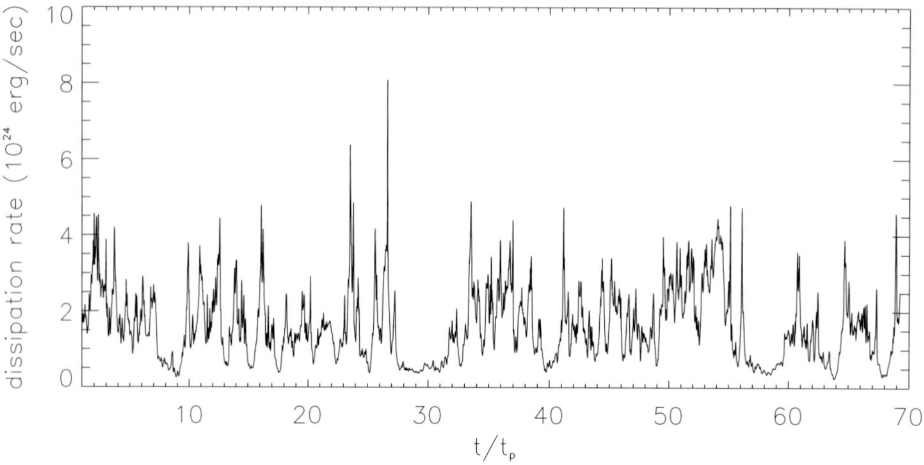

Figure 8. Energy dissipation rate vs time.

by the RMHD equations. Long time-series of the energy dissipation rate are obtained, as the one shown in Figure 8. These time series exhibit the intermittent behavior also shown in Figure 4 for the RMHD case, which is characteristic of turbulent systems.

Parker (1988) proposed that the energy dissipation of the stressed magnetic structures takes place in a large number of small events, which he termed 'nanoflares'. The superposition of a large number of such events would give the global appearance of a spatially homogeneous and stationary heating process. From a turbulent scenario, it seems straightforward to relate this spiky (both in space and time) heating, to the internal intermittency present in all turbulent regimes. We therefore associate the peaks of energy dissipation displayed in Figure 8 to the so-called nanoflares (Dmitruk and Gómez, 1997; Dmitruk, Gómez, and DeLuca, 1998). We estimate the occurrence rate for these nano-events, i.e., the number of events per unit energy and time, $P(E) = \mathrm{d}N/\mathrm{d}E$, so that

$$\mathcal{R} = \int_{E_{\min}}^{E_{\max}} \mathrm{d}E\ P(E) \tag{19}$$

is the total number of events per unit time and

$$\epsilon(2\pi l_p)^2 L = \int_{E_{\min}}^{E_{\max}} \mathrm{d}E\ E P(E) \tag{20}$$

is the total heating rate (in erg s^{-1}) contributed by all events in the energy range $[E_{\min}; E_{\max}]$. We define events simply as the excesses of dissipation with respect to a threshold value ϵ_0 of the order of the heating rate time average. Further details of this derivation can be found elsewhere (Dmitruk and Gómez, 1997; Dmitruk, Gómez, and DeLuca, 1998).

The occurrence rate as a function of energy is found to display a power-law behavior

$$P(E) = AE^{-1.5\pm0.2}\,, \tag{21}$$

in the energy range spanned from $E_{\rm min} \simeq 10^{25}$ erg to $E_{\rm max} \simeq 2\times10^{26}$ erg. We also computed the distribution of events as a function of peak fluxes which is a power law with slope $\alpha_P = 1.7 \pm 0.3$ (Dmitruk, Gómez, and DeLuca, 1998). This slope is consistent with the one derived by Crosby, Aschwanden, and Dennis (1993) (1.68) from X-ray events and somewhat flatter than those reported by Hudson (1991) (1.8). Similar event distributions have been derived by other authors, both from MHD simulations (Georgoulis, Velli, and Einaudi, 1998) and from cellular automata models (Lu and Hamilton, 1991; Lu *et al.*, 1993; Vlahos *et al.*, 1995). Other important results from this statistical analysis are the correlations between the different parameters of these events. For instance, the energy released vs event duration τ can be fitted by a power law $E \sim \tau^2$ (Dmitruk, Gómez, and DeLuca, 1998). This result is consistent with the correlation reported by (Lee, Petrosian, and McTiernan, 1993) from hard X-ray observations.

7. Conclusion

The main results of the present study can be summarized as follows:

(1) TRACE movies show that coronal active regions are extremely dynamic and inhomogeneous objects. This ubiquitous spatial and temporal variability suggests the presence of a rather intermittent heating mechanism for these regions. We propose that the development of MHD turbulence in coronal loops driven at their footpoints will naturally produce intermittent dissipation.

(2) Numerical integrations of the RMHD equations, driven by a stationary pattern of large-scale footpoint motions show that the system reaches a turbulent stationary regime, characterized by a Kraichnan energy power spectrum. Also, the ensuing energy cascade enhances the average dissipation rate to the levels required by the conductive and radiative losses of the coronal plasma. Most of the dissipation occurs in current sheets approximately parallel to the loop axis.

(3) The MHD turbulence developed in coronal loops is magnetically dominated, i.e., the magnetic energy is much larger than the kinetic energy. Turbulent velocities of about 10–20 km s^{-1} are obtained, which are fully consistent with those inferred by various authors from the broadening of EUV spectral lines.

(4) By combining dimensional arguments on the RMHD equations with a sequence of numerical simulations, a scaling law is derived, which relates the turbulent heating rate with the various relevant physical parameters (Dmitruk and Gómez, 1999). This kind of scaling law, arising from theoretical heating models,

can be tested against observational constraints (Mandrini, Démoulin, and Klimchuk, 2000).

(5) The intermittent nature of energy dissipation suggests that the heating mechanism can be interpreted as a stochastic superposition of discrete events or *nanoflares* (Parker, 1988). In this regard, it is important to note that the turbulent scenario is not inconsistent with Parker's scenario (Longcope and Sudan, 1994; Hendrix *et al.*, 1996). Furthermore, it could perhaps be considered as its natural extension to dissipative plasmas, since the tangential discontinuities arising in an ideal plasma will become thin current sheets if a small amount of dissipation is allowed. A statistics of dissipation events obtained from numerical simulations, yields a power law energy occurrence rate. The slope of this distribution is approximately 1.5, which is quite comparable to similar analysis performed for the occurrence of flares (Hudson, 1991; Crosby, Aschwanden, and Dennis, 1993) and microflares (Shimizu, 1995).

Acknowledgements

The authors wish to express their gratitude to the local organizers of the Physics of the Solar Corona and Transition Region Workshop, where this paper was presented, Drs Karel Schrijver, Zoe Frank, and Neal Hurlburt, for such a superb meeting and for their kind hospitality. This work is supported by the University of Buenos Aires (grant UBACYT TX065/98), and the Agencia Nacional de Promoción de Ciencia y Tecnología (grant PICT 02305/97). DG is Member of the Carrera del Investigador Científico of CONICET, Argentina.

References

Biskamp, D.: 1993, *Nonlinear Magnetohydrodynamics*, Cambridge University Press, Cambridge.

Biskamp, D. and Welter, H.: 1989, *Phys. Fluids* **B1**, 1964.

Canuto, C., Hussaini, M. Y., Quarteroni, A., and Zang, T. A.: 1988, *Spectral Methods in Fluid Dynamics*, Springer-Verlag, New York.

Chae, J., Schuhle, U., and Lemaire, P.: 1998, *Astrophys. J.* **505**, 957.

Cheng, C., Doschek, G. and Feldman U.: 1979, *Astrophys. J.* **227**, 1037.

Crosby, N. B., Aschwanden, M. J., and Dennis, B. R.: 1993, *Solar Phys.* **143**, 275.

Dmitruk, P. and Gómez, D. O.: 1997, *Astrophys. J.* **484**, L83.

Dmitruk, P. and Gómez, D.O.: 1999, *Astrophys. J.* **527**, L63.

Dmitruk, P., Gómez, D. O., and DeLuca, E. E.: 1998, *Astrophys. J.* **505**, 974.

Einaudi, G., Velli, M., Politano, H., and Pouquet, A.: 1996, *Astrophys. J.* **457**, L113.

Frisch, U.: 1996, *Turbulence*, Cambridge University Press, Cambridge.

Fyfe, D., Montgomery, D., and Joyce, G.: 1977, *J. Plasma Phys.* **17**, 369.

Galsgaard, K. and Nordlund, A.: 1996, *J. Geophys. Res.* **101**, 13445.

Georgoulis, M., Velli, M., and Einaudi, G.: 1998, *Astrophys. J.* **497**, 957.

Golub, L. and Pasachoff, J.: 1997, *The Solar Corona*, Cambridge University Press, Cambridge.

Golub, L., Herant, M., Kalata, K., Louvas, I., Nystrom, G., Pardo, F., Spiller, E., and Wilczynski, J. S.: 1990, *Nature* **344**, 842.
Gómez, D. O.: 1990, *Fund. Cosmic Phys.* **14**, 361.
Gómez, D. O. and Ferro Fontan, C.: 1988, *Solar Phys.* **116**, 33.
Gómez, D. O. and Ferro Fontán, C.: 1992, *Astrophys. J.* **394**, 662.
Gómez, D. O., De Luca, E. E., and Mc Clymont, A. N.: 1995, *Astrophys. J.* **448**, 954.
Grant, H. L., Stewart, R. W., and Molliet, A.: 1962, *J. Fluid Mech.* **12**, 241.
Harra-Murnion, L. K., Matthews, S. A., Hara, H., and Ichimoto, K.: 1999, *Astron. Astrophys.* **345**, 1011.
Hendrix, D. L. and van Hoven, G.: 1996, *Astrophys. J.* **467**, 887.
Hendrix, D. L., van Hoven, G., Mikic, Z., and Schnack, D. D.: 1996, *Astrophys. J.* **470**, 1192.
Herring, J. and Kraichnan, R. H.: 1972, in M. Rosenblatt and C. van Atta (eds.), '*Statistical Models and Turbulence*', Springer-Verlag, Berlin.
Heyvaerts, J. and Priest, E. R.: 1983, *Astron. Astrophys.* **117**, 220.
Heyvaerts, J. and Priest, E. R.: 1992, *Astrophys. J.* **390**, 297.
Hollweg, J. V.: 1985, in B. Buti (ed.), *Advance Space Plasma Physics*, World Science Publ. Co., Singapore, p. 77.
Hudson, H. S.: 1991, *Solar Phys.* **133**, 357.
Ionson, J. A.: 1982, *Astrophys. J.* **254**, 318.
Kolmogorov, A. N.: 1941, *Dokl. Acad. Sci. URSS* **30**, 301.
Kraichnan, R. H.: 1965, *Phys. Fluids* **8**, 138.
Lee, T. T., Petrosian, V., and McTiernan, J. M.: 1993, *Astrophys. J.* **412**, 401.
Lilly, D. K.: 1969, *Phys. Fluids* **12**, 240.
Longcope, D. W. and Strauss, H. R.: 1994, *Astrophys. J.* **426**, 742.
Longcope, D. W. and Sudan, R. N.: 1994, *Astrophys. J.* **437**, 491.
Lu, E. T. and Hamilton, R. J.: 1991, *Astrophys. J.* **380**, L89.
Lu, E., Hamilton, R., McTiernan, J., and Bromund, K.: 1993, *Astrophys. J.* **412**, 841.
Mandrini, C. H., Démoulin, P., and Klimchuk, J. A.: 2000, *Astrophys. J.* **530**, in press.
Matthaeus, W. H., and Goldstein, G. L.: 1982, *J. Geophys. Res.* **87**, 6011.
Matthaeus, W. H. and Lamkin, S. L.: 1986, *Phys. Fluids* **29**, 2513.
Matthaeus, W. H., Oughton, S., Ghosh, S., and Hossain, M.: 1998, *Phys. Rev. Letters* **81**, 2056.
Meneguzzi, M., Frisch, U., and Pouquet, A.: 1981, *Phys. Rev. Letters* **47**, 1060.
Mikic, Z., Schnack, D. D., and van Hoven, G.: 1989, *Astrophys. J.* **338**, 1148.
Milano, L. J., Gómez, D. O., and Martens, P. C. H. 1997, *Astrophys. J.* **490**, 442.
Milano, L. J., Dmitruk, P., Mandrini, C. H., Gómez, D. O., and Démoulin, P.: 1999, *Astrophys. J.* **521**, 889.
Montgomery, D.: 1983, in M. Neugebauer (ed.), '*Solar Wind V*', NASA Conf. Publ. 2280, p. 107.
Nakariakov, V. M., Ofman, L., DeLuca, E. E., Roberts, B., and Davila, J. M.: 1999, *Nature* **285**, 862.
Narain, U. and Ulmschneider, P.: 1990, *Space Sci. Rev.* **54**, 377.
Narain, U. and Ulmschneider, P.: 1996, *Space Sci. Rev.* **75**, 453.
Nesis, A., Hammer, R., Kiefer, M., Schleicher, H., Sigwarth, M., and Staiger, J.: 1999, *Astron. Astrophys.* **345**, 265.
Ofman, L., Davila, J. M., and Steinolfson, R. S.: 1995, *Astrophys. J.* **444**, 471.
Oughton, S., Ghosh, S., and Matthaeus, W. H.: 1998, *Phys. Plasmas* **5**, 4235.
Parker, E. N.: 1972, *Astrophys. J.* **174**, 499.
Parker, E. N.: 1983, *Astrophys. J.* **264**, 642.
Parker, E. N.: 1988, *Astrophys. J.* **330**, 474.
Politano, H., Pouquet, A., and Sulem, P. L.: 1989, *Phys. Fluids* **B1**, 2330.
Porter, L. J. and Klimchuk, J. A.: 1995, *Astrophys. J.* **454**, 499.
Saba, J. and Strong, K.: 1991, *Astrophys. J.* **375**, 789.
Schrijver, C. J. and 16 co-authors: 1999, *Solar Phys.*, **187**, 261.

Seely, N. *et al.*: 1997, *Astrophys. J.* **484**, L87.
Shimizu, T.: 1995, *Publ. Astr. Soc. Japan* **47**, 251.
Sommeria, J.: 1986, *J. Fluid Mech.* **170**, 139.
Strauss, H.: 1976, *Phys. Fluids* **19**, 134.
van Ballegooijen, A. A.: 1986, *Astrophys. J.* **311**, 1001.
Vlahos, H., Georgoulis, M., Kluiving, R., and Paschos, P.: 1995, *Astron. Astrophys.* **299**, 897.
Withbroe, G. L. and Noyes, R. W.: 1977, *Ann. Rev. Astron. Astrophys.* **15**, 363.
Zirker, J. B. 1993, *Solar Phys.* **148**, 43.

MACS FOR GLOBAL MEASUREMENT OF THE SOLAR WIND VELOCITY AND THE THERMAL ELECTRON TEMPERATURE DURING THE TOTAL SOLAR ECLIPSE ON 11 AUGUST 1999

NELSON L. REGINALD[1] and JOSEPH M. DAVILA[2]

[1] *Department of Physics and Astronomy, University of Delaware, Newark, DE 19716, U.S.A. (e-mail: reginald@stars.gsfc.nasa.gov)*

[2] *Laboratory for Astronomy and Solar Physics, Code 682, NASA's GSFC, Greenbelt, MD 20771, U.S.A. (e-mail:davila@lindsay.gsfc.nasa.gov.)*

(Received 27 September 1999; accepted 23 March 2000)

Abstract. MACS for Multi-Aperture Coronal Spectrometer is a fiber-optic-based spectrograph designed and used to perform global measurement of the solar wind velocity and the thermal electron temperature of the solar corona during the total solar eclipse on 11 August 1999. The motivation for the construction of MACS was provided by the theory formulated by Cram (1976) for the formation of the K-coronal spectrum and a method for determining the radial profile of the thermal electron temperature of the solar corona. Based on this theory a subsequent application was carried out by Ichimoto *et al.* (1996) using a slit-based spectroscopic study during the total solar eclipse on 3 November 1994. We have modified Cram's theory to incorporate the role of the solar wind velocity in the formation of the K-corona and have identified wind and temperature sensitive intensity ratios. Instead of a slit-based spectrograph MACS consists of twenty fiber optic tips placed at the focal plane of the telescope and positioned to see different radii and latitudes of the solar corona. Another fiber is placed at the center of the frame and uses the lunar shadow for a measure of the background signal. The other ends of the fibers are vertically aligned and placed at the primary focus of the collimating lens of the spectrograph thus providing simultaneous spectra from all of the fibers. In this first paper (Paper I) we describe our instrument and the obtained coronal spectra. The final and complete results will be presented in Paper II (Reginald and Davila, 2000).

1. Introduction

The determination of the radial and latitudinal thermal electron temperature and the solar wind velocity profiles of the solar corona are of great importance in understanding the coronal heating mechanism and the dynamics of the coronal features. On this background, our effort has been focused on developing a mechanism and designing for an instrument that would globally and simultaneously determine the thermal electron temperature and the solar wind velocity in the solar corona.

The underlying theory is based on the theory formulated by Cram (1976) for the formation of the K-corona. The K-corona is a result of the Thomson scattering of the photospheric radiation off the free electrons in the solar corona. Cram (1976) in his paper noted that in association with the Fraunhofer lines, notably the Calcium H and K lines and the G-band (a collection of absorption features due predomi-

The physics of the solar corona and transition region. Part II.
Reprinted from Solar Physics **195,** 2000.

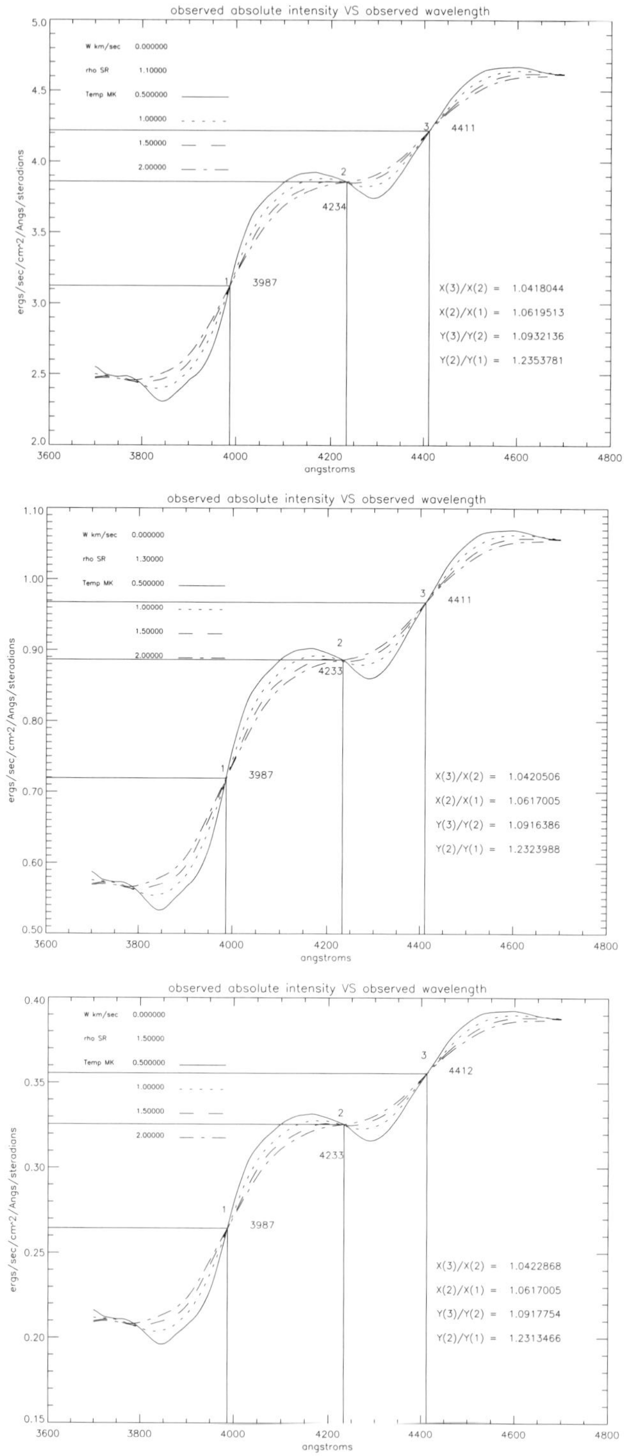

Figure 1. Plots showing (from above) the K-coronal intensity as a function of wavelength at 1.1 $R_\odot$, 1.3 $R_\odot$, and 1.5 $R_\odot$ for zero solar wind velocity.

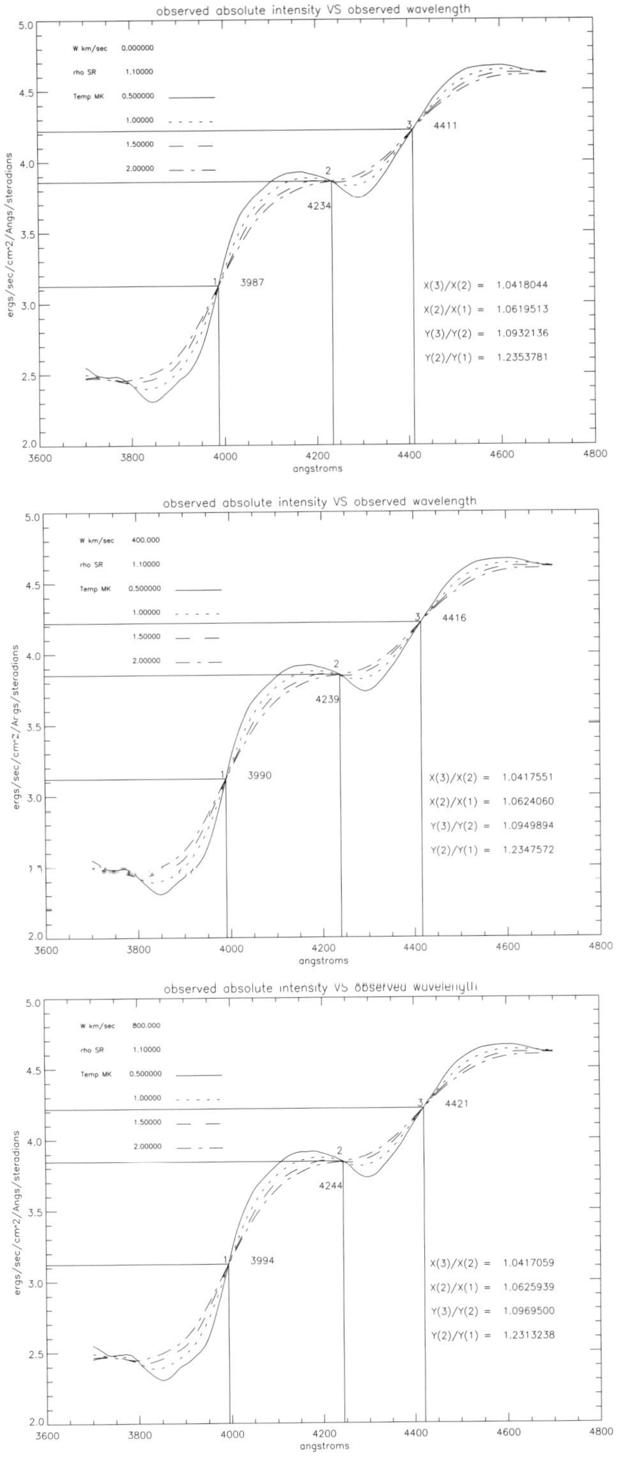

Figure 2. Plots showing (from above) the K-coronal intensity as a function of wavelength at 1.1 $R_\odot$ for 0, 400, and 800 km s^{-1} solar wind velocities.

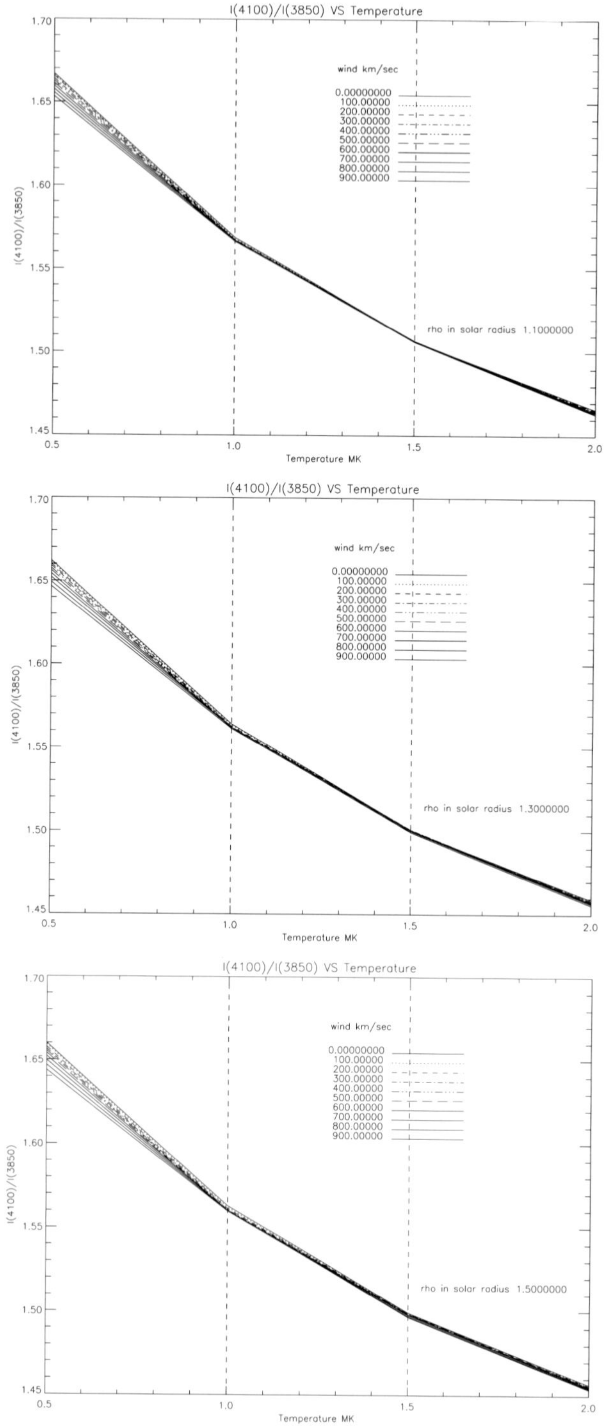

Figure 3. These plots demonstrate the sensitivity of the intensity ratio at 4100 Å and 3850 Å as a function of temperature for a range of solar wind velocities for line-of-sight heights at 1.1 $R_\odot$, 1.3 $R_\odot$, and 1.5 $R_\odot$.

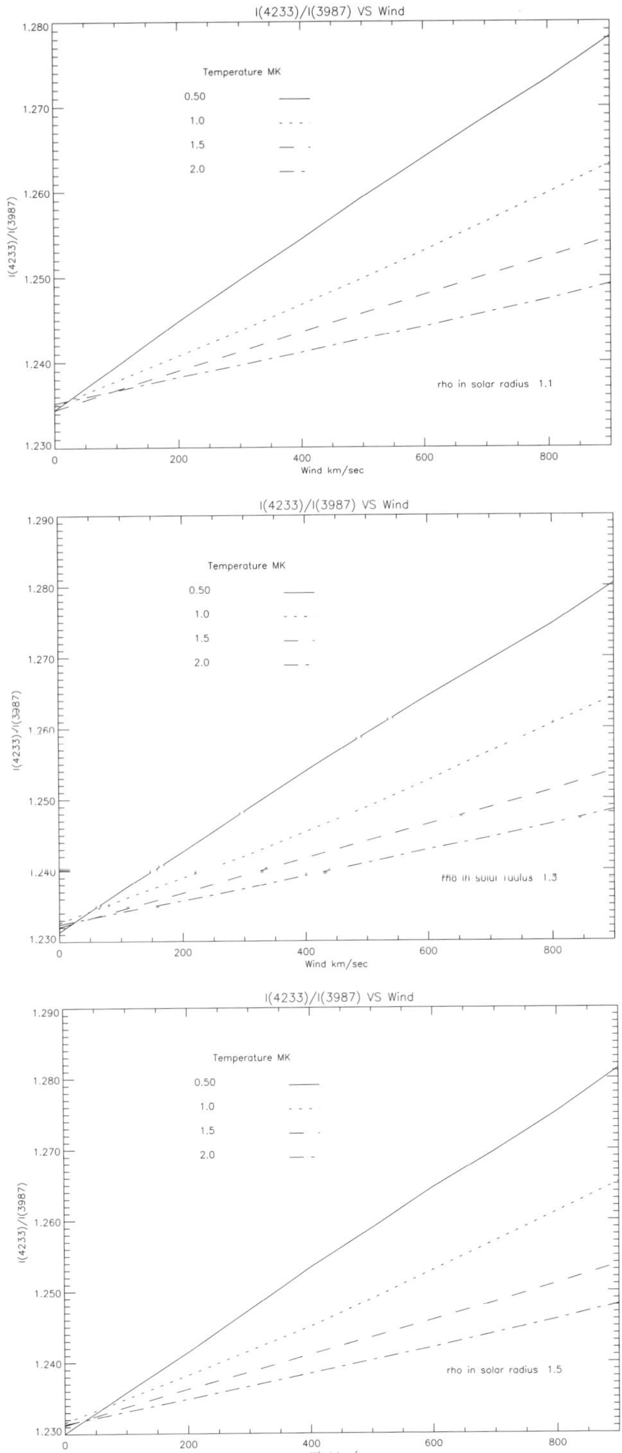

Figure 4. These plots demonstrate the sensitivity of the intensity ratio at 4233 Å and 3987 Å as a function of wind for a range of temperatures for line-of-sight heights at 1.1 $R_{\odot}$, 1.3 $R_{\odot}$, and 1.5 $R_{\odot}$.

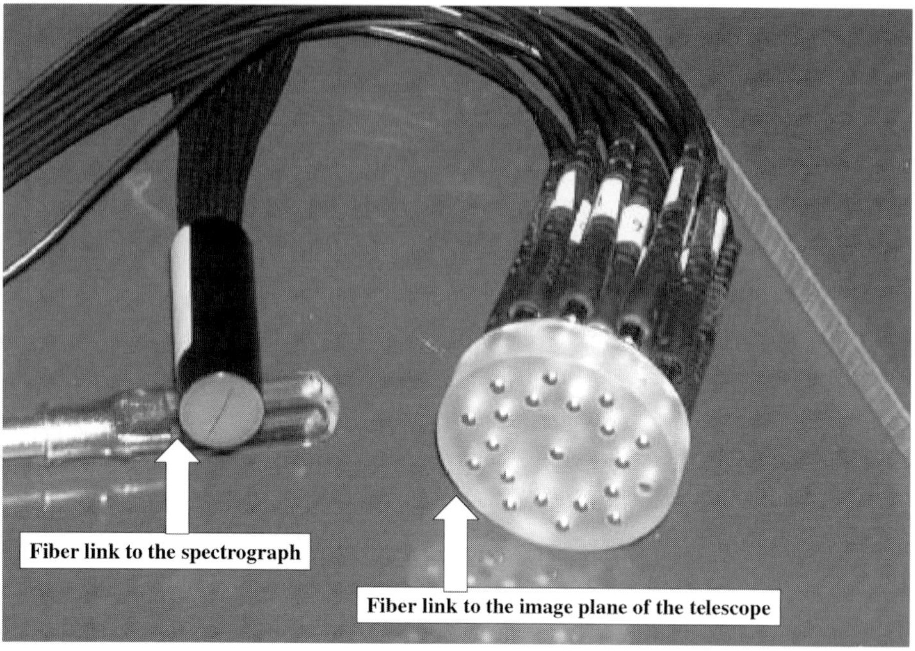

Figure 5. Picture of the fiber link between the image plane of the telescope and the spectrograph. The fiber link at the telescope consists of twenty-one fibers. The inner and the outer loops are at 1.1 $R_\odot$ and 1.5 $R_\odot$ solar radii, respectively, and consist of ten fibers at each altitude. The fiber in the center of the frame uses the lunar shadow during the eclipse to record the background counts. Four other fibers link to the spectrograph and are connected to a mercury calibration lamp.

velocity could be determined to within ± 0.2 MK and ± 200 km s^{-1}, respectively, from the K-coronal spectrum for a given line of sight.

4. The Instrument

MACS (Multi-Aperture Coronal Spectrometer) was designed to obtain simultaneous spectra from multiple locations on the solar corona during the total solar eclipse on 11 August 1999. In this instrument twenty-one fiber optic ends are positioned at various locations on the image plane of the telescope. The distributions of locations are ten fibers, equally spread in azimuth, at 1.1 $R_\odot$ and at 1.5 $R_\odot$. One additional fiber is in the center of the circular frame and uses the lunar shadow to measure the background counts. Four other fibers are attached to a mercury lamp as a calibration source. The other ends of the fiber bundle consisting of twenty-five fibers are arranged vertically at the prime focus of the collimating lens of the spectrograph. Here the separation between each fiber from center to center is 0.4 mm. The fibers have a diameter of 200 μm and a transmission efficiency of 80%, or more, in the wavelength region 3500–4500 Å. The fibers are 1 m in length and produced by Oriel Corporation.

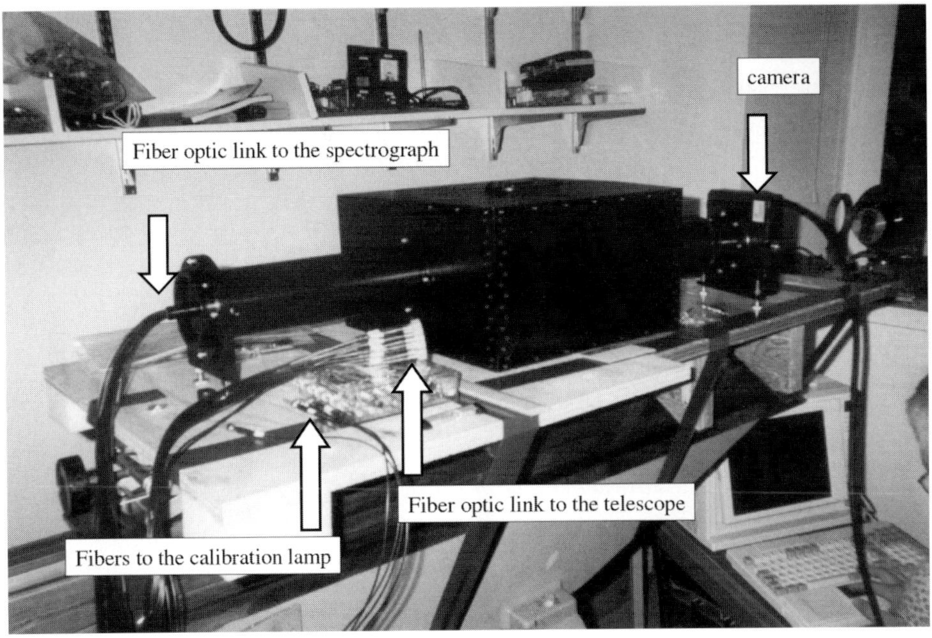

Figure 6. This is a picture of the spectrograph. The link between the spectrograph and the telescope is via the fiber optic cables. This also allows for the telescope and the spectrograph to function as two detached systems.

The image to feed into MACS was produced by a 12-inch, $F/10$ Schmidt–Cassegrain telescope manufactured by the Meade Corporation. The f-number was reduced to $F/6.3$ using a focal reducer. The plate scale of the telescope with the focal reducer in place is 0.107 arc sec μm^{-1}. The telescope features auto tracking and polar alignment.

The collimating and the camera lenses are achromats of 8 cm in diameter with f-numbers of $F/3.9$ and $F/2$, respectively, and produced by Spindler and Hoyer. The grating is a transmission grating of dimension 6×6 cm^2 with a ruling density of 600 lines mm^{-1} and produced by American Holographic. All the optical components are anti-reflective UV coated to enhance the transmission below 4000 Å.

The camera is aligned with the aim to record spectra in the wavelength region 3500–4500 Å in the first order of the grating. The camera consists of a thermo-electric cooler capable of cooling 50 C below ambient and a back thinned SITE - CCD chip with 512×512 square pixels of size 24 μm and produced by the Apogee Instruments Inc. The read out time is $\sim$10 s. The spectral and the spatial resolutions are, respectively, $\sim$2.4 Å pixel^{-1} and $\sim$5.0 arc sec pixel^{-1}.

Figure 7 is a schematic drawing detailing the concept of MACS during totality. Here twenty-one fibers are located at the focal plane of the telescope with ten each at 1.1 and 1.5 solar radii and another is at the center of the frame and uses the lunar shadow for a measure of the background counts. The other ends of these

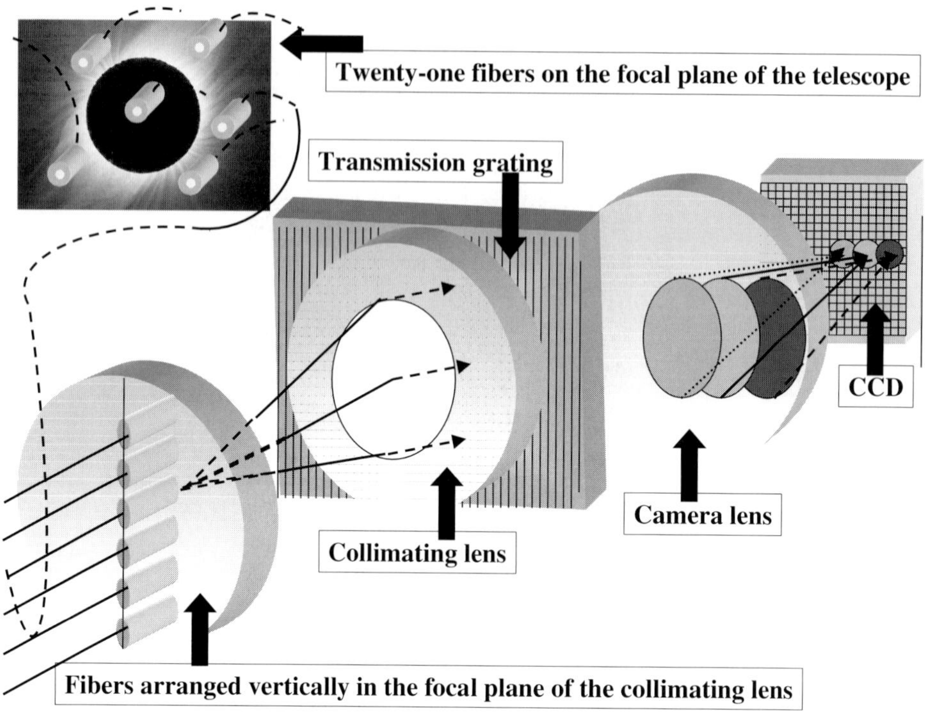

Figure 7. A schematic drawing of MACS showing the dispersion of coronal light from a single fiber located at the focal plane of the telescope. In MACS twenty-one fibers are located in the focal plane of the telescope with ten each at 1.1 and 1.5 solar radii and another placed at the center of the frame. The fiber in the center of the frame uses the lunar shadow during totality, as shown in the drawing, to measure the background counts. Each fiber is imaged on 5 pixels on the detector in the spatial direction with the centers of the neighboring fibers displaced by 8 pixels.

fibers are arranged vertically, that is parallel to the rulings on the grating, at the focal plane of the collimating lens. The collimating lens collimates the light from all of the fibers onto the transmission grating. The grating disperse the light in the direction perpendicular to the rulings on the grating. The camera lens re-focuses the dispersed light from the grating onto the detector, that is at the focal plane of the camera lens. Figure 7 shows the spectral dispersion of one fiber recorded on the detector. The light from the other fibers too will simultaneously be dispersed and recorded on the detector, however, displaced from each other in the spatial direction on the detector. Each fiber is imaged on 5 pixels on the detector in the spatial direction with the centers of the neighboring fibers displaced by 8 pixels.

5. Observational Results

Figure 8 shows the spectra from MACS on the solar corona of the total solar eclipse on 11 August 1999 in Elazig, Turkey. Three sets of spectra were obtained for 4 s,

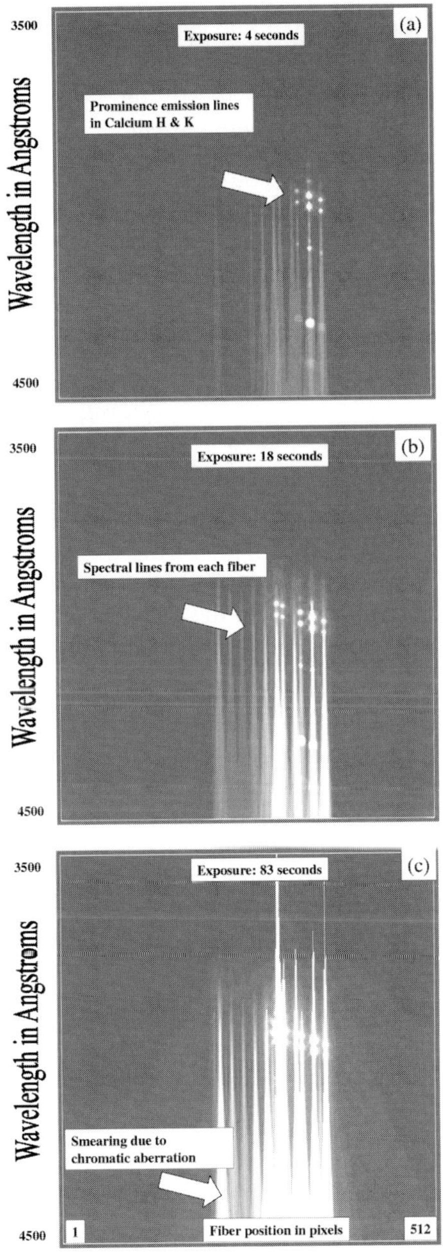

Figure 8. Spectra obtained by MACS for the exposure times of 4, 18, and 83 s are shown by (a), (b), and (c), respectively. The wavelength region covered is from 3500–4500 Å. The individual spectra are produced by the fibers located at the focal plane of the telescope. The top most adjacent bright spots on several of the spectra are due to prominence emission lines in calcium H and K. The tilt of these emission lines results from the tilt of the fiber link at the spectrograph with respect to the grating ruling direction.

18 s, and 83 s exposure times, respectively. The sky was cloudless for the entire duration of the totality. Each vertical line is a spectrum attributed to a single fiber on the image plane of the telescope. The reason for the tilt seen on the prominence emission lines from different fibers is mostly due to the tilt of the fiber link at the spectrograph with respect to the grating ruling direction. This will later be corrected from the calibration spectra obtained soon after totality with the mercury calibration lamp. The tilt was apparent in the calibration spectra too and is a result of a human error.

The broadening of the spectrum at the bottom end is due to chromatic aberration. Nevertheless, the region of interest pertaining to the temperature and wind sensitive intensity ratios, discussed in Section 2 of this paper, remains focused. In the 83 s frame, some smearing is evident due to the over exposure of the fibers located at 1.1 $R_{\odot}$.

The two adjacent emission lines seen in all the frames of Figure 8 result from the calcium H and K lines from prominences. Since these emission lines are not observed in all of the spectra, they could be identified as the locations of the prominences at totality. Also by cross-referencing the positions of those fibers with these emission lines with eclipse images showing prominences, the accuracy of the individual fiber location with respect to the solar north could be verified. This is especially important since at the present stage MACS lacks an imaging camera to simultaneously record the totality. Using imaging data at totality from other space and ground-based solar instruments that are dedicated to observe specific features, it is now possible to infer the solar features that would have crossed the line of sight of each individual fiber and determine the thermal electron temperature and the solar wind velocity associated with those features.

In order to isolate the K-coronal spectrum for our analysis the E-corona and F-corona need to be subtracted. Also the data need to be corrected for the wavelength-dependent sensitivity of the entire instrument.

The final and complete results of the study will be presented in Paper II (Reginald and Davila, 2000).

Acknowledgements

N. L. R. acknowledges NASA research assistantship provided by grant NAG5-8696 administered through the University of Delaware.

References

Allen, C. W.: 1964, *Astrophysical Quantities*, Athlone Press, London.

Baumbach, S.: 1937, *Astron. Nachr.* **263**, 121.

Cram, L. E.: 1976, *Solar Phys.* **48**, 3.

Ichimoto, K., Kumagai, K., Sano, I., Kobiki, T., Sakurai, T., and Munoz, A: 1996, *Publ. Astron. Soc. Japan* **48**, 545.

Reginald, N. L. and Davila, J. M.: 2000, *Solar Phys.* (in preparation).

MULTI-WAVELENGTH OBSERVATIONS OF THE 1998 SEPTEMBER 27 FLARE SPRAY

PETER T. GALLAGHER[1,2,*], DAVID R. WILLIAMS[1,2], KENNETH J. H. PHILLIPS[2], MIHALIS MATHIOUDAKIS[1], RAYMOND N. SMARTT[3] and FRANCIS P. KEENAN[1]

[1]*Department of Pure and Applied Physics, The Queen's University of Belfast, Belfast, BT7 1NN, N. Ireland, U.K.*

[2]*Space Science and Technology Department, Rutherford Appleton Laboratory, Chilton, Didcot, Oxon. OX11 0QX, U.K.*

[3]*National Solar Observatory/Sacramento Peak, Sunspot, NM 88349, U.S.A.*

(Received 28 October 1999; accepted 27 March 2000)

Abstract. We report on observations of a large eruptive event associated with a flare that occurred on 27 September 1998 made with the Richard B. Dunn Solar Telescope at Sacramento Peak Observatory (several wave bands including off-line-center Hα), in soft and hard X-rays (GOES and BATSE), and in several TRACE wave bands (including Fe IX/X 171 Å, Fe XII 195 Å, and C IV 1550 Å). The flare initiation is signaled by two Hα foot-point brightenings which are closely followed by a hard X-ray burst and a subsequent gradual increase in other wavelengths. The flare light curves show a complicated, three-component structure which includes two minor maxima before the main GOES class C5.2 peak after which there is a characteristic exponential decline. During the initial stages, a large spray event is observed within seconds of the hard X-ray burst which can be directly associated with a two-ribbon flare in Hα. The emission returns to pre-flare levels after about 35 min, by which time a set of bright post-flare loops have begun to form at temperatures of about 1.0–1.5 MK. Part of the flare plasma also intrudes into the penumbra of a large sunspot, generally a characteristic of very powerful flares, but the flare importance in GOES soft X-rays is in fact relatively modest. Much of the energy appears to be in the form of a second ejection which is observed in optical and ultraviolet bands, traveling out via several magnetic flux tubes from the main flare site (about 60° from Sun center) to beyond the limb.

1. Introduction

High-resolution ultraviolet (UV) and extreme ultraviolet (EUV) observations of solar flares are now possible with NASA's Transition Region and Coronal Explorer (TRACE) (Handy *et al.*, 1998) and the ESA/NASA Solar and Heliospheric Observatory (Domingo, Fleck, and Poland, 1995) spacecraft. Combining such observations with those from large ground-based telescopes gives us a means of investigating the link between chromospheric and transition region emission during flares. This paper reports on observations of an unusual three component flare accompanied by a large spray event which were taken with the TRACE spacecraft and the Richard B. Dunn Solar Telescope (DST) at the U.S. National Solar

*Current Address: Center for Solar Research, Physics Department, New Jersey Institute of Technology, Newark, NJ 07102, USA.

The physics of the solar corona and transition region. Part II.
Reprinted from Solar Physics **195,** 2000.

TABLE I
Details of the DST observing sequence

Line	Wavelength (Å)	Exposure time (ms)
G band	4305.0	20
Ca II K_3	3933.7	700
Hα	6562.8	300
Hα (+0.65 Å)	6563.5	200
Hα (+1.00 Å)	6563.8	200
Mg I b_1 (−0.40 Å)	5183.2	450
Mg I b_2	5172.7	900
Fe I (LCP)	6302.4	100
Fe I (RCP)	6302.4	100
Continuum	5256.4	100

Observatory, Sacramento Peak, New Mexico. Simultaneous TRACE and optical observations were made for over an hour, from well before the flare impulsive stage, marked by hard X-ray emission seen with the Burst and Transient Source Experiment (BATSE) on board the Compton Gamma-Ray Observatory, to some 50 min later.

In common with other mass ejections, such as jets (e.g., Alexander and Fletcher, 1999) and CMEs (e.g., Lyons and Simnett, 1999), this spray is accompanied by a filament eruption which appears to be the case for of the order of 70% of such events (Munro *et al.*, 1979). The ejection is directly followed by a two-ribbon flare (e.g., Hirayama, 1974; Švestka and Cliver, 1992) which is observed to move away from the original eruption site along curved paths, some of which extend outwards and others towards a neighboring active region, while post-flare loops develop at the flare site itself. A highly complex magnetic field geometry such as that described by Moore *et al.* (1997) and Antiochos (1998) is thus indicated. The second and final components of the flare occur in the penumbral region of the nearby active region AR 8340. In general, flares occurring in the umbral and penumbral regions of large, magnetically complicated spots such as δ or $\beta\gamma\delta$ spots tend to produce M and X class flares such as that discussed by Gaizauskas *et al.* (1998). In this case, and similarly to the 11 November 1993 plasma ejection reported by Ohyama and Shibata (1997), the event had comparatively modest radiative output (C5.2) but seems nevertheless to have been extremely energetic, the bulk of the energy going into the motion of the eruptive material.

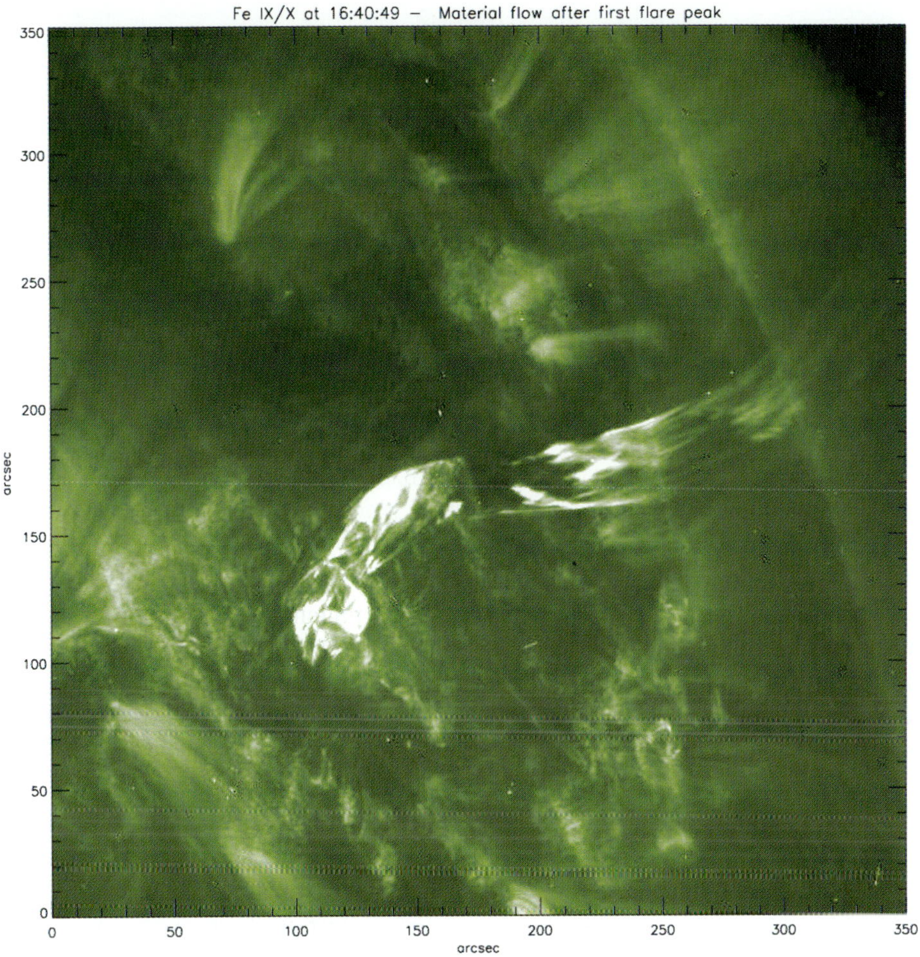

Figure 1. TRACE 350×350 arc sec^2 image in Fe IX/X (171 Å) formed at a temperature of 1.3×10^6 K. This image shows the extent of the ejected material some 15 min after the hard X-ray burst at 16:26 UT.

2. Observations and Analysis

The ground-based data were obtained on 27 September 1998 with the DST. The observations were made at the end of a eleven-day-long campaign (17–27 September) with the TRACE spacecraft to observe the relation between the quiet-Sun chromospheric and transition region network and the photospheric magnetic field. (It was originally intended to involve observations from the SOHO spacecraft too but at the time communications with SOHO were temporarily lost.) The Sun was in fact very active towards the end of this period, with a large active region, NOAA AR 8340, approaching the west limb, and it was decided to use the DST to make observations of it. A smaller active region AR 8342 was located to the west of

TABLE II
TRACE observation passbands

Ion	Wavelength (Å)	Bandwidth (Å)
Fe XII	195	6.5
Fe IX/X	171	6.4
C IV	1550	20
C I/Fe II/UV continuum	1600	275
Continuum (UV)	1700	200
Continuum (visible)	5000	broad

AR 8340, so that there was some degree of magnetic complexity in the Kitt Peak magnetograms made around this period. Both active regions were reported as having delta configurations (opposite magnetic polarity intrusions within the largest sunspot) in the few days leading up to the flare reported here, a good predictor for particularly energetic flares. The larger active region (AR 8340) appears to be the same region (AR 8319) seen on the previous solar rotation during a series of fast-cadence green-line coronal observations made with the Solar Eclipse Coronal Imaging System (SECIS) on the Evans Solar Facility coronagraph at Sacramento Peak on 5 September 1998 (Phillips *et al.*, 2000).

Between 15:06 and 21:43 UT on 27 September, a sequence of 169×169 arc sec^2 images centered on AR 8340 were obtained using an in-series arrangement of the DST Universal Birefringent Filter (UBF: Bonaccini *et al.*, 1989), a Hallé birefringent filter centered on the Ca II K line, and a G-band filter. This sequence thus produced an image every 40 s at each wavelength. The narrow-band regions selected together with their exposure times are given in Table I. Other wave-bands were chosen to image regions of the solar atmosphere that included the photosphere, temperature minimum region, and chromosphere. The two wave-bands centered on the Fe I (6302.4 Å) line measure the right- and left-handed circularly polarized components from which it is possible to extract the photospheric magnetic field; we will use the data from these wave-bands in a later analysis. It was evident by visual inspection of the images that a powerful flare was in progress with several bright and absorbing features being ejected from the flare site, near the large sunspot of AR 8340, by about 16:27 UT. Some of these ejecta were traveling towards the north-west limb, and at 16:39 UT it was decided to slew the telescope to follow these features as they were now almost out of the field of view.

Extreme ultraviolet, ultraviolet, and optical images of active region AR 8340 and its surroundings were obtained with TRACE between 16:04 UT and 19:10 UT. TRACE is a NASA Small Explorer Mission that provides 8.5×8.5 arc min^2 images of the solar chromosphere, transition region and corona with a spatial resolution of 1 arc sec and provides a continuous coverage of the Sun for long stretches of

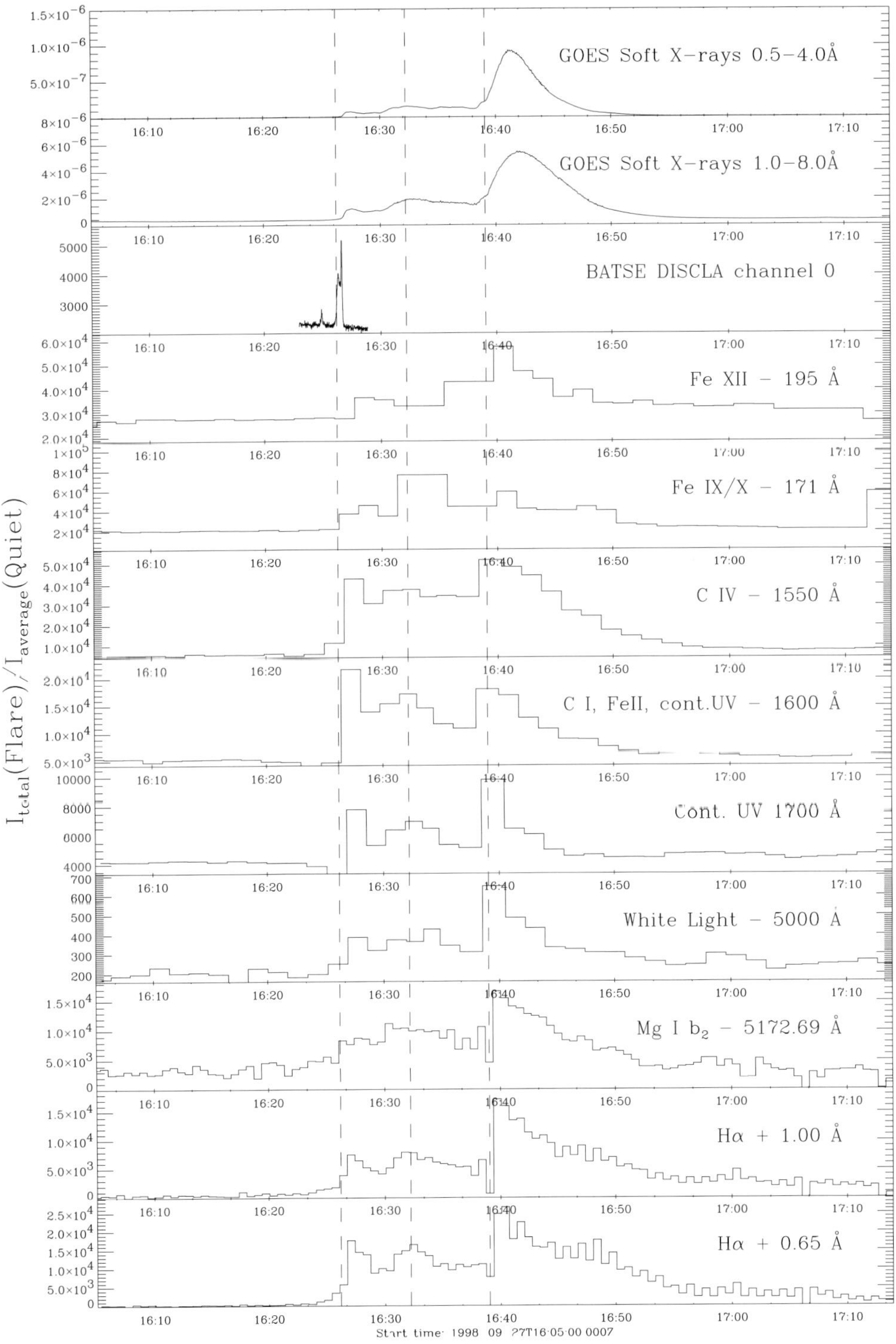

Figure 2. Light curves from GOES, BATSE, TRACE, and the DST. Start time: 1998 Sep 27, 16:05:00 UT.

time because of its Sun-synchronous orbit. For the flare observations reported here, the coronal images obtained in spectral regions centered on 173 Å and 195 Å are particularly relevant. The first includes the intense Fe IX and Fe X emission lines (wavelengths 171, 175 Å; temperature $\sim 1 \times 10^6$ K), and the second includes the intense Fe XII emission line (temperature 1.5×10^6 K). We also used the images in a region centered on 1550 Å which includes the transition-region C IV emission line doublet. The images are obtained by a multi-layer, normal-incidence optical arrangement, and so the bandpasses for each channel include other (mostly much weaker) spectral lines as well. Figure 1 shows one of the Fe IX/X images at a late stage of the flare event, when the flare ejecta are shown most dramatically moving from the active region along a curved path to the solar limb. In Table II we give details for each of the TRACE bandpasses.

Taken together, then, the DST and TRACE observations allow us to observe the flare and more particularly the associated ejecta over a wide temperature range, from a few thousand K to 1.5×10^6 K, including intermediate temperatures characteristic of the transition region.

The reduction and analysis of the TRACE and DST data was carried out using standard routines within the *SolarSoftWare* package written in IDL (Freeland and Handy, 1998). Before a detailed analysis of the images were performed the data were first corrected for cosmic ray events and CCD readout bias and were flat-fielded. There are two intervals in the TRACE data that include small amounts of missing data. Between 16:31 UT and 16:39 UT, three frames were dropped in the TRACE Fe IX/X and Fe XII sequences. These missing frames are preceded in both cases by a rapid increase in the amount of noise-like spiking. Since this is barely noticeable in the UV exposures this image loss may well be due to orbital background effects. This effect occurs again in EUV bands between 17:05 UT and 17:42 UT, causing missing or bad data between these times.

2.1. Flare Light Curves

Figure 2 shows light curves of emission in the various wavelength bands mentioned before. The top two panels show the soft X-ray emission as recorded by GOES which makes total soft X-ray flux measurements of the Sun continuously in two energy bands, 0.5–4 Å and 1–8 Å, with a time resolution of 3 s. The third panel from the top shows hard X-ray emission for a 300-s interval following a flare trigger as recorded by BATSE which unfortunately passed into satellite night before the second and third soft X-ray peaks.

Panels 4 to 9 (from the top) of Figure 2 show data from TRACE. The emission was estimated as follows. As the intensity calibration of both the DST and TRACE is uncertain for this period, we selected a set of quiet-Sun frames taken two days earlier (25 September) during a coordinated TRACE and DST quiet-Sun campaign and calculated the average intensity (I_{av}(Quiet)) for each image. The flare intensity level was then calculated by selecting pixels with intensities greater

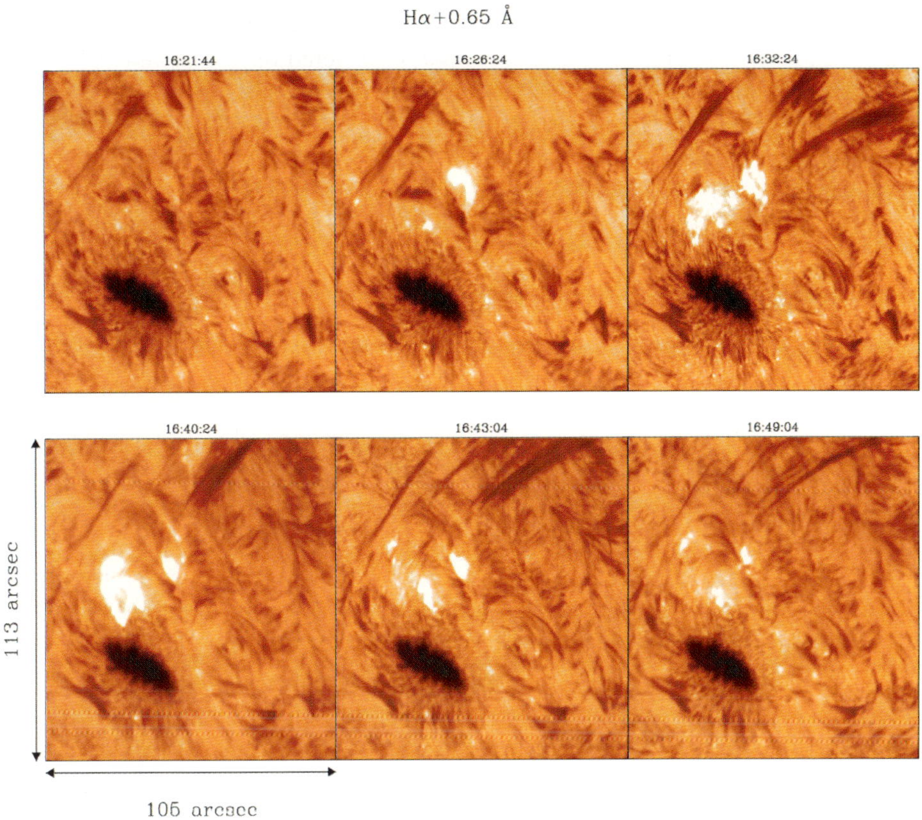

Figure 3. Hα + 1.00 Å images of the 27 September 1998 flare (time (UT) is indicated above each image).

than $I_{av}(\text{Quiet}) + 2\sigma$. These pixels were then co-added. The total intensity of the flare could then be estimated relative to the quiet Sun using $(I_{tot}(\text{flare})/I_{av}(\text{Quiet}))$. The effects of the missing data mentioned earlier can be seen in the 171 Å and 195 Å light curves of Figure 2 as a plateau between the times referred to.

3. Results and Discussion

Figures 3–5 show the development of the flare and associated ejecta at six different times in the two Hα bands and the Mg b_2 band respectively. The six times, from 16:21 UT to 16:50 UT, cover the interval from just before the hard X-ray burst to well into the flare decay in soft X-rays. The image size in each case is 105 × 113 arc sec^2. Figures 6 and 7 show TRACE images in C IV and Fe IX/X at six times that are almost equivalent to those in the optical images and for the same fields of view. The large sunspot associated with AR 8340 is evident in the optical images and can just be made out in the C IV image also from the lack of emission at the

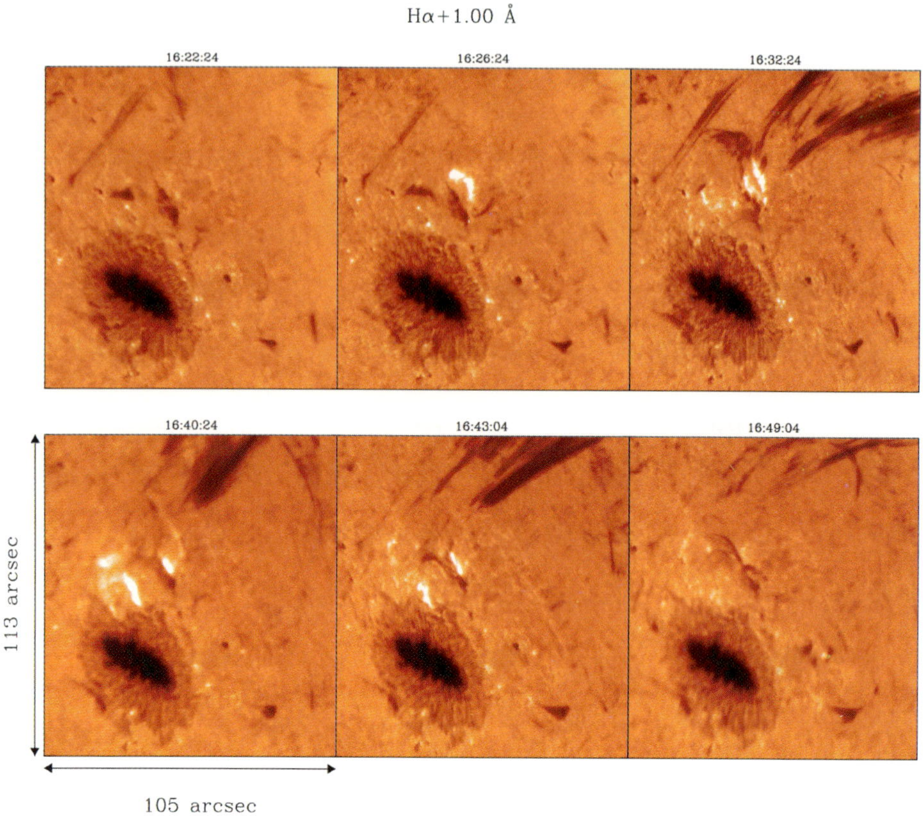

Figure 4. Hα + 0.65 Å images of the 27 September 1998 flare (note that the time (UT) of the last image is slightly different from the last image of Figure 3).

location concerned. The six optical images are a selection of the numerous (1 frame per 40 s) images available from the DST data. The top portion of Figure 8 shows four Hα images, two taken before the telescope slew at 16:39 UT and two after. They show the dark absorbing ejected features moving from the flare site, both towards the north-west limb and towards the neighboring active region AR 8342, to the west. Both trajectories show significant curvature.

Inspection of the two Hα movies which can be viewed on the accompanying CD-ROM and in the figures displayed here show the presence of much moving absorbing material. This is clear within two minutes of the flare onset (marked by the hard X-ray burst at 16:26 UT), the features taking paths which either connect to nearby regions or are directed outwards at locations to the north-west of the large sunspot. One minute before the hard X-ray burst, two foot-point brightenings can be seen to the north-west of the sunspot, visible in Hα, Mg b_2, and particularly in C IV (panel 1, Figure 9). A loop-like feature connecting these two foot-points then brightens rapidly and erupts (panels 2 and 3, Figure 9) ejecting a large volume of absorbing material. After the eruption, the bright 'loop' visible in panel 2 of

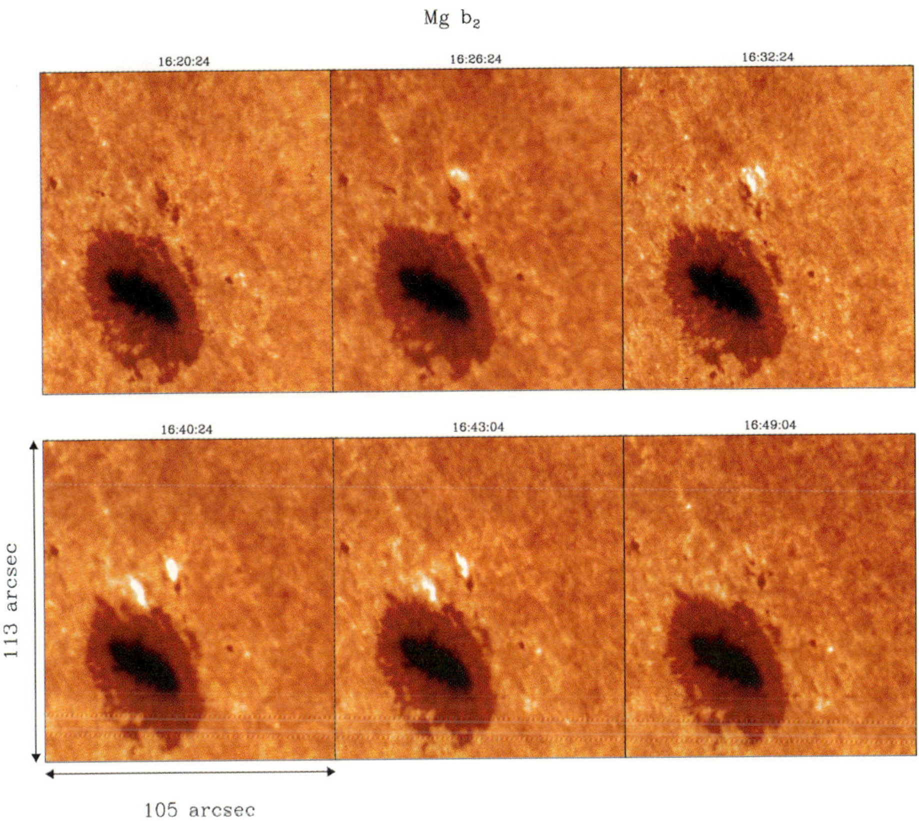

Figure 5. Mg b_2 images of the 27 September 1998 flare (times (UT) are indicated above each image).

Figure 9 separates into two short parallel ribbons clearly visible on panel 4 of Figure 9. This initial component of the flare and the associated mass ejection can be interpreted in terms of the emerging flux model described by Shibata (1997) in which a rising magnetic element reconnects with the overlying ambient field. The mass contained within the rising element is then transferred into the reconnected open flux tube to form a jet or spray (Canfield *et al.*, 1996; Schmieder *et al.*, 1995).

As the two bright ribbons in panel 4 of Figure 9 continue to separate, the ribbon closest to AR 8340 interacts with a new set of ribbons in the penumbra of the sunspot (see frames 19 to 21 of the Hα + 0.65 Å movies at the accompanying CD-ROM). These ribbons intrude deep into the penumbra, almost to the boundary of the umbra and then gradually fade over the following 7 min. In fact, the soft X-ray flux remains almost constant between 16:32 and 16:39 UT indicating that the ribbons cool very slowly during this period. At 16:40 UT the penumbral ribbons are again activated and brighten rapidly to produce a C5.2 class which then decays exponentially over the final 20 min of the flare.

The pattern of the TRACE data recorded from this spray event follows two distinct patterns: one for the UV and continuum white-light observations; and

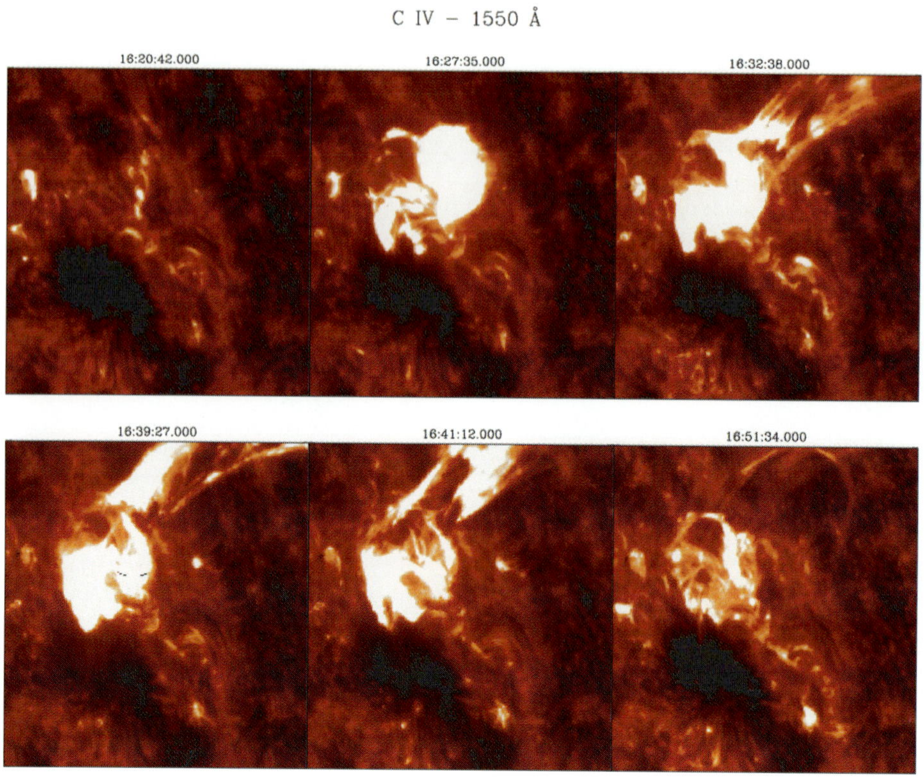

Figure 6. TRACE C IV (1550 Å) images of the 27 September 1998 flare (times, indicated above each image, are near to those in the optical images of Figures 2–4). The bright emission features have temperatures equal to the emission temperature of C IV, viz. 10^5 K.

one for the EUV observations. Although in Fe IX/X and Fe XII the initial peak is simultaneous with the initial peak in the UV and white-light bands, the main peak in intensity is associated with the formation of the post-flare loops, and the rise in intensity is by a factor of no more than two. By contrast, the increase in UV intensity in the 1550 Å band is by a factor of 10. Similarly, the white-light band records a four-fold increase in flux, according to the light-curve algorithm described above. The triple-peaked structure of the light curves is more easy to see in the ultraviolet bands. The first peak occurs almost immediately after the hard X-ray burst recorded by BATSE and the flux rises up from the background extremely rapidly. This first peak appears to be associated with the reconnection point, whereas the peaks at around 16:33 UT and 16:39 UT appear more associated with a loop whose brightening may have been triggered by the initial reconnection event.

The pre-flare activity is perhaps most noticeable in the transition region (C IV -10^5 K). There is evidence of magnetic reconnection as a high-lying transition region loop (loop length $\sim 9 \times 10^4$ km) makes contact with a lower loop (length

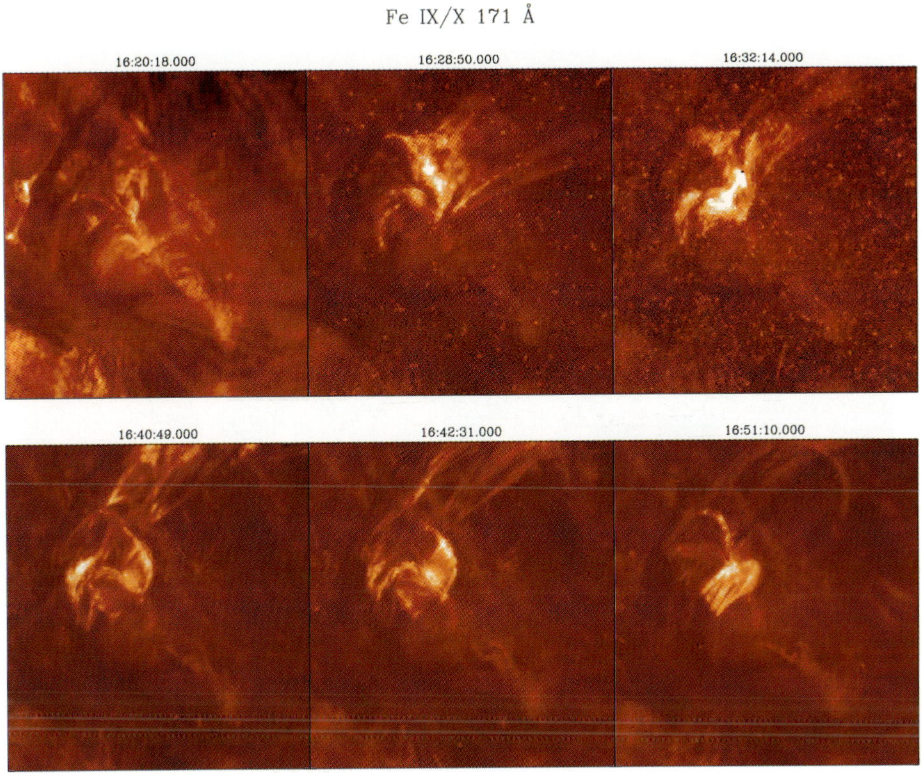

Figure 7. TRACE Fe IX/X (171 Å) images of the 27 September 1998 flare. The bright emission features have temperatures equal to the emission temperature of Fe IX or Fe X, viz. 10^6 K.

$\sim 10^4$ km) on the north-western edge of the sunspot's penumbra. Flow is also in evidence along this higher loop, although its direction is ambiguous due to the restrictions of the time series' cadence. At reconnection (16:27 UT) this point erupts, triggering a rapid expansion of the lower loop, the ejection of material, and the rapid brightening of two nearby loops, one of which appears also to erupt at 16:39 UT, just before a post-flare loop system is seen to emerge in the EUV wavelengths.

At 16:46 UT, this system of post-flare loops is seen to emerge in a bright flash. The loops are significantly smaller ($\sim 3 \times 10^4$ km) than the higher loop system seen before and after the impulsive phase of the flare in the transition region (C IV). As the system re-stabilizes, the loop planes are seen to rotate by approximately $\pi/6$ rad about a roughly radial axis, but the loops themselves stay constant in size. At 17:05 UT the loops can still be seen and have stopped rotating, but have begun to fade appreciably. Once again, noise effects caused dropped frames and heavily deteriorated data in the EUV bands between 17:05 and 17:42 UT, after which time the post-flare loops are no longer visible. Material flow along three main magnetic lanes is also noted in the transition region: one in a loop with both foot-points on

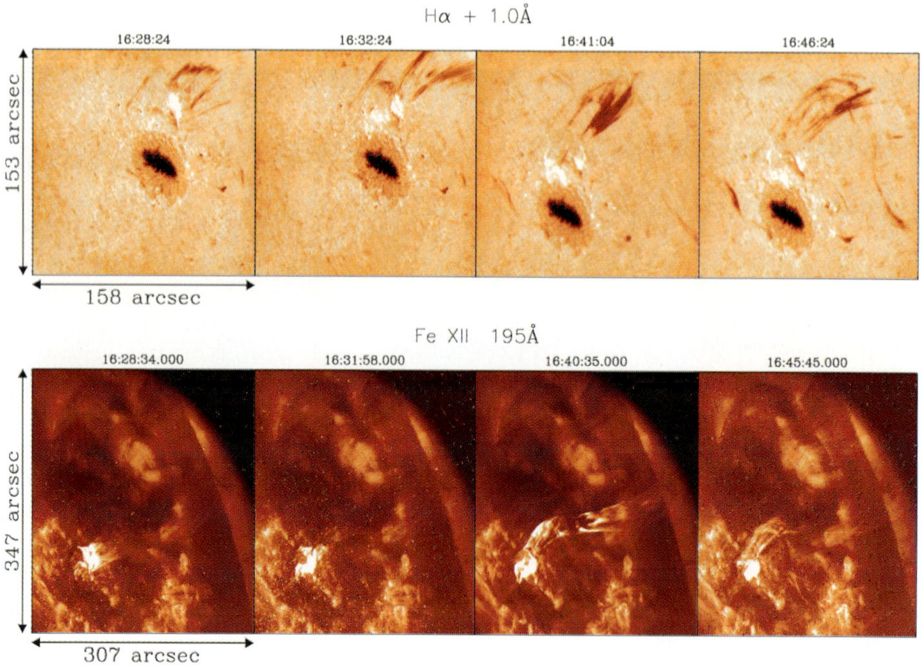

Figure 8. Hα + 1.0 Å and Fe XII images of the 27 September 1998 showing the full extent of the spray event. Note, the shift in position of the sunspot in the DST images was due to the telescope being slewed at 16:39 UT to take in the full extent of the spray.

the northern penumbra of the sunspot; a second in a lane which rises up from the penumbra along with the first loop, then arcs over this loop to run parallel and close to the surface on the region's western side; and a third which is the escaping flow path seen in Figure 1. The lower loop where the initial reconnection occurred remains visible forty minutes after the beginning of the event's impulsive phase.

Figure 2 (third panel) shows a slight increase in hard X-ray flux which coinsides with the initial rise of the flare in the UV, EUV, and optical wavelengths. Interestingly, the main, double-peaked increase in hard X-ray activity recorded by BATSE appears to be nearly *simultaneous* with the rapid increase in flux in all the TRACE and DST passbands.

It is difficult to reconcile this flare with a particular theoretical picture due to its extremely complex nature. Indeed, the light-curve from about 16:21 UT to 16:50 UT has been shown to consist of three distinct brightenings which are accompanied by two separate mass ejections. Thus, the combination of these high-resolution ground-based and complimentary UV/EUV images from TRACE reveals that flares and mass ejections such as jets, sprays and CMEs have a more complicated structure than previously observed and should therefore be subject to further high spatial and temporal investigation.

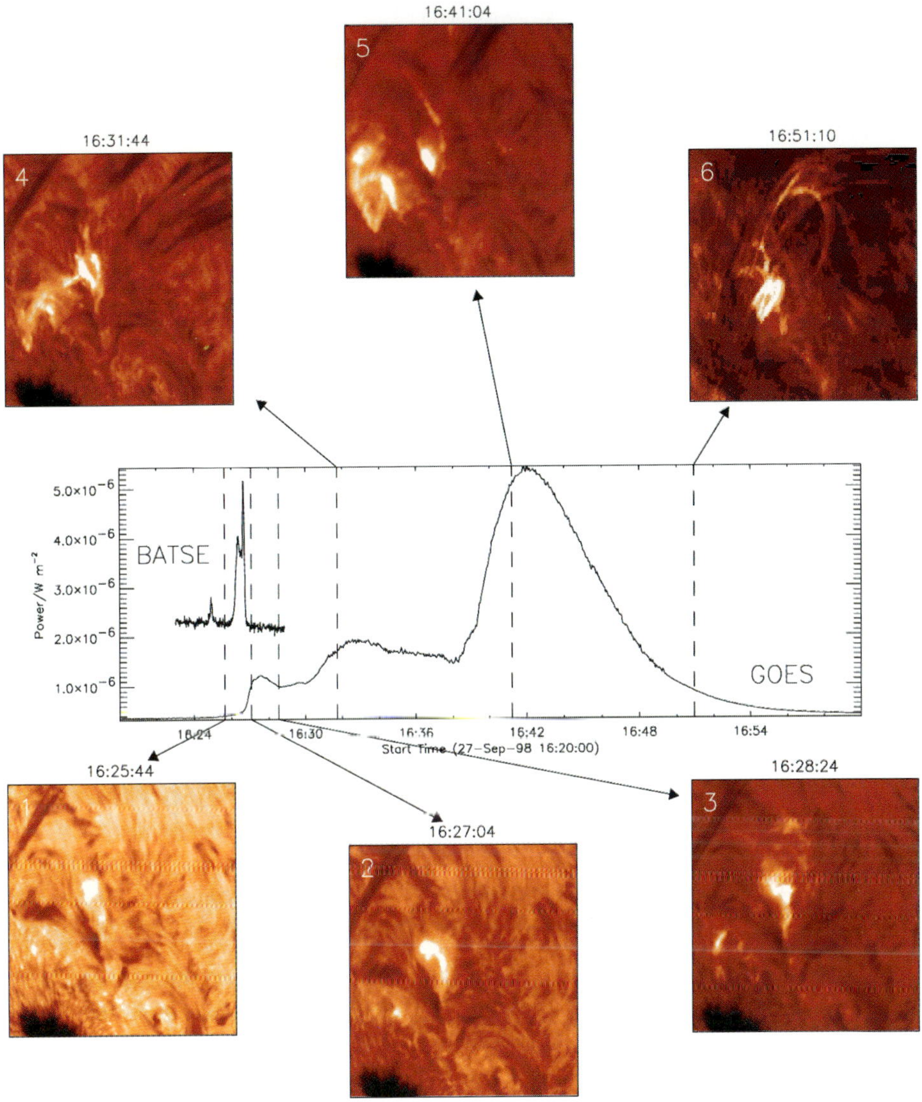

Figure 9. The relationship between hard and soft X-ray light curves and the high-resolution Hα + 0.65 Å (panels 1–5) and Fe IX/X (171 Å) (panel 6) filtergrams.

Acknowledgements

The DST observations were obtained by PTG as a visiting research student at the National Solar Observatory, National Optical Astronomy Observatories, operated by the Association of Universities for Research in Astronomy, Inc. (AURA), under cooperative agreement with the National Science Foundation. DRW and PTG would like to thank the Department of Education for Northern Ireland and the

Rutherford Appleton Laboratory for financial support. Dr Jeongwoo Lee is thanked for useful discussions during the preparation of this manuscript.

References

Alexander, D. and Fletcher, L.: 1999, *Solar Phys.* **190**, 159.

Antiochos, S. K.: 1998, *Astrophys. J.* **502**, L181.

Bonaccini, D., Righini, A., Cavalline, F., and Ceppatelli, G.: 1989, *Astron. Astrophys.* **217,** 368.

Canfield, R. C., Reardon, K. P., Leka, K. D., Shibata, K., Yokoyama, T., and Shimojo, M.: 1996, *Astrophys. J.* **464**, 1016.

Domingo, V., Fleck, B., and Poland, A.: 1995, *Solar Phys.* **162**, 1.

Freeland, S. L. and Handy, B. N.: 1998, *Solar Phys.* **182**, 497.

Gaizauskas, V., Mandrini, C. H., Demoulin, P., Luoni, M. L., and Rovira, M. G.: 1998, *Astron. Astrophys.* **332**, 353.

Handy, B. N., Acton, L. W., Kankelborg, C. C., Wolfson, C. J., Akin, D. J., Bruner, M. E., Caravalho, R., Catura, R. C., Chevalier, R. D. Duncan, W. C., Edwards, G., Feinstein, C. N., Freeland, S. L., Friedlander, F. M., Hoffmann, C. H., Hurlburt, N. E., Jurcevich, B. K., Katz, N. L., Kelly, G. A., Lemen, J. R., Levay, M., Lindgren, R. W., Mathur, D. P., Meyer, S. B., Morrison, S. J., Morrison, M. D., Nightingale, R. W., Pope, T. P., Rehse, R. A., Schrijver, C. J., Shine, R. A., Shing, L., Tarbell, T. D., Title, A. M., Torgerson, D. D., Golub, L., Bookbinder, J. A., Caldwell, D., Cheimets, P. N., Davis, W. N., DeLuca, E. E., McMullen, R. A., Amato, D., Fisher, Maldonado, R. H., and Parkinson, C.: 1999, *Solar Phys.* **187**, 229.

Hirayama, T.: 1974, *Solar Phys.* **34**, 323.

Lyons, M. A. and Simnett, G. M.: 1999, *Solar Phys.* **186**, 363.

Moore, R. L., Schmieder, B., Hathaway, D. H., and Tarbell, T. D.: 1997, *Solar Phys.* **176**, 153.

Munro, R. H., Gosling, J. T., Hildner, E., MacQueen, R. M., Poland, A. I., and Ross, C. L.: 1979, *Solar Phys.* **61**, 201.

Ohyama, M. and Shibata, K.: 1997, *Publ. Astron. Soc. Japan* **49**, 249.

Phillips, K. J. H., Read, P. D., Gallagher, P. T., Keenan, F. P., Rudawy, P., Rompolt, B., Berlicki, A., Buczylko, A., Diego, F., Barnsley, R., Pasachoff, J. M. and Babcock, B. A.: 2000, *Solar Phys.* **193**, 261.

Shibata, K.: 1997, in *The Corona and Solar Wind Near Minimum Activity*, **ESA SP-404**, 533.

Schmieder, B., Shibata, K., van Driel-Gesztelyi, L., and Freeland, S.: 1995, *Solar Phys.* **156**, 245.

Švestka, Z. and Cliver, E. W.: 1992, in Z. Švestka, B. V. Jackson, and M. E. Machado (eds.), *Eruptive Solar Flares*, Springer-Verlag, Berlin, p. 1.